Forc
Matt
and
Energy

GW01460068

Force
Matter
and
Energy

D J Williams BA (Oxon)

THE ENGLISH UNIVERSITIES PRESS LTD

ISBN 0 340 12374 5

First published 1974

Copyright © 1974 D. J. Williams
All rights reserved. No part of this publication may be
reproduced or transmitted in any form or by any means
electronic or mechanical, including photocopy, recording,
or any information storage and retrieval system, without
permission in writing from the publisher.

The English Universities Press Ltd
St Paul's House, Warwick Lane, London EC4P 4AH

Printed in Great Britain by
Butler & Tanner Ltd, Frome and London

Preface

This volume is designed to cover that area of the A-level physics syllabus traditionally known as 'Mechanics and Properties of Matter', and is largely based on the teaching scheme used by the author in the past few years. The aim has been to provide opportunities for the reader to think things out for himself, while at the same time preserving those features of a didactic approach which would seem to be valuable – not least the sense of security and direction which is gained from a logically developed structure. A particular attempt has been made to discuss fundamental concepts more fully than is usual in an A-level textbook, in the hope that if these are clearly understood the applications and developments will be more easily appreciated. Closely integrated exercises interspersed with the text have been included in an attempt to encourage involvement on the part of the reader. The intention is that they should be worked as they are met.

Mathematical description has been included wherever it seems to complement and illuminate qualitative treatment. Broadly speaking, any mathematics which is beyond O-level standard has been introduced with some explanation, not only for the sake of those readers not pursuing mathematics at A-level, but also because of the rather different emphasis which the physicist requires. The mathematical arguments are illustrative and pragmatic, rather than rigorous. It is hoped therefore that the book will also be of assistance to those preparing for the new A-level examination 'Physics and Mathematics'.

Inevitably a book dealing with such a well-established part of school physics will contain very little that is original. I would like to record my gratitude to all those who have consciously or unconsciously contributed by what they have taught me. In particular I am indebted to the series editor, Mr C. B. Spurgin, and to Dr J. W. Warren for his valuable criticism of the manuscript and suggestions for improvement. I should also like to thank Mr J. Pennock, a pupil at Wolverhampton Grammar School, who assisted in checking for errors.

I am grateful to the following examining boards for permission to use questions from past A-level papers:

the Associated Examining Board (AEB)
the Joint Matriculation Board (JMB)
the Oxford and Cambridge Schools Examining Board (O & C)
the Oxford Delegacy of Local Examinations (O)
the Southern Universities Joint Board (SUJB)
the University of Cambridge Local Examinations Syndicate (C)
the Welsh Joint Education Committee (WJEC)

My thanks must also go to the many persons and organisations who generously provided photographs and to the publishers for their encouragement, patience and careful attention to all details.

D. J. W.

Acknowledgements

We would like to acknowledge the following firms, public bodies and individuals who have kindly provided and given permission for the reproduction of the photographs:

Associated Press, p. 255

British Railways, p. 86

Henry Brown and Son Ltd, p. 274

Camera Press, p. 265

Central Electricity Generating Board, p. 95

Central Office of Information, pp. 50, 116, 283

Ronald Chapman, p. 59

De Beers Consolidated Mines Ltd., p. 156

Tony Duffy, p. 110

Esso Petroleum Co. Ltd., p. 79

G.K.N. Ltd., p. 153

Sir Alexander Gibb and Partners, p. 74

Girling Ltd., p. 101

Glacier Metal Co. Ltd., p. 4

Griffin and George, pp. 34, 104, 149

Phillip Harris Ltd., pp. 55, 103, 117, 160, 180, 184, 197, 207, 208, 220, 238

E. D. Lacey, p. 192

Leybold-Heraeus, p. 190

Ministry of Defence (Crown Copyright), p. 235

Mount Wilson and Palomar Observatories, pp. 280, 287, 288

Professor E. W. Mueller, p. 153

N.A.S.A., pp. 43, 51, 82, 187, 198, 234, 307

National Physical Laboratory (Crown Copyright), pp. 7, 9, 10

Professor J. F. Nye (Copyright, The Royal Society), pp. 152, 153, 154, 155

Omega Watch Co. Ltd., p. 33

Paul Popper, pp. 1, 62, 66, 99, 131, 189

Rubery, Owen and Co. Ltd., p. 45

J. K. St. Joseph (Copyright, the Director of Aerial Photography, University of Cambridge), p. 294

Scottish Tourist Board, p. 281

Science Museum (Crown Copyright), p. 105

Shell Petroleum Co. Ltd., p. 127

Sport and General, p. 193

A. E. Stathers, p. 147

Swissair-Photo Ag., p. 225

Peter Tempest, p. 194

U.S. Navy, p. 83

Venner Ltd., p. 20

Contents

1 The business of science: explaining or describing?

This book is mainly about how things move and why they move, or rather, as we shall see, why they do or do not *change* their motions. The snag about trying to explain why anything happens is that the answer usually provokes another question. Why does a stone fall to the ground? The earth pulls it. Why does the earth pull it? Every piece of matter attracts every other piece of matter. Why does every piece of matter attract every other piece of matter? Eventually we are reduced to saying that the stone falls to the ground because it does.

We must say what we mean when we claim to *understand* motion, or light, or the atom. When we look at our falling stone, we can do little more than find some rules which describe its behaviour and use them to guess what will happen next time we let it go. We shall be searching for ways in which the motion produced in one set of circumstances is linked with what happens in a different situation. It may look as though we are settling for quite a modest aim, but in fact it is understanding at this kind of level that we are always seeking in science. We rate our success by the number of different types of phenomenon we can link together, and when we can describe the falling stone, ripples on a pond and the motion of the earth around the sun in terms of a few rules common to all three situations, we feel we have got somewhere. If we like we can ask why there are rules, indeed some of us think we ought to ask that question, but we shall not find the answer in a physics book; scientific method is not adequate to deal with it.

Science and mathematics

Built into every part of science is the notion that the material world can be described in terms of *numbers*; the rules are mathematical rules. Man devises his mathematics, perhaps initially to meet practical needs in commerce and engineering, or perhaps even for sheer entertainment, like a crossword puzzle, and then he finds almost unexpectedly that he can apply it to the behaviour of other aspects of the world around him. The fact that mathematics can be used in this way is so important to the physicist that he comes to take it for granted.

Experiment and theory

A further assumption the physicist makes is that *the complex can be explained in terms of the simple*. If he wants to understand why waves break on a beach, he finds it is best to start by looking at tidy, controlled ripples in a tank. In this book we shall look first at carefully selected simple situations where the rules stand out clearly, and we shall use our mathematics and our sense of what is logical to predict what ought to happen in more complicated cases. Quite often our common sense works well, but it has been found necessary from bitter experience to check each time what actually happens in practice. It is sometimes amazing how wrong we can be. This process, perhaps rather an obvious one to our way of thinking, has come to be called *the scientific method*. We study our phenomenon until we can summarise what happens in one or two simple rules; these are the experimental *laws*. Then we try to make an intelligent guess at a basis for explanation—a hypothesis—and from this we build up a logical, usually mathematical, working out of the consequences—a *theory*. If this gives answers which agree with what actually *does* happen, then our hypothesis and theory are sound so far, but we

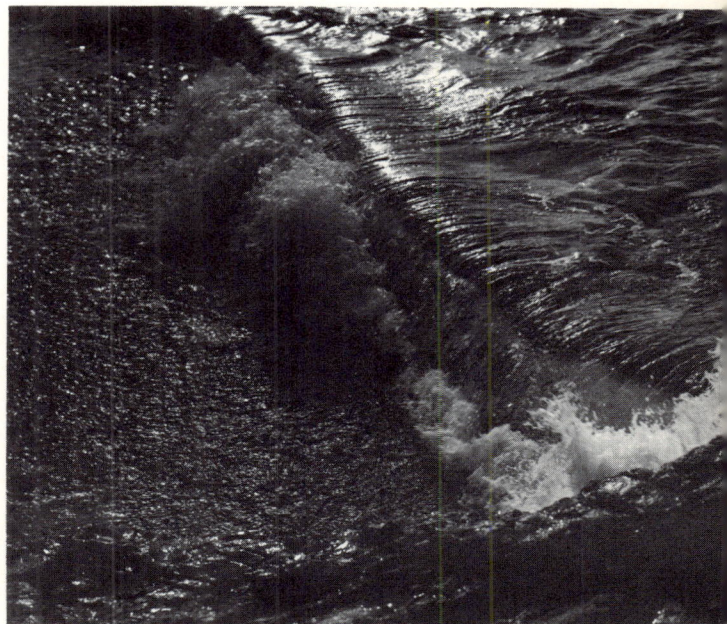

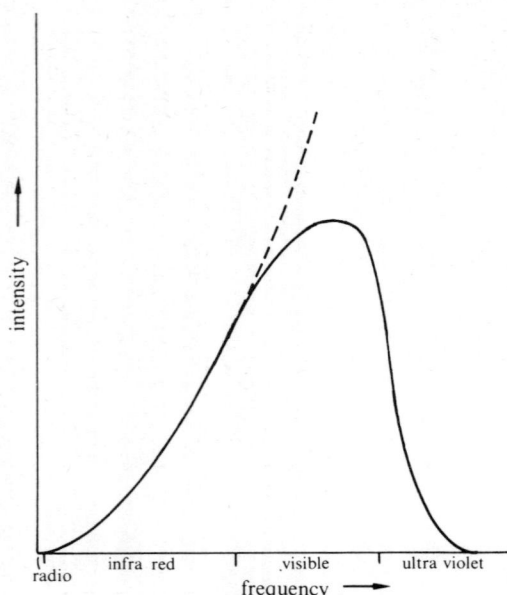

Common sense can sometimes be wrong. The full curve shows the variation of intensity with wavelength of electromagnetic radiation emitted from a hot luminous body. The broken line shows the curve predicted by theory towards the end of the nineteenth century (after Rayleigh and Jeans). This discrepancy between experiment and theory puzzled scientists for a long time, and was eventually removed by the revolutionary ideas of quantum theory.

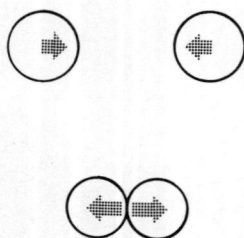

Atoms attract each other at close range and repel at very close range.

must keep trying them out; we must push our ideas to the very edge of their possibilities. As soon as we get a discrepancy between theoretical prediction and experiment, we try to find what has gone wrong. It could be faulty experimental procedure, or badly thought out theory, or it could be that we are working in a situation where our original laws no longer apply. Some laws are trusted more than others, but no law is trusted absolutely in untried situations. Experiment, theory, more experiment, more theory, and so our knowledge increases. That is the way it is supposed to be done, and that is how it works out in most cases. Occasionally a genius has cut a few corners and followed a hunch he had about the way the rules ought to go, but always the ultimate test of his ideas is to look at how the real thing behaves.

Qualitative and quantitative descriptions

Quite often we find that although the complicated case could be worked out in principle, in practice it could not be done without years of work or the help of a fast computer. This is just one reason why we shall often make use of a *qualitative* rather than *quantitative* description. If I say it is a long time to the summer holidays, that is a qualitative description. If I say it is fourteen weeks to the summer holidays, that is a quantitative description. The quantitative description generally conveys more by way of information, but for a complete picture we like to have both–even in our rather basic example the qualitative statement conveyed something which was missing from the quantitative one. There is a sense in which the physicist feels he must put his mathematics into words; he likes to have a picture in his mind of what he is talking about. Speaking loosely, he calls it a *model*, although normally this term implies not only a qualitative visualisation but also a basis for calculation.

Microscopic and macroscopic descriptions

A good example of a model is the picture of the atom which we shall use many times in this book. Physicists are fairly confident that they now know a good number of the rules governing the behaviour of atoms. They are mathematically complicated, they are not what our common sense expects, and their description lies for the most part outside the scope of this book. Nevertheless we shall find talking about atoms an invaluable help as we try to relate the various properties of matter which concern us. We shall get a new slant on the *macroscopic* (large-scale) phenomenon by attempting a *microscopic* description–a description in terms of the behaviour of atoms. For this purpose we shall not need to know everything there is to know about an atom. It will get us quite a long way if we know that atoms can attract each other quite strongly at close range but repel each other still more strongly at closer range. That will tell us for instance why a lump of iron like a poker is hard to compress but can be broken by sufficiently strong forces, and that when it *is* broken it is not good enough merely to fit the bits back together to make the poker whole again. In Chapter 11 we shall look into these ideas a little more closely.

We can try to think of a large-scale model which will have just the properties mentioned–perhaps a golf ball with tiny magnets built in all around its surface. So we might think of an atom as a tiny magnetic golf ball about one ten millionth of a millimetre across. Now that is without doubt a very inaccurate description of an atom, but it is a useful model; it is a picture which brings to mind a number of aspects of the atom which help our understanding of what we can see around us. It is not by any means the best model that can be devised; in fact we shall need to modify it occasionally to bring in a new aspect. A model is a useful tool but it can get in the way of progress when it dominates our thinking too much.

Exercise 1.1

Roughly how many atoms like the one described could be (a) laid on a square of side one centimetre? (b) put in a cubic box of side one centimetre?

Atoms in solids, liquids and gases

We shall take it that all matter consists of atoms and that any object visible to the naked eye contains a very great number of them. There are about a hundred *chemically* different types of *electrically neutral* atom. These are combined in innumerable ways to make up all the materials around us. Water is composed of oxygen atoms and hydrogen atoms, steel is made up mainly of iron atoms with some carbon atoms, a potato consists mostly of carbon, hydrogen and oxygen atoms. In taking part in such combinations the atoms are often changed significantly. They may gain or lose electric charge (charged atoms are called *ions*) and they may become members of close-knit groups called *molecules*, which are likely to have quite different properties from any of their constituent atoms.

Atoms, molecules and ions are not static–they are always on the move. Helium atoms trapped in a container at room temperature do not pile up in a heap at the bottom–they fly around at hundreds of metres per second in all directions, colliding with each other and with the atoms making up the walls. The attractions between the helium atoms are simply not strong enough to subdue this motion and only the walls prevent them from flying away in all directions. This is our microscopic description of a *gas*.

Exercise 1.2

Use the model to describe how a gas can exert a force on a piston, as in an internal combustion engine, and why that force increases when the piston is pushed in to reduce the space in the cylinder.

If we replace the helium by oxygen atoms there is a slight difference–the atoms fly around in pairs. There *is* strong attraction between two oxygen atoms, strong enough to bind them firmly and form an oxygen molecule, but there is a much weaker attraction between this pair and a third atom or another pair. Already our magnetic golf ball is an inadequate representation: perhaps we can think of the attraction as being stronger at some parts of the surface of the ball than others. Most gases consist of molecules rather than single atoms–steam for example is composed of molecules each containing one oxygen and two hydrogen atoms–and it is generally true to say that the forces which bind atoms within an individual molecule are large compared with the attractive forces between molecules.

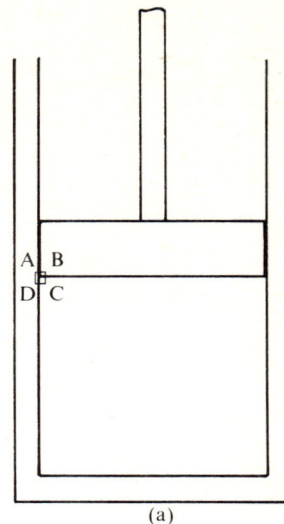

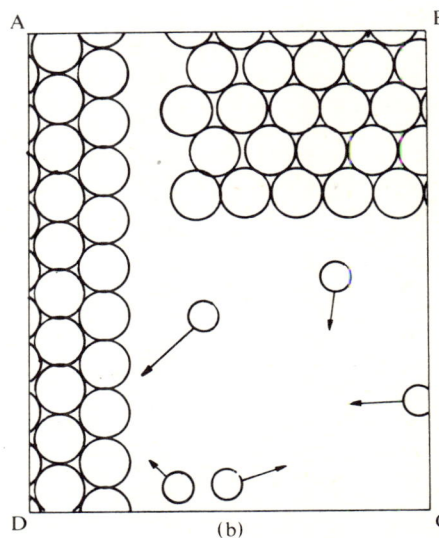

(a) Macroscopic, (b) microscopic representations of a gas trapped in a container by a piston.

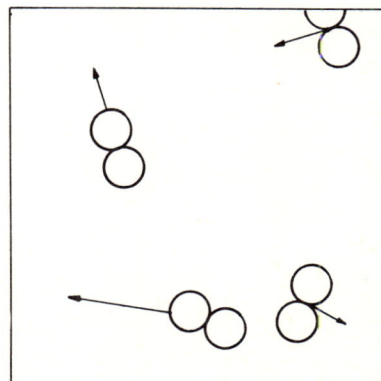

Microscopic representation of oxygen at room temperature.

Exercise 1.3

The oxygen atom is about sixteen times as massive as the hydrogen atom. What fraction of the mass of a cylinder of steam is made up by the oxygen?

If we cool our oxygen gas down, what happens to the molecules? They travel more slowly. There we have our macroscopic and microscopic descriptions complementing each other: macroscopically we say the gas has cooled, microscopically we say the molecules have on average slowed down. It is bad physics to mix the two and say the molecules have cooled down; whenever we take a temperature we are finding out something about the average motion of a very large number of atoms or molecules, and the temperature of just one of them is not a meaningful idea.

Exercise 1.4

Use the model to explain how a rapidly moving piston can change the temperature of the gas it encloses.

As we continue to cool our oxygen, the molecules slow down to such an extent that the weak attractions between the molecules are able to overcome their tendency to move around independently, rather in the way that a harassed teacher's appeals on a school trip may be more effective in drawing the class into a group when the children are tired. The molecules congregate in a pile, each more or less touching the next. A molecule somewhere in the middle of the pile experiences forces from the molecules around it on all sides; these forces tend to counterbalance each other and the molecule can jostle its way through the crowd – it still has quite a bit of freedom. But a molecule on the edge which commences to wander off on its own experiences attraction back towards the pile, and escapes are comparatively rare. Our gas has turned into a *liquid*; it has a definite volume but it can flow.

PTFE bearings used in reconstruction of the Britannia Bridge, Menai Straits. The low-friction surfaces can slide over each other to allow for expansion of bridge members.

If we cool the liquid further the molecules slow down again, until they reach the stage where they are unable to force their way through the gaps between the molecules close to them. Each atom now becomes anchored in its own position; the most it can does to vibrate. If we move some atoms at one end of the pile, the atoms at the other end have to move as well–thousands of millions of individual attractions and repulsions are ensuring that the structure moves as a whole. We now have a solid.

As we moved from liquid to solid we quietly dropped the word molecule. For many solids the term has little meaning. It is worth distinguishing molecules as entities when they exist as close-knit groups which keep reasonably aloof from each other and can move around independently, but when every atom is tightly bound in place by all its immediate neighbours the idea of the family group may lose its significance. That does not mean that it is *never* appropriate to talk about molecules in a solid. The non-stick coating for frying pans, PolyTetraFluoroEthylene, PTFE for short, is a solid composed of molecules which are very like long cylinders; they can rotate although they cannot shift position. A smooth surface slides very easily on a piece of PTFE–it is moving on molecular rollers!

Further Exercises

Exercises 1.5–8

5 Given that iron atoms are about 2×10^{-10} metres across, estimate the number of atoms in a pinhead. (See Appendix 1 for powers of ten.)

6 A kilogram of iron in the shape of a cube has a side of length about 0·05 metre. Estimate the mass of one atom of iron.

7 If the air in a room were liquefied it would take up about one thousandth of the space it occupied as a gas. Estimate the average separation of gas molecules in the room in terms of the diameter of one molecule.

8 What picture does our microscopic model give of the following?
   a) When a stick of chalk is snapped, the two pieces will not stick together again.
   b) A solid heated at one end eventually becomes hotter at the other. (For a more satisfactory development of this model see Chapter 15.)

2 Mass, length and time

'When I use a word,' Humpty Dumpty said, in rather a scornful tone, 'it means just what I choose it to mean–neither more nor less.'

Lewis Carroll, *Through the Looking Glass*

We all know what is meant by length–or do we? The idea is too basic to describe adequately, so there must be some doubt as to whether we all think exactly the same way about it. In physics as in other fields we are concerned to establish as far as possible what we are talking about, and so the physicist goes to some trouble to say what he means by length. His argument goes something like this. Length in the abstract, he is not keen to argue about–he prefers to leave that to the philosophers. What he will do is tell you how to *measure* length–how to find the length of a piece of string. He will give you a few metre rules and a set of instructions concerned with laying them down nose to tail along the string, and if you carry out those instructions properly you will be able to write down the length of the string as so many metres. It is called an *operational definition* and we can be quite sure that everyone understands it in the same way because we can compare answers to see if they agree.

This may not sound very satisfactory–we have defined length merely in terms of another length and a counting process. What *is* length? Perhaps that question has meaning, perhaps not, but we do our physics without answering it in terms other than the comparison of one length with another. Length is an example of a *quantity*, and by a quantity we mean something which can be measured in terms of an example of itself and a counting process. In fact the manner in which we decide to carry out the counting process is the only way we have of saying what is meant by the quantity concerned. *A* is twice as *long* as *B* means that two identical copies of *B* can be laid end to end in such a way that their free ends coincide with those of *A*. Substance *X* has twice the *density* of substance *Y* means that if we make some blocks of identical size out of *X* and *Y*, two blocks of *Y* give the same reading as a single block of *X* on a given balance. It is the *counting* process which defines for us the quantity we are talking about. Suppose we try to say *A* is twice as *hard* as *B*. If that is to have any meaning we must invent a counting process to go with it, perhaps that *B* indents twice as far when pressed in by a needle under certain conditions, and as soon as we have invented the counting process we have defined what we mean by *hard*.

This process of comparison by counting becomes most convenient if we pick a particular example of the quantity concerned and use it again and again in our comparisons. *C* is three times as long as *B*, *D* is four times as long as *B*. It only remains to give the value of the quantity possessed by *B* a distinctive *name*, and we have a *unit* to measure with. We can choose *any* example of the quantity for a unit, but it makes life a lot easier if we can all agree to use the same one. The unit of length now adopted by international agreement is the *metre* (abbreviated to m). All the quantities we shall meet now have internationally agreed units; they are called SI units (Système International d'Unités).

The standard metre

It is of course essential that we all agree about the size of the unit as well as its name, and this is ensured by making a prototype or *standard*. For many years the prototype metre was the distance between two marks on a particular bar of metal kept at Sèvres near

Paris. The marks could have been made in an entirely arbitrary fashion; in point of fact some attempt was made to arrange that the distance between the marks – the metre – should be one ten millionth of the length of that quadrant of the earth's circumference, which starts at the north pole, goes through Paris and ends at the equator. The idea was that there should be some permanent record in case the prototype should be lost or damaged – but the original criterion is now little more than a curiosity. Every metre rule ever made had to conform as closely as possible to the distance between those two marks, and any inconsistencies could be referred back to it to settle the matter.

The standard metre could never be *wrong* – it was *the metre* – but it could be *unreliable* if not chosen carefully. Suppose the bar bent under its own weight one day; we could not logically say that the distance between the marks was no longer one metre, but we should have to call in all metre rules and shave them down a little. Obviously steps were taken to minimise this sort of occurrence, but physicists became dissatisfied with the degree of uncertainty still possible, and in 1960 a new, more reliable standard was adopted, based on the wavelength of a particular light source. For the record, a *metre* is now the length equal to 1 650 763·73 wavelengths in vacuum corresponding to the transition between the levels $2p_{10}$ and $5d_5$ of the krypton-86 atom.

Of course at the time of adoption every care was taken to make the new standard agree with the old, but in any future disagreement between the two the old will be disregarded. It is now no more than a museum piece, though it is still comforting to keep in the back of our minds – most of us are far more at home with two marks on a piece of metal than with a million or so wavelengths of light of a particular colour.

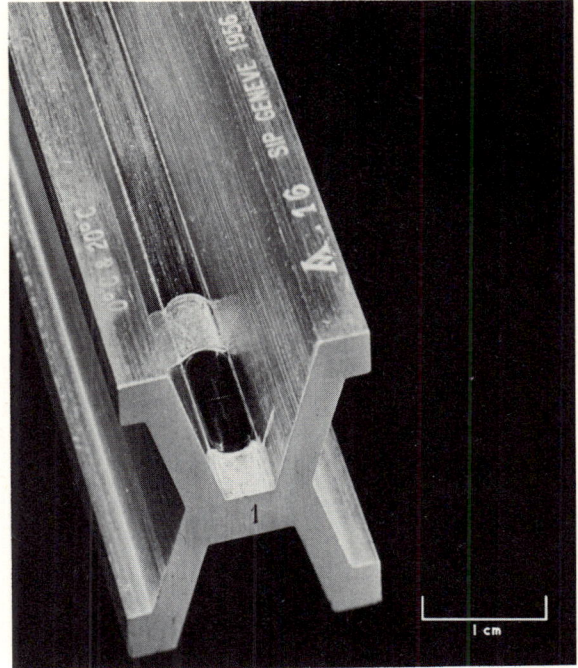

One end of the British National Copy (no. 16) of the pre-1960 standard metre.

Electrical discharge lamp containing krypton-86 for producing the orange spectrum line used as the basis of the new metre standard. On the right is the cryostat in which the lamp is maintained at a temperature of 63 K.

Armed with our metre we can now describe any length we like. Two metre rules, each conforming to the standard, laid end to end in a straight line have a length of two metres, and so on. Obviously we must be able to deal with fractions of a metre; we divide a metre rule into 100 identical parts, and then the length of any one of these is 0·01 m or 1 centimetre (cm). Other submultiples in common use are the familiar millimetre (mm) equal to 10^{-3} m, the micrometre (μm) equal to 10^{-6} m, and the nanometre (nm) equal to 10^{-9} m. Biologists sometimes call the micrometre a micron.

Exercise 2.1

If 2×10^9 iron atoms laid end to end in a line (in a poker for example) occupy a length of 50 cm, what is the separation between the centres of two adjacent atoms, in nm?

Straight lines and curves

Before leaving the measurement of length it is worth saying what we mean by a straight line. We can do it in terms of an operational definition; we give ourselves the task of constructing the straight line between two given points. We start by doing the best we can with a piece of cotton stretched tight between the points and lay our metre rules along it to measure the length. Suppose we have some reason to suspect the line is not in fact straight – perhaps it is such a long line that the cotton is sagging, or curving round the earth's surface, or there may be some other less obvious snag, and we believe we can draw another line which is straighter. We lay our metre rules along the new line and count again; if the answer is less than the previous one, the second line is closer to a straight line than the first. *The* straight line is the shortest one, the one that gives us the smallest answer no matter how many lines we try. Though we need a *rigid* metre rule to carry out this exercise, we need not rely on it to be a *straight* one.

There is nothing to stop us measuring the length of any line, no matter how much it curves and twists. Perhaps the line curves away from our metre rule: then the rule is useless, we cannot lay it along the line properly. Very well, but a curve is indistinguishable from a straight line *if we look at a sufficiently small section of it*. An astronaut looks at the earth approaching him; of course it is round, like an orange – you could never measure around the surface of that with a straight rule! And yet taking measurements with my metre rule on the ground I am hard put to it to detect any curvature at all, because I can look at only a small section at a time.

So if we need to find the length of a curved line, we simply need a shorter rule – a millimetre rule perhaps, or something a good deal shorter still. In practice we might have to lay down very many of them, but the job could be done if required.

Time

Time cannot fail to fascinate us; we are so bound by it. Everything that happens to us, any event, can be catalogued by naming the place and the time – they tie down the event completely. It is a mystery that we have so much liberty to choose the place, to visit a wide range of places at will, and yet that we cannot proceed to and stay at a chosen instant of time. What *is* time? That is another question the physicist is content to leave alone; he can get along with knowing how to compare one time interval with another. Time is another *quantity*, measured in terms of a unit; the SI unit of time is the *second* (s).

For many years the standard second was defined in terms of the rotation of the earth on its own axis, in such a way that there should be exactly 86400 seconds in what is called a *mean solar day*. The idea was simple enough–build a clock which ticks exactly 86400 times from high noon on one day to high noon on the next–but in practice this was fraught with tremendous difficulties. Two clocks built to agree with the sun on a particular day can be made in such a way that they agree closely with each other a fortnight later, but they will no longer agree with the sun; the inescapable conclusion is that days are of different duration. This is partly due to the inclination of the equator to the earth's orbit round the sun, and partly due to the eccentricity of the orbit–the earth is closer to the sun at some times of the year than at others. Measurements on the stars can be made with greater precision; they rise and set like the sun of course. There is the complication that a *sidereal* day, the time for a star to get right round the sky once, as seen from an observatory rotating round the earth, is shorter than a *solar* day by about 1 part in 365.

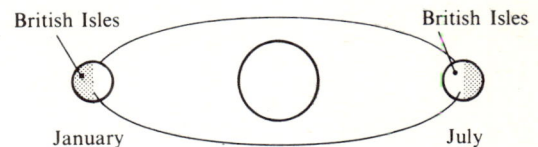

British Isles   British Isles

January   July

Figure 2.1 (Exercise 2.2)

Exercise 2.2

If the earth did not rotate at all on its axis, what interval would be occupied by (a) a solar day? (b) a sidereal day? (Figure 2.1).

So we see that because the earth goes round the sun but does not go round the stars, the number of sidereal days in a year is greater than the number of solar days by one.

The best clocks we can build at present are kept in time by radiation of one particular frequency emitted by excited caesium atoms. Two caesium clocks will agree with each other to an extraordinary degree of precision, and yet they will not keep in step with the earth's rotation as measured by observations on the stars; in fact they indicate that the earth is slowing down by about 0·4 seconds per year. When you have an instrument which can show your standard to be unreliable, it is time to change the standard, and so in 1968 it was agreed that the second should be defined in terms of the caesium clock. Formally stated, *the second* is the duration of 9 192 631 770 periods of the radiation corresponding to the transition between the two hyperfine levels of the ground state of the caesium-133 atom. The second, like the metre, can be divided into a number of identical parts; the millisecond (ms) equal to 10^{-3}s, the microsecond (μs) equal to 10^{-6}s and the nanosecond (ns) equal to 10^{-9}s, are all useful submultiples.

Exercise 2.3

Light travels at nearly 3×10^8 metres per second. How far does it get in a nanosecond?

Mass

Strangely enough, although we are happy to take length and time for granted, when we come to talk about mass we feel we ought to make some attempt at a dictionary definition. It is often said that the mass of a body is the amount of matter it contains. If we had to rely on

Lower end of the National Physical Laboratory caesium beam frequency standard ('atomic clock'). Two such clocks agree to about one part in 10^{12}.

British National Copy (no. 18) of the standard kilogram.

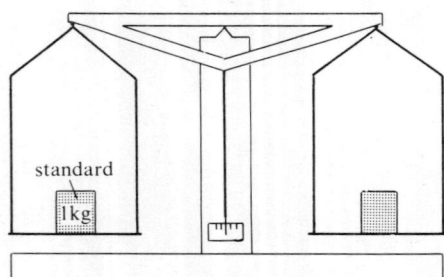

Copying the standard kilogram.

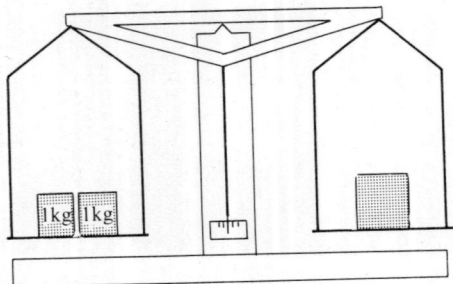

One way of defining 2 kg.

that as a basis we should never be quite sure that it meant the same thing to everybody, so once again we go for an operational definition in terms of a unit and a counting process. The unit of mass is the *kilogram* (kg) and the standard is the mass of a particular lump of platinum alloy kept at the Bureau International des Poids et Mesures, Sèvres, France. We do not pretend to remove *all* possibilities of confusion; for example, an alloy has adhering to it a surface film of air molecules, some of which persist even when the space around is highly evacuated. Do we intend that they should be included as part of the kilogram or not? What about the platinum atoms that are occasionally lost from the surface? The answer to all of these questions is that we shall worry about them when the difference they make is enough to show up in an experiment. It is pointless worrying about something which might affect the twelfth significant figure in a number if the most accurate measurements of mass that we can carry out at present involve only nine or·ten significant figures. Perhaps one day it may be necessary to redefine the kilogram, in terms of a certain number of atoms of a particular element for example, but we have not come to that pitch yet.

This time we need to take a little more care in describing the counting process; one way of doing it is in terms of a beam balance. The balance must be in a vacuum, otherwise, the surrounding air affects the result (see p. 81), and the arms must be symmetrical about the knife-edge on which they rest. When the arrangement balances with the arms horizontal we say the masses in the pans are identical. First we need a few copies of our kilogram; we make each of these by filing down a chunk of metal until it exactly balances the standard. Now we can define a mass of two kilograms – it is the mass of that object placed in one pan which will balance two of our kilogram copies in the other pan. [Although this is the method by which masses are compared in practice, we find it more convenient to define the counting process for mass in a rather different way, considered on p. 58.]

We can make submultiples too, by dividing a kilogram copy made in the shape of a bar of uniform cross section into a number of identical pieces, and so arrive at the gram (g, 10^{-3} kg) and milligram (mg, 10^{-6} kg). Notice there is no question of making a standard gram, a piece of platinum alloy in a glass case. If we did that, one day someone might demonstrate that the standard kilogram had a mass which·was, say, 1000·006 times as big as that of the standard gram and then we should have to live with the fact that there were 1000·006 g in 1 kg. There is no point in looking for trouble; we settle for just one standard.

Mass and weight

As we look at our method for comparing masses we may well come to an unfortunate conclusion about what mass really means: are we to take it that the mass of a body is really the pull the earth exerts on it? That would be a natural conclusion, but it is not correct. We have a name for the pull of the earth on a body but it is not mass; we call it the *weight* of the body and we do not measure it in kilograms. Objection! We used the *weights* of our bits of metal when we said we were comparing their masses! Yes, because two objects with equal masses also have equal weights when they are in the same place. But if I take my kilogram ten kilometres up, the pull of the earth on it will be measurably different, yet it will still be the same chunk of matter, it will still have the same mass. So the mass of a body is controlled by the space it takes up? No, that is not true either – a kilogram of expanded polystyrene occupies a lot more space than a kilogram of platinum alloy. But there *is* something about a mass of one kilogram which is the same wherever it is and whatever it is composed of; we shall look into it in detail in Chapter 5, but we can state the basic point here. *It has the same degree of objection to changing its motion.* It is easier to think about something big, like a 1000 kg trolley. It is difficult to get moving *and once it is moving it is difficult to stop or deflect.* The trolley would experience less pull towards the earth in an aircraft flying at 10 km than at ground level, but we would have an identical amount of difficulty in getting it to start rolling along the floor, given that the bearings were well lubricated. The name we give to this reluctance which bodies have to change of motion is *inertia*.

Two phenomena are involved therefore when we talk about mass – gravitational attraction, and inertia. As far as we can tell, if two objects have the same inertia, they also experience the same gravitational pull at the same point in space, and so we can use either effect to compare masses. But although the gravitational pull depends on the mass, it depends on other things as well (it is different in different places) and that is why we have to distinguish between the mass and the pull it experiences.

Area

Though all measurements in physics involve a unit and a counting process, not all units are defined by a standard in the same way as the metre, the second and the kilogram. Take a familiar quantity like area. We could if we wished set up a unit of area by making a standard; it could be a square of plastic like a floor tile. We would need a name for it, the slab perhaps, and then we could measure any area by laying down copies of the slab and counting. Soon a very significant feature would emerge. Suppose it happened that a square with sides one metre long needed sixteen of our tiles to cover it – it had an area of sixteen slabs. We should then find that a rectangle with edges of length one metre and three metres would need forty-eight tiles, and a square of side two metres would need sixty-four tiles, and so on. All the examples we tried could be summarised in the following statement. Area of rectangle in slabs = length in metres × breadth in metres × 16.

Now we have a formula connecting the quantities we have been measuring. We may think we can get at it by pure logic, but the modern view would be that it is a matter for experiment to decide. It is a simple example of a *law*.

It would not take long to find formulae for areas of more complicated shape, and then we should rapidly become tired of laying down tiles every time we wanted an area – we should find it quicker to measure lengths and use the formula. The factor 16 would soon become tiresome and we should all agree to scrap the slab and start with a new unit, corresponding to a standard tile one metre square. But that standard tile would be open to the same objection as the standard gram; one day someone might succeed in demonstrating that the tile had a side only 0·999997 m long, and we should be stuck with a number far worse than 16 in our formula. So we do not bother with a standard tile; we define the unit of area as a square of side exactly one metre.

Our formula is now

$$\text{Area} = \text{length} \times \text{breadth}$$

or in symbols

$$A = lb,$$

and because we are multiplying metres by metres we call the unit of the resulting quantity the *square metre* (m^2). It is called a *derived* unit because it does not have its own standard – it is expressed entirely in terms of other units. Units which are defined in terms of standards or prototypes are called *basic* units; the metre, the second and the kilogram are all basic units.

Exercise 2.4

How many (a) square centimetres, (b) square millimetres are there in a square metre?

Volume

We can follow through a very similar argument for volume. This time we shall end up with the formula for a box with perpendicular faces:

$$\text{volume} = \text{length} \times \text{breadth} \times \text{height}$$

or

$$V = lbh,$$

and so we call the unit of volume the *cubic metre* (m^3).

Exercise 2.5

How many (a) cubic centimetres, (b) cubic millimetres are there in a cubic metre?

Actually in this case something very like the story of the slab did occur. For many years there were two ways of measuring volume, one based on the cubic metre and one on the *litre*. Originally the litre was simply $1/1000\,m^3$, but because an attempt was made to build the prototype kilogram so that it should have the same mass as $1/1000\,m^3$ of water under specified conditions, the litre was re-defined in 1901 in terms of the volume of a kilogram of water. Needless to say, it has been established that the prototype kilogram does *not* have a mass exactly equal to that of $1/1000\,m^3$ of water, and so for a long period scientists had to put up with the fact that there were not quite 1000 litres in a cubic metre. This source of confusion was removed in 1964 by going back to the old definition of $1/1000\,m^3$ for the litre.

Density

Another derived quantity which we find very useful is the *density* of a material – the mass of unit volume of it. The formula this time is

$$\text{density} = \frac{\text{mass}}{\text{volume}}$$

or in the usual symbols

$$\rho = \frac{m}{V}$$

and so the unit is kilogram per cubic metre ($kg\,m^{-3}$). (Italic symbols are used for quantities, roman symbols are used for units, so *m* denotes a mass and m refers to a metre.)

Exercises 2.6,7

6 a) In view of the fact that a litre of water has a mass of almost exactly one kilogram, find the density of water in $kg m^{-3}$.

b) If the mass of $50 cm^3$ of mercury is $680 g$, express its density in $kg m^{-3}$.

c) The mass of an initially evacuated 2·0 litre flask increases by 2·4 g when air is let into it. Find the density of the air in $kg m^{-3}$ and the mass of the air in a room measuring 4·0 m × 3·0 m × 2·5 m.

d) Given that the density of a substance is $X g cm^{-3}$, express its density in $kg m^{-3}$. (Usually the symbol for a quantity stands for the *magnitude with its unit*, for example $m = 3 kg$, and it would generally be incorrect to write $m kg$. Throughout this book, therefore, the symbols X, Y and Z are reserved for *numerical values* of quantities – hence $X g cm^{-3}$ in the question.)

7 Most of the mass of an atom is found to be concentrated in a tiny central core called the nucleus, about $10^{-15} m$ across. Taking a typical density of $1000 kg m^{-3}$ for solid material, and using the fact that atoms are about $10^{-10} m$ across, find what value this gives for the density of nuclear matter in $kg m^{-3}$.

Coherent system

Though it is usually more convenient to form a derived unit when we can, it occasionally pays to start fresh and introduce another basic unit. The ampere for electric current and the kelvin for temperature are further examples of basic units in the SI system. Once the basic units have been decided on, the arithmetic becomes a lot easier if all the other units are derived from them without introduction of any numerical factors except unity; we chose the square metre rather than the 'slab' for the unit of area, for example. A set of units constructed in this way is called a *coherent system*; SI units form a coherent system. It is not the only possible system, indeed there is nothing sacrosanct about the basic units, and there would be nothing to stop us making up a system which had area as a basic quantity defined in terms of a standard, and length as a quantity derived from it.

Exercise 2.8

In a possible new system of units, density and volume are adopted as the basic quantities, and the symbols for their *units* are d and V respectively. Find units for length, area and mass in terms of d and V.

The radian

We must mention one more so-called *supplementary* unit which is incorporated into the SI system but which is a little different in character from those we have considered so far, it is the unit for measurement of *angle*. We can measure angles in degrees of course but there is a unit which is more useful in many cases. Very often in physics we use an angle as a convenient measure of how far we have travelled round the circumference of a circle, and so it suits us very well if there is a simple connection between the angle and the arc of

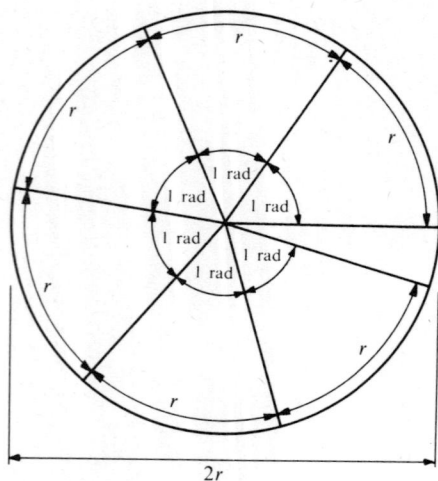

Figure 2.2

$360° = 2\pi\,\text{rad}$, or approximately $6\cdot283\,\text{rad}$.

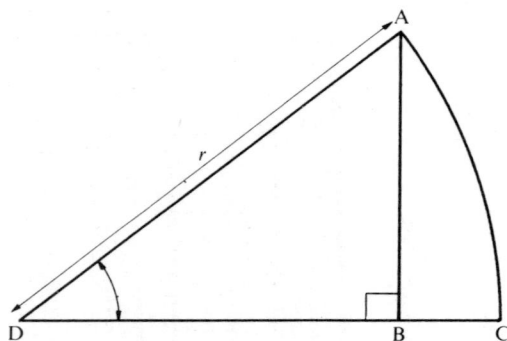

Figure 2.3 (Exercise 2.10)

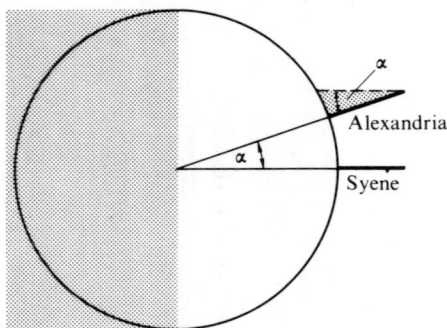

Figure 2.4

Eratosthenes' determination of the radius of the earth (Exercise 2.12)

the circle it defines. If we use degrees we do not get a very simple connection.

The unit angle we choose is the one which defines a length of arc equal to the radius of the circle (Figure 2.2), and it is called one *radian* (rad). Two radians will give us a length of arc equal to twice the radius, and so on. We can summarise all possible cases by the formula:

$$\begin{array}{c}\text{Distance round that part of the}\\\text{circumference defined by a given angle}\end{array} = \begin{array}{c}\text{radius}\\\text{of arc}\end{array} \times \begin{array}{c}\text{the angle}\\\text{in radians}\end{array}$$

or

$$\text{length of arc} = r \times \theta$$

This formula is *not true if we measure θ in degrees*. When we write a formula in symbols it looks deceptively simple, but it has to be treated with care; we have to know exactly what each symbol means and the circumstances it refers to before we are safe to use it. To reduce possible confusion we shall take it from now on that the symbol θ refers always to an angle measured in radians.

It is useful to know how to convert from degrees to radians. The total circumference is $2\pi r$ so we can get right round it in 2π arcs of length r, or in other words 2π radians of angle. So $2\pi\,\text{rad} = 360°$; a radian is just less than $60°$, or to five significant figures $57\cdot296°$.

Exercises 2.9–12

9 Find the formula connecting the length of an arc of a circle with its angle when the angle is in degrees.

10 In Figure 2.3, AC is an arc of a circle with centre at D.
   a) Write down expressions for the following in terms of r, AB, AC (measured along the arc) and DB: (i) θ (in radians), (ii) $\sin\theta$, (iii) $\tan\theta$, (iv) $\cos\theta$.
   b) Read the values of all four from tables when the angle ADB is $10°$, and infer what approximations can be made for $\sin\theta$, $\tan\theta$, and $\cos\theta$ when θ is small.
   c) Find the fractional difference $(\theta - \sin\theta)/\theta$ (i) when angle ADB is $5°$, (ii) when it is $10°$.
   d) Find the ratio between the length of arc AC and the line AB when $\theta = 0\cdot5\,\text{rad}$.
   e) Evaluate BC/DC when the angle ADB is (i) $1°$, (ii) $5°$. Hence consider whether it is important that rules should be truly vertical when measuring heights.

11 A and B are two points on the earth's surface $2000\,\text{km}$ apart, as measured along a line laid on the surface. Find the length of the straight line linking A and B. Take the radius of the earth to be $6400\,\text{km}$.

12 Eratosthenes (275–194 BC) found a value for the radius of the earth by a method indicated in Figure 2.4. At noon on Midsummer day a vertical pole erected at Syene produced no shadow; at noon on the same day at Alexandria the angle α subtended from the top of a vertical pole by its shadow was found to be $1/50$ of a full circle. Eratosthenes estimated that Alexandria was 5000 stades due north of Syene. Given that 10 stades $= 1\cdot6\,\text{km}$, find the value this information gives for the radius of the earth.

3 The interpretation of experimental data; errors and graphs

[Hiram] . . . made the Sea of cast metal; it was round in shape, the diameter from rim to rim being ten cubits; it stood five cubits high, and it took a line thirty cubits long to go round it.

I Kings 7: 23 *New English Bible*

Suppose we have a hypothesis from which we are able to develop some theory and arrive at a mathematical formula. We devise an experiment which gives us measurements to fit into the formula to see if it works, and we find that our numbers almost fit but not quite: they are out by, say, one part in twenty, a very common experience! The question we have to answer is which is wrong, the hypothesis or the results? Hypotheses always bow to experiment! But was that experiment carried out properly, could the results be in error? It becomes essential in every experiment not only to get a result but also to establish how much trust we can place in it, how much error may be involved. It is convenient to talk about three types, *systematic*, *calibration* and *random* errors.

Systematic error

Systematic errors are those which turn up again to the same extent when an experiment is repeated, even when the instruments used have been accurately calibrated. A lot of them are due to insufficient matching of the experiment to the theory it is supposed to test. In Chapter 15 we shall work out some theory which will predict the period of a pendulum bob swinging on the end of a thread. The period is the time for one complete swing, for example from A to C and back to A, or from B to A to C and back to B (Figure 3.1). When we are measuring periods we usually time from the centre of the swing because that is where the bob is travelling most quickly and we can be most certain about the instant it passes through. In our theory (p. 207) we take it that the angle θ which measures the extent of the swing is equal to $\sin\theta$. This is never true in practice, but Exercise 2.10 shows that it is wrong by about one part in a thousand if θ is no more than 0.1 rad ($\approx 5°$). In addition we do not write into the theory the fact that the thread has some mass, nor that the bob takes up some space, nor that the surrounding air may affect the motion. We pretend we have an ideal case of a thread which is not there and a bob which consists of a lot of matter all concentrated at one point. This product of the imagination is called a *simple pendulum*.

We do our theory in this way not only because it would be difficult to bring in the other things but also because we want to concentrate on the most significant aspects, and we believe we can set up a real bob on a real thread which is so close to the ideal that the effects we have not dealt with will affect the period by an insignificant amount. In our experimental test therefore we had better not use a tennis ball on the end of a thick rope; that would be a sure way to introduce systematic error. We want a small dense bob and fine thread. Since the length of the thread is of great interest we shall have to be careful about it; we measure from the middle of the bob to the point of suspension, and the point of suspension has to be a *point*. We need an arrangement like that in Figure 3.2a, so that the length remains the same throughout the swing, which we shall keep below 5°. All these points are potential sources of systematic error, and the trouble is that we are never *sure* we have thought of everything.

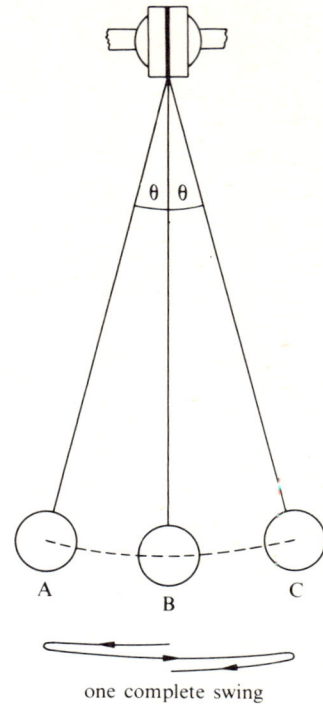

one complete swing

Figure 3.1

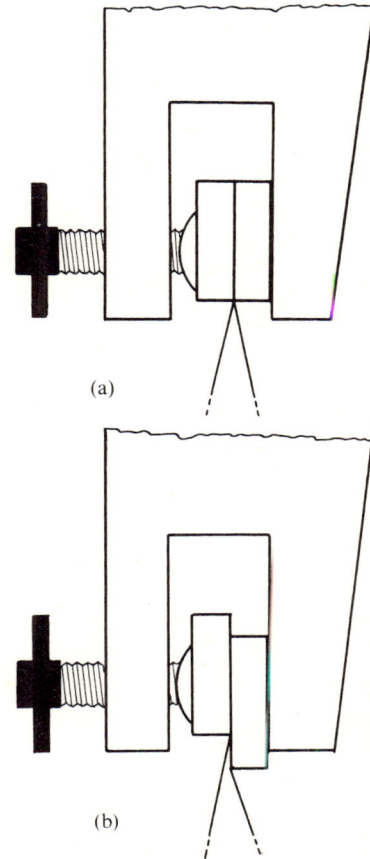

(a)

(b)

Figure 3.2
(a) a simple pendulum support which ensures constant length throughout the swing, and (b) one which does not!

Calibration errors

We can find another fertile source of error in the instruments themselves; we need a metre rule and a clock, and they may not match up to the quality we should like. The rule may have irregular divisions, or the clock may run fast or slow, and such *calibration errors* will not be eliminated by repeating a reading; in this respect they form a special class of systematic errors. Metre rules, clocks and ammeters must be checked against reliable copies of the standards, called substandards–a confusing name, since the precision with which they are made is anything but substandard! In this respect we have to make the distinction between an *accurate* instrument and a *sensitive* one; accuracy is a word describing how nearly the instrument corresponds to the agreed standards, and sensitivity has to do with how small a change can be detected. A good stopwatch reading to the nearest 0·1 s may not be as sensitive as an electronic device registering to the nearest 0·001 s, but over a period of several minutes it may be more accurate.

Even if the clock runs at the correct rate, it may have a *zero error*. For example it may start 0·3 s late every time we press the button. Instruments should be checked for zero error as a matter of course, though in fact it is often possible to detect a zero error by a careful look at the tabulated results, as we shall show.

Random errors

By contrast, random errors are different each time a reading is repeated; we measure the length of the pendulum and record it as 46·35 cm, and when we measure it again later we think it is 46·37 cm. Of course we are not going to think differently on the second occasion unless we remove and replace the rule in the meantime, in which case we are unlikely to put it back in exactly the same place. Many random errors, though not all, are due to the fallibility of human judgement. They are much more manageable than systematic errors because they can be both estimated and minimised by standard procedures.

Estimation of error from scale

In some cases it is appropriate to be guided by the *size of the smallest scale division*. The smallest division on a metre rule for example is usually 1 mm, and so it is difficult to be certain of a length to within anything better than perhaps one fifth of a division, 0·2 mm, either way. It would be appropriate therefore to quote the length of the pendulum as (46·36 ± 0·02) cm (this is read as 'plus or minus nought point nought two centimetre'). The uncertainty or estimated error is ±0·02 cm, the estimated fractional error is plus or minus two parts in five thousand (it is pointless quibbling about the odd 364) or ±1/2500 and the estimated percentage error is ±(1/2500) × 100 % or ±0·04 %.

Exercise 3.1

Write down the estimated percentage errors for the following results: (28·3 ± 0·3) kg, (84·6 ± 0·1) cm, (35 ± 1) s.

If we were measuring something about 5 cm long instead of 50 cm, the error would be the same but the fractional error would be ten times as great. Often the fractional error is the more significant of the two and it is worth taking steps to minimise it by choosing as large a value as possible for a given quantity. For this reason we time not one period of the pendulum but many–fifty or a hundred. The stopwatch we use advances in 0·1 s jerks, so there is no question of estimating parts of a division! If we relied on the information given by the watch scale we should quote the error as ±0·1 s, but there are more dependable criteria available to us.

Estimate of error from range

Table 3.1 shows the results of twenty determinations of the time t for fifty swings of the pendulum. The symbol (t/s) at the head of the table is an example of standard SI notation; a column of results or a graph axis is labelled by the symbol for the quantity concerned, a solidus, and the unit in which the quantity is measured. In fact the solidus denotes a division process; the idea is that t stands for the quantity including its unit, for example $t = 68·5$ s. If we divide this by the unit, a second, we get a pure number, 68·5, which appears in Table 3.1.

The best estimate we can make of the true value of a quantity from a reliable set of readings is the *arithmetic mean*–the sum of all the readings, divided by the number of readings. The arithmetic mean of the numbers in Table 3.1 is 68·555, and they range between 68·3 and 68·8, so if there is no systematic error or calibration error we can say fairly safely that the answer lies between these extremes. We could if we wished write the result as (68·56 ± 0·25) s but that would be rather presumptuous. There is no point in pretending that the last figure means very much either in the reading or in the error and it is probably better to settle for (68·6 ± 0·2) s or (68·6 ± 0·3) s; there is little to choose between these on the rough basis we are using.

Histogram and normal distribution

A look at the range of the results is quite adequate for estimating error if just a few readings are available, but as more and more readings are taken, it is possible not only to estimate the random error more closely but also to reduce it. A very good way of presenting the data in Table 3.1 at a glance is to draw a *histogram* (Figure 3.3). Each possible value is given its own rectangle, and the height of the rectangle represents the number of readings of that value obtained. Now suppose we took not twenty readings but a very large number, and at the same time we used a clock with very much smaller scale divisions, each of which was given its own vertical column in the histogram. The top edges of the rectangles would form such minute steps that the histogram would begin to look like a continuous curve. The peak of the curve would indicate the value occurring most frequently; strictly this is called the *mode*, but for most practical purposes it can be taken to be equal to the mean as long as a large number of readings are involved. Statisticians tell us that as more and more readings are taken the shape of the curve will correspond more and more closely to a theoretical curve which they can represent by quite a simple mathematical formula; it is called the *normal distribution curve* or random error curve, and it applies not just to clock readings but also to a host of other situations. The peak of the normal distribution curve shows the true value of the quantity concerned, corresponding to the mean of an infinite number of readings.

Table 3.1

Twenty determinations of the time t for 50 periods of a pendulum.

	t/s		
68·6	68·8	68·3	68·4
68·4	68·4	68·7	68·5
68·5	68·5	68·6	68·8
68·7	68·6	68·6	68·5
68·6	68·5	68·5	68·6

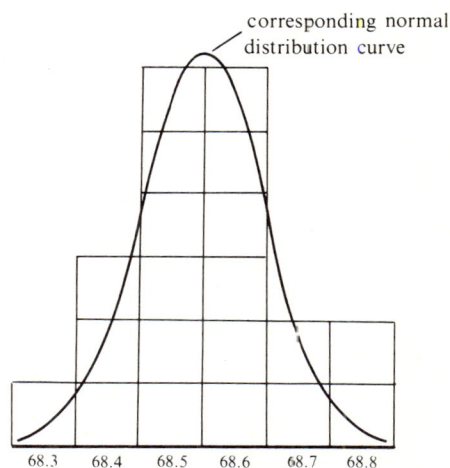

corresponding normal distribution curve

68.3   68.4   68.5   68.6   68.7   68.8

Figure 3.3
Histogram for Table 3.1.

Standard deviation and standard error

Now although the mean of twenty readings need not by any means coincide with the peak of the normal distribution, it has a better chance of being close to it than does a single reading chosen at random. Statistical theory can tell us what chance there is that the correct result lies within given limits on either side of the calculated mean. The arguments are rather long and involved so we shall just quote the formula; to do that we need to understand what is meant by *standard deviation*.

The deviation of any reading from the mean is, as we should expect, merely how far it is from the mean. To find the standard deviation of the readings in Table 3.1 we have first to write down all twenty deviations and square each one. Then we add the twenty squared deviations together and divide by twenty: the result is called the *mean squared deviation* or *variance*. Finally we take the square root of the mean squared deviation and obtain the *standard deviation*, for which we use the symbol σ.

Example

Table 3.2

Reading	39·5	40·0	40·5
Frequency*	2	5	3

*number of occurrences

Find the mean and standard deviation of the readings in Table 3.2.

$$\text{Mean} = \frac{\text{sum of readings}}{\text{number of readings}}$$

$$= \frac{2 \times 39{\cdot}5 + 5 \times 40{\cdot}0 + 3 \times 40{\cdot}5}{10}$$

$$= \frac{400{\cdot}5}{10} = 40{\cdot}05.$$

$$\text{Variance} = \frac{\text{sum of squares of deviations from mean}}{\text{number of readings}}$$

$$= \frac{2(40{\cdot}05 - 39{\cdot}5)^2 + 5(40{\cdot}05 - 40{\cdot}0)^2 + 3(40{\cdot}05 - 40{\cdot}5)^2}{10}$$

$$= \frac{2(0{\cdot}55)^2 + 5(0{\cdot}05)^2 + 3(0{\cdot}45)^2}{10}$$

$$= 0{\cdot}1225.$$

Standard deviation $= \sqrt{(0{\cdot}1225)} = 0{\cdot}35.$

Though any set of results can be processed in similar fashion, the work involved can often be reduced by making use of a working mean (sometimes called an arbitrary origin)–an initial guess at the answer. It is shown in books on statistics that

$$\text{mean} = \text{working mean} + \frac{\text{sum of deviations from working mean}}{\text{number of readings}}$$

and

$$\text{variance} = \frac{\text{sum of squares of deviations from working mean}}{\text{number of readings}} - \left[\frac{\text{sum of deviations from working mean}}{\text{number of readings}}\right]^2$$

and we shall use these relationships to work the same problem again, selecting a working mean of 40·0.

Table 3.3

Reading	Frequency	Deviation of reading from working mean, x'	$(x')^2$	$x' \times$ frequency	$(x')^2 \times$ frequency
39·5	2	−0·5	0·25	−1·0	0·5
40·0	5	0	0	0	0
40·5	3	0·5	0·25	1·5	0·75
			Totals	0·5	1·25

From Table 3.3,

$$\text{Mean} = 40 + \frac{0·5}{10} = 40·05$$

$$\text{Variance} = \frac{1·25}{10} - \left(\frac{0·5}{10}\right)^2 = 0·1225$$

$$\text{Standard deviation} = 0·35.$$

Exercise 3.2

a) Find the standard deviation of the results in Table 3.1 (p. 17).
b) Find the mean and standard deviation of the results in Table 3.4.

Table 3.4

Reading	49·7	49·8	49·9	50·0	50·1	50·2	50·3	50·4
Frequency	1	3	5	6	3	2	1	1

We can express the definitions for mean and standard deviation in symbols but they look rather formidable at first; if we have taken n readings, a typical one being x_i (the suffix i labels the reading x, telling us which of the n readings it is) then the mean $\bar{x}$ is given by

$$\bar{x} = \frac{\sum_{i=1}^{n} x_i}{n}$$

(read this as 'x bar equals sum of the x_i's from i equals 1 to n, divided by n') and the standard deviation

$$\sigma = \sqrt{\left(\frac{\sum_{i=1}^{n}(x_i - \bar{x})^2}{n}\right)}.$$

Exercise 3.3

Make use of the above symbols to write down similar expressions for the mean and variance in terms of the working mean m.

Once we have found the standard deviation σ we can write down all there is to know about the likely error; here we shall simply quote three useful results.

a) There is about one chance in three that the mean will be further than $\sigma\sqrt{n}$ from the correct result.

b) There is about one chance in twenty that the mean will be further than $2\sigma\sqrt{n}$ from the correct result.

c) There is about one chance in five hundred that the mean will be further than $3\sigma\sqrt{n}$ from the correct result.

The number $\sigma\sqrt{n}$ is called the *standard error* of the mean and is usually quoted as 'the error'. Clearly the standard error decreases as the number of readings increases; if we take eighty readings instead of twenty we shall halve the error. An interesting consequence is that a watch moving in 0·1s jerks can be used to quote a result with a standard error considerably *less* than 0·1s if sufficient readings are taken.

Exercise 3.4

Write down the standard errors for the results in Tables 3.1 (p. 17) and 3.2 (p. 18). In the latter case how many readings would be required to give a standard error of $\pm0·01$, assuming σ did not change?

An example of the use of graphs: investigation of free fall
We can often take in the significance of a set of results at a glance if we plot a graph. In Chapter 4 we shall be interested in the way the time of fall of an object depends on the height through which it drops. Figure 3.4 shows suitable apparatus for an investigation, making use of an electric stop clock reading to 0·01s; the object to be timed is a steel ball. There are two difficulties to be faced; we must ensure that the clock reads zero at the instant the ball starts and we must be able to stop the clock as the ball passes a given point. The switch-off is the easier to arrange; the current for the clock passes through a switch consisting of a metal gate, just held in position by a small magnet. The ball crashes through the gate at the

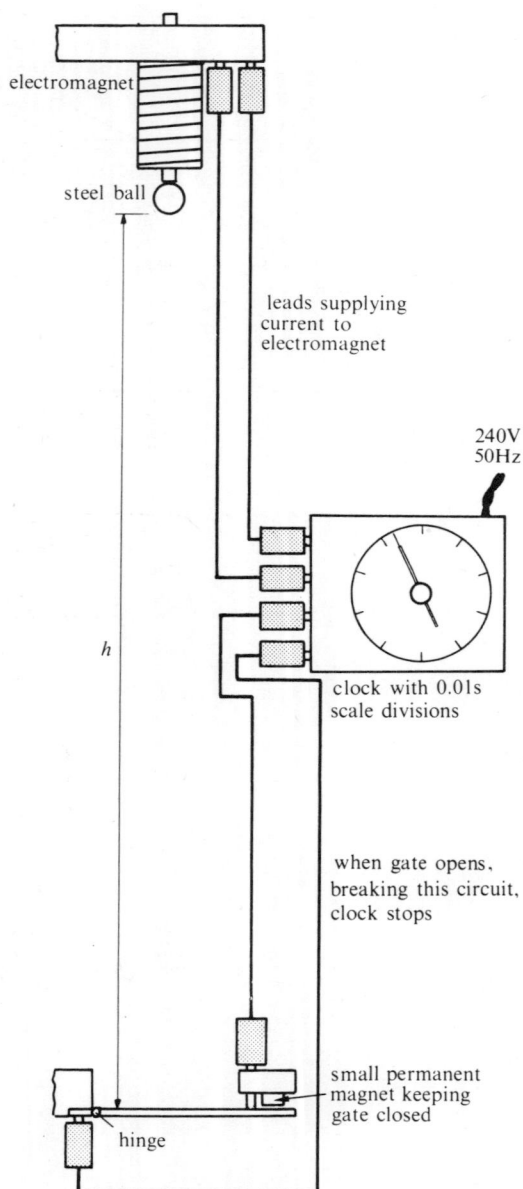

electromagnet

steel ball

leads supplying current to electromagnet

240V 50Hz

h

clock with 0.01s scale divisions

when gate opens, breaking this circuit, clock stops

small permanent magnet keeping gate closed

hinge

Figure 3.4

Electric stop clock reading to 0·01s.

end of its fall and stops the clock. We can control the starting time by releasing the ball from an electromagnet and the clock is so designed that the current for the electromagnet is switched off just as the clock pointer passes through zero–a rotating arm inside the clock breaks a set of contacts for a few hundredths of a second each second, and the current for the electromagnet is carried through these contacts. The departure timetable is very simple, therefore; the ball can start every second, on the second.

Exercise 3.5

What information is needed in addition to the vertical separation of magnet and gate to find how far the ball drops?

Table 3.5 shows a typical set of times recorded for different distances; the errors have been estimated as ± 0.01 s and ± 0.5 mm respectively.

Figure 3.5 shows a graph of (T/s) against (h/m). It does not really matter whether we plot (T/s) along the y-axis or the x-axis (the conventional names for 'vertical' and 'horizontal' respectively) but we usually reserve the x-axis for the variable over which we have direct control, called the *independent* variable, as opposed to the *dependent* variable. We could set up a drop of exactly one metre without too much trouble but we could not arrange for a time of exactly one second without a good deal of trial and error. We have plotted vertical bars instead of points to indicate the error in the timing; the length of the bar corresponds to the estimated error. We could in the same way use a horizontal bar to represent the error in the length of the drop, but on this occasion it is too small to indicate.

Exercise 3.6

What fraction of one division on the horizontal axis (0.5 units) would a horizontal bar have to be in order to indicate the error in (h/m)?

The curve we have drawn is a guess at the many points for which there are no data. We use a smooth curve through the plots rather than a series of straight lines joining them because we do not believe that the plots for which we happened to take readings are special: there is no more reason for thinking that the line suddenly kinks at these points than at any others.

We may be tempted to continue the curve back to the origin; after all, we can be quite sure that it takes no time to go no distance. However, it is quite possible that we have some zero error–not in the length of drop, we can be fairly certain of that, but in the timing. There may for example be some small inherent delay before the release of the ball after the switch has opened, so that the release does not coincide exactly with the transit of the pointer past zero. To take account of this possibility we shall distinguish between the *clock reading* T and the *time taken for the drop* t; if the ball leaves the electromagnet a short time T_0 after the pointer passes zero, T and t will be related by

$$T - T_0 = t.$$

Table 3.5

Time T registered by clock for length of drop h

h/m ± 0.0005	T/s ± 0.01
0.395	0.30
0.851	0.45
1.277	0.53
1.614	0.60
1.963	0.66

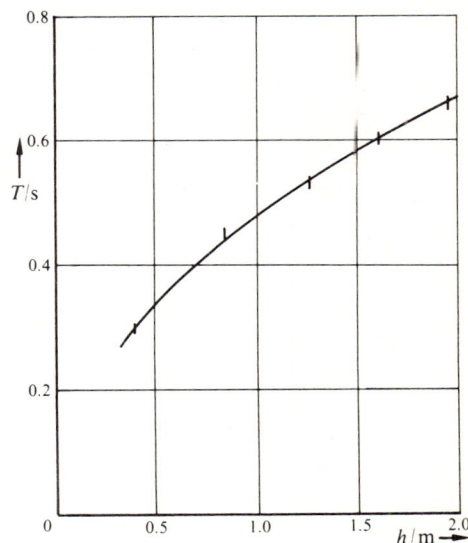

Figure 3.5

Graph of clock reading T against length h of drop for free fall of a steel ball.

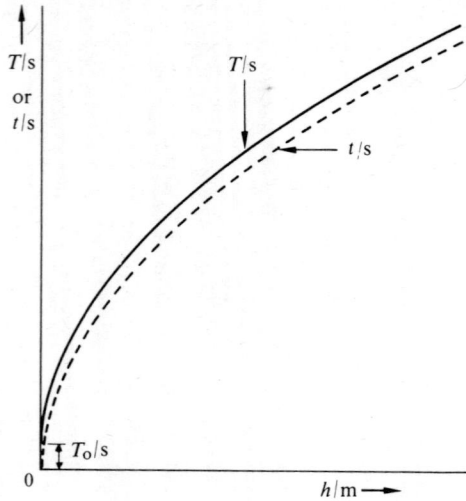

Figure 3.6

Comparison of clock reading T and time t of fall assuming a small zero error T_0.

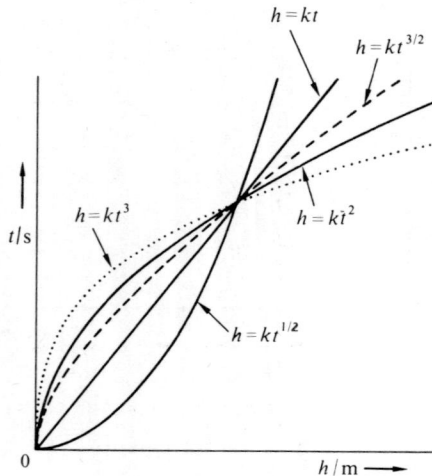

Figure 3.7

Curves which would result if h and t were linked by the formulae indicated.

This means that a graph of (t/s) against (h/m) would have the same shape as the curve in Figure 3.5 but would be shifted down by an amount T_0/s (Figure 3.6). Though we have little idea from Figure 3.5 how large T_0 may be, the treatment which follows will enable us to make an estimate of it.

We are really interested in whether or not there is some *simple mathematical formula* linking t and h. In particular in Chapter 4 we shall want to know whether the formula happens to be

$$h = kt^2,$$

where k is some number which does not vary (a constant). If h and t are linked in this way then a graph of (t/s) against (h/m) will be a *parabola*, and so therefore will a graph of (T/s) against (h/m). Though the graph in Figure 3.5 looks as though it *could* be a parabola we cannot be sure; we are not very good at distinguishing a parabola from other curves similar to it by eye. Figure 3.7 shows what we are up against; it indicates the shapes of curves which would result if h and t were connected by various formulae. All but one look quite like parabolas over the range drawn, but the exception is very important to us–the *straight line*, corresponding to

$$h = kt.$$

The straight line: slope and intercept

Because the straight line is the only one we can recognise at sight with any safety (and we can easily check with a ruler) we try to plot quantities which lead to a straight line wherever possible. The great thing about the straight line is that it shows increases in y which are proportional to the corresponding increases in x; if we go one division along the x-axis the line takes us, say, three divisions up the y-axis, if we go two divisions along the x-axis we are taken six divisions up the y-axis, and so on. A very convenient formula for the straight line is

$$y = mx + c,$$

and Figure 3.8 shows its significance. We have drawn a line corresponding to $y = x$; every point on that line must have a y-value equal to its x-value and so the line passes through the origin and goes up one division as it goes along one division. We have also drawn the line $y = 2x$; it goes up two as it goes along one, and it looks *steeper*. The line $y = 1{\cdot}5x$ lies between the two. The *gradient* or *slope* of the line is a measure of its steepness and we define it as the number of divisions a line climbs in the y-direction per division moved in the x-direction. The slopes of the three lines are therefore $1{\cdot}0$, $2{\cdot}0$ and $1{\cdot}5$ respectively; in general we use the symbol m for slope.

Also shown is the graph of $y = 2x + 3$. Now the only difference between $2x$ and $2x + 3$ is that three has been added! So every y-value which used to be $2x$ is now $2x + 3$. The line shifts up the y-axis by three divisions, but is otherwise unchanged, and its slope is still $2{\cdot}0$. In particular, it hits the y-axis at $y = 3$; this is called the *intercept on the y-axis* and is usually given the symbol c. *The line*

$$y = mx + c$$

is a straight line with slope m and intercept c on the y-axis.

Exercise 3.7

Write down the equations of lines 1 to 6 in Figure 3.9.

Choice of quantities for a straight line

The free-fall experiment with the steel ball presents us with some delicate points when we come to select the best graph to draw. We want to know if the length of the drop is proportional to the square of the time taken, and if we could be absolutely sure that the clock and release mechanism were free from zero error we could therefore simply test whether T^2 and h were proportional to each other; we could do that by plotting a graph of (h/m) against $(T/s)^2$ and looking for a straight line through the origin. But if there is a zero error T_0, we shall really want to test if

$$h = k(T - T_0)^2 \qquad \ldots(1)$$

that is if $\qquad h = k(T^2 - 2TT_0 + T_0^2)$.

If we plotted (h/m) against $(T/s)^2$, we would get an intercept on the (h/m)-axis (because in equation (1) h is not zero when $T = 0$) and we would also get a line with a slight curvature, because if h is proportional to $(T^2 - 2TT_0 + T_0^2)$ then it is *not* proportional to T^2. For this reason it is better to rewrite equation (1) in the form

$$\sqrt{h} = \sqrt{k}(T - T_0)$$

or $\qquad\qquad T = \dfrac{\sqrt{h}}{\sqrt{k}} + T_0. \qquad \ldots(2)$

We can now make a direct comparison with

$$y = mx + c.$$

Term for term, variable y corresponds to variable T
variable x corresponds to variable $\sqrt{h}$
constant m corresponds to constant $1/\sqrt{k}$
constant c corresponds to constant T_0.

If we plot (T/s) against $\sqrt{(h/m)}$, the intercept on the (T/s)-axis will give the magnitude of T_0, and if we have a correct mathematical representation of the situation we shall get a straight line with gradient numerically equal to $1/\sqrt{k}$. The graph is plotted in Figure 3.10. The units of k can be worked out from equation (2); $\sqrt{(h/k)}$

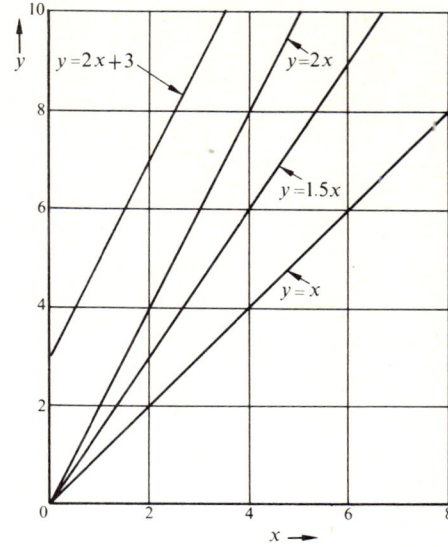

Figure 3.8

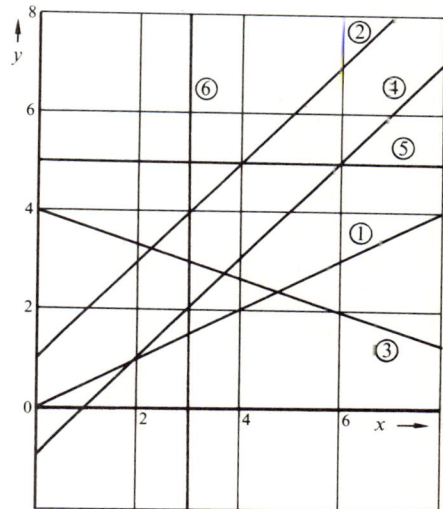

Figure 3.9 (Exercise 3.7)

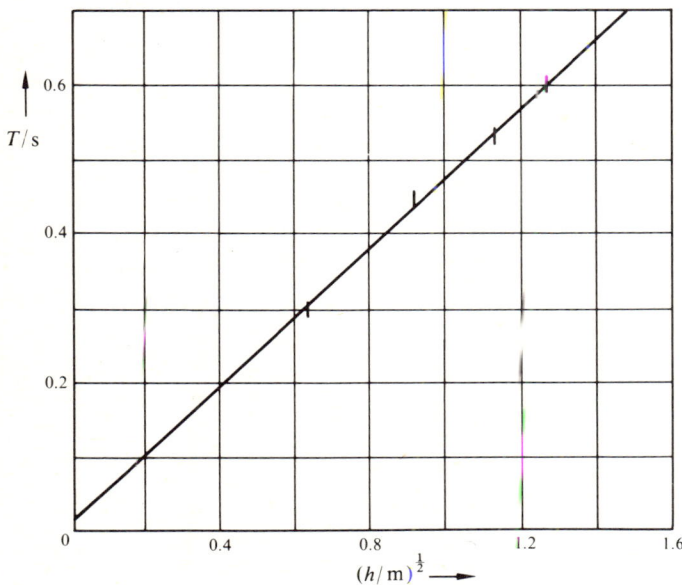

Figure 3.10
Graph of clock reading T against $\sqrt{}$(length h of drop) for free fall of a steel ball.

must have the same units, seconds, as T_0 because the two terms are added together in the equation, so (h/k) has units s^2. This means that the units of k are ms^{-2}.

Exercises 3.8, 9

8 Use the graph to write down the zero error T_0, and decide whether the ball left before or after the clock pointer passed through zero. Show also that the numerical value of $1/\sqrt{k}$ is 0·46.

9 What graphs would you choose to test whether two quantities p and q with units m and s were connected by

a) $p^2 = kq^3$
b) $p = kq^2 + r$
c) $p^2 + kq^2 = r$
d) $\dfrac{1}{p^2} + \dfrac{1}{kq^2} = r,$

where k and r are constants? In each case give the units of k and r and relate their numerical values to m and c, the gradient and intercept of the resulting line, assuming that it turns out to be straight. (See also Chapter 9.)

Estimation of error from the graph

The line drawn in Figure 3.10 is an attempt at the best one we can draw, the one which takes full account of all the plots and puts no more reliance on one than another. It is useful to get an idea of the possible error involved in drawing the line by defining the limits within which we can reasonably expect it to lie. These limits are drawn as (a) and (b) in Figure 3.11. The line (c) is one we could *not*

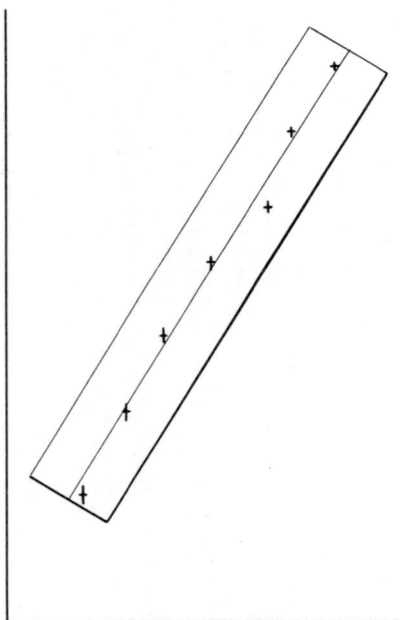

A transparent rule with a central line helps to find the best straight line.

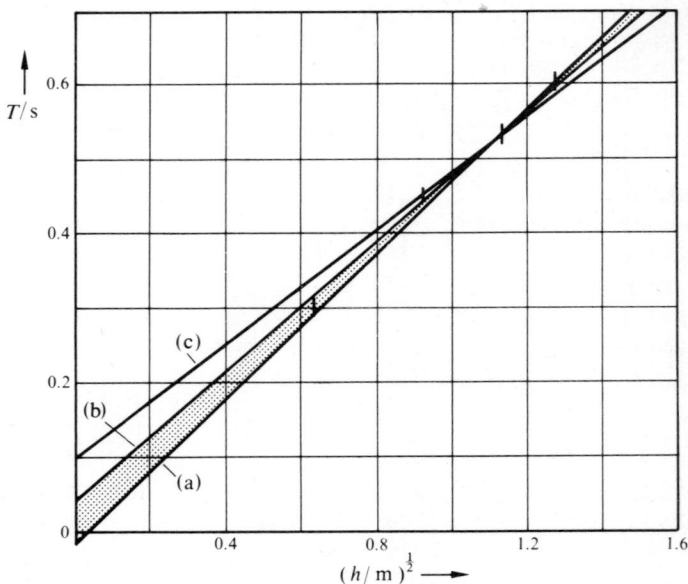

Figure 3.11

reasonably expect; it goes through two of the points but pays no attention to all the evidence provided by the other three. A transparent rule with a central line is an invaluable aid here so that all the plots are continually in view.

Exercise 3.10

Write down the slopes of (a) and (b) and thus quote $1/\sqrt{k}$ with its estimated error. Write down also the zero error of the clock together with its estimated error.

Method of least squares

There is in fact a standard method of arriving at the best straight line without any guesswork, but it is rather tedious and not always worthwhile when the points themselves are not particularly accurate. The criterion selected is that the *mean of the squares of the deviations of the points from the line* should be as small as possible; the line which makes that mean the least value is the best line. We shall merely quote the result; it applies only when all the points are equally trustworthy. If we have n pairs of readings for x and y, x_i and y_i being typical ones, then the best straight line is

$$y = mx + c,$$

where m and c are given by the simultaneous equations

$$\sum_{i=1}^{n} y_i = m \sum_{i=1}^{n} x_i + nc$$

and

$$\sum_{i=1}^{n} x_i y_i = m \sum_{i=1}^{n} x_i^2 + c \sum_{i=1}^{n} x_i.$$

The sums of the x_i's, y_i's, x_i^2's and $x_i y_i$'s can all be worked out mechanically using a desk calculator or computer and the simultaneous equations can be solved to find m and c.

Exercise 3.11

The readings in Table 3.6 are taken for two quantities x and y which are believed to be linearly related, but an accident prevents further readings.

Use the method of least squares to find the most likely straight line and draw it on a graph on which the three points are plotted.

Table 3.6

x	1·5	3·0	5·5
y	2·0	5·5	8·5

Logarithmic graphs

If we are completely in the dark about the relationship between the quantities we are measuring it may be some time before we hit on the functions which give a straight line, if indeed there are *any* functions of the quantities which will do so. In such a case there is an alternative approach which is often very useful. We suppose that h and T are linked by the formula

$$h = kT^n$$

where n is some constant number. It may be that no such link exists, but if that is the case it soon becomes apparent (the method is not able to cope with a zero error T_0 for example; we are relying on the assumption that it is very small). We now proceed by taking logarithms: strictly this can be done only with pure numbers, and h, k and T are all quantities with units. We can overcome this in the following manner. If for example

$$T = 5·0\,\text{s},$$

then

$$T/\text{s} = 5·0\,\text{s}/\text{s} = 5·0,$$

and this is a pure number with a meaningful logarithm, 0·6990. (T/s) means 'the numerical value of T when measured in seconds' and if we can arrange our equation so that it consists of purely numerical terms like (T/s) and (h/m) we shall be able to take logarithms successfully. We therefore write

$$\frac{h\,m}{m} = k\left(\frac{T\,s}{s}\right)^n \qquad \ldots(3)$$

thus leaving both sides unchanged. Equation (3) can be rearranged to give

$$\frac{h}{m} = \left(\frac{T}{s}\right)^n \times \left(\frac{k\,s^n}{m}\right),$$

and since (h/m) and $(T/s)^n$ are pure numbers, $(k\,s^n/m)$ has no option but to be a pure number also. We can now take logarithms of both sides. By the usual rules,

$$\log(h/m) = n\log(T/s) + \log(k\,s^n/m)$$

and if we compare this with $y = mx + c$ we can match the two equations term by term.

$$\log(h/m) \text{ corresponds to } y$$
$$\log(T/s) \text{ corresponds to } x$$
$$\text{constant } n \text{ corresponds to } m$$
$$\text{constant } \log(k\,s^n/m) \text{ corresponds to } c.$$

We hope n and k are both going to be constant, and if they are we see that we can expect a straight line when $\log(h/m)$ is plotted against $\log(T/s)$; n will be the slope and $\log(k\,s^n/m)$ will be the intercept on the y-axis. The graph is plotted in Figure 3.12. We have drawn the two limits within which we can be reasonably certain that the line should lie; their gradients are 1·9 and 2·15, so we can say that $n = 2·0 \pm 0·15$. We are so accustomed to finding the rules simple that we immediately put $n = 2$; even if n does deviate from 2 slightly this experiment is not sensitive enough to show it.

The intercept lies between 0·62 and 0·70, so

$$\log(k\,s^n/m) \text{ lies between } 0·62 \text{ and } 0·70.$$

From antilogarithm tables therefore,

$$(k\,s^n/m) \text{ lies between } 4·2 \text{ and } 5·0,$$
$$k = (4·6 \pm 0·4)\,\text{m}\,\text{s}^{-2},$$

which is consistent with the value of $1/\sqrt{k}$ obtained by the previous method (Exercise 3.10).

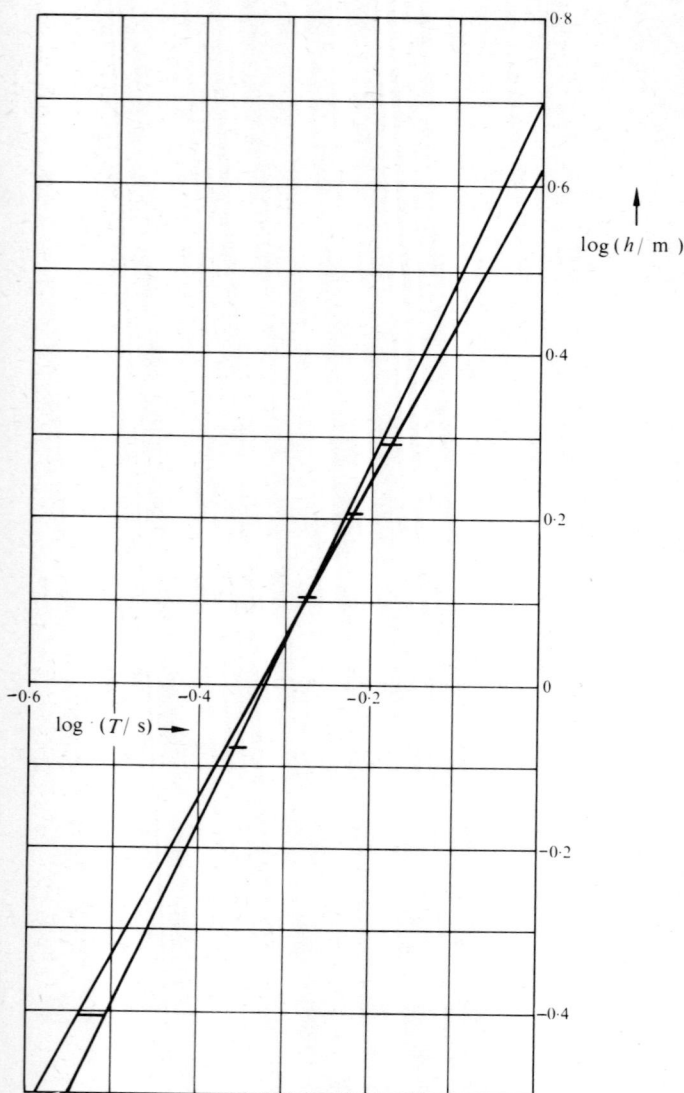

Figure 3.12

Logarithmic graph for free fall of a steel ball; T is the clock reading and h the length of the drop.

Exercises 3.12, 13

12 The readings in Table 3.7 are taken for two quantities p and q with units s and m respectively, and it is believed that they are linked by the expression $p = kq^n$. Plot a graph to confirm this and find k and n.

Table 3.7

q/m	0·200	0·400	0·800	1·200	1·600
p/s	0·895	1·27	1·79	2·19	2·53

(Remember $\log 0·200 = \bar{1}·6990 = -1 + 0·6990$).

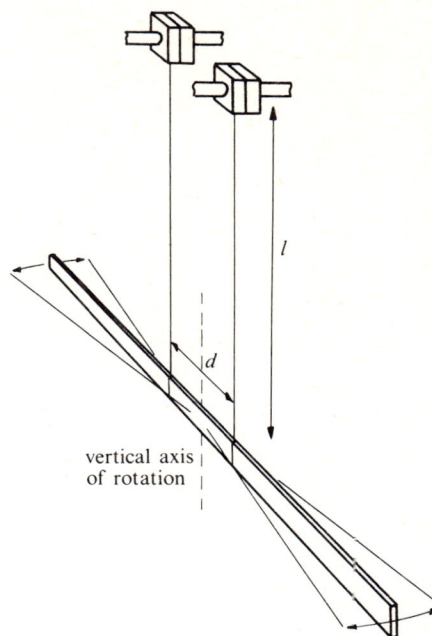

13 Figure 3.13 shows a *bifilar suspension* for a metre rule. Set up such an arrangement with l about 50 cm and d about 10 cm and measure the period T of small-angle oscillations of the rule about a vertical axis through its centre as indicated. Repeat for various values of d between 5 cm and 80 cm. Plot a suitable graph to test whether a relationship of the form

$$T = kd^n$$

exists between T and d, and find n if this is the case.

Combination of errors

It is one thing to estimate what error is involved in a reading; it is quite another to estimate how much error there will be in the final answer after that reading has been used in a formula. We need to know how the errors behave when we manipulate the numbers. There are several common situations.

a *Multiplication by a constant*

We have already met this case on p. 16. Suppose we have a reading x with error $\pm\delta x$. The symbol δx (delta x) merely means a certain small bit of x, an increment–it does *not* mean delta times x. The fractional error will be $(\delta x/x)$ and the percentage error $(\delta x/x \times 100)\%$. If for some reason we need to multiply our reading by, for example 500, the answer could clearly be anywhere between $(500x + 500\delta x)$ and $(500x - 500\delta x)$, that is $500x \pm 500\delta x$. So when the reading is multiplied by a constant the error is multiplied by the same constant. The fractional error is now $\pm 500\delta x/500x$; it has not changed.

b *Addition of two independent readings*

Exercise 3.14

Suppose we wish to add two readings $x \pm \delta x$ and $z \pm \delta z$. Write down the biggest and smallest values that the sum of the readings could take.

The exercise shows that the answer must be quoted as

$$x + z \pm (\delta x + \delta z).$$

When two independent readings are added their errors must be added to find the maximum error.

c *Subtraction of two independent readings*

Exercise 3.15

Find the biggest and smallest values that could result when $z \pm \delta z$ is subtracted from $x \pm \delta x$.

It follows that the result is $x - z \pm (\delta x + \delta z)$. *When one reading is subtracted from another independent reading the errors still have to be added to find the maximum error.*

d *Multiplication of two independent quantities*

If the two readings have to be multiplied, the results could be as much as $(x + \delta x)(z + \delta z)$ and as little as $(x - \delta x)(z - \delta z)$. Taking the

Figure 3.13 (Exercise 3.13)

top limit we can write it as

$$xz + x\delta z + z\delta x + \delta x \delta z$$

that is, it deviates from xz by an amount $x\delta z + z\delta x + \delta x \delta z$. This corresponds to a fractional error

$$\frac{x\delta z + z\delta x + \delta x \delta z}{xz} \quad \text{or} \quad \frac{\delta z}{z} + \frac{\delta x}{x} + \frac{\delta x \delta z}{xz}.$$

Suppose the fractional errors in z and x (respectively $\delta z/z$ and $\delta x/x$) are each about 0·01. Then $\delta x \delta z/xz$ will be about 0·0001, sufficiently small compared with $\delta z/z$ or $\delta x/x$ to be ignored. So the fractional change in xz is the sum of the fractional changes in x and z. We can follow through a similar process for the bottom limit, and conclude that *when two independent readings are multiplied, the maximum fractional error of the result is the sum of the individual fractional errors.*

e Division of two independent quantities

If we divide one reading by the other, the biggest answer we can get is $(x + \delta x)/(z - \delta z)$ and the smallest is $(x - \delta x)/(z + \delta z)$. Dealing again with the top limit we can find how much it deviates from x/z. The difference is

$$\frac{x + \delta x}{z - \delta z} - \frac{x}{z}$$

or

$$\frac{(x + \delta x)z - x(z - \delta z)}{z(z - \delta z)}$$

which simplifies to

$$\frac{z\delta x + x\delta z}{z(z - \delta z)}.$$

We can afford to ignore δx or δz in this expression when they are added to a much larger quantity, such as z or x. Thus although we cannot ignore δx or δz in the numerator (if we did we would get zero) we can ignore δz in the denominator, because $z(z - \delta z)$ is so close to z^2. So the expression approximates to

$$\frac{z\delta x + x\delta z}{z^2}.$$

The bottom limit can be followed through in the same way, and we conclude that the result we should quote is

$$\frac{x}{z} \pm \frac{z\delta x + x\delta z}{z^2}.$$

The fractional error is much easier to work with; it is

$$\frac{z\delta x + x\delta z}{z^2} \div \frac{x}{z} \quad \text{or} \quad \frac{\delta x}{x} + \frac{\delta z}{z}$$

(N.B. plus, not minus).

So *when two independent readings are divided, the fractional errors add to give the maximum fractional error.*

f Any function of a quantity

Suppose we have a reading $x \pm \delta x$ and we need to square it in our formula. The top limit is $(x + \delta x)^2$, that is

$$x^2 + 2x\delta x + \delta x^2$$

and the deviation is

$$2x\,\delta x + \delta x^2,$$

which is approximately $2x\,\delta x$ if δx is small compared with x. The fractional error in x^2 is therefore $2x\,\delta x/x^2$ or $2\,\delta x/x$, which is double the fractional error in x.

   We can deal with other powers of x along the same lines, but if we are prepared to use standard results of differentiation (see Appendix 2) the job can be done more quickly. We use the fact that

$$\frac{\mathrm{d}y}{\mathrm{d}x} \approx \frac{\delta y}{\delta x}$$

as long as δx and δy are small enough. So if the error δy in some number y is required, and y can be written in terms of x, then

$$\delta y \approx \frac{\mathrm{d}y}{\mathrm{d}x}\delta x$$

and the fractional error is given by

$$\frac{\delta y}{y} \approx \frac{\mathrm{d}y}{\mathrm{d}x}\frac{\delta x}{y}.$$

Some of the uses to which these expressions can be put are indicated in the following examples.

Example

The percentage error in x is 2%. What is the percentage error in x^4?

   In this case
$$y = x^4$$
$$\frac{\mathrm{d}y}{\mathrm{d}x} = 4x^3.$$

So

$$\frac{\delta y}{y} \approx 4x^3\frac{\delta x}{y}$$

$$\approx 4x^3\frac{\delta x}{x^4}$$

$$\approx \frac{4\,\delta x}{x}.$$

So the percentage error in x^4 is 8%.

Exercise 3.16

If $y = kx^{\frac{1}{2}}$, what percentage error in y is caused by a 2% error in x?

Example

An angle α is measured as $35° \pm 0.5°$. What is the fractional error in $\sin\alpha$?

   Though the answer could be found simply by looking up $\sin 35°$ and $\sin 35.5°$ in tables, we shall use the previous method again for illustration.

$$y = \sin\alpha$$
$$\frac{\mathrm{d}y}{\mathrm{d}\alpha} = \cos\alpha \qquad\qquad \ldots(4)$$

So

$$\frac{\delta y}{y} \approx \frac{\delta \alpha}{y} \cos \alpha \approx \frac{\delta \alpha \cos \alpha}{\sin \alpha};$$

but there is the snag that equation (4) applies only when α is in radians.

So

$$\frac{\delta y}{y} \approx \left(0{\cdot}5 \times \frac{\pi}{180}\right) \times \frac{\cos 35°}{\sin 35°}$$

$$\approx 0{\cdot}0125$$

and the percentage error is $\pm 1{\cdot}3\%$, probably better quoted as $\pm 1\%$.

Exercises 3.17, 18

17 Show that if the percentage error in x is $p\%$ the percentage error in x^n is $np\%$.

18 Readings are taken of quantities connected by the following formulae, and the percentage errors are given. Find in each case the percentage error involved in the calculation of the unknown quantity.

a) $Y = \dfrac{FL}{\pi r^2 l}$ $\begin{aligned}&F \pm 0{\cdot}1\% \quad L \pm 0{\cdot}1\% \\ &r \pm 0{\cdot}5\% \quad l \pm 1\%.\end{aligned}$

b) $V = \dfrac{\pi p r^4}{8 \eta l}$ $\begin{aligned}&V \pm 2\% \quad p \pm 2\% \\ &r \pm 1\% \quad l \pm 0{\cdot}5\%.\end{aligned}$

Sometimes we have to cope with a mixture of additions and multiplications. In such cases we have to work with actual errors while adding or subtracting, and convert to fractional errors when dividing or multiplying, converting back again if necessary.

Example

The following measurements are taken in an experiment to determine the length d of the diagonal joining opposite corners of a rectangular box with sides a, b and c (Figure 3.14):

$$a = (10{\cdot}0 \pm 0{\cdot}1)\,\text{cm}$$
$$b = (5{\cdot}0 \pm 0{\cdot}1)\,\text{cm}$$
$$c = (2{\cdot}0 \pm 0{\cdot}1)\,\text{cm}$$

Find the percentage error in d.

The formula is $d = \sqrt{(a^2 + b^2 + c^2)}$ and the working can be tabulated as in Table 3.8.

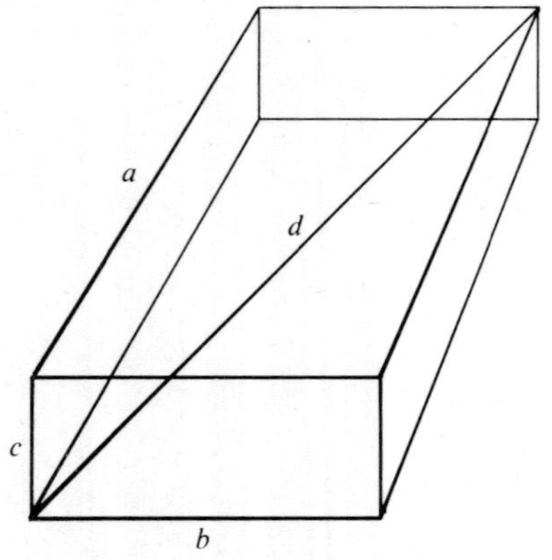

Figure 3.14

Table 3.8

Quantity	Magnitude	Percentage error	Error
a	10·0 cm	1	
b	5·0 cm	2	
c	2·0 cm	5	
a^2	100 cm²	2	2 cm²
b^2	25 cm²	4	1 cm²
c^2	4 cm²	10	0·4 cm²
$a^2 + b^2 + c^2$	139 cm²		3·4 cm²

So percentage error in $(a^2 + b^2 + c^2) = \dfrac{3\cdot4 \times 100}{139}\%$

$$= 2\cdot4\%.$$

Making use of the result quoted in Exercise 3.17, therefore,

percentage error in $\sqrt{(a^2 + b^2 + c^2)} = 1\cdot2\%,$

which we shall quote as 1%.

Exercise 3.19

What would be the percentage error in d if the errors in b and c were halved but the error in a doubled?

The rules derived in this chapter apply *only* when all the terms in the formula are *independent*. Suppose for example that we want to find the fractional error in $(x + 1)/x$ when $x = 50 \pm 1$. If we tried to use the rule for the division of one reading by another we would in effect be crediting x with the value 51 in the numerator and 49 in the denominator, and we would get an answer of 4% error. It is clear however that we must allow x to keep the same value throughout the calculation, and $(x + 1)/x$ must lie somewhere between $(52/51)$ and $(50/49)$. These ratios differ by 8 parts in 10000, giving a percentage error of $\pm0\cdot04\%$.

Summary of rules for combination of errors: Table 3.9

	Errors	Fractional or percentage errors
Quantity multiplied by constant n	multiplied by n	remain the same
Quantity raised to power n	no simple relationship	multiplied by n
Independent quantities added or subtracted	added	no simple relationship
Independent quantities multiplied or divided	no simple relationship	added

Exercises 3.20–23

20 Figure 3.15 shows an experimental arrangement for investigating the variation of the intensity of γ-rays with distance from the small source S. Since measurements cannot be made directly from the exact position of the source to the window of the Geiger–Muller tube, the distance d is taken from two convenient reference points as marked, and should be reduced by an amount x to give the corrected distance.

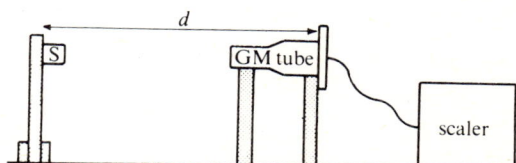

Figure 3.15 (Exercise 3.20)

Table 3.10

d/cm	15	20	25	30	35	40
Counts per minute	2527	1135	652	427	301	235

In an experiment the results shown in Table 3.10 were obtained.

The background count of the tube was 27 per minute.

The intensity of the γ-rays is thought to be inversely proportional to the square of the corrected distance. Express this mathematically, and show how the results should be plotted to give a graph which will be a straight line if this is true. Draw the graph and from it find the value of x.   (o)

21 At very low temperatures the specific heat capacity C of a metal varies with the absolute temperature T according to the equation

$$C = AT + BT^3$$

where A and B are constants.

Find, by means of a suitable straight-line plot, the values of A and B from the experimental results shown in Table 3.11.

Table 3.11

T/K	1·0	1·5	2·0	2·5	3·0	3·5	4·0	4·5	5·0
C/mJ mol^{-1} K^{-1}	0·78	1·28	1·92	2·74	3·78	4·88	6·72	8·71	11·1

(WJEC)

22 The diameter of a cylinder and its height are measured. Given that the maximum error in each of these measurements is $\pm 2\%$ write down, with your reasons, the approximate maximum percentage error in the volume of the cylinder as calculated from these measurements.   (o)

23 In an experiment with a reversible pendulum the following expression occurs:

$$C = (T_1{}^2 - T_2{}^2)/(l_1 - l_2).$$

If the errors in l_1 and l_2 are both negligible but T_1 and T_2 are both subject to uncertainties $\pm \Delta T$, what is the maximum permissible value of ΔT if C is to be accurate to 1%?   (o & c)

4 Motion

How fast can Smith run? On the track he has a best time of
$(13{\cdot}0 \pm 0{\cdot}1)$ s for 100 metres.
So in 13·0 s Smith covers 100 m
and in 1 s he covers $(100/13{\cdot}0)$ m or 7·7 m
giving a speed of 7·7 metres per second $(m\,s^{-1}) \pm 1\%$.

Average speed

This is hardly good enough as an answer to our question however.
We know very well that Smith takes some time to get up to top
speed – we are sure he runs the last ten metres faster than the first ten.
All we have found so far is his *average* speed over the hundred
metres; *we define average speed as (distance gone/time taken)*, with
units $m\,s^{-1}$.

Speed

How can we find his actual speed at any instant? More funda-
mentally, how can we *say what we mean* by the actual speed at any
instant? We need to take a more detailed look at the motion, and
this time we position timekeepers at ten-metre intervals along the
track; the times they record are shown in Table 4.1.

Exercise 4.1

Work out Smith's average speed for (a) the slowest, (b) the fastest
10 metre section.

This is at least a step in the right direction; we can now get some
idea of how the speed varies. We should get an even better idea if we
could position a clock every metre, though it would have to read to
0·01 s to be of much value since Smith covers a metre in a little more
than 0·1 s. We define the speed at any instant by taking this idea to
extremes. Suppose at some time t after the start Smith has covered a
distance x, and he gets a further short distance δx in an extra time
interval δt. The average speed over the extra distance is $\delta x/\delta t$, and
we say that the smaller we make δt, the nearer $\delta x/\delta t$ approaches v,
the speed at time t. In rather more formal language, as δt tends to

Table 4.1

Time t taken by Smith to get distance x from the start

x/m	t/s $\pm$ 0·1
10	2·3
20	3·8
30	5·1
40	6·3
50	7·4
60	8·5
70	9·6
80	10·7
90	11·8
100	13·0

zero, $\delta x/\delta t$ tends towards the speed v at time t. This is an *operational* definition for speed; it tells you what to measure, and how you ought to do it.

Calculus notation

We have a symbol for '$\delta x/\delta t$ as δt tends to zero'. It is written as $\mathrm{d}x/\mathrm{d}t$ (and read as '$\mathrm{d}x$ by $\mathrm{d}t$'); see Appendix 2.

$$\text{So} \quad v = \frac{\mathrm{d}x}{\mathrm{d}t} = \frac{\text{limit}}{\text{as } \delta t \to 0} \left(\frac{\delta x}{\delta t}\right)$$

$$= \text{rate of change of } x \text{ with respect to } t.$$

The idea of a limit takes some getting used to; surely, we might say, as δt approaches zero, δx will also approach zero, and as zero divided by any number is also zero, how can $\mathrm{d}x/\mathrm{d}t$ have a finite, and at that a perfectly definite value? The answer is that $0/0$ need not be zero at all, in fact it could be practically anything, and in any case, we have not said δx is zero, only that it *approaches* zero. We must be careful not to go all the way to zero in our definition; we can go as close as we like so long as we do not actually get there. This is merely saying that we cannot write down anything about the speed if we know just one position and its corresponding time; we must know the position at another time as well, though the interval between the two can be very small. Even apparently instantaneous speed recorders, car speedometers or radar speed trap devices, do not work instantaneously; they need information about the change of position over a short time interval.

Uniform speed

If we take a look at Smith's progress between 40 and 90 metres we see that the average speed over each 10 metre section is the same, $10/1\cdot1\,\mathrm{m\,s}^{-1}$ or $9\cdot1\,\mathrm{m\,s}^{-1}$. It would seem that the speed does not vary much over this section, and we may be tempted to say that it is constant. This may or may not be true; he could have covered the first five metres of a given section in 0·5 s and the next five metres in 0·6 s, and we should need more timekeepers stationed at smaller intervals to find out. From this idea we compose an operational definition for *uniform speed*: *an object is moving at uniform speed if it travels equal distances in equal intervals of time, no matter how small those intervals may be.*

Distance–time graph; gradient

Figure 4.1 is a graph of distance against time for Smith's run; we have made an exception to our usual rule and plotted the independent variable x along the vertical axis. The useful characteristic of this graph is that it enables us to see the way the speed is changing at a glance. Where the line is steeply inclined the runner is covering a large distance in a short time. At the start the line is less steep because Smith has not yet got up to full speed. We can put this idea on a quantitative basis, and we shall imagine first that Smith had a flying start and ran so as to produce a straight line graph (Figure 4.2). Since this line rises by the same amount in each second it indicates that Smith ran with uniform speed, and to find that speed we have only to look at how much the line rises in one second. In other words we need the *gradient m*. Defining it as a pure number, we have

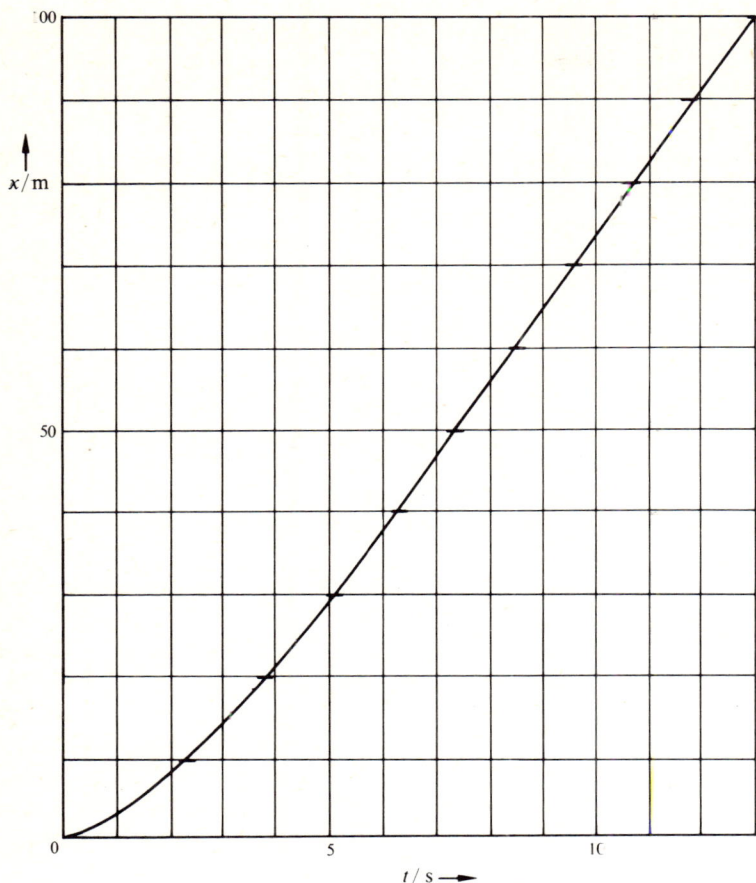

Figure 4.1

Graph of distance x against time t for Smith's run.

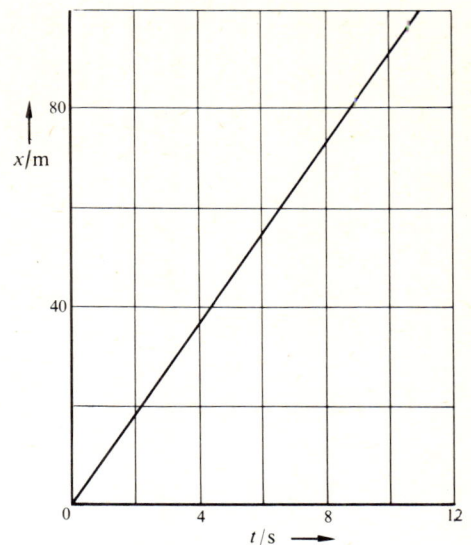

Figure 4.2

A flying start for Smith.

$$m = \frac{\text{increase in } (x/\text{m})}{\text{corresponding increase in } (t/\text{s})}$$

$$= \frac{(\delta x/\text{m})}{(\delta t/\text{s})}$$

(we have to take extra care here in distinguishing gradient m and metre m).

$$m = \left(\frac{\delta x}{\delta t}\right) \bigg/ \text{m s}^{-1}$$

so the speed v, because it is uniform in this case, is given by

$$v = \frac{\delta x}{\delta t} = m\, \text{m s}^{-1}.$$

Exercise 4.2

Write down the speed indicated by Figure 4.2.

How can we modify this approach to deal with a distance–time graph which is curved, as in Figure 4.3a? It is not good enough to

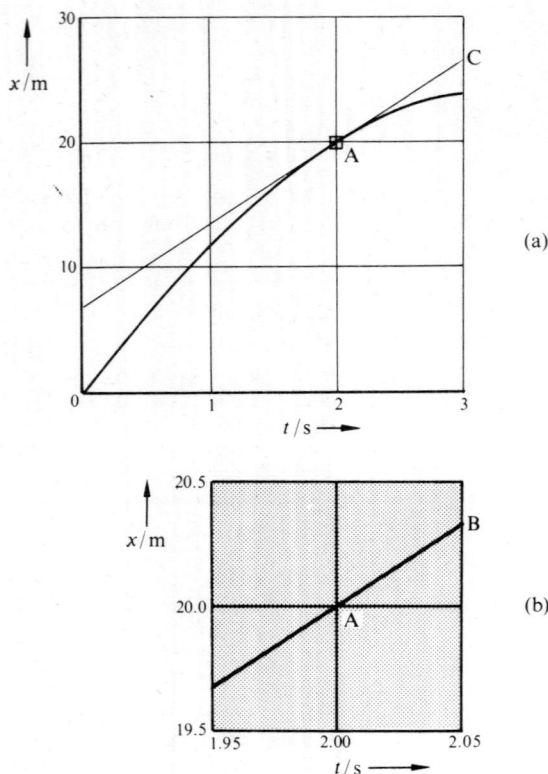

(a)

(b)

Figure 4.3

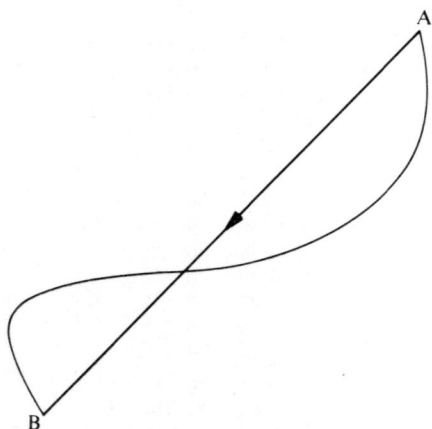

Figure 4.4

talk in terms of the number of vertical divisions the line rises through in one horizontal division; this will give us merely the average speed over the corresponding time interval. However, any curve looks like a straight line *as long as we look at a sufficiently small section of it*. In Figure 4.3b a small part of the main curve has been magnified, and since it is indistinguishable from a straight line we can measure its gradient by the usual method.

Exercise 4.3

Write down the gradient of the line *AB* in Figure 4.3b.

Imagine now that the line *AB* is carried on as a *straight* line. Between *A* and *B* it could be straight or it could belong to the curve, but outside the region *AB* the straight line deviates from the curve and heads for the point *C*. Clearly the gradient of *AC* is the same as the gradient of *AB*, and if we imagine *B* infinitesimally close to *A*, the line *AC* becomes the *tangent* to the curve at point *A*. If we want the gradient of a curve at a given point, therefore, we measure the gradient of the tangent to the curve at that point. (See also Appendix 2.) The gradient of the curve in Figure 4.1 increases at first, indicating a build-up of speed; Smith is accelerating from rest.

Exercise 4.4

Use Figure 4.1 to estimate Smith's speed 20 m from the start.

Scalars and vectors

Before going further we must mention a distinction made in physics between *speed* and *velocity*. To specify the velocity of an object we must give not only its speed but also the direction in which it is moving; a vehicle rounding a bend at constant speed will have a velocity which is changing at every instant. Sometimes it is only the speed which interests us–how fast can Smith run?–but if we are taking a journey on a motorway it is quite important to establish whether we are heading north or south: in this case the *velocity* is significant. In a similar way we distinguish between distance moved and *distance moved in a given direction*; the latter is called *displacement*. A person getting from *A* to *B* by the devious route shown in Figure 4.4 has gone a distance equal to the total length of the wavy line, and that is a useful quantity to know if we want to judge how tired he is or how much petrol he has used. But if we are interested only in whether he reached his desired destination, it is the *displacement*, the length and direction of the straight line *AB*, which concerns us. Whereas *speed* is *the rate of change of distance at a particular instant*, *velocity* is defined as *the rate of change of displacement at a particular instant*.

Quantities such as velocity and displacement, which must be defined by their size and direction, are called *vector* quantities; distance and speed on the other hand are examples of *scalar* quantities. We shall use bold-face type to denote vectors; *v* refers to a velocity, defined in size and direction. If at any stage we want to refer simply to the size of the vector we shall revert to ordinary italics, so that *v* means 'the magnitude of *v*'. Though strictly we should always call *v* 'the magnitude of the velocity' we shall refer to it simply as 'the velocity' in any situation where the direction of *v* is obvious.

Acceleration

We have made the distinction between vectors and scalars at this point in order to introduce the term *acceleration*, which is a measure of how quickly the velocity is changing. The point is that rate of change of speed is *not* in general the same as rate of change of velocity (a vehicle rounding a bend with no change of speed is changing its velocity), and of the two, rate of change of velocity is the more significant and useful quantity. This is by way of an advance warning. We shall look first at the special case of motion confined to a straight line, for which the changes in speed are equal to the magnitudes of the corresponding velocity changes. Smith's run is not the most convenient example to consider; in Table 4.2 we have a record of his progress in a straight line on a bicycle, starting from rest. His position is recorded every 0·02s by means of a *ticker-tape timer* (Figure 4.5). The device works from an a.c. source of frequency 50 Hz, and the diode allows current to pass in one direction only so that the soft iron core inside the coil becomes magnetised and demagnetised fifty times a second. The soft-iron lever above it is attracted every time the current flows but pulled back by the spring whenever the current is switched off, so that fifty times every second the hammer at the end strikes a piece of carbon paper and puts a small dot on the strip of ticker tape shown. If we attach one end of the tape to Smith's bicycle at the same vertical height as the vibrator and let him pull the tape through the timer we get a record of the amount δx by which the cycle advances in a straight line in each 0·02s time interval (we shall call the direction of that line the 'x-direction').

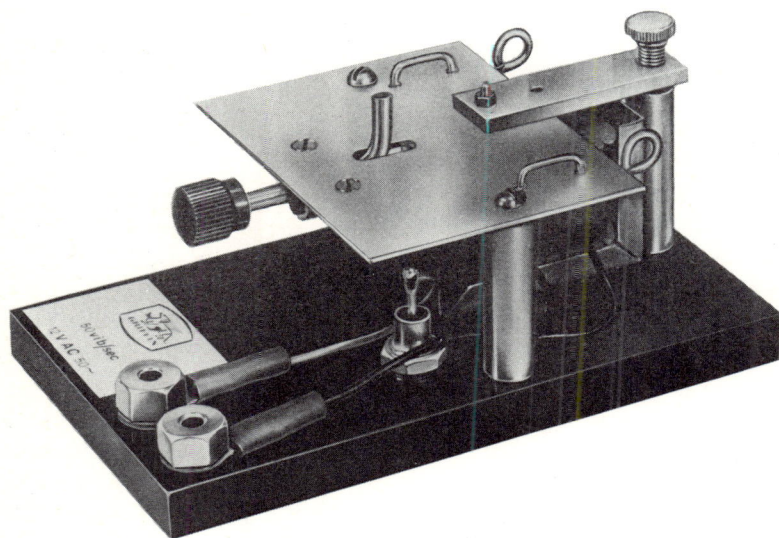

Figure 4.5
Ticker-tape timer.

We can use Table 4.2 to estimate Smith's velocity at any instant, such as 1·00s after the start, as follows. [We use vector rather than scalar terms in this situation because if for some reason Smith's cycle started going backwards we should want to reckon the x-value as decreasing, and clearly while the quantity 'distance gone' is indifferent to a change of direction, 'displacement' is not. This 'plus or minus' property of a vector confined to one direction is called its 'sense'.] Smith's velocity is changing all the time, so we know we have

Table 4.2

Sizes of changes of displacement δx experienced by Smith's bicycle in successive 0·02s intervals as time t increases.

(t/s)	$(\delta x/\text{m})$
0	0
0·02	0·001
0·04	0·002
0·06	0·003
0·08	0·004
0·10	0·006
0·12	0·007
0·14	0·009
0·16	0·010
0·18	0·011
0·20	0·012
0·22	0·013
0·24	0·013
0·26	0·014
0·28	0·015
0·30	0·015
0·32	0·016
0·34	0·017
0·36	0·018
0·38	0·018
0·40	0·019
0·42	0·020
0·44	0·021
0·46	0·022
0·48	0·023
0·50	0·024
0·52	0·024
0·54	0·025
0·56	0·026
0·58	0·027
0·60	0·028
0·62	0·029
0·64	0·030
0·66	0·031
0·68	0·032
0·70	0·033
0·72	0·034
0·74	0·035
0·76	0·036
0·78	0·036
0·80	0·037
0·82	0·037
0·84	0·038
0·86	0·038
0·88	0·038
0·90	0·039
0·92	0·039
0·94	0·039
0·96	0·040
0·98	0·040
1·00	0·040
1·02	0·040
1·04	0·041

to look at how far he gets in a *short* time interval. Suppose we pick an interval 0·04s long having 1·00s right in the middle of it; between 0·98s and 1·02s he travels 0·080m, so his average velocity during this short time interval is of magnitude $0·080/0·04 \, ms^{-1}$ or $2·0 \, ms^{-1}$. We take this to be a good estimate of his actual velocity half-way through the time interval, that is, at 1·00s.

Now we should like to find the *rate of change of velocity*–the *acceleration*. Since the initial velocity was zero (Smith started from rest) there has been a velocity change of $2·0 \, ms^{-1}$. But the change did not occur in a regular fashion; if it had done so we should expect the size of the velocity at 0·50s to be $1·0 \, ms^{-1}$, the speed at 0·25s to be $0·5 \, ms^{-1}$, and so on.

Exercise 4.5

Use Table 4.2 to estimate the size of Smith's velocity at 0·50s and at 0·25s.

All we have found so far is the average magnitude of the rate of change of velocity over the first second, which is called the *average acceleration*.

$$\text{average acceleration} = \frac{\text{velocity change}}{\text{time taken for change}} \qquad …(1)$$

The units are (ms^{-1}) per s, or ms^{-2}, *metres per second squared*.

Exercise 4.6

Find the size of the average acceleration (a) between 0s and 0·50s, (b) between 0·50s and 1·00s.

Now just as we developed the idea of speed from the average speed during a small time interval, so we can define the acceleration, the rate of change of velocity at any instant, as the limiting case of the average acceleration measured over a small time interval. In symbols, if the velocity increases from v at time t to $v + \delta v$ at time $t + \delta t$, the average acceleration over the time interval δt is of magnitude $\delta v/\delta t$, and the nearer δt gets to zero the closer $\delta v/\delta t$ approaches the size a of the acceleration at time t:

$$a = \frac{\text{limit}}{\text{as } \delta t \to 0} \left(\frac{\delta v}{\delta t} \right). \qquad …(2)$$

We define acceleration as a vector; in the situation we are considering it points in the x-direction, and we shall deal with a more general case on p. 46 when we have considered the rules by which vectors can be combined.

Using calculus notation, the right-hand side of equation (2) can be written as dv/dt, and because $v = dx/dt$ we often write

$$a = \frac{d}{dt}\left(\frac{dx}{dt}\right) \quad \text{or} \quad \frac{d^2x}{dt^2}$$

(read this as 'd two x by dt squared').

Exercises 4.7, 8

7 Use the method you employed in Exercise 4.5 to estimate the size

of Smith's velocity at 0·46 s and 0·54 s, and hence find the average acceleration between these two times. Because the time interval δt is so short, this gives in theory a good estimate of the acceleration half-way through the time interval at 0·50 s, but of course the fractional error is large because the error in the calculation of δv is considerable.

8 Use the same procedure to estimate the acceleration at 0·10 s.

Uniform acceleration

The answers to Exercises 4.7 and 4.8 confirm that Smith's acceleration is not constant. We can devise an operational definition for uniform acceleration along the lines we used for the definition of uniform speed: *a body is moving with uniform acceleration if it experiences equal changes in velocity in equal time intervals no matter how small those intervals may be.* Whereas motion with uniform *velocity* must be motion in a straight line (because if the displacements occurring in equal times are to be equal in all respects they must have the same direction) uniform acceleration will produce motion along a *curved* path unless the acceleration lies along the same straight line as that of the initial velocity. (See Chapter 13.)

Velocity–time graph

Figure 4.6 is a displacement–time graph of the cycle run, and Figure 4.7 is the corresponding graph of *velocity* against time. Although the graphs can indicate magnitudes only, we use the words displacement and velocity rather than distance and speed because the graph enables us to show the sense of these quantities: positive or negative. (See p. 37.)

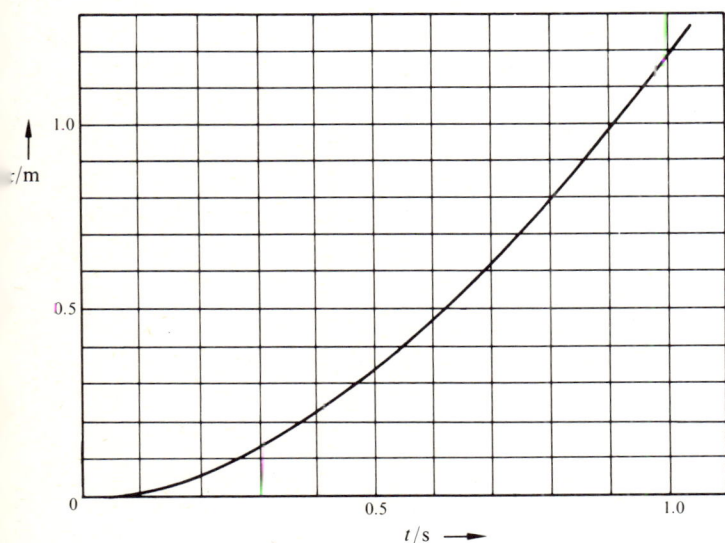

Figure 4.6
Graph of displacement x against time t for Smith's cycle ride.

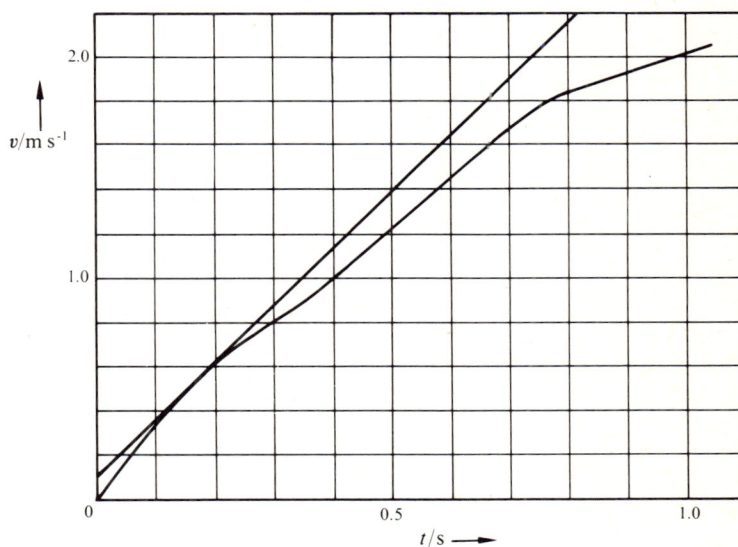

Figure 4.7
Graph of velocity v against time t for Smiths's cycle ride.

The information for the velocity–time graph could be obtained by finding the velocities at various times from Table 4.2 in the way we have already used, or it could be found by drawing tangents at suitable intervals on the displacement–time graph and measuring their gradients.

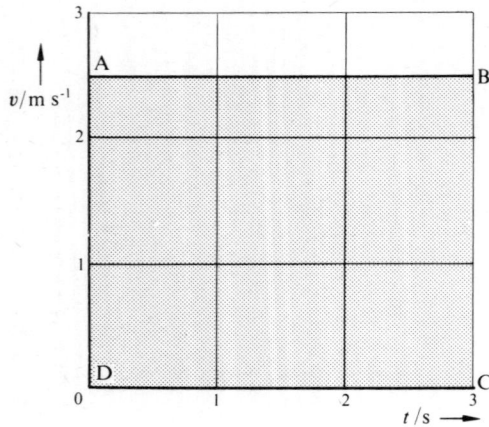

Figure 4.8

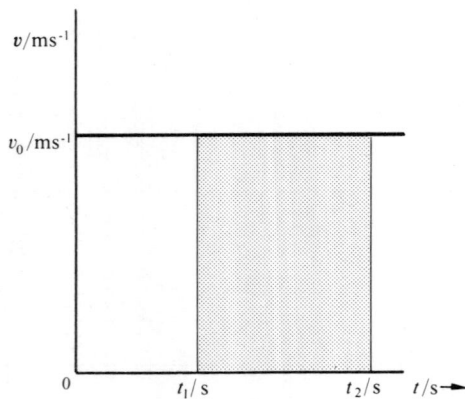

Figure 4.9

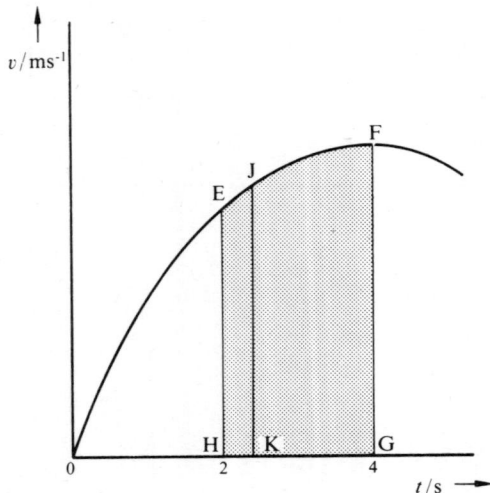

Figure 4.10

Find the velocity at $t = 0.5$s from the gradient of the displacement–time graph (Figure 4.6) and compare it with your answer to Exercise 4.5.

Gradient of a velocity–time graph

Just as the gradient of a displacement–time graph gives the velocity, so *the gradient of a velocity–time graph gives the acceleration*. The tangent drawn to the curve in Figure 4.7 rises 6.5 divisions in 5 horizontal divisions, or 1.3 vertical divisions per horizontal division. This corresponds to an increase of 0.26ms^{-1} per 0.1s: that is, to an acceleration of magnitude 2.6ms^{-2}. The tangent grazes the curve at $(t/s) = 0.20$, so 0.20s after the start the acceleration is 2.6ms^{-2}. If Smith's bicycle had been accelerating uniformly, this gradient would have been maintained; *a straight line on a velocity–time graph indicates uniform acceleration*.

Find the acceleration at $t = 0.50$s from the graph.

Displacement from a velocity–time graph

When we are given a velocity–time graph of Smith's ride, we know the velocity at any instant, and this dictates completely how far he gets. Locked up somewhere in the velocity–time graph therefore there should be complete information about how far he has got at a given time; we shall now consider a method of calculating displacements from a velocity–time graph. We start with a very simple graph – a horizontal line (Figure 4.8).

From the graph in Figure 4.8, write down the velocity at $t = 1.0$s and $t = 3.0$s. What type of motion does this horizontal line represent? Show that the object will travel 7.5m in the first three seconds.

Significantly the rectangle $ABCD$ has an area of 3×2.5 or 7.5 square divisions. Now this rectangle is defined by four rather special lines: the (t/s)-axis, the graph line itself and the vertical lines through $(t/s) = 0$ and $(t/s) = 3.0$. We call its area 'the area under the line between $(t/s) = 0$ and $(t/s) = 3$' and it is a measure of the displacement occurring within the time interval concerned. In general, if we consider a velocity–time graph corresponding to a uniform velocity v_0, the displacement in the time interval defined by t_1 and t_2 will have magnitude $v_0(t_2 - t_1)$ (see Figure 4.9) and the area under the line between t_1 and t_2 would measure (v_0/ms^{-1}) divisions by $[(t_2 - t_1)/s]$ divisions.

$$\text{Area } S \text{ under the line} = (v_0/\text{ms}^{-1}) \times (t_2 - t_1)/s$$
$$= \frac{v_0(t_2 - t_1)}{\text{ms}^{-1}\text{s}},$$

so

$$v_0(t_2 - t_1) = S\text{m}.$$

Will the same rule work for non-uniform velocity? Can we say that the area $EFGH$ under the curve between $t = 2$s and $t = 4$s in Figure 4.10 gives the size of the displacement occurring in that time interval? We will now concentrate on a thin vertical slice of that

area; we have shown it as just one line *JK* in Figure 4.10, but in
Figure 4.11 we have expanded the scale of the (t/s)-axis considerably
(while keeping the scale of the (v/m s^{-1})-axis the same as before) and
the slice now appears as *JMLK*. It is almost a rectangle, and the
thinner the slice the nearer it will approach a rectangle, because the
line will be able to rise only an infinitesimal amount in the time
available. So we have the same case as before; the area under the
curve between *J* and *M* is a measure of the displacement between
2·400 s and 2·402 s.

This is fine for a thin slice but what about a thick slice? Well,
what is true for one thin slice is true for another thin slice, and we
can think of a thick slice as being made up of a large number of thin
ones. All the tiny displacements represented by the areas of the thin
slices add up to the total displacement, and all the tiny areas add up
to the area of the thick slice. So, *the area under a velocity–time curve
between two given times is numerically equal to the size of the displacement occurring in the time interval concerned.* The area can be found
by counting squares, or by cutting out the area and comparing its
mass with the mass of one square division.

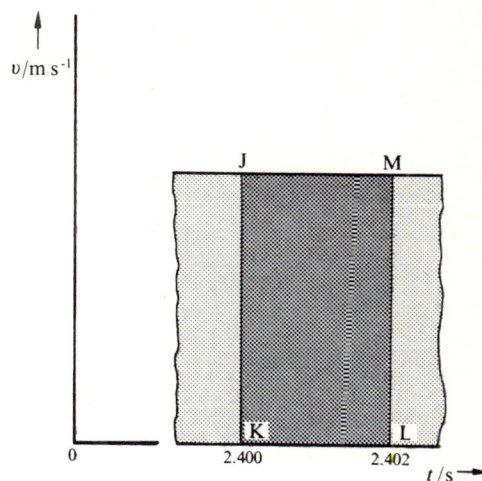

Figure 4.11

Exercise 4.12
Use the velocity–time graph of the cycle run (Figure 4.7) to estimate
how far Smith travelled in the first 0·30 s. Compare your answer with
the result given by Figure 4.6.

There is one piece of information given by a displacement–time
graph which is lacking in a velocity–time graph: namely that the
velocity–time graph does not tell us from where the object started.
We have always taken it that $x = 0$ m when $t = 0$ s (there has been
no reason to do otherwise) but a situation might arise where it would
be convenient to start from, say, $x = 2·0$ m. We could indicate it on
a displacement–time graph, but it would not appear on the velocity–
time graph; we have to adopt an arbitrary zero for x unless we are
given further information.

Calculus notation for area under a curve
If the velocity at time t is v, the extra displacement experienced in
the short time interval δt immediately following will be of size δx,
given by

$$\delta x \approx v \delta t.$$

The approximation sign is needed because v is not necessarily constant, and even in a short time δt it changes a little. But the smaller
we make δt the less opportunity there is for this to occur, and the
better is the approximation. The total displacement between times
t_1 and t_2 is of magnitude x given by

$$x = \sum \delta x$$

$$\approx \sum_{t_1}^{t_2} v \delta t,$$

where the sum is over all the tiny intervals δt between t_1 and t_2. As
the intervals get smaller and smaller the sum approaches the true
value of x more and more closely, and we can write

$$x = \lim_{\text{as } \delta t \to 0} \sum_{t_1}^{t_2} v \delta t,$$

for which we have the symbol

$$x = \int_{t_1}^{t_2} v\,dt$$

(This is read as 'integral from t_1 to t_2 of $v\,dt$'; see Appendix 3.)

Equations of motion for uniform acceleration

The symbols we have written down are all very well, but so far they are no more than a compact way of expressing what in many cases is likely to be a very laborious arithmetic process; they do not actually work anything out for us. If we are prepared to look at very special, simple types of motion we can find some rules which enable us to write down very easily the magnitudes of velocity and displacement at given times. We shall look at the simple case of motion with uniform acceleration in a straight line, not only because it *is* simple, but also because a number of important real situations involve accelerations which are at least approximately uniform. The object we are going to consider has a constant acceleration a throughout its journey, which lasts a time t. The velocity is u at the start and v at the end, and the size of the corresponding displacement is x. In this situation the average acceleration will be *the* acceleration; there is no question of any variation. So, making use of equation (1) (p. 38), we have

$$\text{acceleration} = \text{velocity change/time taken}$$
$$a = (v - u)/t$$
$$v = u + at. \qquad \ldots(3)$$

The average velocity will be the mean of the initial and final velocities, with magnitude $(v + u)/2$; the object will spend half its time with a velocity less than this and the other half with a greater velocity. So we can write

$$\text{displacement} = \text{average velocity} \times \text{time taken}$$
$$x = \frac{(v + u)}{2} \times t \qquad \ldots(4)$$

A glance at these equations will show that of the five quantities x, u, v, t, a, equation (3) does not contain x and equation (4) does not contain a. We can in fact form three more equations, each of which is short of one of the quantities. By writing t as the subject of equation (3) for example, and using the resulting expression in place of t in equation (4), we get

$$x = \frac{(v + u)}{2} \times \frac{(v - u)}{a}$$

or

$$2ax = v^2 - u^2 \qquad \ldots(5)$$

Exercise 4.13

Write v as the subject of equation (3) and use it in equation (4) to obtain

$$x = ut + \tfrac{1}{2}at^2 \qquad \ldots(6)$$

These four equations are very useful indeed when dealing with cases of uniform acceleration; the fifth, obtained by eliminating u between equations (3) and (4) has less frequent application.

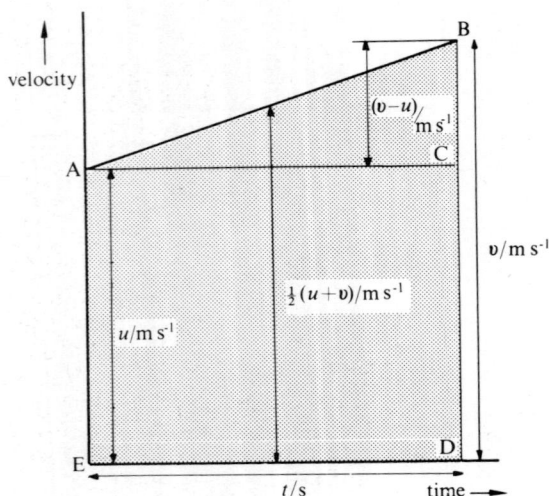

Velocity–time graph illustrating equations (3) and (4)
$a/\text{ms}^{-2} = \text{gradient} = BC/AC = (v - u)t^{-1}/\text{ms}^{-2}$
$x/\text{m} \quad = \text{area } AEDB = \tfrac{1}{2}ED(AE + BD)$
$\quad\quad = \tfrac{1}{2}(v + u)t/\text{m}.$

Exercise 4.14

Find the fifth equation.

Any problem involving acceleration can be solved if we are given three of the five quantities x, u, v, t, a by use of the appropriate equation. If for example we are given x, a and u and asked to find v we should need equation (5).

Example

A train travelling at $30\,\mathrm{m\,s^{-1}}$ loses speed at a uniform rate of $0{\cdot}5\,\mathrm{m\,s^{-2}}$ until it comes to rest. How far will it travel while slowing down?

$$x = ?$$
$$u = 30\,\mathrm{m\,s^{-1}}$$
$$v = 0$$
$$t$$
$$a = -0{\cdot}5\,\mathrm{m\,s^{-2}}$$

$$2ax = v^2 - u^2$$
$$2 \times -0{\cdot}5\,\mathrm{m\,s^{-2}} \times x = (0 - 900)\,\mathrm{m^2\,s^{-2}}$$
$$x = 900\,\mathrm{m}.$$

Note that we put $a = -0{\cdot}5\,\mathrm{m\,s^{-2}}$, not because the train is slowing down but because its acceleration is in a direction opposite to that of its displacement. We could work the same problem by assuming the initial velocity to be in the direction of *decreasing* x, in which case we should write $u = -30\,\mathrm{m\,s^{-1}}$ and $a = +0{\cdot}5\,\mathrm{m\,s^{-2}}$, and obtain $x = -900\,\mathrm{m}$.

Exercises 4.15, 16

15 A sports car can achieve a speed of $30\,\mathrm{m\,s^{-1}}$ from rest in $12\,\mathrm{s}$. What uniform acceleration does this correspond to?

16 A space vehicle during re-entry must slow down from $11\,\mathrm{km\,s^{-1}}$ to rest in a distance of about $1500\,\mathrm{km}$. What uniform rate of change of speed does this correspond to?

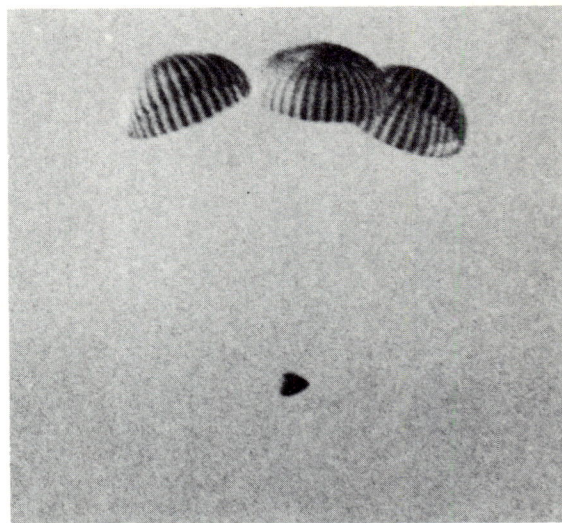

Apollo 13 after re-entry.

Acceleration of free fall: g

In Chapter 3 we described an experiment to investigate the motion of a falling body. We should now like to consider whether the results show that the acceleration of free fall is uniform. If it *is* uniform we can use $x = ut + \frac{1}{2}at^2$ to describe the motion. Transferring to the symbol h used for the length of the drop (p. 21) and putting u equal to zero, we can write this equation as

$$h = \tfrac{1}{2}at^2 \qquad \ldots(7)$$

so we can infer that the acceleration is constant if h is proportional to t^2. This point was settled by Figures 3.10 and 3.12 (p. 23 and p. 26 respectively). They provide two alternative methods of confirming that, within the accuracy of the experiment, the motion of a freely falling steel ball over a range of about two metres from rest can be described by the formula

$$h = kt^2. \qquad \ldots(8)$$

By comparing equations (7) and (8) we can write
$$a = 2k.$$

The usual symbol for the acceleration of free fall is g. Exercise 3.10 (p. 25) gives us that $1/\sqrt{k} = (0{\cdot}46 \pm 0{\cdot}03)\,\mathrm{s\,m^{-\frac{1}{2}}}$, so

$$k = (4{\cdot}8 \pm 0{\cdot}6)\,\mathrm{m\,s^{-2}},$$
$$g = (9{\cdot}6 \pm 1{\cdot}2)\,\mathrm{m\,s^{-2}}.$$

There are more accurate methods of finding the magnitude of g, and in Chapter 20 we shall find that it varies a little from place to place, but for anywhere on the surface of the earth the value $9.8 \, \mathrm{m \, s^{-2}}$ is correct within 0.3% and it is the same for all bodies. This does not mean that every falling body will continue with an acceleration of $9.8 \, \mathrm{m \, s^{-2}}$ throughout its fall, because the faster it goes the more air resistance it meets; this question is taken up more fully in Chapter 16.

In the following exercises take g as $10 \, \mathrm{m \, s^{-2}}$ and ignore air resistance.

Exercises 4.17–20

17 How long will a stone take to drop (a) 1 m (b) 10 m from rest?

18 A stone is dropped down a well and the splash is heard after three seconds. How deep is the well? Ignore the short time taken for the sound of the splash to return.

19 A ball is thrown vertically up at $20 \, \mathrm{m \, s^{-1}}$. How long will it take to return to the starting level?

20 A paper-weight is dropped from the roof of a block of multi-storey flats, each storey being three metres high. It passes the ceiling of the twentieth storey at $30 \, \mathrm{m \, s^{-1}}$. How fast will it go past the ceiling of the tenth storey? How many storeys does the block have? (The ground floor is numbered as the first storey.)

Addition of displacements

If I walk 4 km due south and then 3 km due east my legs will tell me I have covered 7 km, but I shall be only 5 km from my starting point. Because of their directional property, displacements do not add like ordinary numbers; we have to use a little geometry. AB and AC each represent a displacement (Figure 4.12) and if an object receives both displacements then its total or *resultant* displacement will be represented in size and direction by the diagonal AD of the parallelogram. Figure 4.12 is a *vector diagram*, and we think of the lines in the following way; although the length and direction of each line is significant and represents the corresponding aspect of the displacement, the actual *position* of the line is *not* significant. The lines AB and CD are equally good representations of a displacement of corresponding length and direction, and if we use the symbols a, b and c to represent the lines AB, AC and AD respectively we shall write

$$a + b = c$$

though of course this does *not* mean that $a + b = c$.

Exercise 4.21

Find the resultant displacement, in magnitude and direction, of a displacement 5 km due east added to one 3 km south-west.

Addition of velocities

The rule for addition of displacements may have seemed fairly obvious, but we can use the same idea to add two *velocities* which are not in the same direction. Think of a gantry crane in a steel works; the hook can be hoisted vertically and at the same time the trolley from which it hangs can run horizontally along the gantry.

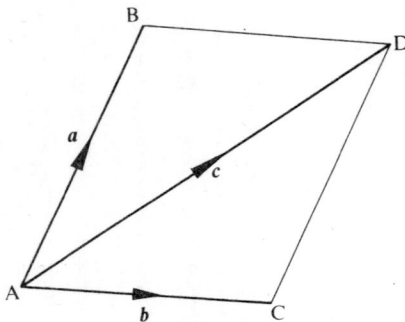

Figure 4.12

The hook can have horizontal and vertical velocities simultaneously, and these combine to form the resultant velocity which is not simply the sum of their magnitudes and is not in the direction of either of them. The hook will move through space along a definite line making an angle with the floor. The rule for addition of velocities becomes very plain if we remember that velocities are merely *displacements per unit time interval* (though the time interval may have to be made infinitesimally small if the velocity is changing). Talking first in terms of uniform velocities, if we combine the vertical displacement of the hook in one second with its horizontal displacement in one second we shall get its resultant displacement in one second – its resultant velocity. We know how to combine these displacements: we need a vector diagram like Figure 4.12, with suitable lengths and directions. If the velocities are not uniform, we merely consider a much smaller time interval than one second, so that the velocities do not have time to change appreciably. So velocities will combine according to a *parallelogram law*, which can be stated as follows: *If two velocities which an object possesses simultaneously are represented in magnitude and direction by the adjacent sides of a parallelogram, the resultant velocity will be represented in magnitude and direction by the diagonal through their point of intersection.*

Example

A ferry man steers his boat (which can move at $1.5 \mathrm{m\,s^{-1}}$ in still water) straight for the far bank of the river, but he is swept downstream by a current of $1.0 \mathrm{m\,s^{-1}}$.

a) How far down the river does he hit the opposite bank if the river is 50 m wide.
b) At what angle should he steer to hit the bank at a point opposite his starting place?

a) Figure 4.13a is the vector diagram for part (a); v_1 is the velocity of the boat relative to the water and v_2 the current velocity. The resultant velocity v makes an angle α with the normal to the bank

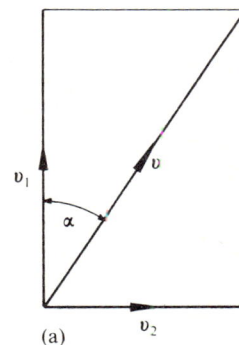

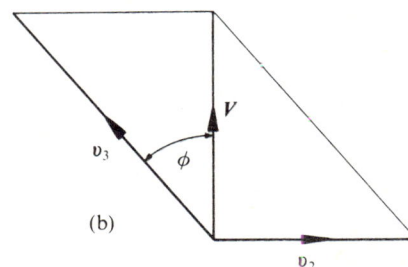

(a)

(b)

Figure 4.13

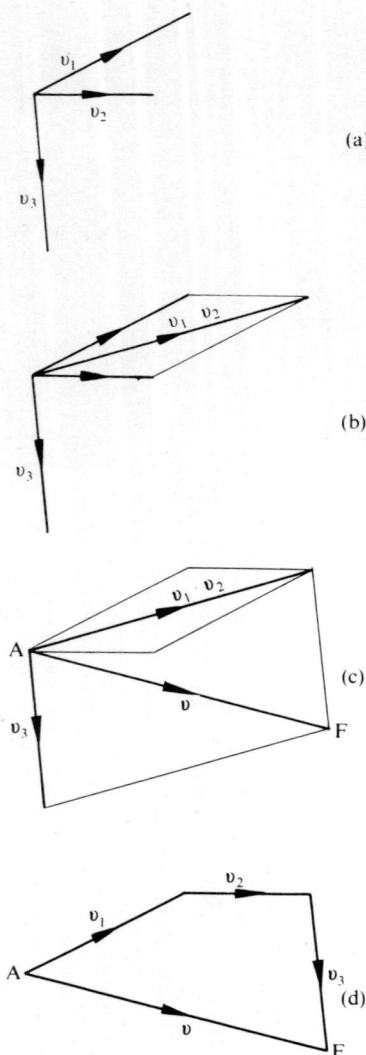

Figure 4.14
(a) Three velocities to be combined. (b) v_1 and v_2 combined. (c) Their resultant combined with v_3. (d) An alternative representation of the addition process.

such that

$$\tan\alpha = \frac{v_2}{v_1} = \frac{1}{1\cdot 5}.$$

In moving 50 m across the river therefore he will move a distance d downstream given by

$$\frac{d}{50\,\text{m}} = \frac{1}{1\cdot 5}$$

$$d = 33\cdot 3\,\text{m}.$$

b) Referring to Figure 4.13b, the new velocity of the boat relative to the water, v_3, must be in a direction making some angle ϕ with the normal to the bank such that the new resultant velocity V is directed along the normal. This time

$$\sin\phi = \frac{v_2}{v_3} = \frac{1}{1\cdot 5}$$

$$\phi = 41\cdot 9°.$$

Exercise 4.22

How long will it take to cross the river in each case?

If we need to combine three different velocities v_1, v_2, v_3, we can do it in two stages, combining two of the velocities and using their resultant to form another parallelogram with the third velocity (Figure 4.14).

The line AF then represents the resultant velocity v. The same line would result however if we merely laid down end to end the three lines representing the three constituent velocities (Figure 4.14d). It is not even necessary for the velocities to be in one plane; going back to our crane, we can have the velocity of the gantry along the shop floor added to the other velocities, and then we should need a drawing depicting three dimensions for a vector diagram (Figure 4.15).

Addition of accelerations
The parallelogram rule works for displacements and velocities, but will it also work for accelerations? We can show that it does; *if an object simultaneously possesses two accelerations represented in size and direction by the adjacent sides of a parallelogram, the diagonal through their point of intersection represents the resultant acceleration in size and direction.* The proof that follows is not difficult, but it is rather long and may well be omitted from a first reading.

Figure 4.16a is a vector diagram for two velocities v_1 and v_2 which an object possesses at one and the same time. After a short interval δt both velocities have changed, by δv_1 and δv_2 respectively, to v_1' and v_2' (Figure 4.16b) where

$$v_1' = v_1 + \delta v_1$$

and

$$v_2' = v_2 + \delta v_2.$$

(Remember these equations do *not* mean $v_1' = v_1 + \delta v_1$ or $v_2' = v_2 + \delta v_2$ unless the new velocities happen to be in the same directions as the old ones respectively.) The object is therefore experiencing two simultaneous accelerations, a_1 roughly of magnitude $\delta v_1/\delta t$ in the direction δv_1, and a_2 of approximate magnitude $\delta v_2/\delta t$ in the direction of δv_2.

$$a_1 \approx \frac{\delta v_1}{\delta t} \quad \text{and} \quad a_2 \approx \frac{\delta v_2}{\delta t}$$

in fact we can even write

$$\boldsymbol{a}_1 \approx \frac{\boldsymbol{\delta v}_1}{\delta t} \quad \text{and} \quad \boldsymbol{a}_2 \approx \frac{\boldsymbol{\delta v}_2}{\delta t},$$

to show that the accelerations are in the directions of the corresponding small velocity changes.

The net velocity change $\boldsymbol{\delta v}$ is given by

$$\boldsymbol{v} + \boldsymbol{\delta v} = \boldsymbol{v}',$$

where $\boldsymbol{v}$ is the initial, $\boldsymbol{v}'$ the final, resultant velocity. The change $\boldsymbol{\delta v}$ is represented by the line AC in triangle ABC, where AB is drawn in the direction of $\boldsymbol{\delta v}_1$ and BC is in the direction of $\boldsymbol{\delta v}_2$. A certain amount of labour with congruent triangles will show that

$$AB = EF \quad \text{and} \quad BC = GH;$$

in other words AB and BC are perfectly good vector representations of $\boldsymbol{\delta v}_1$ and $\boldsymbol{\delta v}_2$ respectively. This means that the separate velocity increments add to give the resultant velocity change according to the parallelogram rule, and if we divide $\boldsymbol{\delta v}_1$, $\boldsymbol{\delta v}_2$ and $\boldsymbol{\delta v}$ each by δt and approach the limit ($\delta t \rightarrow 0$) we can say that the *accelerations* $\boldsymbol{a}_1$ and $\boldsymbol{a}_2$ also add according to the parallelogram rule to give the resultant acceleration $\boldsymbol{a}$.

The above arguments can be put very compactly if we are prepared to make full use of vector notation. In what follows, an equation such as

$$\boldsymbol{a} + \boldsymbol{b} = \boldsymbol{c}$$

is to be interpreted as follows: if we draw a parallelogram with sides representing $\boldsymbol{a}$ and $\boldsymbol{b}$ in size and direction, then the diagonal through the point of intersection represents $\boldsymbol{c}$ in size and direction. Further,

$$\boldsymbol{a} + \boldsymbol{b} + \boldsymbol{c} = \boldsymbol{d}$$

says that when $\boldsymbol{a}$ and $\boldsymbol{b}$ are combined according to the parallelogram rule, and their *resultant* is combined with $\boldsymbol{c}$ by a second application of the rule, the final resultant is equal to the vector $\boldsymbol{d}$ in size and direction. Starting with the parallelogram of velocities we can write all of the following:

$$\boldsymbol{v}_1' = \boldsymbol{v}_1 + \boldsymbol{\delta v}_1$$
$$\boldsymbol{v}_2' = \boldsymbol{v}_2 + \boldsymbol{\delta v}_2$$
$$\boldsymbol{v}' = \boldsymbol{v} + \boldsymbol{\delta v}$$
$$\boldsymbol{v}_1 + \boldsymbol{v}_2 = \boldsymbol{v} \qquad \qquad \dots(9)$$
$$\boldsymbol{v}_1' + \boldsymbol{v}_2' = \boldsymbol{v}'. \qquad \qquad \dots(10)$$

Substituting the first three equations into equation (10) we get

$$\boldsymbol{v}_1 + \boldsymbol{\delta v}_1 + \boldsymbol{v}_2 + \boldsymbol{\delta v}_2 = \boldsymbol{v} + \boldsymbol{\delta v}$$
$$\boldsymbol{\delta v}_1 + \boldsymbol{\delta v}_2 + (\boldsymbol{v}_1 + \boldsymbol{v}_2) = \boldsymbol{v} + \boldsymbol{\delta v}.$$

So

$$\boldsymbol{\delta v}_1 + \boldsymbol{\delta v}_2 + \boldsymbol{v} = \boldsymbol{v} + \boldsymbol{\delta v}$$

using equation (9). We can safely add the same vector to each side, so presumably we can also subtract $\boldsymbol{v}$ from each side, giving

$$\boldsymbol{\delta v}_1 + \boldsymbol{\delta v}_2 = \boldsymbol{\delta v}$$

and then it follows that

$$\underset{\text{as } \delta t \rightarrow 0}{\text{limit}} \left(\frac{\boldsymbol{\delta v}_1}{\delta t} \right) + \underset{\text{as } \delta t \rightarrow 0}{\text{limit}} \left(\frac{\boldsymbol{\delta v}_2}{\delta t} \right) = \underset{\text{as } \delta t \rightarrow 0}{\text{limit}} \left(\frac{\boldsymbol{\delta v}}{\delta t} \right).$$

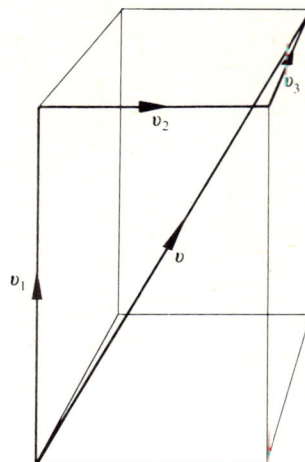

Figure 4.15

Vector addition in three dimensions.

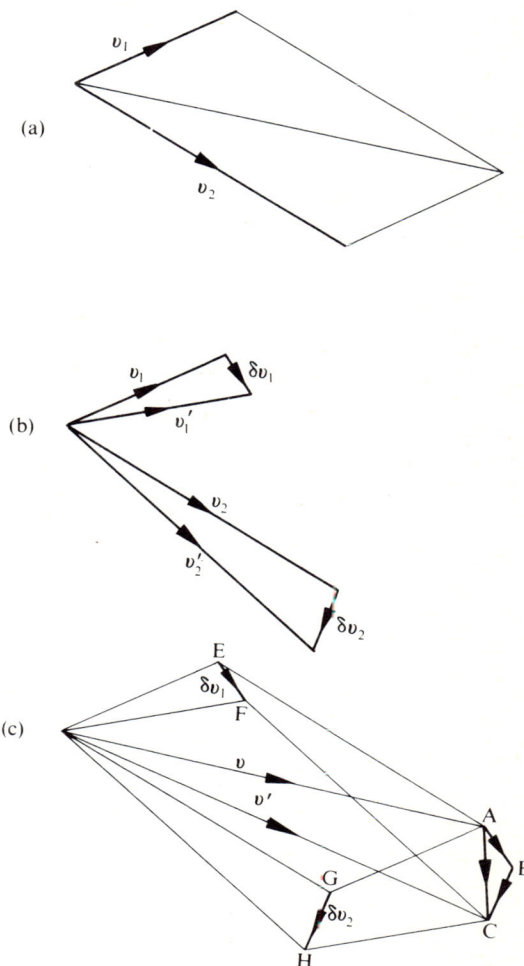

Figure 4.16

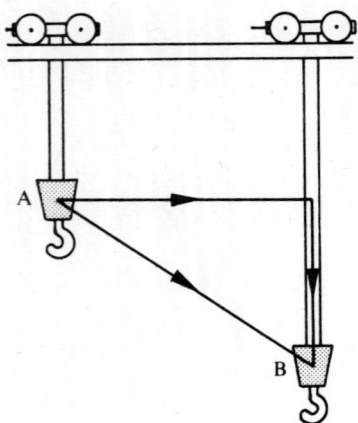

Horizontal and vertical components of displacement AB.

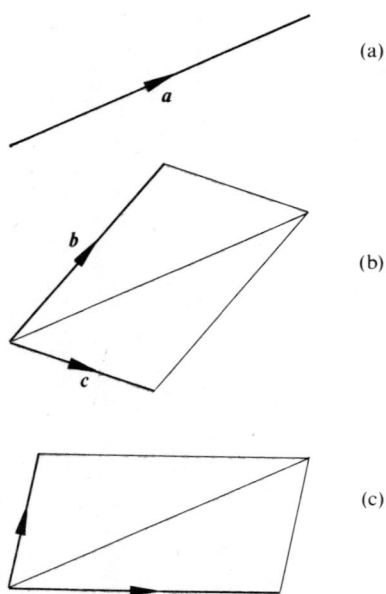

Figure 4.17
Two sets of components of a.

Resolution of vectors

We have considered what happens when the hook of a gantry crane is given two different velocities at the same time, and we have seen that the resultant velocity can be found by using the parallelogram rule. The crane driver must think about the problem in reverse; he has the job of getting the hook from point A to point B, and he must judge how much horizontal and how much vertical motion is needed – he must split the total velocity required into the horizontal and vertical parts which will make it up.

This process of breaking a vector down into two other vectors which together are equivalent to it is called *resolution*, and we shall find it a very useful technique. In Figure 4.17b for example we say that the vector a has been resolved into the two components b and c. Of course these are not the only vectors into which a can be resolved: Figure 4.17c shows an alternative pair; the process can be carried out in an infinite number of ways. Almost invariably it is most useful to resolve into two vectors which are *perpendicular* to each other; the phrase 'resolved into components' will imply perpendicular components throughout this book unless otherwise stated. Even when we restrict ourselves to perpendicular components resolution can be carried out in an infinite number of ways, some of which appear in Figure 4.18. In any particular case, however, one set of components turns out to be of greater use than any other. For the crane, for example, the obvious components to think about are the horizontal and vertical velocities.

Once we have settled on the directions of the components their magnitudes are out of our control. If we decide to resolve a in Figure 4.19 into horizontal and vertical components, they *have* to be x and y as shown. We cannot make x a little longer and then make y a little shorter to compensate; no matter how we juggle with their magnitudes we can never make a up differently.

The magnitudes of the components

From Figure 4.19 we can write down some useful relationships. We know that $AB = AC\cos\alpha$, and $AD = AC\sin\alpha$. But AC, AB and AD represent a, x and y to scale, so

$$x = a\cos\alpha$$
$$y = a\sin\alpha$$

If any vector v is resolved into perpendicular components, that component which makes an angle α with v will be of magnitude $v \cos \alpha$, and the component perpendicular to it will be of magnitude $v \sin \alpha$.

Exercises 4.23–26

23 Resolve a velocity of $7.0\,\mathrm{m\,s^{-1}}$ 35° east of north into components running north-east and north-west.

24 A projectile is to be fired to a vertical height of 3000 m above the point of projection. What must its initial *vertical* velocity be? (Ignore air resistance.)   (o & c)

25 A train initially at rest stands on a straight track. In a hypothetical case, it accelerates at a constant rate of $0.6\,\mathrm{m\,s^{-2}}$ for 30 s, then moves with constant speed for 20 s, and finally decelerates uniformly to rest in the next 60 s.

Draw (a) acceleration–time, (b) speed–time graphs for the motion. Determine, from the latter graph, the distance travelled by the train in the time interval between 20s and 80s. Deduce the average speed during the whole motion. (WJEC)

26 A particle A is projected vertically upwards from the base of a tower at the same time that a particle B is dropped from rest from the top of the tower, the paths of the two particles being in the same straight line. The initial velocity of A is $30\,\text{ms}^{-1}$ and the tower is 120m high.

What is the initial relative velocity of the two particles? What is it at a slightly later time? Will the particles collide before either reaches the ground? If so, when and where does the collision occur? (WJEC)

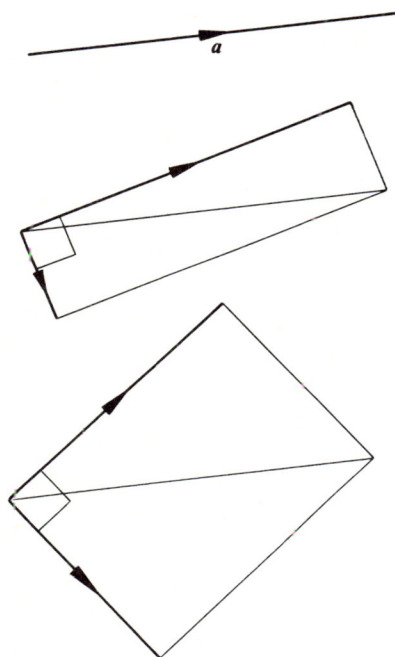

Figure 4.18

Two sets of perpendicular components of a.

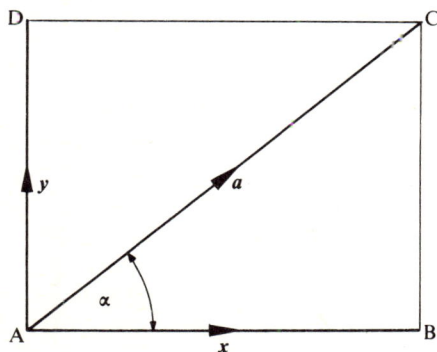

Figure 4.19

5 Force

Force [L. *fortia*, n. of quality L. *fortis* strong.] 1. Physical strength. Rarely in *pl.* 2. Strength, impetus, violence, or intensity of effect. 3 (a). Power or might; *esp.* military power. (b). In early use, the strength (of a defensive work, etc.) Subseq., the fighting strength (of a ship). 4 (a). A body of armed men, an army. In *pl.* the troops or soldiers composing the fighting strength of a kingdom or of a commander. (b). A body of police: often absol. the *force* = policemen collectively. 5. Physical strength or power exerted upon an object; *esp.* violence or physical coercion. . . .

Extract, *Shorter Oxford English Dictionary.*

It may be no surprise to learn that the range of meaning implied by the word *force* is far too wide to be of much use in a scientific context; to make it more serviceable to us we pare it down and restrict its meaning fairly drastically. Suppose some respectable citizen walking purposefully along the pavement suddenly finds himself sprawling in the gutter; no doubt he will want to ascertain who or what was responsible. We say he was acted on by a force, and such an unfortunate incident illustrates three aspects of what we mean. First, it is something which we feel, often painfully, a pull or a push, and second, it alters our motion, and third, it involves someone or something else.

Of course we need to put the idea of force on a quantitative basis, we need a unit and a counting process. One way of establishing this would be in terms of the stretch of a spring, which involves two equal and opposite forces as indicated in Figure 5.1. We could say that when a certain spring is stretched one centimetre, each end of it is being pulled by a force of one unit. Two such springs side by side, each stretched one centimetre, would need at each end two such unit forces, which we call a force of two units, and so on. [Although in practice multiples of the unit force can conveniently be obtained by using springs, force is not in fact defined in this manner–see p. 58.]

Now we find a happy accident; provided the spring is not over-stretched in the process, equal and opposite forces, each of magnitude two units, applied to the ends of one of the springs on its own, stretch it two centimetres. That cannot be worked out by logic, it is a matter for experiment, and a specific example of *Hooke's law* (see Chapter 11). It makes it easy for us to calibrate a spring balance in arbitrary units. We just need equally spaced divisions on the scale.

A unit of force defined in this way would be a *basic* unit; it would be an arbitrary fresh start. We shall find later in the chapter that we can devise a *derived* unit for force, and we have seen that derived units are usually preferable when we can get them. Nevertheless the spring balance is a very useful instrument to have in mind when we are fixing our ideas, and we shall use it to arrive at the most appropriate definition of the derived unit.

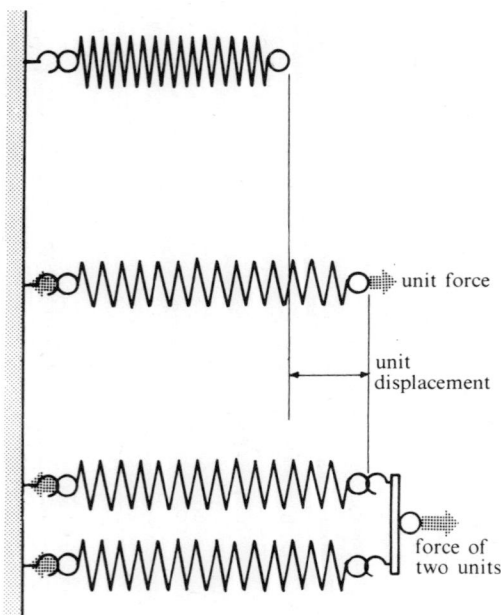

Figure 5.1
A possible way of defining a counting process for force.

Newton's first law of motion

The spring balance emphasises the pull–push aspect of force; now we want to concentrate on the other aspects and think of a force as an interaction between two bodies changing their motion. This statement contains the germ of two fundamental laws of motion stated by Isaac Newton (1642–1727) in his *Philosophiae Naturalis Principia Mathematica* (mathematical principles of natural knowledge). The substance of his first law can be stated as follows: *any object will continue in its state of rest or uniform motion in a straight line unless acted on by an external force.*

At first sight this just appears to be a definition of the word force.

Newton's first law is more obviously acceptable in this environment.

We have simply lumped all the objects in the world into two categories; the first category is labelled 'all bodies at rest or in uniform motion in a straight line' and the other is labelled 'all other bodies'. Newton has stuck on two fresh labels, which read 'no force' and 'force acting' respectively. Now anyone can stick on labels and make definitions; what makes a definition into a *law* is the fact that those labels are useful and significant in other situations. For example, every time my motion is changed I have a totally independent way of describing it; I say I feel a push. [This would not apply if I were an astronaut moving freely with motors turned off in the vacuum between earth and moon. At most points on that journey I should be accelerating under the action of gravitational forces, but since they would be applied to every individual atom of my body, and not simply to atoms in my surface layer, I would not experience any distortion and would therefore not 'feel' the force.] It is because forces have properties independent of the one specifically mentioned in Newton's first law that we call it a law, not a definition.

Everyday experience would suggest that Newton's first law is *not* true; things which are moving on the earth slow down and stop if left to themselves. This is because things which are moving on the earth are always experiencing forces, in particular frictional forces, which arise whenever one object slides on another, and air resistance. We can take great pains to minimise these forces, and the more trouble we take the more convinced we become about the first law. The puck shown in Figure 5.2 was given an initial push on a smooth horizontal surface and illuminated by a xenon lamp flashing briefly at regular intervals. Friction was reduced to a very small value because the puck was floating on a cushion of carbon dioxide, supplied at a steady rate from a lump of 'dry ice' contained inside it.

Exercise 5.1

Take measurements from Figure 5.2 to determine the velocities near the extremes of the recorded path, given that the frequency of

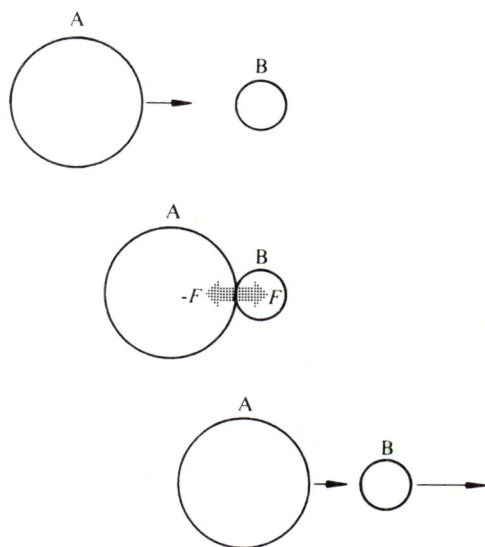

Figure 5.2

Newton's third law: the force which *A* exerts on *B* is at every instant equal and opposite to the force which *B* exerts on *A*.

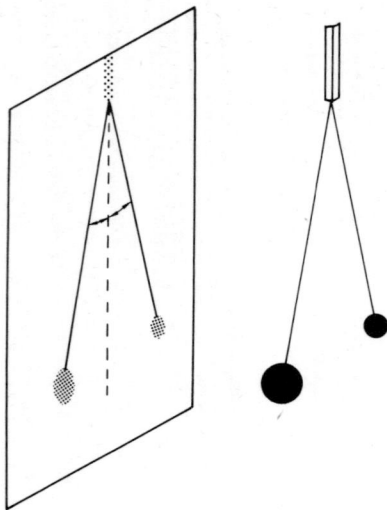

Figure 5.3
Possible test of Newton's third law in a particular situation.
The charged polystyrene spheres are illuminated by a small
light source at the other end of the room.

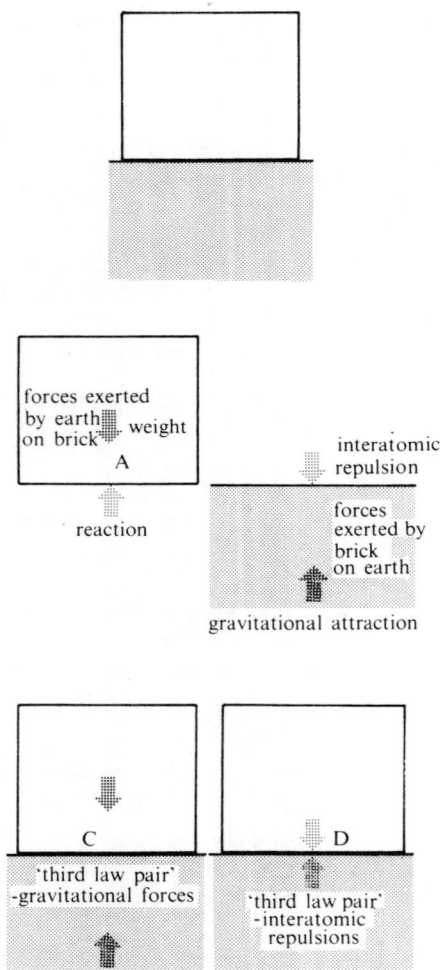

Figure 5.4

flashes is 10 per second. Can you tell which way the puck moved,
assuming a perfectly horizontal table? Is there any evidence of
deviation from a straight line?

Newton's third law of motion

Before dealing with Newton's second law, we shall look at the idea
that a force always involves *two* bodies; this appears in his *third* law.
It can be stated as follows: *the force which body A exerts on body B is
exactly equal and opposite to the force which body B exerts on body A.*
If I am standing on a small trolley and walk off, the trolley moves off
in the other direction–whatever force the trolley exerts on me, I
exert an equal one on the trolley in the opposite direction. If I
collide with someone we both get deflected from our original path,
and we both suffer indignity, if not physical pain. Figure 5.3 shows a
possible arrangement for a quantitative test, making use of the
electrostatic repulsion between two charged spheres of expanded
polystyrene. Measurment of the deflections of the threads and
masses of the spheres enables us to calculate the force each is
experiencing from the other; the ways in which this can be done are
explored in Chapter 6.

When one of the bodies involved is very much larger than the other,
the law is still true, though the effect may be harder to see. If I
drop a brick out of a window, the entire earth exerts a gravitational
force on the brick and it accelerates to the ground–so much is
obvious. It is also true to say that the brick is attracting the earth, and
the force the brick exerts on the earth is exactly equal and opposite
to the force the earth exerts on the brick. While the brick is accelerating to the earth, the earth is being accelerated towards the brick. It
does not come up and meet the brick halfway of course–the earth is
a lot bigger and would need a much larger force to produce appreciable motion in the time available. But though the brick would not
produce any measurable effect on the earth, a somewhat larger
object like the moon can produce significant acceleration–this point
is discussed in Chapter 20.

Newton's third law is often stated in the form, 'action and reaction are equal and opposite', but this can lead to misunderstanding.
In the preceding sentence the words *action* and *reaction* mean 'the
force which *A* exerts on *B*' and 'the force which *B* exerts on *A*'
respectively. Unfortunately the word *reaction* is commonly used in
quite a different way: when an object is resting on the ground, the
force the ground exerts on the body, preventing it from falling
through, is often called simply 'the reaction of the ground on the
brick'. This can lead to a certain amount of confusion; in this book
we shall refer to the third law always in terms of *A* acting on *B* and
B acting on *A*, and the word 'reaction' will be reserved for the force
one object exerts on another object resting against it. Microscopically the force is made up of interatomic repulsions in the region of
contact.

We need to take a careful look at our brick when it has hit the
ground to see how the third law operates then. Usually we show the
forces in this situation as (i) the gravitational pull of the earth on the
brick (its weight), and (ii) the push or reaction of the earth's
surface on the brick (labelled *A* in Figure 5.4). If the brick is not
accelerating (that is, sinking in at a non-uniform rate), these two
forces are equal and opposite, but it is a misunderstanding to conclude
that this is an example of Newton's third law. The fact is that the

third law deals with corresponding forces on *different bodies*, but the two forces we have mentioned so far (labelled *A* in the figure) are both exerted by the earth on the brick. To see the operation of the third law we have to consider *four* forces, two 'third law pairs'. The four forces are:

first pair (labelled *C*)
- gravitational pull of earth on brick (weight of brick)
- gravitational pull of brick on earth

always equal and opposite; Newton's third law

second pair (labelled *D*)
- push of earth's surface up on brick
- push of brick down on earth's surface

always equal and opposite; Newton's third law

The equalities indicated are true (i) when the brick is still falling through the air (when both members of the second pair are zero), (ii) when it is accelerating downwards through liquid or quicksand (when the members of the second pair are less than those of the first), or (iii) when it is perched on a compressed spring which has just been released (when the members of the second pair are greater than those of the first). The only situations for which the members of the first pair are equal to those of the second pair are when the brick rests on a firm unyielding surface or sinks in with uniform velocity, and the fact that they *are* equal in these circumstances has nothing to do with Newton's third law.

Force mass and acceleration

Forces alter motion: what sort of acceleration do they produce? We find it simplest to look first at what a constant force will do.

Figure 5.5
Trolley accelerated by a constant force.

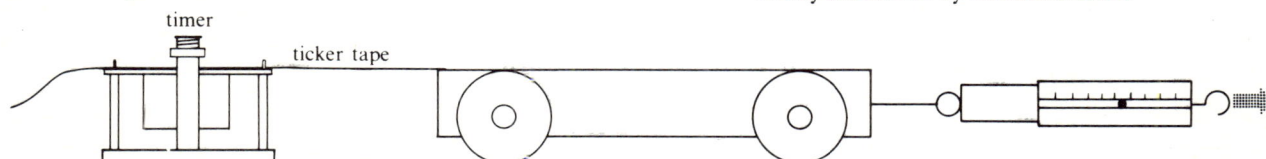

Figure 5.5 shows an experimental arrangement which attempts to approach the ideal of a constant net force acting on an object. The horizontal cord towing the trolley along the runway is attached to a spring balance, and the idea is to hold the spring balance in the hand and pull in such a way that its reading stays constant – it takes a certain amount of practice to manage this. An obvious point, but an important one, is that we need to pull in the direction in which the trolley is facing. Force is another *vector* quantity, and the direction of the force is as important as its size. In Chapter 6 we shall be concerned with forces acting in different directions, but at present we shall concentrate on the *magnitudes* and consider situations where the force and the changes of motion it produces are confined to one direction.

We can use ticker-tape to record the details of the motion (see Chapter 4), and Figure 5.6 shows the velocity–time graph for a typical run. The plots lie sufficiently close to a straight line to suggest that a constant force produces a constant acceleration.

Exercise 5.2

Find the acceleration indicated by Figure 5.6.

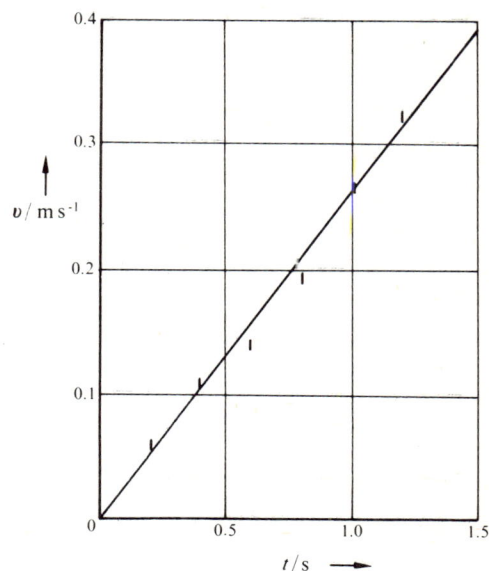

Figure 5.6
Velocity–time graph for the accelerated trolley.

Table 5.1

Acceleration a of trolley produced by force F

F / arbitrary units	(a/ms^{-2})
1	0.26 ± 0.05
2	0.49 ± 0.05
3	0.75 ± 0.05
4	0.93 ± 0.05

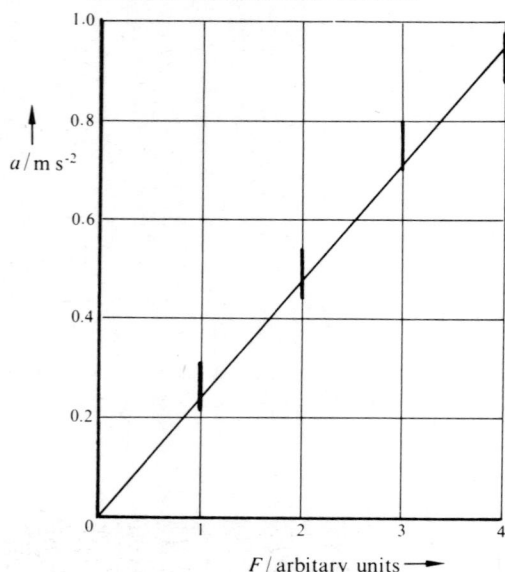

Graph of acceleration a against applied force F for the results in Table 5.1.

Table 5.2

Acceleration a produced on a trolley of mass m by a constant force

(m/kg)	(a/ms^{-2})	(m/kg)$^{-1}$
11.7	0.95 ± 0.05	0.0855
16.7	0.66 ± 0.05	0.0598
21.7	0.48 ± 0.05	0.0460
26.7	0.36 ± 0.05	0.0374

To those who have done some physics this result will be no surprise: we have come to expect this sort of thing. It is worth pausing a moment therefore to try and take in the significance. Here we have two quantities which started off quite independently; one is to do with the stretch of a spring, the other is to do with how much further a trolley gets between two ticks of a clock than it did between two previous ticks, and now we find that a constant value of one produces a constant value of the other. That is by no means trivial, and it certainly could not be predicted from logic.

In Table 5.1 we have the accelerations produced by one, two, three and four arbitrary units of force. Before we carry out a run some attempt can be made to compensate for frictional drag by tilting the runway slightly until the trolley moves forward at constant speed when given a push to start it.

There is enough evidence here to *suggest* that acceleration is proportional to the force producing it, or in symbols

$$a \propto F \text{ (when the mass is constant).} \qquad \ldots(1)$$

What happens if we load the trolley? Table 5.2 shows a set of accelerations produced by the same force – the trolley has a different mass for each run.

Clearly there is no direct proportion this time, but if we plot acceleration a against reciprocal of mass $1/m$ we get the graph shown in Figure 5.7 indicating that

$$a \propto 1/m \text{ (when the force is constant).} \qquad \ldots(2)$$

Exercise 5.3

Work out the product ma for each line of results in Table 5.2 and confirm that it is (approximately) constant. This is an alternative way of testing whether $a \propto 1/m$.

The acceleration is doubled if we double the force or if we halve the mass. If we double the force *and* halve the mass we shall get the acceleration increased by a factor of four: equations (1) and (2) can be combined into a single statement

$$a \propto F/m.$$

In fact we hardly need to load the trolley to find how the acceleration depends on the mass; we can draw on results from a rather more precise experiment. We have been provided with a very convenient force which is constant to a very satisfactory degree and produces a constant acceleration – the pull of the earth on an object. The steel ball used in the experiment on p. 20 fell with an acceleration of $9.8\,\mathrm{m\,s}^{-2}$; it is a matter of common experience that a steel ball with twice the mass falls with the *same* acceleration. But the force is *double* in the second case; there is twice as much pull on the larger mass – surely the acceleration should have doubled? No, because when we doubled the force, we could not help but double the mass at the same time. Doubling the force *and* doubling the mass brings us back to the same acceleration, so if the acceleration is proportional to the force it must also be inversely proportional to the mass.

Proportional signs are not the most convenient symbols to handle; therefore we usually replace them by equalities as soon as possible, and we can do this by introducing a constant into the equation. We rewrite

$$a \propto F/m$$

as

$$ka = F/m$$

where k is some constant, or

$$F = kma. \qquad \qquad ...(3)$$

If we continue to measure forces in terms of an arbitrary unit defined by a mark on a particular spring balance, the value of k is a matter for experiment, but the presence of such a constant in this equation is a sure sign that we have an opportunity to dispense with our arbitrary basis and define a derived unit. In order to get a coherent system (see p. 13) we make k the simplest number possible – one – and this then fixes the size of our force unit. *It is that force which will give a mass of one kilogram an acceleration of one metre per second squared*, and it is called a *newton* (N). Equation (3) now becomes

$$F = ma,$$

and this is the only form which need be remembered. Both the unit of force and the counting process can be defined by this equation: a force of two newtons will give a 1 kg mass an acceleration of $2\,\mathrm{m\,s^{-2}}$. It need hardly be said that the formula will give F in newtons *only* if m is in kg and a in $\mathrm{m\,s^{-2}}$.

We get the most direct indication of the size of one newton by considering a falling body. A 1 kg mass allowed to fall freely has an acceleration of approximately $9{\cdot}8\,\mathrm{m\,s^{-2}}$; we can use $F = ma$ to work out the force which is producing this acceleration.

$$F = 9{\cdot}8 \times 1\,\mathrm{N}$$

and so a mass of 1 kg is pulled towards the earth with a force of approximately $9{\cdot}8\,\mathrm{N}$; the precise value will vary from place to place. More generally, a mass m placed in a region where the acceleration of free fall is g has a weight of mg, m being in kg, g in $\mathrm{m\,s^{-2}}$ and the weight in N.

Mass as inertia

We can now be much clearer about that aspect of mass mentioned briefly on p. 11, namely its aversion to change of velocity, or *inertia*. We cannot change the velocity of any object instantaneously; if we put a force on it, it 'sizes up' the force, and changes its velocity gradually at a rate consistent with $F = ma$. Inertia does not imply reluctance to *move* unless the object happens to be stationary; rather it indicates resistance to *change* of motion.

So mass has two quite distinct properties – gravitational attraction and inertia. What is more, experiment shows that the two are proportional, at least to an accuracy of 1 part in 10^3. That is rather remarkable. Suppose we pick up two chunks of matter from the ground, different substances, shapes and sizes, and we trim one down until the two pieces have identical inertia; that is, they accelerate at the same rate when subjected to the same force. If we now measure their weights at a given position on the earth's surface we shall find them to be precisely equal. This is known as the *principle of equivalence*, and we may well wonder whether there is a reason why it should be so. Albert Einstein also wondered whether there was a reason, and from his thinking emerged general relativity theory.

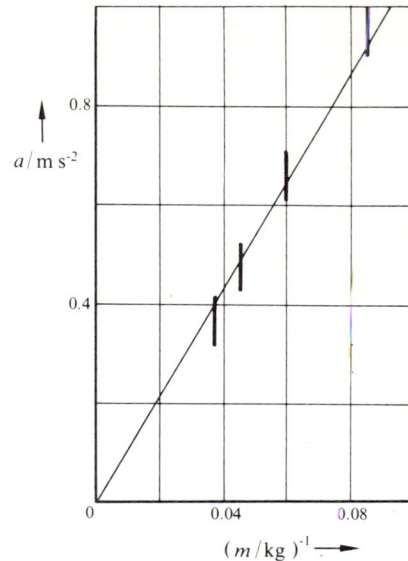

Figure 5.7
Graph of acceleration a against reciprocal of mass $1/m$ for a loaded trolley acted on by a constant force.

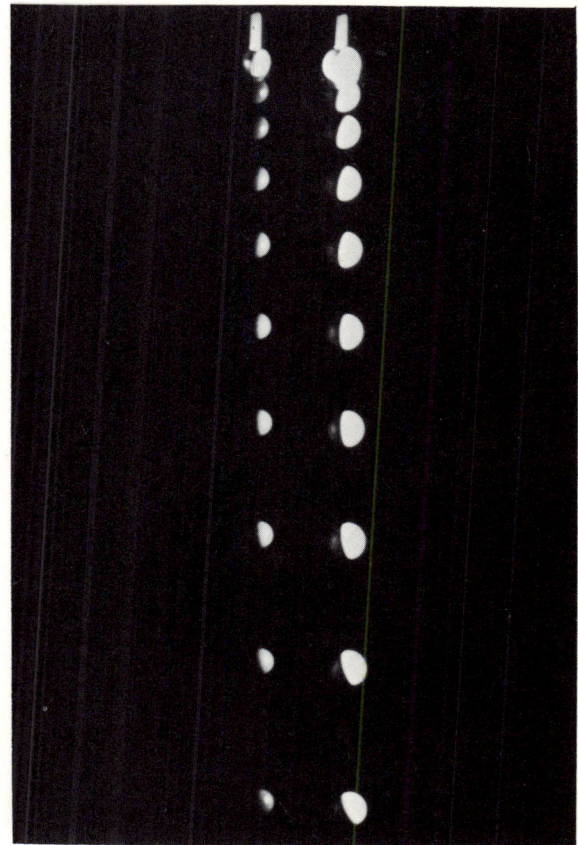

g is independent of mass: falling objects released simultaneously and illuminated by a stroboscopic lamp.

The importance of $F = ma$

Much of the rest of this book is an exploration of the implications of this one simple equation $F = ma$ in various circumstances. The evidence we have given for its accuracy is so far slender. The real test of its reliability is not a collection of rather doubtful results from one experiment with a trolley, nor even many experiments with different trolleys, but rather the fact that it can account with great precision for the motions which occur in a countless number of more complex cases. It was in fact a consideration of the behaviour of heavenly bodies and what keeps them in their orbits which led Newton to formulate the laws which bear his name. Speaking loosely, $F = ma$ summarises Newton's second law of motion; to state the law more precisely we must introduce a new quantity, the *momentum* of a body, defined as the product of its mass and velocity. So far we have found it appropriate to write

$$F = m \times \text{rate of change of velocity.}$$

For a system having constant mass, we could equally well write
$$F = \text{rate of change of (mass} \times \text{velocity)}$$

or

$$F = \text{rate of change of momentum}$$
$$= \mathrm{d}(mv)/\mathrm{d}t \text{ in symbols.}$$

At speeds approaching the speed of light it turns out that $[m \times$ rate of change of $v]$ is no longer equal to [rate of change of (mv)], and we have to decide which to adopt for the definition of force. The second alternative is preferred, and Newton's second law is usually stated as follows:

When a force acts on a body, the rate of change of momentum it produces is proportional to the force, and the change takes place in the direction of the force.

This is our link between cause and effect, between forces and the changes of motion they produce. In most cases discussed in this book, Newton's second law can be summarised by the formula $F = ma$. To discover the details of the changes of motion likely to occur in any given circumstance, we shall write down the forces on the objects concerned, reduce them to the simplest form and then use $F = ma$. This is worked out in some simple cases in the following examples.

Example

A car of mass 800 kg travelling on a level road accelerates forwards at $2 \, \text{m s}^{-2}$. If it is experiencing a drag of 100 N at the time, find the total forward force which must be provided.

If the total forward force required is F_1, with magnitude X newtons, and the drag force is F_2, net forward force $= F_1 - F_2$.
This is the force referred to in the formula $F = ma$, so

$$F_1 - F_2 = ma$$
$$X - 100 = 800 \times 2$$
$$X = 1700$$

so
$$F_1 = 1700 \, \text{N}.$$

Exercise 5.4

If as a result of an increase in speed the drag force increases to 500 N, the total driving force remaining at 1700 N, what acceleration will be achieved?

Example

A lift journey to the top of a building commences from rest with a uniform acceleration of $2 \cdot 0 \, \text{m} \, \text{s}^{-2}$ to a top speed of $4 \cdot 0 \, \text{m} \, \text{s}^{-1}$; this speed is maintained until a uniform retardation of $2 \cdot 0 \, \text{m} \, \text{s}^{-2}$ brings the lift to a halt. A spring balance attached to the roof of the lift carries a mass of 1 kg and is damped so that oscillations die away almost immediately. Describe the variations which occur in the reading during the journey, taking g to be $9 \cdot 8 \, \text{m} \, \text{s}^{-2}$.

We can take it that any acceleration which the lift experiences is also experienced with negligible delay by the 1 kg mass. During any stage of the journey,

$$\text{net force on the 1 kg mass} = F_0 + W$$

(note this is a *vector* sum). F_0 is the upward force exerted by the spring, with magnitude equal to the balance reading, and W is the force on the 1 kg mass due to the attraction of the earth, that is, $-9 \cdot 8 \, \text{N}$. Writing the mass as m and its upward acceleration as a.

$$F_0 - 9 \cdot 8 \, \text{N} = ma$$

so
$$F_0 = ma + 9 \cdot 8 \, \text{N}.$$

For the first stage of the journey $a = 2 \, \text{m} \, \text{s}^{-2}$

so
$$F_0 = 11 \cdot 8 \, \text{N}$$

For the second stage,
$$a = 0$$

so
$$F_0 = 9 \cdot 8 \, \text{N}$$

and for the third stage
$$a = -2 \, \text{m} \, \text{s}^{-2}$$

so
$$F_0 = 7 \cdot 8 \, \text{N}$$

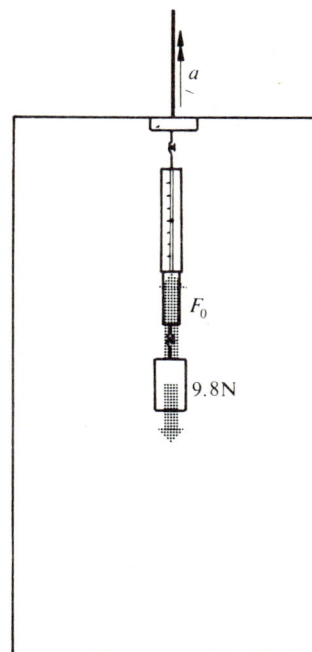

Forces on an object suspended from a spring balance in an accelerating lift.

Exercise 5.5

During one stage of a launch from the earth's surface a team of astronauts experiences a vertical acceleration of $25 \, \text{m} \, \text{s}^{-2}$. If a spring balance were used to record their 'weights' in these conditions, how would its readings compare with those recorded before the journey commenced?

Law or definition?

We shall pause at this point to take stock, and in particular to ask whether we have any right to call $F = ma$ a *law*, since it is now tied up with how we *define* force. We recall that whenever we define a quantity we need to specify both the unit and the counting process. Certainly $F = ma$ is what we have in mind when we define the *unit*: a newton is that force which gives a mass of one kilogram an acceleration of one metre per second squared. Whether $F = ma$ defines the *counting process*, that is, defines 2 N, 3 N and so on, is another matter; we have a choice to make. On p. 50 we defined the counting process for forces in terms of springs exerting equal forces and acting in parallel, and if we stick to this we can then discover $F = ma$ as a *law*, because it is a *summary of the results* we get when we measure the accelerations of objects such as trolleys acted on by different forces (see p. 54). It would be fair to say that this is how the counting process is defined in *practice* – we calibrate a spring balance by hanging a load from it, then two equal loads, then three

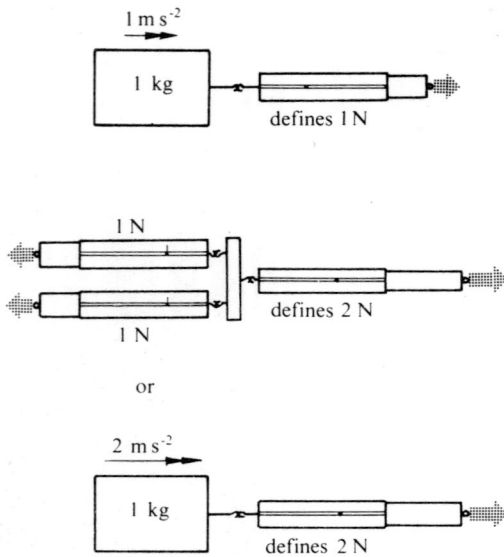

Two possible ways of defining 2N: experiment must decide whether they are equivalent.

Two possible ways of defining 2kg: experiment must decide whether they are equivalent.

and so on. (This happens to result in equally spaced divisions, but that is purely fortuitous as far as the present argument goes.)

There is a tendency to take a different view however. The relation $F = ma$ is taken to be the *basis of the counting process*, that is, a force of two newtons is defined as the force which causes a mass of 1 kg to accelerate at $2\,\mathrm{ms}^{-2}$. This makes $F = ma$ a *definition* of force. But we cannot leave it like that. If we decide to turn a law into a definition by re-defining our terms, somewhere we must get another law, another summary of how things behave, because we are not making up our equations in a vacuum. We are making equations to fit the material world we have been measuring, real trolleys pulled by real springs. The law we get now is a rather strange one; it can be illustrated by the following experiment. Let us set up two spring balances A and B in such a way that each one can exert a force of 1 N as defined in an acceleration process by $F = ma$. Similarly we shall set up another spring balance C so that it exerts a force of 2 N, again defined by $F = ma$. Now we hook up A and B in parallel to pull against C–and lo and behold, they balance! That is an *experimental result* and could *not* have been predicted from logic if we had started with $F = ma$ as the basis for defining force. We have found that 1 N plus 1 N in parallel equals 2 N; in other words *parallel forces combine according to the rules of ordinary numbers*. That is the resulting law if $F = ma$ is made the definition of force.

The argument so far has involved masses of 1 kg only. Further complications arise if we consider masses of other sizes, and of differing sizes, because the definition of mass, with *its* counting process, is also involved. If we subject two different masses m_1 and m_2 to equal forces F (the forces can be of any size, but they must be equal) we find that they accelerate at different rates a_1 and a_2 given by

$$\frac{m_1}{m_2} = \frac{a_2}{a_1} \qquad \ldots(4)$$

or

$$m_1 a_1 = m_2 a_2$$

(each side of this equation being equal to F). On p. 10 we used combinations of identical lumps of matter to define 2 kg, 3 kg and so on, that is, to define the counting process for mass. On this basis, equation (4) is a law; it summarises the results of the experiment with trolleys of different mass on p. 54. However, complications which arise in more advanced work make it more convenient to *define* the counting process in terms of equation (4), so that, for example, a mass of 2 kg is that mass which experiences half as much acceleration as a 1 kg mass in similar circumstances. Once again if we decide to turn a law into a definition we shall get a different law in place of the original definition. The question to ask this time is, what is the mass of two 1 kg masses put together? We cannot just *say* it is 2 kg, because a 2 kg mass is now something which is decided by a measurement of *acceleration*. We have to *measure* the mass of our pair of 1 kg masses; we have to subject them to a force, as a result of which we *find* that the acceleration is half that which either of them experiences when it alone is acted on by the force. So their mass *is* 2 kg–by experiment. The law which summarises the results of all such experiments states that when a number of masses are placed in contact, their combined mass is the sum of the constituent masses; masses add according to the rules of ordinary numbers.

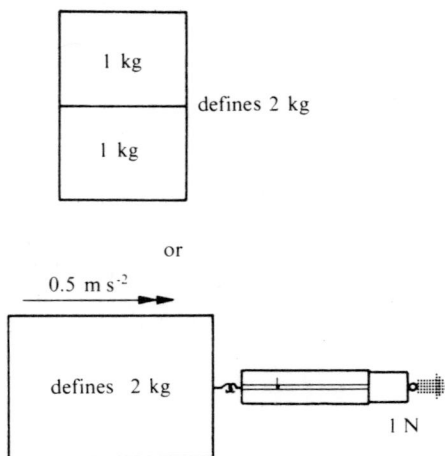

Begging the question as to whether there is something *absolutely* right about the equation $F = ma$, there is no doubt that it is a very *convenient* relationship. It would certainly not have been a happy choice for a basis of definition if the ensuing laws of addition of masses and of parallel forces had not corresponded to the rules of addition of ordinary numbers. Finally, it may be fair to ask why it is necessary to worry about such a point, since, for example, either definition of mass produces equations which agree with each other in the end: the two possible definitions are *consistent* with each other. On a practical level the answer is: none whatever, *until we come to a situation where they are not consistent*. And the world in which we measure is a strange place, because for masses travelling at speeds approaching the speed of light, peculiar things happen, and we then have to ask ourselves exactly what we mean by our definitions.

Varying mass

Like many formulae in physics, $F = ma$ is not merely a relationship between three numbers. Very often we shall find it connects *variables* – quantities which change as time goes on. We shall have to deal with situations where the force is changing, but at any and every instant $F = ma$ will hold exactly. We can tell this is so because the motion we work out on that basis can be experimentally verified. It is also possible for the mass m to vary, for example, in the case of a rocket expelling its products of combustion, and the question then arises whether we should use $F = ma$ or $F =$ rate of change of (mv), as these two expressions are liable to give different answers; $\mathrm{d}(mv)/\mathrm{d}t$ does not equal $m\,\mathrm{d}v/\mathrm{d}t$ if m is changing. Such cases must be treated with care; the rocket problem is deferred until p. 91 where it is shown that the thrust F exerted by the motors produces an acceleration a given by $F = ma$. This means that the statement 'force = rate of change of momentum' is reliable only when matter is not being added to or taken from the system considered. It *is* however applicable to *some* such situations involving changing mass; a horizontal conveyor belt fed by a hopper is a good example. If the belt has a constant speed of $Y\,\mathrm{m\,s^{-1}}$ and the load, which we assume has no horizontal velocity initially, arrives at the rate of $X\,\mathrm{kg\,s^{-1}}$, we can say that during each second a mass X kg is accelerated from rest to a speed of $Y\,\mathrm{m\,s^{-1}}$. We take it that the force required to achieve this is that which would be needed to give a rigid body of mass X kg an acceleration of $Y\,\mathrm{m\,s^{-2}}$, that is, the force needed is XY N. The appropriate formula for the force needed to keep the belt moving at speed v is therefore

$$F = v \times \text{rate of arrival of mass}$$
$$= v\,\mathrm{d}m/\mathrm{d}t$$

which in this case happens to be $\mathrm{d}(mv)/\mathrm{d}t$ because v is constant.

Conveyor belt.

Example

A hose can deliver $0.5\,\mathrm{m^3}$ of water per minute from a hole of diameter 2 cm. If the jet is played horizontally against a wall, what is the force on the wall? Assume that water which has hit the wall drops vertically.

We write the volume of water emerging per second as Q and the

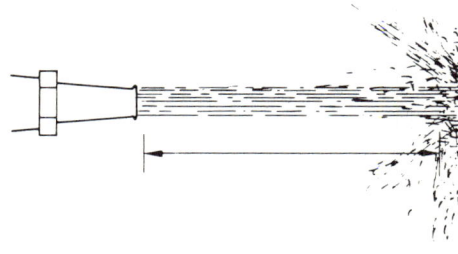

The column of water indicated by the arrow, of length numerically equal to v, contains a volume of water numerically equal to Q.

whole area as A. The water velocity can be found by the following useful argument. Each second a column of water with volume numerically equal to Q and cross-sectional area A emerges; the length of this column is therefore numerically equal to Q/A. Since the leading edge of such a column travels a distance equal to the length of the column in one second, the water velocity v is Q/A. The force F on the wall is given by $F = v\,dm/dt$

that is, $F = v \times$ rate at which mass is leaving the column as a result of being stopped at the wall.

The rate at which mass is leaving $= \rho Q$, where ρ is the density of water. So

$$F = v\rho Q = Q^2\rho/A.$$

Using the data given, and remembering that the equation is related to $F = ma$ and will give F in newtons only if lengths are in metres, masses in kilograms and times in seconds,

$$F = \frac{(0\cdot5)^2}{3600} \times 1000 \times \frac{1}{\pi(0\cdot01)^2}\,\text{N}$$
$$= 220\,\text{N}.$$

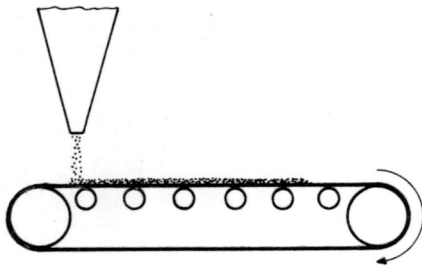

The force needed to keep this conveyor belt moving at constant speed is equal to the product of the belt speed and the rate at which mass is added (ignoring frictional losses).

Exercise 5.6

Sand drops vertically from a hopper at $2\,\text{kg s}^{-1}$ on to a horizontal conveyor belt moving at $1\cdot5\,\text{m s}^{-1}$. What force is needed to pull the belt along in the absence of any other retarding mechanisms?

Further exercises

Exercises 5.7–9

7 a) An object of mass m rests on the floor of a lift which is ascending with acceleration a. Draw a diagram to show the external forces acting on the object, and write down its equation of motion. How do these forces arise? Show graphically how their magnitudes vary with the acceleration of the lift. What force constitutes the second member of the action–reaction pair in the case of each of these external forces?

b) Five identical cubes, each of mass m, lie in a straight line, with their adjacent faces in contact, on a horizontal surface, as shown in Figure 5.8.

Suppose the surface is frictionless and that a constant force P is applied from left to right to the end face of A. What is the acceleration of the system and what is the resultant force acting on each cube? What force does the cube C exert on cube D?

Given that friction is present between the cubes and the surface, draw a graph to illustrate how the total frictional force varies as P increases uniformly from zero. (WJEC)

8 A helicopter of total mass $1000\,\text{kg}$ is able to remain in a stationary position by imparting a uniform downward velocity to a cylinder of air below it, of effective diameter $6\,\text{m}$. Assuming the density of air to be $1\cdot2\,\text{kg m}^{-3}$, calculate the downward velocity given to the air. (JMB)

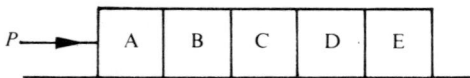

Figure 5.8 (Exercise 5.7)

9 A bag containing a mass M of sand is suspended from a hook on the arm of a balance at a height h above the balance pan. At a time $t = 0$, the sand starts to pour from a hole at the bottom of the bag, falling onto the pan beneath, and continues at a constant rate r (mass per unit time) until the bag is empty.

a) Find the mass required on the other pan to maintain balance under steady conditions when a continuous stream of sand is falling from the bag. (A simple balance with equal arms is envisaged.)

b) Show graphically the variation with time of the mass required to maintain balance throughout the experiment, indicating by suitable labelling the quantities involved. Assume ideal conditions, under which air resistance and balance inertia and damping effects may be ignored.

c) Describe how you might attempt to verify your results and discuss what allowances, if any, you would make because of the impossibility of realising these ideal conditions in practice. (C)

6 Forces in combination

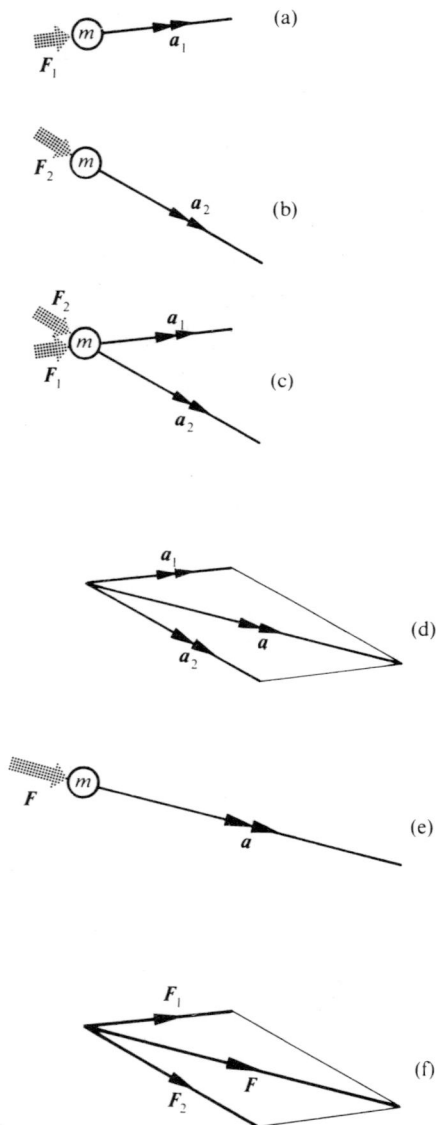

Figure 6.1

Addition of forces: (a) and (b) F_1 and F_2 produce accelerations a_1 and a_2 respectively. (c) F_1 and F_2 act together. (d) The accelerations combine according to the parallelogram law. (e) The net acceleration could also be produced by a single force F as shown. (f) Because the forces are proportional to the corresponding accelerations, F_1 and F_2 must be related to F by the parallelogram law.

Newton's second law not only gives the *size* of the acceleration produced in a given set of circumstances, it deals also with its *direction*; $F = ma$ is a *vector* relation. Figure 6.1 illustrates some of the implications. In the first two diagrams we have two forces F_1, F_2 respectively, acting in turn on the mass m and producing corresponding accelerations a_1, a_2 in the directions shown. Figure 6.1c shows both forces acting simultaneously; now the mass experiences both accelerations at once. Well, we already know how to combine two accelerations – we use the parallelogram law – so the net acceleration a will be as indicated in Figure 6.1d. But this acceleration could equally well have been produced by a single force F of appropriate size and direction. We call F the *resultant* of F_1 and F_2; it is the single force to which they are equivalent. That leads us to an important conclusion. The sizes of the accelerations are proportional to the forces, and each acceleration has the same *direction* as the force producing it, so if the accelerations are linked by the parallelogram law, it looks as though the forces must be linked by the same rule. Forces combine according to the parallelogram law.

It does not pay to put too much reliance on logical processes of that kind however; fallacies are sometimes difficult to spot, and scientists have been caught out more than once in the past. We need to carry out an experimental test. Figure 6.2 shows a simple arrangement which is suitable. Each of the three lengths of thread knotted together at P passes over a pulley with low-friction bearing and is loaded underneath by a suitable weight. We want to know if we can replace the forces F_1 and F_2 exerted on the knot by two of the threads with a single force F given by a parallelogram law of addition. If we set up the weights so that the knot remains stationary, F will have to balance F_3 exactly. Common experience tells us what F must be like to do this; it must have size equal to F_3 and it must act in a direction exactly opposite to F_3. This is our test then; if when we try the parallelogram rule on F_1 and F_2 we get a resultant exactly equal and opposite to F_3, that will indicate we have used the right rule of addition. Table 6.1 shows some typical results (see Figure 6.3).

We can use a scale drawing to produce the parallelogram, or we can use some trigonometry; if the parallelogram law holds, then F is given by

$$F^2 = F_1{}^2 + F_2{}^2 - 2F_1F_2\cos\beta$$

Also,

$$\widehat{APB} = \alpha_2 + \gamma,$$

where

$$\frac{F_2}{\sin\gamma} = \frac{F}{\sin\beta}.$$

Exercise 6.1

Show that the values given in Table 6.1 are consistent with the parallelogram law.

The above test is not complete because it has been carried out in special circumstances, namely, an *equilibrium* (that is, unaccelerated) situation. A more thorough test would involve measurement of the *acceleration* produced by a system of forces acting on an object. The real evidence for the parallelogram law is that it is found to give correct results in every situation to which it is applied. To state the law formally: *if two forces acting simultaneously on a body are represented in size and direction by the adjacent sides of a parallelogram, their resultant is represented in size and direction by the diagonal through their point of intersection.*

Line of action; moment of a force

The parallelogram rule plays a very important part in the way forces combine, but it is not the whole story. Hit a ball fair and square in the middle and it will do what you want it to; hit it with the same force in the same direction but near one edge and quite a different result will occur. We need to take into account the *line of action* of a force as well as its size and direction. Where we hit the ball decides how much it will spin – the line of action of a force defines its *turning effect* about any axis that may interest us. A very simple experiment gives us the rule for evaluating this turning effect. In Figure 6.4 we have a metre rule balanced at its midpoint with weights W_1 and W_2 hung at different distances x_1 and x_2 from the pivot on either side. It soon emerges that the metre rule will balance horizontally only when $W_1x_1 = W_2x_2$; it is the quantity *force $\times$ distance from the pivot* which must be the same on both sides. A slight modification (Figure 6.5) by which one of the forces acts along the line AB, inclined to the vertical, shows that it is the *perpendicular* distance of the force from the pivot which matters; the condition for balance in this case is $W_1p = W_2x_2$.

The product of the force and its perpendicular distance from the axis considered is called the moment of the force about that axis and it is a measure of the turning effect which the force has about it; the

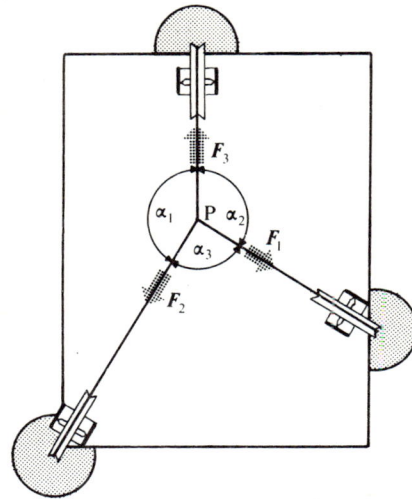

Figure 6.2

Table 6.1

$F_1 = 1{\cdot}70\,\text{N}$	$\alpha_1 = 121°$
$F_2 = 1{\cdot}50\,\text{N}$	$\alpha_2 = 131°$
$F_3 = 1{\cdot}90\,\text{N}$	$\alpha_3 = 108°$

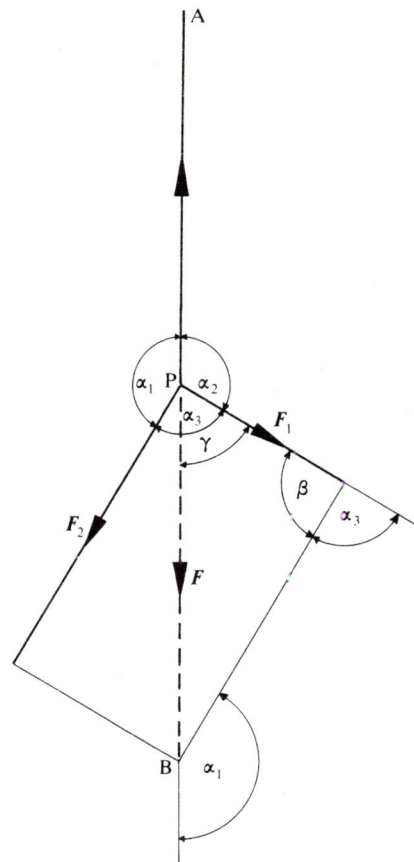

Figure 6.3

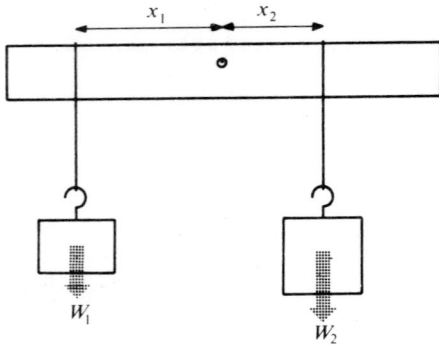

Figure 6.4

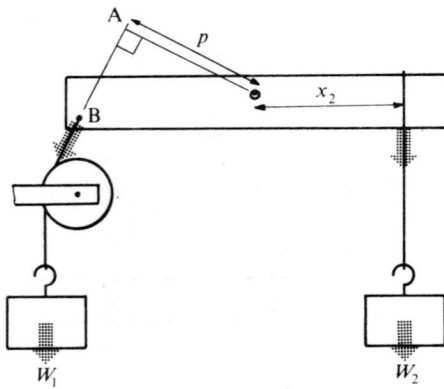

Figure 6.5

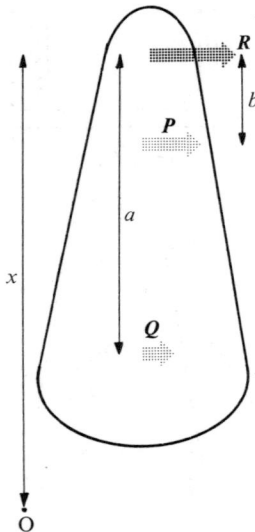

Figure 6.6

What line of action must R have so that it can replace P and Q?

unit is the newton metre, N m. (Notice that it is not enough to define moment as 'the turning effect of a force about a given axis'. We have to supply an *operational* definition, giving instructions how to measure it: 'find the product of force and perpendicular distance'.) With the turning effect in mind we need to take a rather more thorough look at how forces combine, and we start with the special case of forces having parallel lines of action.

Combination of two like parallel forces

Figure 6.6 represents two forces P and Q parallel and facing in the same direction ('like' parallel forces). Of course the object they act on cannot be localised at a point; it must be large enough to give both forces a foothold; we call it an *extended* object. We want to know if we can replace P and Q by a single force which reproduces most of the effects of P and Q. It needs to be said that we cannot reproduce *all* the effects. In practice P and Q act directly on perhaps a few million atoms each; these atoms in turn interact with atoms around them, and so on, until by an enormous number of mutual attractions and repulsions the effects are felt by the entire object. Now, if we replace P and Q by a different force system the details of these interactions may be different, and if the body has any tendency to distort we shall be able to *see* the difference. The whole problem of change of shape under stress is dealt with in Chapter 11; here we shall deal with a figment of the imagination known as a *rigid body*– one which does not distort at all. Many familiar objects around us are rigid bodies to a good approximation.

Though we do not pretend to be able to find a single force which can produce the same internal atomic attractions and repulsions, we *can* find one which will cause the object to accelerate in the same way as a whole. There are two aspects to look at. One is the transfer of the object from place A to place B–*translation*. The other is reorientation of the object–*rotation*. As far as translation goes, the single force R which replaces P and Q can hardly be in any direction other than parallel to both of them, and of size equal to $(P + Q)$. The question is whether we can find a line of action for R such that by itself it produces the same rotational effect as P and Q acting together. We have got a measure of the rotating effect of a force–its moment. Perhaps we can arrange R so that its moment is the same as the sum of the moments of P and Q, but about what axis? Presumably about *any* axis, if R is to replace P and Q entirely in this respect. This seems rather a tall order. In Figure 6.6 we have drawn R a distance b from P and a distance a from Q; we want to find what a and b must be and we take moments about an axis through O at a distance x from the line of action of R.

For the P, Q system, the clockwise moment about O is

$$Q(x - a) + P(x - b)$$

and for the R system it is xR, or $x(P + Q)$. We have to insist these are equal.

Exercise 6.2

Show that as a result we can write

$$\frac{b}{a} = \frac{-Q}{P}.$$ ...(1)

The significance of the negative sign is that R lies between P and Q as in Figure 6.7.

Well, we have managed it: we have matched the moment of R with that of P and Q with respect to *any* axis, because the answer is independent of x; it does not matter where O is.

It pays to confirm even the most straightforward piece of theory by experiment, and the arrangement in Figure 6.8 provides a suitable check. Parallel forces P and Q are balanced by another force S; the assumption is that to achieve this, S must be equal and opposite to R and have the same line of action. P, Q, a and b can be measured and used to test equation (1).

Combination of two unlike parallel forces

Thus encouraged we tackle the case illustrated in Figure 6.9 where two forces P and T are in opposite directions. In fact we have no need to go through the whole process again; all we need say is that T has magnitude T to the left, or magnitude $-T$ to the right. Then we can use the previous treatment, replacing Q by $-T$.

Equation (1) will now read

$$\frac{b}{a} = \frac{T}{P}, \qquad \ldots(2)$$

so that a and b now have the same sign; this means that the line of action of R will no longer come between P and T. The magnitude of R will of course be $(P - T)$. There is still one question to decide – will R be nearer P or T? Equation (2) tells us that if P is greater than T, a must be greater than b, and this can be so only if we make R closer to P than to T, as in Figure 6.9.

This result can be obtained even more directly from the case involving like parallel forces as follows. Referring to Figure 6.7, we already have that

$$R = P + Q$$

that is,

$$-R + P = -Q$$

which expresses in symbols the fact that two unlike parallel forces $-R$ and P can be replaced by a single force $-Q$. Since we have established that R must be between P and Q, it follows that $-Q$ is *outside* the combination $-R$ and P.

Couples

A snag occurs in equation (2) when $P = T$. This gives $a = b$, and as a and b are supposed to be *different* by an amount equal to the distance between P and T this is a little difficult to arrange! The only way round it would be to say that the line of action of R for this special case is an infinite distance from P and T. What is more, the magnitude of R will have to be zero since $R = P - T$. The fact is that two unlike parallel forces of exactly equal size but different lines of action *cannot* be replaced by a single force. They produce no net translation, they can only *rotate* the object on which they act, and a force system of this type is called a *couple*. The obvious way to define the size of a couple is by means of its moment. In Figure 6.10 we have a couple with forces of size F separated by a distance p (p is called the *arm* of the couple) and we want to write down the moment, or *torque*, of the couple about some chosen axis O, a distance x from one of the forces.

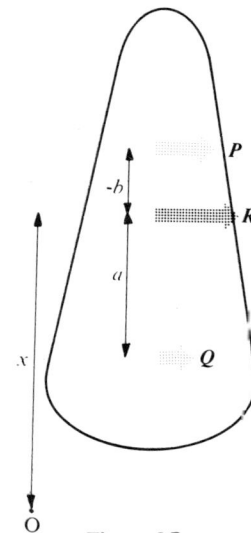

Figure 6.7

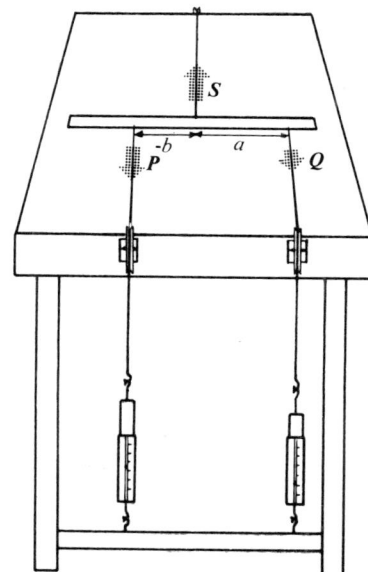

Figure 6.8

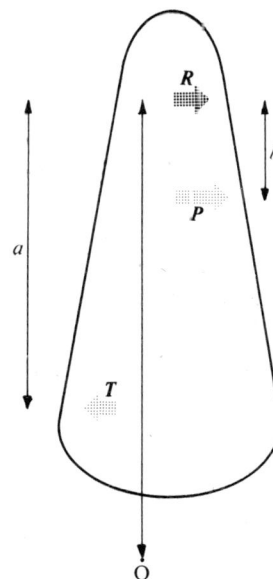

Figure 6.9
Combination of unlike parallel forces.

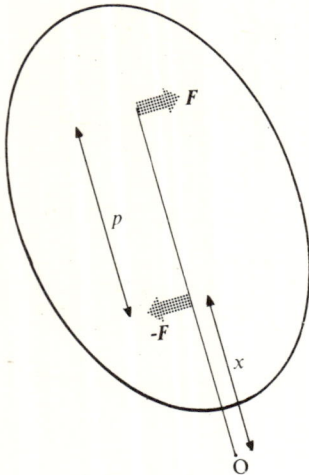

Figure 6.10
A couple.

Torque about $O = F(p + x) - Fx = Fp$.

In other words it does not matter how big or small x is, *the torque of the couple about the axis O is equal to the product of one of the forces and the arm. The moment of a couple is the same about any axis perpendicular to the plane containing the constituent parallel forces.*

Couples abound in everyday experience; in all cases where pure rotation commences or stops a couple is present. If we turn on a water tap, or use a spanner on a nut, a couple is brought into play; in the latter case it looks as though there is only one force, but the other one is provided by the reaction of the bolt on the nut. The longer the spanner, the greater the torque of the couple and the more we can tighten the nut, or strip the thread! Some spanners have a mechanism causing them to slip when a chosen torque is exceeded, to avoid the latter possibility. Steering wheels, cogs, clutches, armatures in electric motors, galvanometer coils, and shafts of every kind all commence or cease to rotate as a result of the operation of a couple.

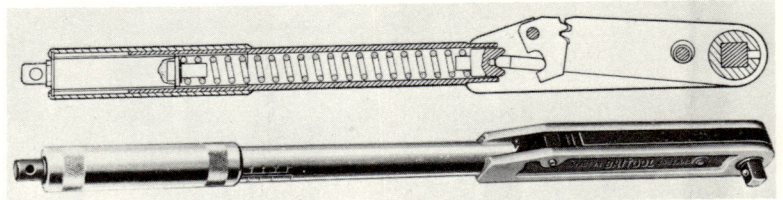

Torque wrench, designed to slip when the chosen torque is exceeded.

Parallel steel suspender wires hanging from the suspension cables support the deck of the Forth Road Bridge.

Several parallel forces; centre of gravity

If we can add two parallel forces we can of course add three or more; we use the rule to add two of the forces and then use it again to add the resultant to the third force, and so on if necessary. This has great importance whenever we have to deal with the weight of an extended object–that is, any object! Every tiny part, every atom, is being pulled to the earth, and the weight of the body is really made up of a huge collection of tiny forces, all with different lines of action, and all parallel, so long as the body is not of gigantic proportions.

We have claimed we can replace a force system like that by a single force, as long as we are not interested in the internal stresses in the object. This single force is what we generally have in mind when we speak of the object's weight, W. Obviously W is the sum of all the constituent forces and acts vertically downwards; this ensures that it produces the same translating effect. We must also see to it that it has the same rotating effect; its line of action must be such that it has the same moment about any chosen axis as the sum of the moments of all the tiny constituent forces about that axis. In particular, if we happen to choose an axis through a point on the line of action of W, this sum will have to be *zero*; that is often useful in fixing the line of action.

We can express this in symbols as follows. If δW is the weight of a typical tiny part of the object, acting a perpendicular distance x from the axis considered, then $x\delta W$ is its moment about this axis, and we can write the sum of all such moments as $\sum x\delta W$. If the net weight W acts a perpendicular distance $\bar{x}$ from the same axis, we have to insist that

$$\sum x\delta W = \bar{x}W$$

(sum of constituent moments equals moment of W about the same axis). But

$$W = \sum\delta W$$

(net weight equals sum of constituent weights).
So

$$\sum x\delta W = \bar{x}\sum\delta W$$

that is,

$$\bar{x} = \frac{\sum x\delta W}{\sum\delta W} = \frac{\text{sum of constituent moments about chosen axis}}{\text{sum of constituent weights}}$$

Figure 6.11 illustrates another point needed to complete the picture. If we change the orientation of the body, all the tiny forces making up the weight will swing round and point in a new direction relative to the body. The line of action of W will also swing round but one point in the body will be common to both lines of action. In fact that point will be common to *all* lines of action of W for every orientation; it is the one point in the body at which all the weight of the body can be considered to act, and it is called the *centre of gravity* of the body. [It is only in a *uniform gravitational field* that the centre of gravity is a fixed point in the object (see Chapter 20, p. 295). The gravitational field is uniform to a very good approximation over any local region of the surface of the earth.]

Calculating the position of the centre of gravity can be very tedious unless the object possesses symmetry of some kind, in which case it must lie on the axis of symmetry. The following examples show how it can be found in some typical cases.

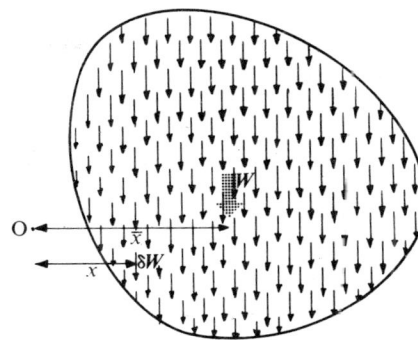

Establishing the line of action of the weight of an object

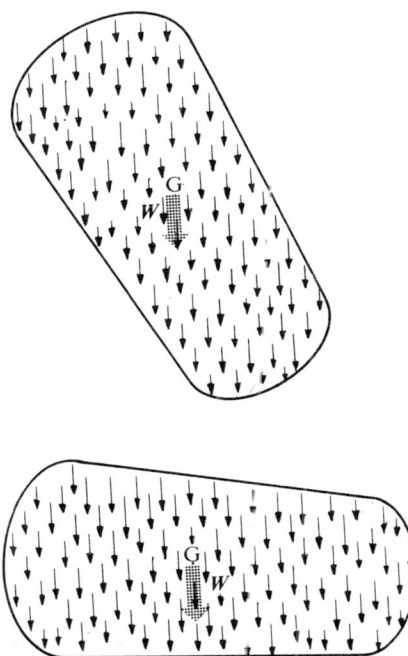

Figure 6.11

There is one point G through which the line of action of the weight always passes–the centre of gravity.

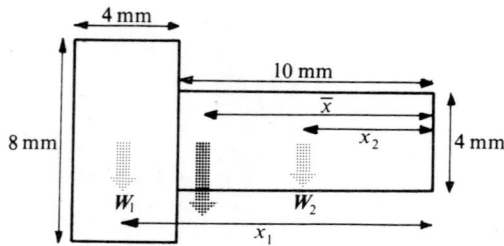

Figure 6.12

Example 1

A particular copper rivet consists of two coaxial cylinders in contact. One has a length of 4 mm and diameter 8 mm; the other has a length of 10 mm and diameter 4 mm. How far is the centre of gravity from the thinner end?

Each cylinder has a high degree of symmetry in its own right, with centre of gravity at its centre. The system can therefore be replaced by two point forces W_1 and W_2 acting at distances x_1 and x_2 from the thin end (Figure 6.12); x_1 is 12 mm and x_2 is 5 mm. If the centre of gravity of the entire rivet is a distance $\bar{x}$ from the thin end, we know that

$$W_1(x_1 - \bar{x}) = W_2(\bar{x} - x_2) \qquad \text{...(3)}$$

because the net moment of W_1 and W_2 about an axis through the centre of gravity must be zero.

Now W_1 and W_2 will be proportional to the volumes of the respective cylinders, that is,

$$\frac{W_1}{W_2} = \frac{\pi 4^2 \times 4}{\pi 2^2 \times 10} = 1 \cdot 6.$$

So putting $\bar{x} = X$ mm in equation (3) we get

$$1 \cdot 6 \,(12 - X) = (X - 5).$$

This gives $X = 9 \cdot 3$, so that $\bar{x} = 9 \cdot 3$ mm.

Example 2

A solid roller from a roller bearing is built in the shape of a right circular cone of height h and base radius a. How far is the centre of gravity from the apex?

We can think of this problem as an extension of the previous one; this time we are dealing with a very large number of coaxial cylinders, each one very thin and infinitesimally larger than the next. We look at one of these discs a distance x from the apex, and take it to be of width δx and radius y. If the density of the roller material is ρ, we can write the weight of this disc as approximately $\rho \pi y^2 \, \delta x$; the smaller we make δx the better will be the approximation.

In this problem it is simplest to take moments about an axis through the apex and perpendicular to the axis of the cone. The moment of our disc about such an axis is approximately

$$\sum \rho \pi y^2 x \, \delta x$$

summed up over all the discs, or, exactly,

$$\int_0^h \rho \pi y^2 x \, dx. \qquad \text{...(4)}$$

The moment of the equivalent single force about this axis is $W\bar{x}$ where $\bar{x}$ is the distance of the centre of gravity from the apex and W is the total weight; we shall need W in terms of ρ. The sum of the weights of the constituent discs will be approximately

$$\sum \rho \pi y^2 \, \delta x$$

or exactly, in integral form,

$$\int_0^h \rho \pi y^2 \, dx. \qquad \text{...(5)}$$

We cannot evaluate either of these integrals until we find the connection between y and x, which is defined by the geometry of the

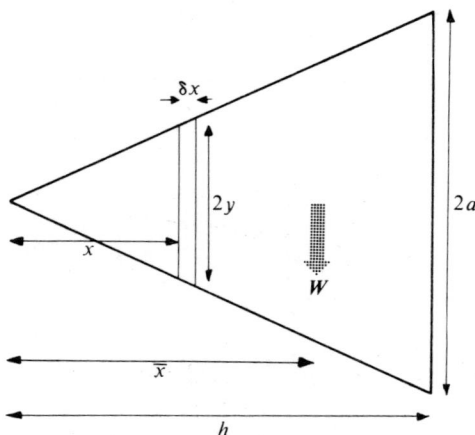

Conical roller (see Example 2)

cone; x and y are both zero at the apex and increase proportionally, until when $x = h$, $y = a$. So

$$y/x = a/h$$
$$y = ax/h.$$

Expression (4), giving the moment of the entire cone about the chosen axis. now becomes

$$\int_0^h \frac{\rho\pi x^3 a^2\,\mathrm{d}x}{h^2} = \frac{\rho\pi h^4 a^2}{4h^2},$$

and expression (5), giving the weight W, becomes

$$\int_0^h \frac{\rho\pi x^2 a^2\,\mathrm{d}x}{h^2} = \frac{\rho\pi h^3 a^2}{3h^2}.$$

Since the moment is $W\bar{x}$, we can write

$$\frac{\rho\pi h^4 a^2}{4h^2} = \frac{\rho\pi h^3 a^2}{3h^2} \times \bar{x}$$

$$\bar{x} = 3h/4.$$

Exercises 6.3, 4

3 The aerial shown in Figure 6.13 has elements of mass $6m$, $3m$, m, m and m respectively, all mounted on a uniform rod of mass $4m$ and length l. They can be taken to be equally spaced. Find the distance of the centre of gravity from the heaviest element.

4 The nose-cone of a rocket consists of a thin open-ended conical shell of uniform thickness; the radius of the base is a and the height h. Find the distance of the centre of gravity from the apex.

Combination of forces which meet at a point; polygon of forces

The size and direction of the resultant of two forces which meet at some point are taken care of by the parallelogram law, and its line of action must pass through the meeting point, as the apparatus in Figure 6.2 will soon show; any attempt to get equilibrium in a situation where the three forces do not share a common point meets with little success. Actually, the fact that the resultant *does* pass through the common point means that it has the same moment about any chosen axis as the sum of the moments of the separate forces making it up; this claim is justified in Appendix 4 with the aid of a little geometry. Of course we can go on to find the resultant of any number of forces as long as they act through a common point; we simply combine two of them by the parallelogram rule, then use the rule again to combine the resultant with force number three, and so on. This is worked out for four forces in Figure 6.14a–c. In Figure 6.14d we have a simpler representation; lines representing the forces in size and direction have simply been laid down nose to tail, with no attempt to represent the correct line of action this time. The straight line joining the ends of the chain then represents the resultant; it can be seen that diagram d contains just some of the lines drawn in diagram c but achieves the same result. If the chain ends up in the same place as it started then there is no resultant; the forces produce equilibrium. This is formally stated in the *polygon of forces law*: *if any number of forces acting at a point can be represented in size and direction by the sides of a polygon, the forces are in equilibrium.*

As long as the forces all meet at a point, there is no need to insist that they should all be in one plane; we can add them by this method

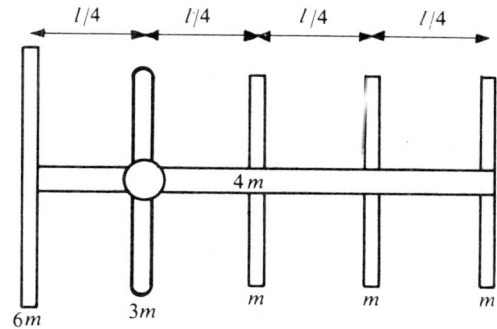

Figure 6.13 (Exercise 6.3)

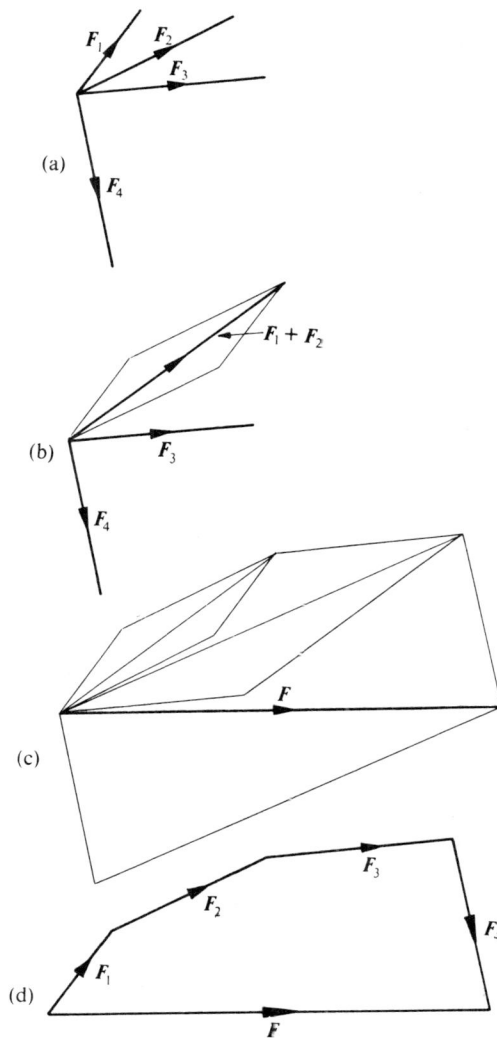

Figure 6.14
Combination of four forces acting at a common point.

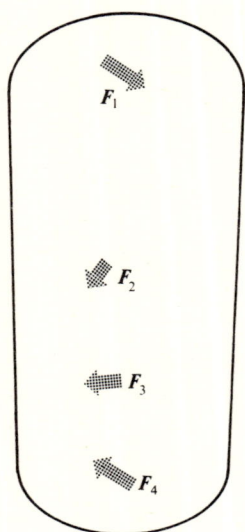

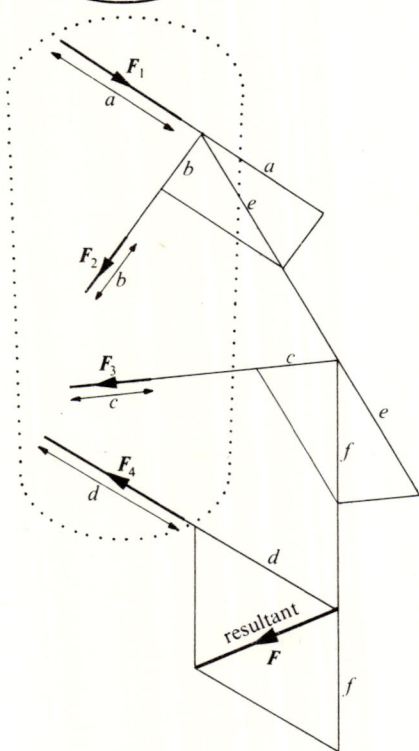

Figure 6.15

Combination of forces not in the same plane, but acting at a common point.

even if some of the forces are pointing 'out of the page', as illustrated in Figure 6.15.

Coplanar forces not sharing a common point of action

When we come to deal with the effect of forces which do not share a common point of action, we find things can get quite complicated unless the forces all lie in one plane, and it is only systems of this kind which we shall be dealing with here. Again, in principle, there is nothing to stop us going ahead with a stage-by-stage application of the parallelogram law. The only difference is that the line representing the resultant at each stage (this time in size, direction *and* line of action) must be extrapolated until it intersects the line of action of the next force to be dealt with; the new resultant will act through that point of intersection. This process is carried out for four forces in Figure 6.16. The only snag that can arise is that it may be necessary to combine two parallel forces at some stage, and this must be done according to the rules already established (p. 64). The entire system will eventually be reduced either to a single force, or possibly a couple.

Resolution of forces

The methods we have discussed provide one way of tackling the problem of the combined effect of a system of forces, but it is often not the most convenient approach. The method we generally find most helpful seems at first absurd – we actually *increase* the number of forces we have to deal with, replacing each force in the original system by a pair of forces producing the same result. In Chapter 4 (p. 48) we dealt with the replacement of a *velocity* by an equivalent pair of velocities; the same arguments will apply to forces. Any force can be replaced by two different forces; we merely need to choose them so that the lines which represent them form the sides of a parallelogram which has a diagonal representing the original force on the same scale. There are an infinite number of ways of doing this

Figure 6.16

Combination of coplanar forces not having a common point of application.

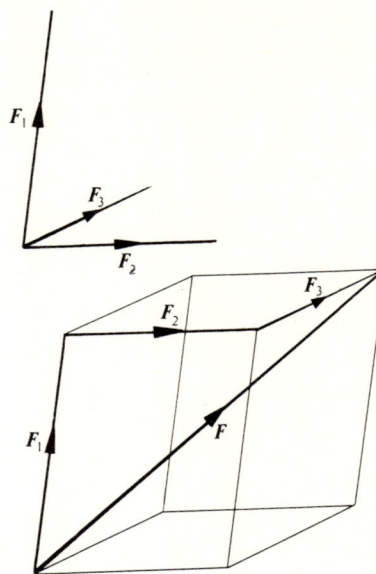

in any particular case–two suitable combinations which replace the force F are shown in Figure 6.17. Of all these combinations, the ones we find of greatest use are those for which the two constituent forces are perpendicular to each other. Even so, we still have an infinite number of combinations to choose from (Figure 6.18). Usually one pair will turn out to be more useful than any other in the particular circumstances we happen to be considering. The process of splitting a force into two constituent forces perpendicular to each other is called *resolution* of the force into *perpendicular components*. Throughout this book we have adopted the convention that whenever *components* are referred to, the vector concerned has been resolved into *perpendicular* components.

With all the different pairs of components which could be chosen for any given force the whole process appears very divergent; it begins to look as though we can do just what we like, and the possibility that resolution might make things simpler seems rather remote. In fact we have very much less freedom of choice than we might think; once the directions of the component forces have been decided on, the size of each of them is fixed. Referring to Figure 6.18, there are many rectangles having AC as diagonal but there is one and only one rectangle $ABCD$ which has AC as diagonal and AB inclined at a particular angle α to AC.

We can use this diagram to establish the sizes of the resolved components P and Q of the force F represented by AC.
Since

$$AB = AC\cos\alpha \quad \text{and} \quad AD = AC\sin\alpha,$$

and since the length of each line is proportional to the force it represents, it follows that

$$P = F\cos\alpha, \quad Q = F\sin\alpha.$$

This result will be in such continual use that it is essential to be able to quote it without any specific reference to the implied vector diagram.

The resolved component of a force F along a direction making an angle α with F is $F\cos\alpha$; the component perpendicular to this is $F\sin\alpha$.

We use the technique of resolution to help us find the resultant of a system of coplanar forces as follows. First we need to pick a convenient pair of axes x and y perpendicular to each other and both in the plane concerned, and then we systematically resolve each force into an x component (its component along the direction of the x-axis) and a y-component (Figure 6.19). Now, since we have a whole collection of vectors pointing in the x-direction we can do some straight addition, paying attention to any contributions pointing in the $-x$-direction, which will have to be counted negative. We can do the same for the components in the y-direction, and as a result we shall end up with just two forces, perpendicular to each other. If we wish, these can be combined according to the parallelogram law, but we often find it more convenient to work with them as separate components.

This process fixes the magnitude and direction of the resultant; the line of action can be settled by looking at the question of moment. We can pick any axis perpendicular to the plane containing the forces in the original system and find the sum of their moments about it. We then have to choose the line of action of the resultant so that

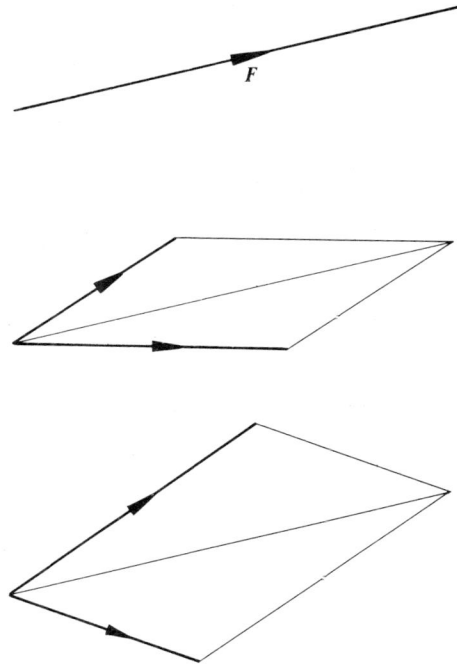

Figure 6.17
Two different sets of forces which could replace F.

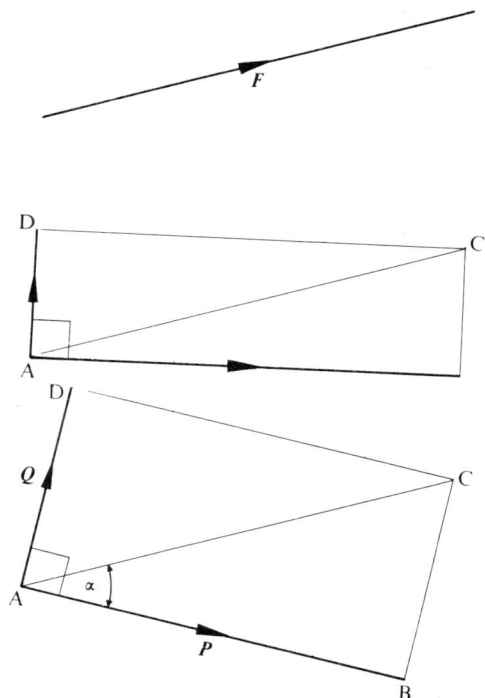

Figure 6.18
Two different sets of perpendicular components of F.

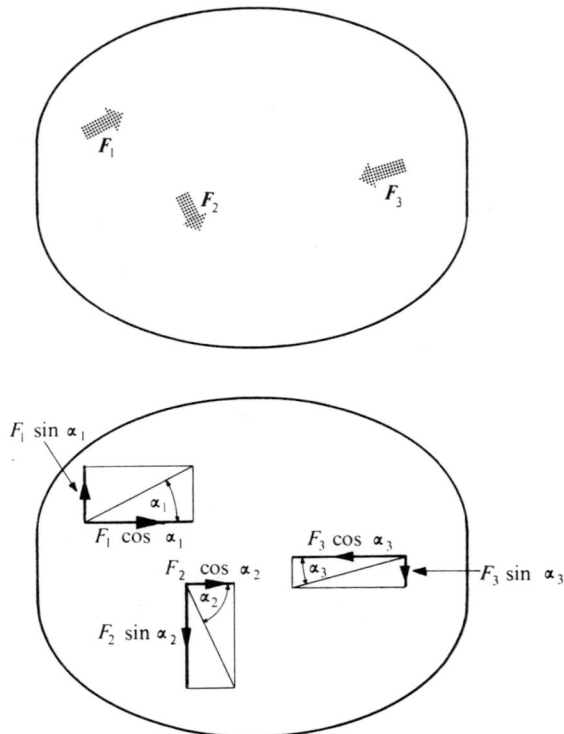

Figure 6.19

Combination of forces by resolution into components:
net horizontal force = $F_1\cos\alpha_1 + F_2\cos\alpha_2 - F_3\cos\alpha_3$ to the right,
net upward vertical force = $F_1\sin\alpha_1 - F_2\sin\alpha_2 - F_3\sin\alpha_3$.

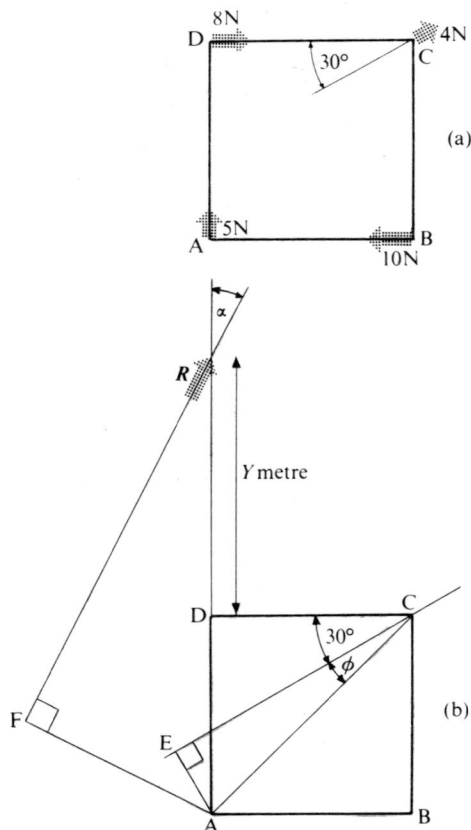

Figure 6.20

its moment about that axis has the same value: in this way we make sure that it produces the same turning effect as the system it is supposed to replace. Objection! We have matched the turning effect for just one axis—what about all the other axes we could have chosen? Have we matched the turning effect for all these too? It seems too good to be true, but it works out that if it is right for one axis, it is right for them all; this is illustrated in Appendix 5. Here we shall show it working out in practice in a specific example.

Example

The square tile $ABCD$ of side 0·2 m is acted on by the four forces shown (Figure 6.20a). Find the size of the resultant, the angle it makes with AD and how far from D it crosses AD.

We shall suppose the resultant to have magnitude R, and to make an angle α with AD, crossing AD at a distance Y metres from D as shown (Figure 6.20b). It will be simplest to work with axes along the directions of the sides of the square. The resolved components along the direction of AB are 8 N, $(4\sqrt{3}/2)$ N and -10 N, remembering that $\cos 30° = \sqrt{3}/2$.

So the sum of components along the direction of $AB = (2\sqrt{3} - 2)$ N.

Sum of components along direction $AD = (5 + 4\sin 30°)$ N
$$= 7\,\text{N}$$

Since R is the resultant of these two perpendicular forces,
$$R^2 = [(2\sqrt{3} - 2)^2 + 7^2]\,\text{N}^2$$
$$= (12 + 4 - 8\sqrt{3} + 49)\,\text{N}^2$$
$$= 51\cdot1\,\text{N}^2$$
$$R = 7\cdot15\,\text{N}$$

Also
$$\tan\alpha = \frac{2\sqrt{3} - 2}{7} = 0\cdot209$$
$$\alpha = 11\cdot8°.$$

Now we have to take moments about a convenient axis; the line through A perpendicular to the plane of the square is suitable. Figure 6.20b shows the perpendicular distances of the various forces from A, which must be found by means of a little geometry.
$$EA = AC\sin\phi$$
$$= 0\cdot2\sqrt{2}\sin 15°\,\text{m}$$
$$= 0\cdot0732\,\text{m}$$
and
$$FA = (Y + 0\cdot2)\sin\alpha\,\text{m}$$
$$= (Y + 0\cdot2) \times 0\cdot205\,\text{m}$$

So the sum of moments of original force system about A
$$= (8 \times 0\cdot2 + 4 \times 0\cdot0732)\,\text{N m}$$
$$= 1\cdot89\,\text{N m}$$

and the moment of R about $A = R \times FA$
$$= 7\cdot15 \times 0\cdot205\,(Y + 0\cdot2)\,\text{N m}.$$

So if R is to replace the original system,
$$1\cdot89 = 7\cdot15 \times 0\cdot205(Y + 0\cdot2)$$
$$Y = 1\cdot09$$

and the line of action of R must cross AD 1·09 m from D.

Exercise 6.5

Show that the sum of the moments of the original force system, and the moment of R, about an axis through D, are both equal to $1 \cdot 60 \, \mathrm{N \, m}$.

The processes of resolution and taking moments really come into their own when we are dealing with an object acted on by a system of forces in equilibrium, that is, when the resultant is zero. The equations we get in such a situation can often enable us to work out the sizes and directions of any of the constituent forces we do not know about; some of the ways in which this can be done are discussed in Chapter 10.

Components of force and acceleration

We began the chapter by saying that $F = ma$ is a *vector* relationship. A force on a particle in a given direction will produce an acceleration in that direction, and will not affect what goes on in any other direction. Now if we care to resolve the force F into two components F_x and F_y in the x- and y-directions respectively, we can say that F_x affects the acceleration in the x-direction only, and F_y the acceleration in the y-direction only. In symbols

$$F_x = ma_x = m \frac{d^2 x}{dt^2}$$

$$F_y = ma_y = m \frac{d^2 y}{dt^2},$$

and if we need to talk about three dimensions, we can also have

$$F_z = ma_z = m \frac{d^2 z}{dt^2}.$$

These equations give us a very useful way of dealing with a particle which is not moving in a straight line; we shall make use of them in Chapter 13 particularly. In more advanced work it is customary to write them in the condensed form

$$F = ma.$$

Because this is a vector equation it implies not one but *three* relationships simultaneously, namely

$$F_x = ma_x$$
$$F_y = ma_y$$
$$F_z = ma_z.$$

(a)

(b)

(a) Components of a force along three mutually perpendicular axes
(b) Corresponding accelerations which they produce.

Exercise 6.6

A light wire is stretched horizontally and a pulley carrying a load runs on the wire, which sags slightly. Draw a diagram of the forces acting on the pulley when it is part way to the centre of the wire, and explain how these forces cause an acceleration of the pulley. (JMB)

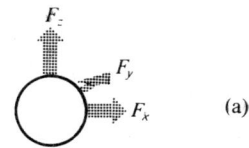

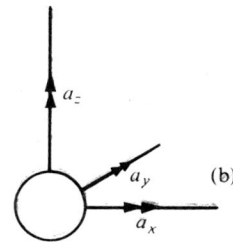

7 Forces in fluids

The methods we have developed so far for dealing with the combined effects of a number of forces on an object will be successful only if the body is rigid. When we talk about a metre rule balanced on a knife-edge we generally have in mind just two forces: the reaction at the support, and the weight acting through the centre of gravity. This is the macroscopic view, and because both of these forces are caused by objects outside the metre rule (the knife-edge and the earth respectively) we call them *external* forces. Microscopically of course each atom experiences forces from all its neighbours. An atom near the end of the rule is pushed down by the weight of all the atoms above it but supported by the atoms below it. The vertical slice of which this atom forms a part is held in place by attractive forces from the atoms in the adjacent slice. As we move in towards the knife-edge the demands made on these attractive forces become greater as more and more of the rule has to be supported. The atoms in the vicinity of the knife-edge experience very large repulsive forces; they have to receive from the knife-edge, and give to the atoms close to them, sufficient force to hold up the entire rule. If the knife-edge is sharp, this force will be shared out over comparatively few atoms and the forces involved will be enough to cause a minute indentation.

All these interatomic attractions and repulsions occur in equal and opposite pairs; whatever atom A does to atom B, B does to A. These forces do not involve any objects outside the metre rule and are therefore called *internal* forces. There *are* also external forces on the atoms – each atom has its own infinitesimal weight and the atoms very near the knife-edge experience repulsions from it–but these are precisely the forces which we have combined into two macroscopic forces, the total weight and the reaction.

Now, in the case of a liquid, the microscopic situation is different, and so it is not surprising that the most convenient macroscopic description is also different. There are, it is true, attractive and repulsive forces between molecules, but the attractive forces are insufficient to bind one molecule permanently to the next. Vertical slices will not hold each other up this time, and the liquid cannot be supported on a knife-edge! It must be supported by the floor and walls of a container. Even less can a gas be supported by forces acting on just a few constituent molecules–it must experience forces from all sides, including the top.

If a liquid or gas experiences forces from every part of the container in contact with it, it is not going to be very useful to attempt to replace all these forces by a single resultant. Instead we accept the reality of the situation and incorporate it in our macroscopic description by talking about the *force per unit area of surface, normal to the surface*. It is the force which unit area of wall surface has to exert on the fluid to keep it in, and also it is the force which unit area of wall surface experiences from the fluid: it is called the *pressure* on the surface. Formally, *the pressure at a given point in a fluid is the force per unit area normal to a small area of surface including the point considered*. The unit of pressure will therefore be newton per square metre, Nm^{-2}, to which is given the name *pascal*, Pa.

We can use the microscopic picture to argue that the pressure at a point is the same in all directions–the orientation of the small area does not matter. In the case of a liquid, any one molecule in the bulk of the liquid will experience and therefore exert equal repulsions on all sides on average, and our small surface experiences a force which is made up of such repulsions, from all molecules in contact with it. In the case of a gas the molecules are moving randomly in all directions and our test surface will experience continual bombardment, the nature and extent of which will be independent of the way it is facing.

The pressure due to a column of liquid

We want to evaluate the pressure caused by a column of liquid at a depth h below its surface. It is simplest to choose a horizontal test surface, the area of which we shall call δA. We want to write down the force it experiences from above. Microscopically that force is made up of the repulsions from the liquid molecules in contact with it and we may well wonder how it can ever be evaluated. The macroscopic view gives the answer immediately–our surface has to support the entire weight of the liquid column vertically above it (Figure 7.1). We shall ignore for the present any force exerted by the atmosphere on the free liquid surface. Most of the attractions and repulsions experienced by molecules in the column are internal forces, cancelling in pairs; the only ones that do not cancel are those contributed by the surrounding liquid, and we shall assume that we can think of these as a large number of horizontal forces acting in from all sides. The only vertical external forces on the column will therefore be the upward force exerted by our test surface and the weight of the column itself, and these must be equal. The weight of the column is $h\delta A\rho g$ where ρ is the density of the liquid, and so

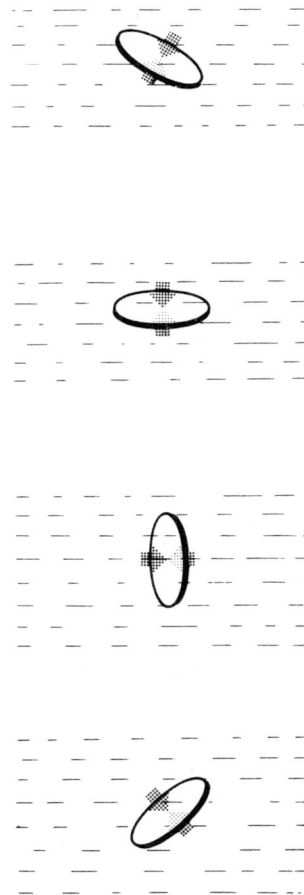

The pressure on a small surface immersed in a fluid is independent of its orientation.

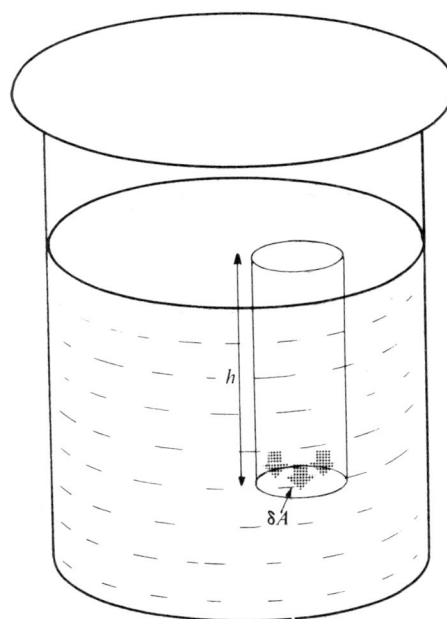

Figure 7.1

Pressure due to a column of liquid.

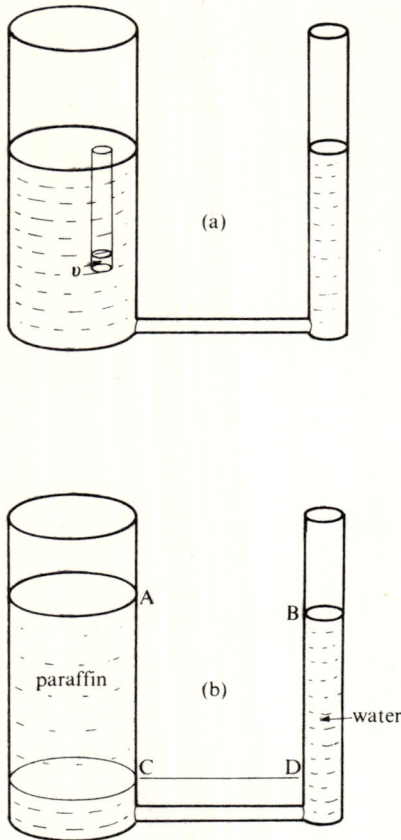

Figure 7.2

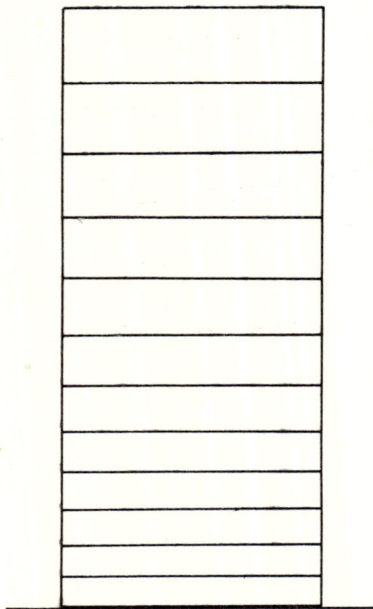

Foam mattresses in a pile are progressively more compressed the nearer they are to the bottom. The density of the atmosphere varies with height in a similar way. (For a discussion of variation of pressure with height, see p. 246)

the force per unit area acting down on the test surface is $h\delta A\rho g/\delta A$, that is,

$$\text{pressure } P \text{ due to the column} = \rho g h$$

What is the use of pressure, from a macroscopic point of view? Figure 7.2a provides one answer. The two columns of water have very different cross-sectional areas; surely the wide column will push the narrow one round? Not so: the condition for equilibrium is to have not equal weights on each side, but equal *pressures*, and $P = \rho g h$ tells us that that means equal heights. The point is that the columns will be in equilibrium when any small volume of water, such as the one marked v in the figure, experiences equal forces on all sides, and it is only that liquid vertically above the surface of v which is effective in pushing it down. It is the *force per unit area* that matters. Perhaps someone may feel that pressure is an unnecessary concept in this situation and that we have here merely a case of 'liquid finding its own level'. He should consider Figure 7.2b, in which paraffin has been poured into the wide column; it sits above the water in an immiscible layer. The levels in the two columns are no longer equal; the paraffin level is higher because paraffin is less dense than water, and a greater height of it is needed to produce equal pressures on both sides.

Exercise 7.1

Given that in Figure 7.2b BD is 12 cm, and the density of paraffin is 800 kg m^{-3}, find the height AC.

Pressure due to a column of gas

As long as we are careful, $P = \rho g h$ can equally well be applied to a gas; the arguments employed in the derivation are still valid, but there is the complication that the density cannot be relied on to remain constant throughout the height of the column. The gas at the base of the column is supporting more weight than the gas higher up, and as gases are readily compressible, the density is correspondingly greater further down. We see this most clearly when we think of the entire belt of atmosphere surrounding our planet. Balancing that particular column of gas against a column of liquid is a routine procedure and provides the most direct measurement of atmospheric pressure; Figure 7.3 shows the familiar simple mercury barometer. The space above the mercury contains nothing but a small quantity of mercury vapour, so that the pressure of the atmosphere, often called simply 'one atmosphere', is balanced entirely by a relatively short column of mercury.

Exercises 7.2–4

2 The height of the mercury column at sea level is always within a few centimetres of 0·76 m. Show that this corresponds to a pressure of almost exactly 10^5 Pa. The density of mercury is 13 600 kg m^{-3}.

3 If you attempted to use a water column instead, what is the minimum height of tube you would select to register 10^5 Pa? Remember that the density of water is 1000 kg m^{-3}.

4 Supposing that the air density did not fall off with increasing altitude, show that an atmospheric belt a little over 8 km high

would be needed to provide a pressure of 10^5 Pa at sea level. Take the air density to be $1\cdot2\,\mathrm{kg\,m^{-3}}$. (In Chapter 17 some attempt is made to take account of varying density.)

Exercise 7.4 shows that there is seldom need to worry about the density variation over differences in height of perhaps 50 m near sea level. If the top of a column of air 50 m high is being compressed by so much atmosphere above it, then the extra 50 m supported by the bottom of the column does not count for much, and we can take the density as being constant throughout the column.

Exercise 7.5

At the bottom of a block of flats 50 m high the town gas pressure exceeds atmospheric pressure by 1800 Pa. What is the excess pressure of the gas supply at the top of the building? (Density of air = $1\cdot2\,\mathrm{kg\,m^{-3}}$, gas supply density = $0\cdot2\,\mathrm{kg\,m^{-3}}$.)

A small difference in pressure between two gases is most conveniently measured by a *manometer*, a U-tube containing water or

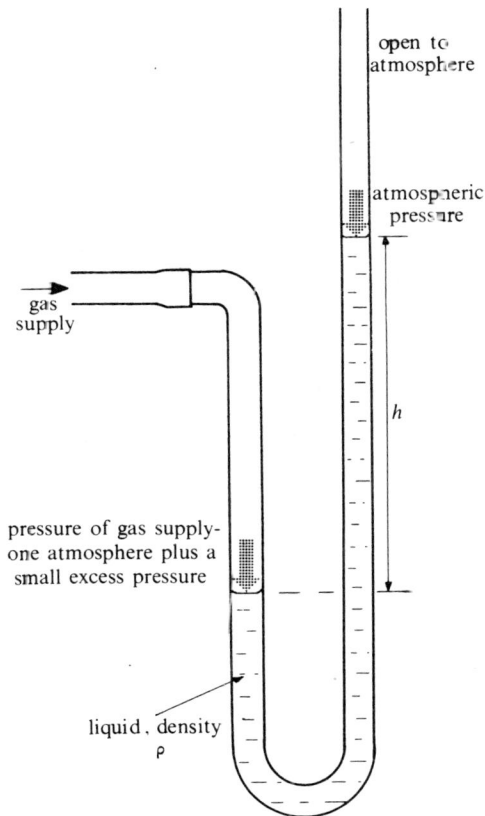

Torricellian 'vacuum'
contains trace of mercury vapour

strong glass tube at least 0·8 m long

this column of mercury – about 0·76 m high balances atmospheric pressure

atmosphere pushes down on free surface of mercury

Figure 7.3
Simple mercury barometer.

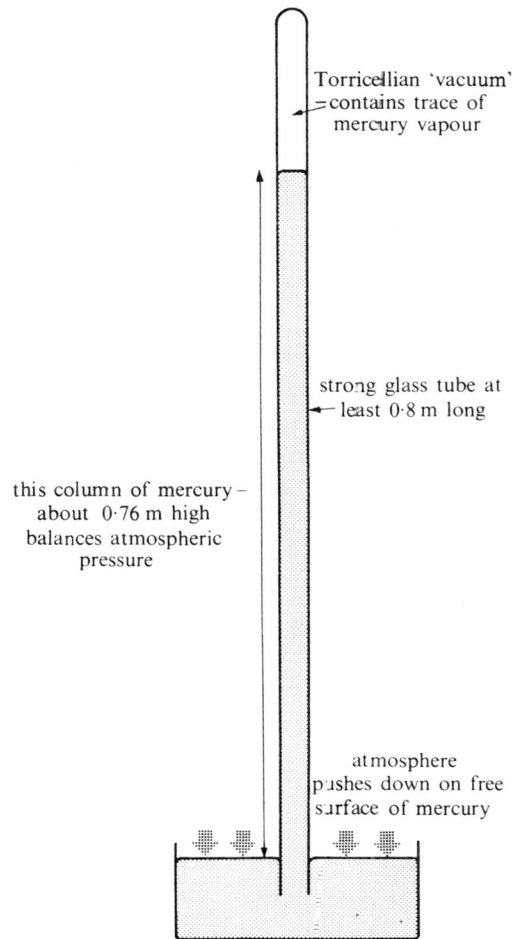

open to atmosphere

atmospheric pressure

gas supply

h

pressure of gas supply-one atmosphere plus a small excess pressure

liquid, density ρ

Figure 7.4
Manometer. Excess pressure of gas supply = $\rho g h$.

light oil. One limb of the manometer shown in Figure 7.4 is connected to the gas supply, the other is left open to the atmosphere. The gas forces the liquid column round until the extra height in the open limb is sufficient to balance the excess pressure of the supply.

Exercise 7.6

a) If a water manometer were used to measure the excess pressure of the gas supply referred to in Exercise 7.5, what difference in heights would be recorded (i) at the bottom, (ii) at the top of the building?

b) What would be the result of using a manometer employing oil of density $800 \, \text{kg m}^{-3}$ instead?

The air inside a tyre has an excess pressure of between two-thirds and four atmospheres, depending on the vehicle; a water manometer able to cope with a pressure of such magnitude would be inconveniently tall, though *mercury* manometers are used at filling stations in some European countries. Figure 7.5 shows an alternative type of

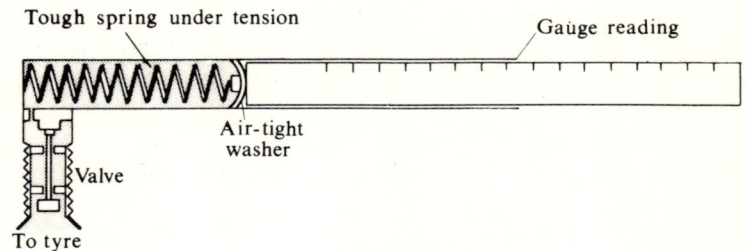

Tyre-pressure gauge.

pressure gauge; there is no column of liquid this time. A tough spring provides the force needed to keep the washer from flying straight out of the tube when the valve is pushed open and air flows in from the tyre. The extension of the spring is an indication of the excess pressure in the tyre; this is often called the *gauge* pressure, as opposed to the *absolute* pressure.

absolute pressure = gauge pressure + atmospheric pressure

Exercises 7.7, 8

7 If the gauge indicates a pressure of $1.5 \times 10^5 \, \text{Pa}$ and has a barrel of internal diameter 8 mm, what force must the spring provide?

8 Town gas with an excess pressure corresponding to a manometer reading of 15 cm of water is stored in a gas holder of circular cross section and diameter 30 m. The holder consists of two concentric overlapping cylinders; the lower cylinder contains water to act as a seal, and the upper cylinder is closed at the top but open at the bottom, where it dips into the water. It rises and falls as the demand fluctuates, and is entirely supported by the gas pressure, apart from an upthrust experienced by the part immersed in the water see p. 81). Neglecting this upthrust, find the weight of the upper cylinder. (Ignore the possibility of differences of gas pressure at different heights within the holder.)

Perhaps we should mention a possible source of confusion here. Suppose we have a quantity of gas kept under pressure, the air in the tyre, or an oxygen cylinder containing gas at a pressure of perhaps $10^7 \, \text{Pa}$. Has $P = \rho g h$ anything to do with a case like this? Very little. If we stand the cylinder vertical it will certainly be true that the pressure at the base will exceed that at the top by an amount equal to $\rho_G g h$, where ρ_G is the density of the gas – but this difference will be so small compared with $10^7 \, \text{Pa}$ that it can safely be ignored.

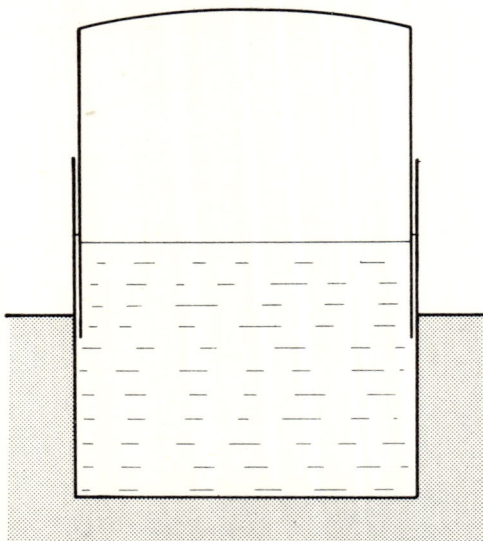

Gas holder (see Exercise 7.8)

Exercise 7.9

The density of oxygen kept at a pressure of 10^7 Pa is in the region of $100 \,\mathrm{kg\,m^{-3}}$. Estimate the percentage difference between the pressures at the top and bottom of a gas cylinder 1 m high.

If it is not the weight of the gas column that is responsible for the force on the floor of the cylinder, what is? Clearly the answer must be that the top surface of the cylinder is pushing down, and this force is transmitted to the bottom surface by the mutual repulsions of the gas molecules during their continual collisions with each other. We have only to open the valve at the top to realise that we are depending on *all* the walls of the cylinder to keep the gas in.

Testing fire-fighting equipment on a high-pressure spherical gas-storage vessel.

Influence of the pressure of a gas on its density

Figure 7.6 shows an almost traditional arrangement for investigating the way in which the density of a gas changes when it is compressed. The sample of gas concerned is trapped in the left-hand limb between the tap and the top of the mercury column; its volume V can be read directly off the graduations on the glassware. We can move the right-hand limb up or down to produce various vertical separations h of the two mercury levels so that the total pressure P of the gas sample, equal to $(H + h)\rho g$, can be varied; here H is the height of the mercury barometer at the time of the experiment and ρ is the density of mercury. For all gases at high

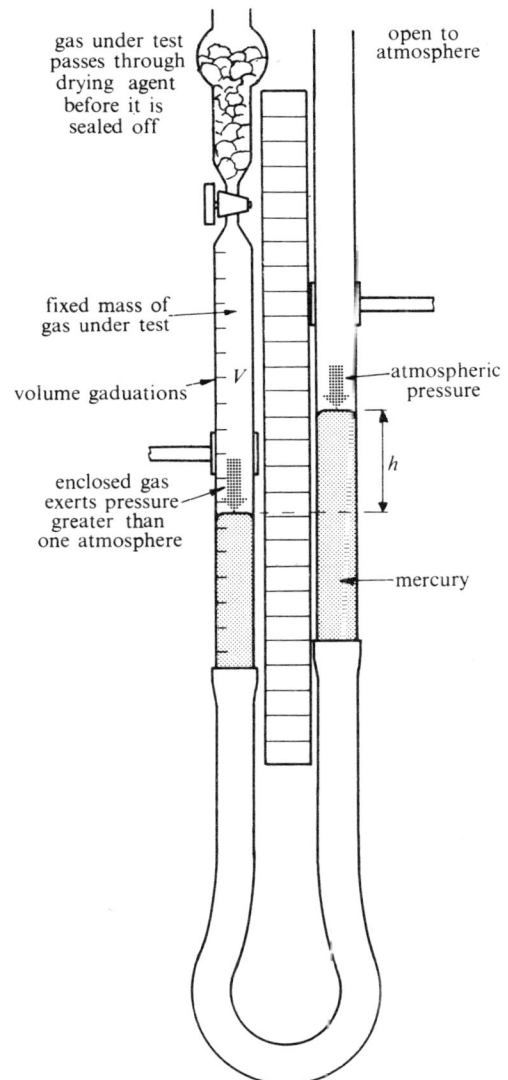

gas under test passes through drying agent before it is sealed off

open to atmosphere

fixed mass of gas under test

volume graduations

V

atmospheric pressure

enclosed gas exerts pressure greater than one atmosphere

h

mercury

Figure 7.6

enough temperatures and sufficiently low pressures, we find that the relationship

$$(H + h) \propto 1/V$$

is obeyed very closely indeed so long as the temperature remains constant. This is an example of *Boyle's law*, which states that *the pressure of a fixed mass of gas at constant temperature is inversely proportional to its volume*. Even a sample of gas approaching lique-faction obeys the law to a fair degree of approximation, but as soon as a condition is reached where further reduction of volume results in conversion of some of the gas to liquid, we are no longer dealing with a fixed mass of gas, and the law is no longer relevant. For this reason it is important that the gas under test be free of water vapour.

It is often convenient to express Boyle's law in the form

$$P_1 V_1 = P_2 V_2$$

where the suffixes indicate that a fixed mass of gas of volume V_1 and pressure P_1 has changed to volume V_2 and pressure P_2. Since density = mass/volume, density and volume are inversely propor-tional when the mass is fixed, so

$$\frac{P_1}{\rho_1} = \frac{P_2}{\rho_2}$$

where ρ_1 and ρ_2 are the initial and final densities.

Boyle's law, as we have indicated, is not always precisely obeyed, especially at high pressures and low temperatures, but it is very useful in many everyday situations and serves as a *rough* guide even for high pressures and when close to liquefaction.

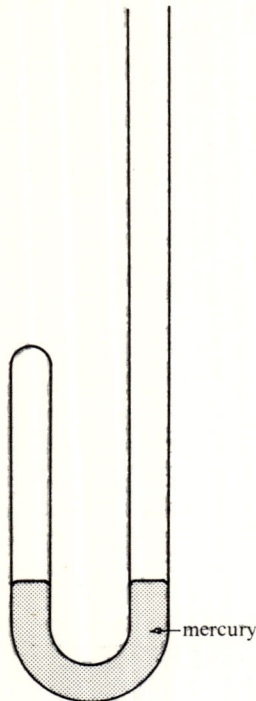

Figure 7.7 (Exercise 7.10)

Exercises 7.10, 11

10 Figure 7.7 shows a J-tube with uniform cross-sectional area equal to $2\,cm^2$. $16\,cm^3$ of dry air is trapped in the closed limb by a mercury column, and the mercury levels in the two limbs are the same. What volume of mercury must be poured into the tube to reduce the volume of trapped air to $12\,cm^3$? The height of the mercury barometer is $760\,mm$.

11 On a day when the height of the mercury barometer is $750\,mm$, $15\,cm^3$ of dry air is trapped in the left-hand limb of the apparatus shown in Figure 7.6. The internal cross-sectional area of each limb is $1\,cm^2$, and the mercury levels are initially equal. The right-hand level is then raised $500\,mm$ above its original position. How far will the left-hand level rise?

Force on an extended surface immersed in a fluid

Since   pressure = force per unit area on a small area
it follows that   force = pressure × area,

but if the area concerned is so large that different parts of it lie at different depths we shall have to cope with the fact that the pressure varies from point to point. (The calculation of the force on a dam wall or lock gate involves just such a difficulty.) As this is typical of problems met with in many branches of physics we shall tackle the specific job of evaluating the total force experienced by one wall of a trough due to the liquid it contains.

Example

A rectangular trough has vertical sides each of length a and contains a depth h of liquid of density ρ. Find the total force exerted by the liquid on one wall.

A simple approach to this problem would be to write

$$\text{total force} = \text{average pressure} \times \text{area of wall covered}$$
$$= \text{pressure at depth } (h/2) \times \text{area}$$
$$= \frac{\rho g h}{2} \times ah.$$

This is in fact the correct answer, but it is safer to arrive at it by a rather more careful approach as follows.

Pressure at depth $y = \rho g y$, so force on a small horizontal strip of width δy at that depth $= \rho g y a \delta y$, where we consider δy so small that there is negligible pressure variation between its upper and lower edges (Figure 7.8).

Total force on the wall = sum of forces on all such strips

$$= \Sigma \, \rho g y a \, \delta y$$

or in integral form

$$\int_0^h \rho g a y \, \mathrm{d}y,$$

which is $\frac{1}{2}[\rho g a y^2]_0^h$ or $\frac{1}{2}\rho g a h^2$, as before.

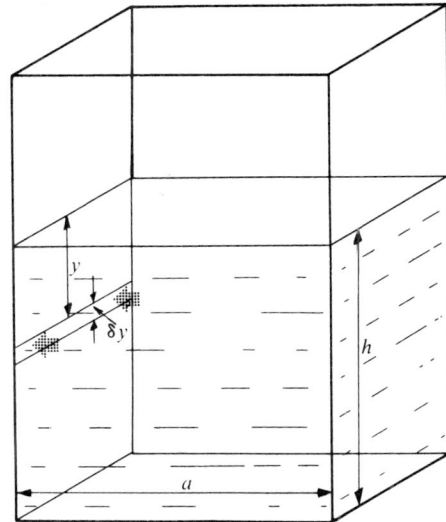

Figure 7.8

Exercise 7.12

Show that the total force on the bottom half of the immersed part of the wall is three times the force on the top half. This is one reason why dam walls are built thicker at the base than at the top.

Upthrust

Any object immersed in a fluid experiences forces from all the molecules surrounding it, and these forces *do not cancel each other completely*: they produce a net upward force, or *upthrust*, on the object. Figure 7.9 shows a cube of side a with its top face a depth h below the surface of a liquid of density ρ. All six faces experience forces from the liquid; we can expect the forces acting on the vertical sides to cancel each other because of the symmetry of the arrangement, but the forces on the top and bottom faces do not balance in this way.

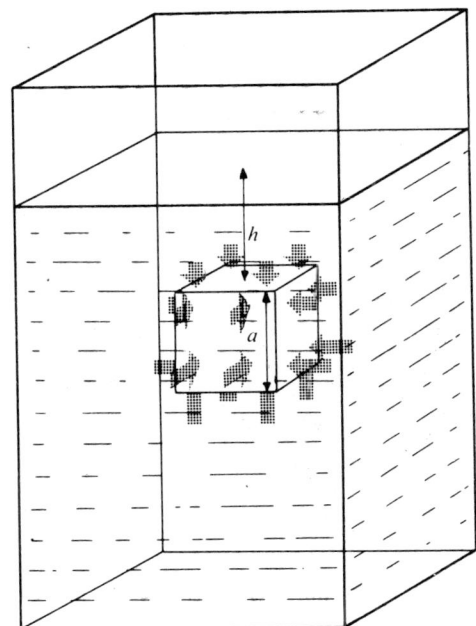

Figure 7.9

Hydrostatic forces on a cube immersed in a liquid.

Exercise 7.13

Write down the pressure and the force on (i) the top surface, (ii) the bottom surface of the cube and show that as a result there is an excess upward force of $a^3 \rho g$ exerted by the liquid.

The expression $a^3 \rho g$ has a direct interpretation: it is equal in magnitude to the weight of liquid which would have been in the space now occupied by the cube – the weight of liquid displaced. This is *Archimedes*' famous principle, and it is by no means limited to cubes. In more general terms it can be stated as follows: *if a body is immersed wholly or partly in a fluid it experiences an upthrust equal to the weight of fluid displaced.* We can use a simple, logical argument to back it. Take any object and put it in a fluid: what force will the

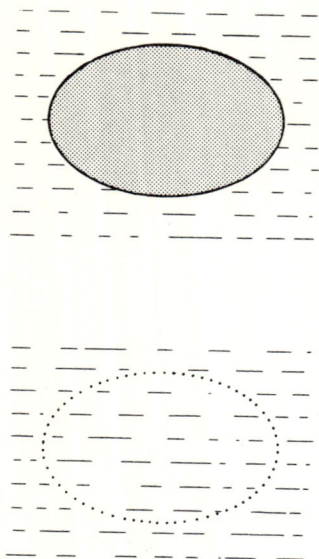

An argument supporting Archimedes' principle. The object immersed in the fluid experiences an upthrust which is identical to that which was experienced by the fluid it displaced, and which was just sufficient to support the weight of that fluid.

Apollo 11 after splash-down, complete with flotation collar and three flotation balloons to right it if needed.

fluid exert on it? Well, what if we remove the object but exactly fill the space it leaves by more of the fluid, separated from what is already there by a thin membrane? Naturally the surrounding fluid will just support the new fluid, membrane or no membrane, but the forces exerted on the outside of the membrane will not be altered by what is inside it, be it fluid or our object. So the object must have experienced an upthrust just sufficient to support the fluid it displaced. The argument applies equally well to liquid or gas surrounding the object though of course the upthrust produced by a gas is smaller, corresponding to a density of about 1/1000 that of a liquid at normal pressures.

Exercises 7.14–16

14 A test of Archimedes' principle was carried out as follows. A metal cylinder suspended in air from a spring balance by a fine thread produced a reading of 48·4 N. It was then immersed completely in water of density $1000 \, \mathrm{kg \, m^{-3}}$ and the new reading noted. If the height and diameter of the cylinder were each recorded as 100 mm, what should the new reading be? Take g to be $9·80 \, \mathrm{m \, s^{-2}}$.

15 In the previous exercise the spring balance used was able to read to $\pm 0·2 \, \mathrm{N}$ and the estimated error in each of the linear measurements was $\pm 0·5 \, \mathrm{mm}$. Within what limits must the final spring balance reading lie if it is to confirm Archimedes' principle?

16 If the density of the air is $1·2 \, \mathrm{kg \, m^{-3}}$, what difference will the upthrust of the air make to the recorded weight of the cylinder in Exercise 7.15? Is this difference likely to be detectable on the balance described?

Principle of flotation

We can go a stage further when we are dealing with an object which floats on the surface of a liquid. Archimedes' principle still applies, but it is also true that the liquid is supplying enough upthrust to support the entire weight of the body: no less, or the body would sink in, and no more, or it would rise. If we couple this with Archimedes' principle we can say that *a floating body displaces a weight of fluid equal to its own weight.* This is known as the *principle of flotation.* We must mention one complication; if the object is fairly small, forces due to surface tension become comparable with the other forces involved, and account must be taken of them (see p. 164).

Exercise 7.17

What fraction of the volume of a floating object lies above the surface of water of density $1000 \, \mathrm{kg \, m^{-3}}$ if the density of the object is (a) $500 \, \mathrm{kg \, m^{-3}}$, (b) $900 \, \mathrm{kg \, m^{-3}}$, (c) $1000 \, \mathrm{kg \, m^{-3}}$?

When we come to apply the principle of flotation to gases we have to be careful to qualify what we mean by a 'floating' object, we must insist that the object is not accelerating up or down. Floating on the surface of water is a relatively simple matter because the upthrust is self-adjusting; if you sink in too far, it increases. Keeping a free helium-filled balloon stationary is quite another matter, as any balloonist will testify–this time the upthrust is *not* self-adjusting and if its magnitude departs only slightly from the total weight of the balloon, there is acceleration–in one direction or the other!

'Skyhook' instrument balloon at launch. The balloon is partially filled to allow for expansion, so that the density of helium inside can remain below the density of the atmosphere at high altitude. This balloon ascended to a height of 35 km.

Exercise 7.18

A balloon is made from a polythene bag having a mass of 80 g. What volume of air does it displace if it is just able to float when partially filled with helium? Take the densities of air and helium at room temperature and pressure to be $1 \cdot 2 \, \text{kg m}^{-3}$ and $0 \cdot 17 \, \text{kg m}^{-3}$ respectively.

Further exercises

Exercises 7.19–28

19 State the principle of Archimedes and explain why, if a glass stopper be weighed in water at two different temperatures, its weight appears to change.

A ship sailing from the sea into a river sinks X mm, and on discharging her cargo rises Y mm; on proceeding again to sea she rises another Z mm. Find an expression for the density of sea-water in terms of X, Y and Z. The ship's side may be assumed vertical at the water-line. (O & C)

20 Explain what is meant by the *pressure at a point* in a fluid, and state the unit in which it is measured. Find an expression for the pressure due to a liquid of density ρ at a point h below the free surface. Describe briefly how you would verify your result by experiment.

A diving bell of internal volume 6 m³ is lowered into sea-water of density $1030 \, \text{kg m}^{-3}$ until the volume of the contained air is reduced to 4 m³. Assuming that the temperature is unchanged, calculate how far the surface of the water in the bell is below the surface of the sea, the pressure of the atmosphere being equal to that due to 760mm of mercury. (Density of mercury = 13 600 kgm^{-3}.)   (o & c)

21 Explain the meaning of the term 'pressure at a point in a fluid'. Prove that the pressure at a point in a liquid at rest is the same for all points in the liquid at the same horizontal level, and that the difference in pressure between two points, not at the same horizontal level, is given by dgh, where d is the density of the liquid and h the vertical distance between the points. Why does this latter result not apply in the case of a gas?

A body of mass 96 g is put to float in water contained in a beaker. What will be the increase in the thrust on the base of the beaker? If the area of the base is 210 cm^2 what is the increase in the height of the water level?

A piece of ice at 0°C floats on water at 0°C in a beaker. If the ice melts, without change of temperature, does the level of the water surface rise or fall? Give reasons for your answer. (WJEC)

22 Describe some form of hydraulic lift or press and explain the principle of its action. Why, in practice, is the efficiency of such a device not 100 %?

A solid incompressible cylinder is suspended on a string with its axis vertical. The cylinder is gradually lowered into an incompressible liquid until it is some distance below the surface. Draw a graph showing how the tension in the string varies with the position of the cylinder. Explain the form of this graph. How would the form of this graph be modified if the liquid were compressible, i.e. if its density increased significantly with pressure? Explain.   (o & c)

23 Show that, as one moves vertically downwards a distance h in a liquid of density ρ, the pressure increases by an amount ρgh. Hence deduce Archimedes' principle.

A spherical buoy of 1 m external diameter is made of steel plates 5 mm thick. It is anchored in sea-water so that it floats half submerged with its cable vertical. Find the tension in the cable where it joins the buoy.

Discuss in general terms whether the tension in the cable is uniform, and whether the force on the anchor is greater or less than the force on the buoy. (Density of iron is 7700 kgm^{-3} and of sea-water is 1030 kgm^{-3}.)   (o & c)

24 State the principle of Archimedes.

A section of a wreck lying under sea-water (density 1025 kgm^{-3}) is to be raised with the help of hollow cylinders which can be lashed to it and then emptied of water by pumping in air. Given that the total mass of the wreck and cylinders is 2×10^5 kg, and the mean density of the metal of which they are made is 7800 kgm^{-3}, calculate (a) the 'lift' in newtons required to raise the wreck to the surface, (b) the volume of sea-water which must be pumped out of the cylinders to give this lift.

A measuring cylinder full of water falls freely under gravity with its open end upwards. What will be the hydrostatic pressure at a depth h below the surface of the water in the cylinder? (o & c)

25 A large plastic sphere floats on water with one-sixteenth of its volume above the surface. Oil is poured on to the water until the sphere is just covered by the layer of oil, when it is found that one-half of the sphere is below the water–oil interface. Find the density of the plastic and the density of the oil.

Why, if the sphere were small, would the results you obtain be only approximate? (O & C)

26 Criticise the following as a statement of Archimedes' principle. 'The upthrust on an object immersed in a liquid is equal to the weight of liquid which it displaces.'

A cylinder of mass M and of average density d_1 is lowered into a vessel of cross-sectional area A, half full of liquid of density d_2 (less than d_1) until it is completely covered.

a) By how much has the surface of the liquid risen?
b) What is the increase in pressure at the bottom of the vessel?
c) What is the increase in thrust there?
d) Can you use this calculation to *prove* Archimedes' principle (i) in this particular case, (ii) in general?

The cylinder is further lowered until it rests on the bottom of the vessel. What is the upthrust of the liquid on the cylinder and the upthrust of the base on the cylinder in the two cases when the axis of the cylinder is (i) vertical, (ii) horizontal?

Do your results suggest that the above statement of the principle should be modified? (SUJB)

27 A hydrometer consists of a hollow cylindrical glass bulb, volume V, to one end of which is attached a stem of uniform cross-section a. It is so weighted that it floats upright. When the hydrometer floats in water a length L of the stem is submerged. Calculate the length submerged when the hydrometer floats in a liquid whose density is ρ.

A body, which weighs 600 N, floats in water with $0.015\,\text{m}^3$ above the surface. Calculate the volume of the body. (O & C)

28 Explain how you would make a correction for buoyancy when weighing a body of density ρ in air of density σ, using 'standard weights', of density ρ', which are correct *in vacuo*.

If the buoyancy correction is neglected and ρ' has the value $8\,\text{gcm}^{-3}$, in which of the two following cases would the greater percentage error be made:
i) weighing a gold alloy of density $20\,\text{gcm}^{-3}$
ii) weighing an aluminium alloy of density $2\,\text{gcm}^{-3}$?
(Density of air $= 0.0012\,\text{gcm}^{-3}$.) (WJEC)

8 Momentum and energy

Bridge demolition using explosives. Rapid conversion of chemical energy to internal energy and translational kinetic energy.

In Chapter 5 we established the link between a change of motion of any familiar object and the cause of such change, $F = ma$. It can be successfully applied to footballs, pistons, muscles, conveyor belts, cars, cruisers, aircraft, rockets, tides, and the earth moving round the sun. In carefully chosen situations we can also apply it to the motions of atoms and sub-atomic particles (electrons and ions) but with the warning that in this region there are situations for which $F = ma$ is not suitable; in fact the whole idea of force becomes very much less useful on such a scale. The equations which replace it in the atomic realm lie beyond the scope of this book; they form the basic tools of *quantum mechanics*.

This limitation apart then, we find that $F = ma$ will, in principle, solve any problem of motion which may confront us. Very often, unfortunately, the amount of computation needed is astronomical. The journey to the moon and back is fully covered by $F = ma$, but as F and a vary continuously, and m varies whenever the motors are operating, a very fast computer is needed to work out an enormous amount of arithmetic, and then correct it quickly enough to modify the flight schedule while the journey is progressing.

Now although $F = ma$ has such a wide application, it is not the *only* rule for writing down what happens, and in a number of cases it is not even the most convenient one; there are two other complementary approaches, one involving *momentum* and the other involving an equally useful quantity called *energy*. We shall approach the rules governing momentum and energy from a consideration of Newton's laws, and at first we can regard them as alternative ways of expressing the facts summarised by $F = ma$ about the way things behave. The modern view, however, is to regard momentum and energy as the really basic quantities, obeying universal laws as far as we can tell, and to see $F = ma$ as a rather restricted rule arising from them. Within the atom for example, they are far more useful and significant than any notions of force and acceleration.

Momentum

The momentum of an object has already been defined in Chapter 5 as the product of its mass and velocity; it is a vector with a direction parallel to that of the velocity concerned. Newton's first law tells us that the momentum of an object will not change unless an external force acts on it, and $F = ma$ can tell us how big a change to expect, and in what direction.

Exercise 8.1

Consider an object of mass m with initial velocity u undergoing a uniform acceleration (in line with u) for a time t under the influence of an external force F (Figure 8.1). As a result it acquires a final velocity v. Show that the momentum change ($mv - mu$) is given by

$$Ft = mv - mu. \qquad \ldots(1)$$

The product Ft is called the *impulse* of the force; *the impulse of a constant force is the product of that force and the time for which it acts,* and we see that the momentum change it produces is equal to its impulse. This means that momentum can be measured either in $\mathrm{kg\,m\,s^{-1}}$ (corresponding to mv) or Ns (corresponding to Ft). We have written down the formula for a force which does not vary during the time t; if it is varying it can be dealt with as follows.

Using the expression (5) of Chapter 5 for Newton's second law, we have

$$F = \frac{\mathrm{d}(mv)}{\mathrm{d}t},$$

or writing mv as p

$$F = \frac{\mathrm{d}p}{\mathrm{d}t}.$$

So

$$\int_{t_1}^{t_2} F\mathrm{d}t = \int_{p_1}^{p_2} \mathrm{d}p = p_2 - p_1 = m(v - u) \qquad \ldots(2)$$

where p_2 is the final momentum, p_1 the initial momentum, and t_1 and t_2 the limits of the time during which the force is acting.

Equations (1) and (2) emphasise that if we want to change the velocity of an object, using a small force, then that force must act for a relatively long time, so that the product Ft (or the integral $\int F\mathrm{d}t$) is sufficiently large. If on the other hand we want the object to stop quickly, the force required will be correspondingly big. We pack fragile articles in foam plastic for protection–if the package is dropped there is a little more time in which to remove the momentum because the foam takes some time to distort. Of course we can argue in terms of $F = ma$ and come to the same conclusion–the pliable covering makes for a more gradual retardation, which requires a smaller force.

We can modify equation (1) to take care of the case for which F is not in the same direction as u. The simplest course is to resolve F into two components F_x and F_y in appropriate directions. If we also

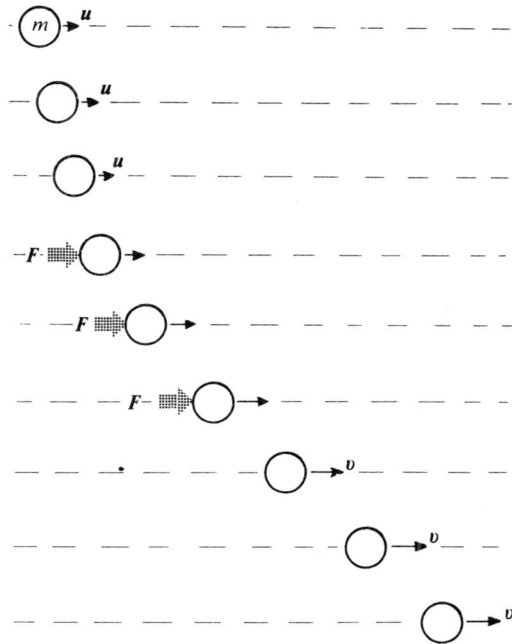

Figure 8.1 (Exercise 8.1)

resolve the initial and final velocities into components u_x and v_x in the x-direction, and u_y and v_y in the y-direction, we can write

$$F_x t = m(v_x - u_x) \quad \text{and} \quad F_y t = m(v_y - u_y).$$

This is simply saying that a force in a given direction affects only the component of the momentum in that direction, and has no influence on the momentum in a perpendicular direction (this point will be considered at greater length in Chapter 13). The size and direction of the *resultant* momentum are affected, and the change, as Newton's second law tells us, is in the direction of the applied force. The impulse still equals the change in momentum: we can express this in vector notation as

$$\boldsymbol{F}t = m(\boldsymbol{v} - \boldsymbol{u})$$

(remember $\boldsymbol{v} - \boldsymbol{u}$ implies a subtraction according to the rules of *vector* combination).

Exercise 8.2

A 2 kg mass having initially a velocity of 10 m s^{-1} in a horizontal line experiences a net vertical force of 20 N for 1·5 s. Find (a) the vertical momentum, (b) the size and direction of the final resultant momentum.

Conservation of momentum

Sometimes equations (1) and (2) are useful in their own right, but the real value and significance of the momentum change emerges only when we bring into the picture the 'third law partner' of the force $\boldsymbol{F}$. Figure 8.2 illustrates a specific case, a collision between two trolleys moving in the same straight line. They have masses m_1 and m_2, and initially the velocities are u_1 and u_2; clearly u_1 must be greater than u_2 if we are to have a collision. As a result of the collision they have velocities v_1 and v_2 in the same direction as before (if it turns out that m_1 is moving to the left it will simply mean that v_1 is negative). To simplify the argument we shall at first make a completely unwarrantable assumption, one that we shall remove at the earliest opportunity, namely that during the short time t that the two bodies are in contact they each experience a *constant* force. Newton's third law enables us to write these forces as $-\boldsymbol{F}$ and $\boldsymbol{F}$ respectively.

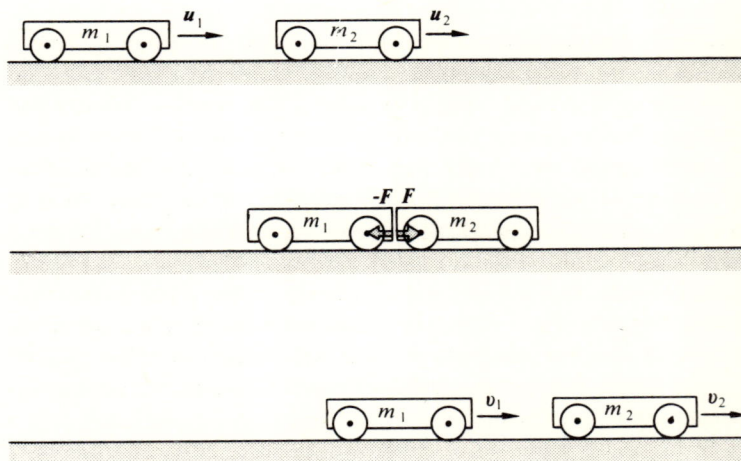

Figure 8.2

Now we write down for each trolley the fact that impulse equals momentum change.

Trolley A: $m_1v_1 - m_1u_1 = -Ft$
Trolley B: $m_2v_2 - m_2u_2 = Ft$.

This means we can put $m_1v_1 - m_1u_1 = -(m_2v_2 - m_2u_2)$
or more significantly $m_1v_1 + m_2v_2 = m_1u_1 + m_2u_2$

that is, total momentum after impact = total momentum before impact.

Of course the repulsive force between the trolleys is not really going to be constant! No, but though it is changing continuously, the force on trolley A is instant by instant equal and opposite to the force on trolley B, so we can get over the difficulty quite neatly. We split the collision time into a large number of sections, each perhaps a millionth of a second in duration, or even shorter if we please. Looking at just one of these tiny intervals at some stage of the collision, it is safe to say that the force is not going to change very much; we can afford to do our 'constant force' treatment and conclude that the total momentum at the end of the section is equal to the total momentum at the start of it. But we can say this of every section in turn, right through the entire collision; instant by instant the total momentum is at all times exactly conserved by the system.

A more compact way of expressing the argument is to write:

$$m_2v_2 - m_2u_2 = \int_{t_1}^{t_2} F\,dt$$

$$m_1v_1 - m_1u_1 = \int_{t_1}^{t_2} (-F)\,dt = -\int_{t_1}^{t_2} F\,dt.$$

So

$$m_1v_1 + m_2v_2 = m_1u_1 + m_2u_2$$

This is the real importance of momentum–the total momentum of the two trolleys is not changed as a result of their collision.

Exercise 8.3

Trolley B is of mass $0.5\,kg$ and initially stationary; trolley A comes in from the left at $0.4\,ms^{-1}$, and an automatic linking mechanism ensures that A and B move off together after the collision. Find the final velocity if A has mass (a) $0.1\,kg$, (b) $0.5\,kg$, (c) $2.5\,kg$.

The point about these calculations is that we do not need to know any details about the way the force between the trolleys builds up and decays during the impact. If we had to use $F = ma$ we should have no mean problem on our hands. Even if we could find a mathematical expression to represent the rapidly varying force, that would simply give us a non-uniform acceleration, from which we should have to attempt to work out the velocity change over the duration of the impact. All this is avoided if we use the fact that the total momentum remains constant. It is not even necessary that the two bodies should come in contact; Figure 8.3 shows an arrangement where magnetic repulsion is the only force between the bodies, and still the total momentum remains unchanged. Formally defined, the *principle of conservation of momentum* states that *for a closed system of bodies the total momentum is unaffected by any interaction between the bodies*; further, the sum of the *resolved components* of momentum of the bodies in the closed system in any chosen direction is also unaffected by any interactions between the bodies. The phrase 'closed system' needs explaining. If we look at the momentum of just *one*

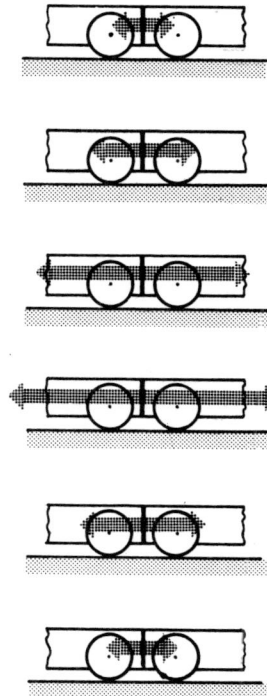

The repulsive forces vary during a collision, but Newton's third law is true at every instant.

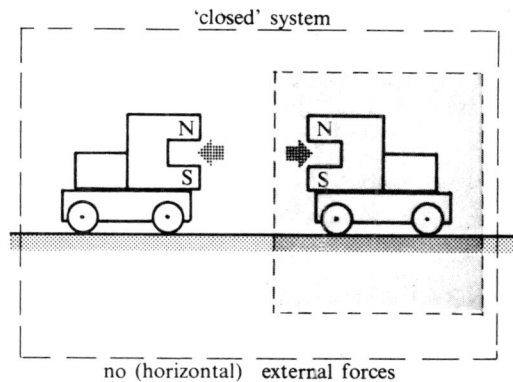

'closed' system

no (horizontal) external forces

Figure 8.3

Magnetic repulsion provides a 'soft' collision between the trolleys. Considering horizontal forces only, the larger rectangle defines a closed system. The shaded part is not a closed system, since one member of a 'third-law pair' is accelerating an object outside the system.

of the trolleys in Figure 8.3 we shall find that the momentum changes, but the reason is that we have not brought in all the objects involved. You can tell when you have not brought in enough objects because if that is the case, you are always left with an external force – in this case it is the force provided by the other trolley. One member of a 'third-law pair' is acting on the system as it stands (shaded in the figure), but its opposite number is acting on some other object as yet outside the system, and that is the object which must be brought in to make the system 'closed'. A 'closed system' means 'all the bodies taking part in the interactions', in other words there must be no external forces left over.

Sometimes we have to range far and wide to get all the bodies involved. Suppose for example one of our trolleys runs into a cupboard. It will certainly undergo a momentum change, but that is because it has experienced an external force, provided by the cupboard. So, we need to bring the cupboard into the system; but the cupboard did not accelerate, it did not suffer any momentum change! Very well, but we are not finished yet; it, too, experienced an external force – friction between the cupboard base and the floor. In other words before we can get a closed system this time we have to bring in a rather large body, the *earth*, and it *does* suffer a momentum change, though because of its huge mass there is of course no perceptible acceleration.

In fact there is rather more to this than meets the eye: the earth *had already experienced* a momentum change while we were accelerating the trolley, but in the opposite direction. The idea is illustrated in the following exercises.

Exercises 8.4, 5

4 A trolley of mass 1 kg stands on a platform of mass 2 kg, itself on wheels and able to move freely on a horizontal surface (Figure 8.4). Initially both trolley and platform are stationary, and points A on the trolley and B on the platform are vertically in line with a point C on the runway. A spring in the trolley is then released, and this pushes a plunger against the stop at one end of the platform, as a result of which the trolley almost instantaneously acquires a steady speed of $0.2 \, \text{m s}^{-1}$ with respect to C. After three seconds it hits the other stop and remains in contact with it. Find
   a) the speed of B with respect to C before and after the trolley hits the stop, and
   b) the positions of A and B with respect to C at the instant that the trolley comes to rest against the stop.

5 With a similar arrangement (but with a different distance between the stops on the moving platform) the trolley again starts from rest at one stop but moves off with a speed of $0.2 \, \text{m s}^{-1}$ with respect to the *platform*, coming to rest against the other stop after 3s. Find
   a) the speeds of A and B with respect to C before and after the trolley hits the stop, and
   b) the positions of A and B with respect to C at the instant that the trolley hits the stop.

Experimental illustration of momentum conservation
We have derived momentum conservation from Newton's laws but it is as well to check with an experiment. Of course no single experiment

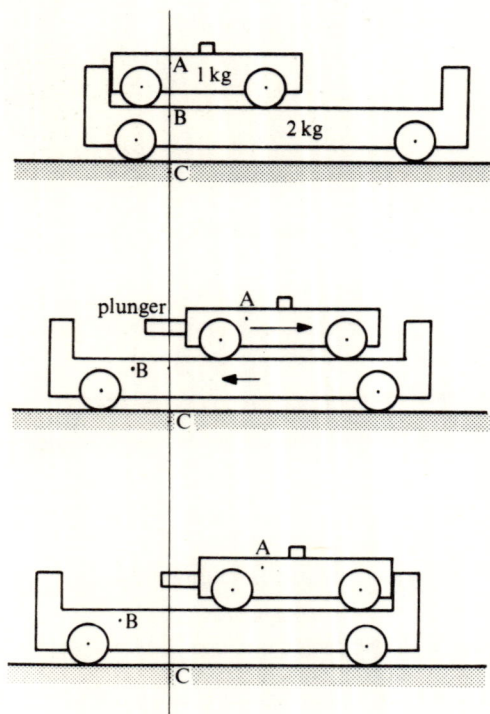

Figure 8.4 (Exercises 8.4 and 8.5)

we do will *prove* the principle; the most we can do is to confirm that it has worked yet again, and the real test is that no exception has yet been found.

A simple example is shown in Figure 8.5; we have here two trolleys with different masses, both initially stationary. One of them has a built-in compression spring of the type mentioned in Exercise 8.4; a sharp tap on the release button causes the trolleys to fly apart.

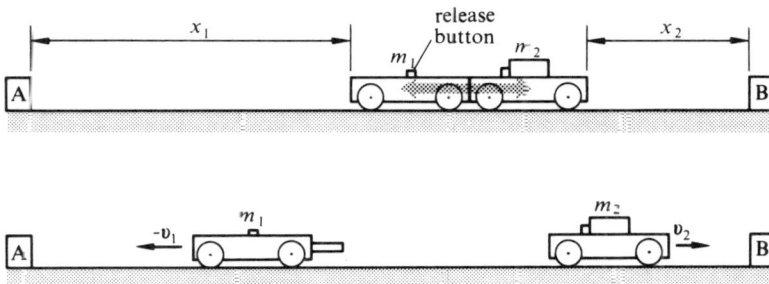

Figure 8.5

We can use ticker-tape to find their velocities, or more simply we can adjust the positions of two blocks A and B until the trolleys hit their respective buffers simultaneously. As long as the running surface is horizontal and smooth, the ratio x_1/x_2 of the distances travelled is then the same as the ratio of the trolley speeds.

The total momentum before the 'explosion' is zero, so the momentum conservation principle insists that the total momentum after the explosion must also be zero, that is,

$$m_1 v_1 + m_2 v_2 = 0,$$

where v_1 and v_2 are the respective trolley velocities, both measured to the right (v_1 will of course be negative).

So the principle is confirmed in this instance if

$$m_1/m_2 = -v_2/v_1 = x_2/x_1.$$

Example involving rocket propulsion

A space vehicle moving in a region where gravitational attractions are negligible is accelerated by a rocket motor which emits gases at a rate Q (in kg s^{-1}), at speed V with respect to the vehicle. V will be constant when the motor is operating normally because the fuel awaiting combustion has experienced the same accelerations as the rest of the vehicle, that is to say, it is at rest with respect to the combustion chamber. Find the acceleration of the vehicle at the instant that it has mass M, and the thrust provided by the motor.

Suppose the mass is M at time t, and that after a very short time interval δt a mass δm of gas has been ejected, in other words, $Q = \delta m/\delta t$. In effect, a mass δm has altered its momentum by an amount $V\,\delta m$ in leaving the vehicle and becoming part of the exhaust gases. This means that the rest of the vehicle must experience an

The trolley on the right is fitted with a spring-loaded plunger, released by tapping the knob in the centre of the trolley.

equal and opposite momentum change of size $V \delta m$ if momentum is to be conserved. If the consequent increase in speed of the vehicle is δv, we can write the momentum change as $(M - \delta m)\delta v$, or $M \delta v$ to a good approximation. So

$$V \delta m \approx M \delta v,$$
$$V \delta m / \delta t \approx M \delta v / \delta t,$$
$$VQ \approx M \delta v / \delta t,$$

or in the limit as δt approaches zero, $\mathrm{d}v/\mathrm{d}t = VQ/M$.

The thrust of the motor can be worked out by analogy with the conveyor belt problem on p. 59; every second a mass numerically equal to Q experiences a velocity change V as it joins the exhaust stream, and the force F required to achieve this is VQ. The vehicle experiences an equal and opposite force VQ, which can also be written as $M \, \mathrm{d}v/\mathrm{d}t$.

Oblique collisions

The momentum conservation principle does not simply say that the *size* of the total momentum of a closed system does not change; the *direction* stays the same too, and that means that if we resolve the total momentum into components, those *components* will not be affected by any interactions within the system. This gives us a very straightforward way of dealing with oblique collisions, as the following example shows.

Example

A particle of mass m and velocity $\boldsymbol{u}$ along the x-axis is struck by another particle of mass M and velocity $\boldsymbol{U}$ in a direction making $150°$ with the x-axis (Figure 8.6). After the collision the particles stick together and move on as one body. Find the size and direction of their velocity.

Suppose the final velocity has size V in a direction making an angle α with the x-axis.

Momentum in x-direction before impact $= mu + MU\cos 30°$
after impact $= (m + M)V\cos\alpha$

Momentum in direction perpendicular to x-axis
before impact $= MU\sin 30°$
after impact $= (m + M)V\sin\alpha$

The momentum conservation principle tells us that both components remain unchanged, so

$$mu + MU\sqrt{3}/2 = (m + M)V\cos\alpha$$
$$\tfrac{1}{2}MU = (m + M)V\sin\alpha$$

Dividing, we obtain

$$\tan\alpha = \frac{MU}{2mu + MU\sqrt{3}}.$$

We can find V by remembering the useful formula

$$\sin^2\alpha + \cos^2\alpha = 1.$$

This gives

$$\frac{M^2 U^2}{4(m + M)^2 V^2} + \frac{(mu + MU\sqrt{3}/2)^2}{(m + M)^2 V^2} = 1$$

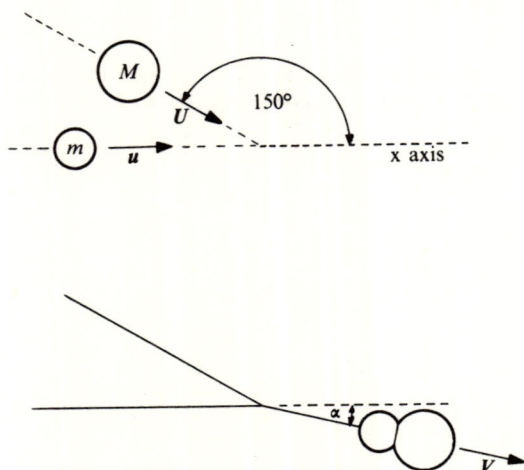

Figure 8.6

or

$$V^2 = \frac{M^2U^2 + (2mu + \sqrt{3}MU)^2}{4(m + M)^2}$$

Exercise 8.6

A drop of liquid moving vertically down at $2\,\mathrm{m\,s^{-1}}$ is hit by another drop, half the size, moving horizontally at $1\,\mathrm{m\,s^{-1}}$. The drops coalesce and move on together. Find the size and direction of their velocity just after the impact.

Energy

The product of force and time led us to the important idea of momentum; the product of force and *distance* leads us to an equally fundamental concept. Of course we have met such a product before–the moment of a force about an axis is the product of the force and its *perpendicular* distance from the axis. This time however we have in mind a different distance, one *in the direction* of the force, and the quantity we get as a result is called *work*. Formally, work is done when the point of application of a force is moved along its line of action, and *the work done is the product of the force and the distance moved by its point of application along its line of action*. The unit of work is the *joule* (J); one joule of work is done when the point of application of a 1 N force is displaced 1 m along its line of action. (Although the moment of a force is also the product of a force and a length, the word 'joule' is not used in connection with moments, which are measured in newton metres). Work is a *scalar*, in spite of the fact that it is the product of two vectors. It does not matter in which direction the force acts; if the point of application moves in that direction, work is done. If it moves a distance x in a direction not in line with the force F, but at an angle α to it, we take the component of F in the direction of the motion, and the work done is $xF\cos\alpha$ (Figure 8.7).

Though we are entitled to define as many new quantities as we please, and units to measure them in, they may not all turn out to be useful. The work done by a force turns out to be an exceedingly useful quantity, in fact in many ways it has more significance than the force itself. For a start, we can argue that work as defined above is indicating something which we recognise as 'hard work' in everyday experience. If I have a heavy trunk to drag from A to B I reckon myself as having some work to do, and if I am asked either to drag two identical trunks over the same distance or to drag one trunk twice as far, I consider myself put upon to twice the extent–and no doubt expect twice the monetary return if that is how I make my living. On the other hand, merely exerting a force on the trunk without moving it–just leaning on it–is *not* hard work.

From the viewpoint of physics, dragging the trunk along the ground is a rather wasteful way of doing work, whereas lifting a trunk on to a table is rather more thrifty. Once you have got your trunk on to the table, you can rig up a pulley block and link trunk X to another trunk Y still on the floor by a rope passing over the pulley (Figure 8.8). If Y is lighter than X, even only just lighter, X can *do work* on Y, it can pick up Y as it moves down. *The work which was originally done on X can be re-used.* In practice not all of it can be regained, of course; the weight of the rope and friction in the pulley block prevent us from raising quite as much weight as is being

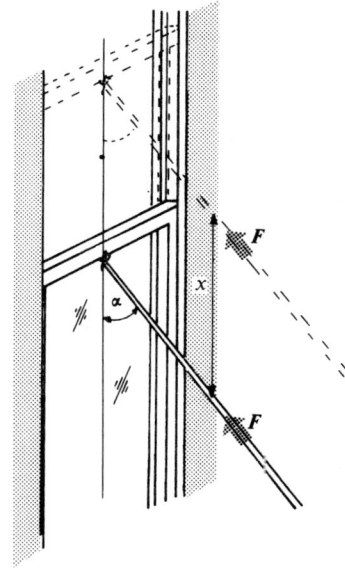

Figure 8.7
No work is done by the horizontal component of F.

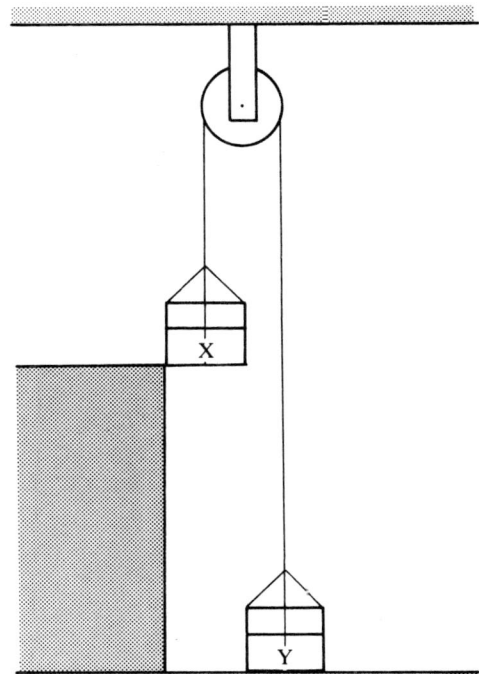

Figure 8.8

lowered, but in principle we can have a string with infinitesimal weight and a pulley block with negligible friction, and get the weight of X as close as we like to that of Y.

The trunk on the table has something which the one on the ground lacks: the *capacity for doing some work*. We call that *energy*; energy is 'stored work', the capacity for doing work, and logically it is, like work, a scalar, and has the same unit, the joule. The trunk has energy on account of its *position* relative to the ground; we call that *potential energy, p.e., energy which bodies possess on account of their positions relative to other bodies*. The bodies in this case are the trunk and the earth; strictly speaking they share the p.e. as a *system*.

Exercise 8.7

A 20 kg trunk is lifted on to a ledge 1·5 m above the floor. Taking g as 9·8 m s^{-2} write down

a) the weight of the trunk,
b) the force needed to lift it,
c) how much work is needed to lift it on to the ledge,
d) how much energy is stored as a result.

Potential energy of a body of mass m height h above the ground $= mgh$

If the trunk in Exercise 8.7 is in a room which has a trap door to the cellar below, we can obviously let it do a lot more work – we can lower it below floor level. This emphasizes an important feature of potential energy; we measure it with respect to an arbitrary zero level, and we have to be careful to say where the zero is. If we take our zero at ground level, then an object of mass m placed on the ground has zero potential energy; if it is lifted a height h it has potential energy equal to mgh, and in the cellar it has *negative* potential energy.

Energy conservation

Figure 8.8 has given us an idea of the significance of energy, but it can hardly indicate what an important part the quantity plays in physics. Energy is to physics what money is to business; it is one of the great unifying ideas of the subject. Like money, it can be stored in all sorts of ways, it can be turned into different forms and back again, it can be used to alter the state of things, and it can be wasted or converted into a form in which it is no longer useful. We have here in fact another *conservation* principle: *energy cannot be created or destroyed, though it can be converted from one form to another*. We shall be looking at the implications during the rest of the chapter.

We may get the impression that this is squeezing rather a lot out of the fact that trunk X will pick up trunk Y because it is heavier! Well, consider the situation in Figure 8.9. Here we have an object A of mass m on a table height h above the ground – it has potential energy mgh. On the ground is an object B of mass $2m$, and we want to get it on to the other table, of height $h/2$. It needs potential energy $2m \times g \times \frac{1}{2}h$, or mgh, to do it. Object A has just enough energy for the job, so we throw a rope over a pulley to link A and B, and of course nothing happens. Does this mean all our ideas about energy are useless? They certainly would be if anything *did* happen,

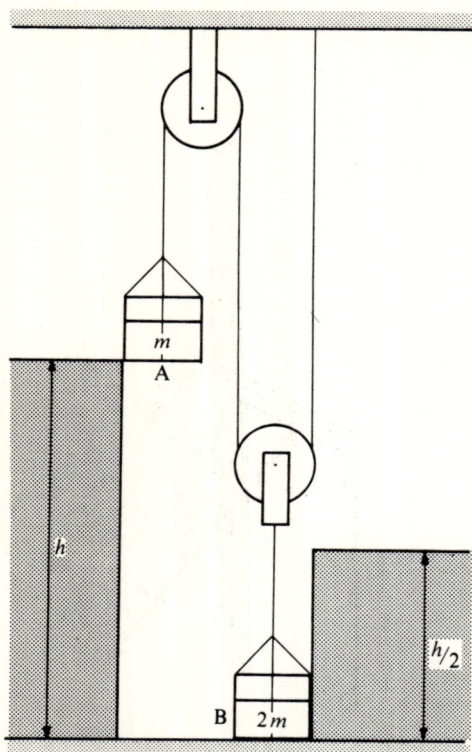

Figure 8.9

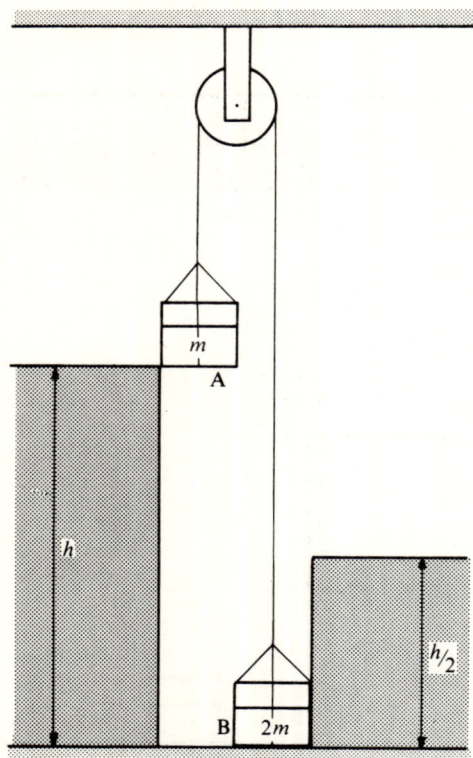

Figure 8.10

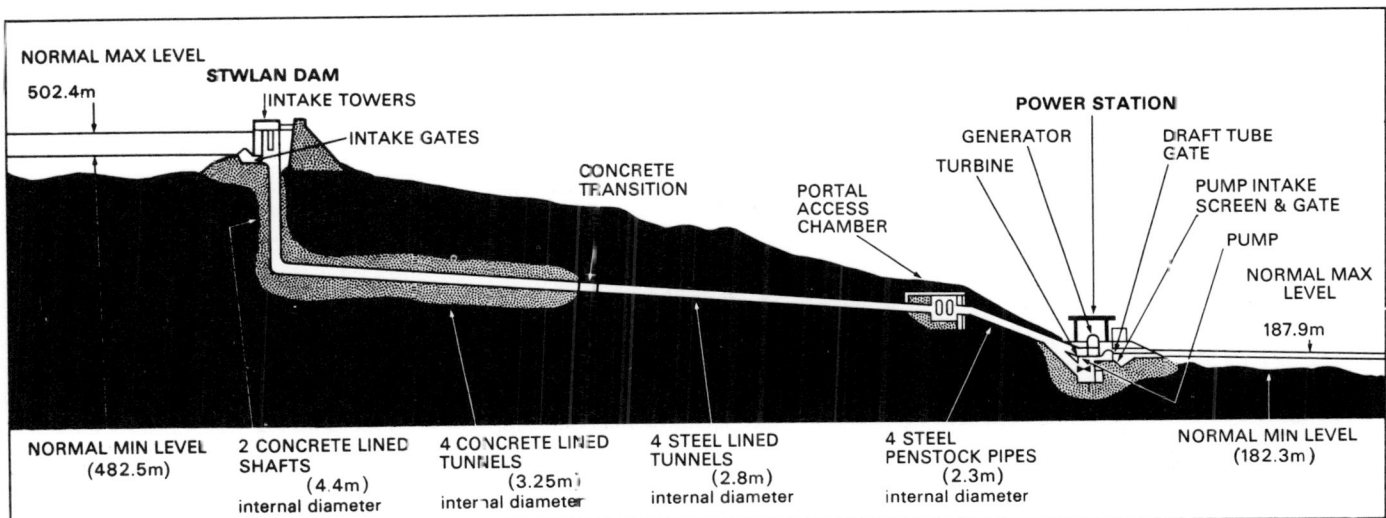

Energy conservation at Ffestiniog power station. At night when demand for electric power drops below the optimum rate for economic production, power is used to pump water to the top level, indicated by the dam in the photograph. During the day the water is allowed to run down again, working a turbine and feeding power back into the national grid.

because by the time B got to its destination A would be only half-way down – it would have given up only half its energy. Somehow we have to find a way of letting A get right down to the bottom while B rises half-way; we need an arrangement like that in Figure 8.10, a two-pulley system. Now A can lose all its potential energy during the ascent of B, and if our ideas about energy are correct, A should be able to lift B. In practice A needs a mass slightly greater than m, chiefly because it has to lift the moving pulley as well, but as long as this is kept as light as possible, the ratio of the mass of B to the mass of A just needed to lift it is convincingly close to two.

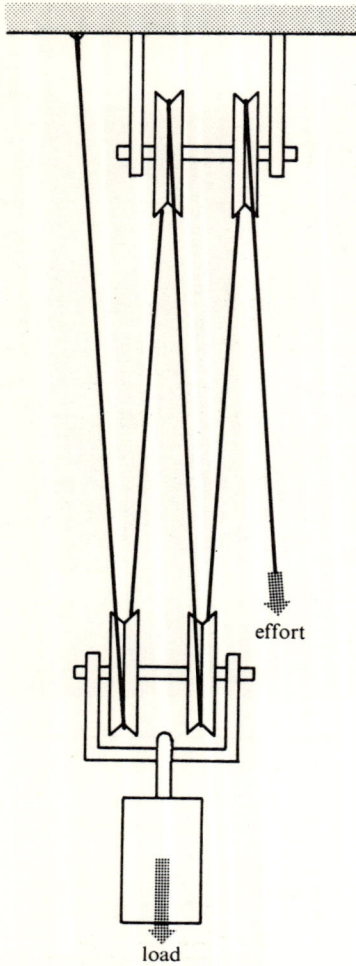

Figure 8.11 (Exercise 8.9)

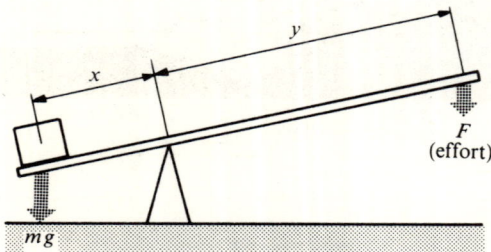

Figure 8.12 (Exercise 8.10)

Machines

Figure 8.10 is an example of a machine, a device which enables the work done by one force–the *effort*–to be done conveniently on a different force, the *load*. (Convenience is the essential reason for using a machine–it makes work easier to do.) Usually the machine is designed to work against a large load using a small effort, though this is not always the case. The ratio (load/effort) is called the *mechanical advantage* (m.a.). Of course we have to pay for the privilege of moving a large load with a small effort by moving the effort further; the ratio

$$\frac{\text{distance moved by effort}}{\text{corresponding distance moved by load}}$$

is called the *velocity ratio* (v.r.). Some machines waste more work than others; the effort has to work against frictional forces and lift parts of the machinery. The *efficiency* of a machine is defined by the expression

$$\text{efficiency} = \frac{\text{work done on load}}{\text{corresponding work done by effort}}$$

and is often expressed as a percentage. By the energy conservation principle no machine can be more than 100% efficient.

Exercises 8.8–10

8 Show from the definitions of m.a. and v.r. that the efficiency of a machine = (m.a./v.r.)

9 The pulley system shown in Figure 8.11 can lift a load of 80 N with an effort of 30 N. Write down the m.a., v.r. and efficiency of the system.

10 Figure 8.12 shows a lever being used as a machine. Write down the v.r. and use the energy conservation principle to find the best possible value for the m.a. Show that this ratio is consistent with that obtained from the equilibrium condition given by the principle of moments if the weight of the lever is negligible.

Kinetic energy

To get an object moving from rest requires a force, and that force must move its point of application as the object moves, so work must

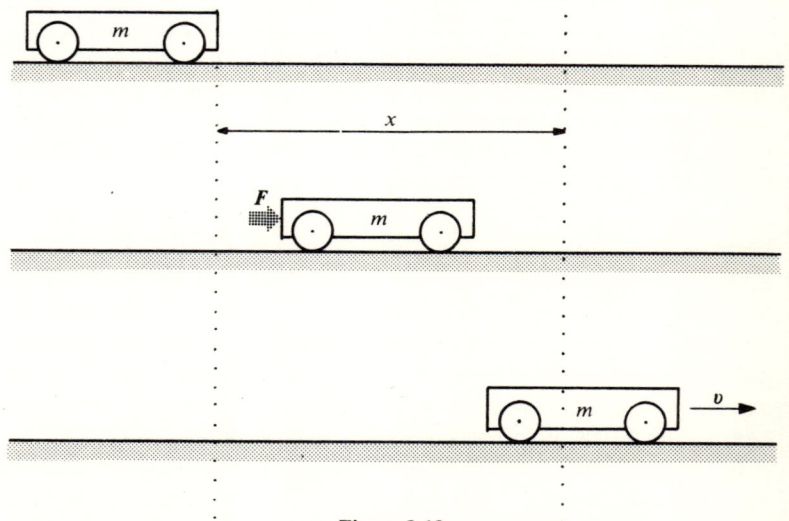

Figure 8.13

be done. Once the object *is* moving it can do all sorts of useful things while slowing down; it can climb a hill or start something else moving–the work we did is made available again. Energy possessed by objects as a result of their motion is called *kinetic energy* (k.e.). Suppose we exert a net force F on a trolley of mass m to accelerate it from rest, and we continue pushing until a distance x has been covered (Figure 8.13). The work done is Fx, but we would like to be able to write this in terms of the final speed v achieved.

Exercise 8.11

Use $2ax = v^2 - u^2$ to show that $Fx = \frac{1}{2}mv^2$.

Now this expression ($\frac{1}{2}$ mass $\times$ velocity^2) is also the work done *by* the trolley in stopping. If the trolley has an initial velocity **u** and is opposed by a constant force $-F$ producing an acceleration $-a$ (**F**, **u** and **a** all having the same direction) then the trolley exerts a forward force **F** on whatever is opposing it, and if it comes to rest in a distance x it does an amount of work Fx. By making the appropriate adjustment to the sign of a in the formula $2ax = v^2 - u^2$ we obtain

$$-2ax = 0 - u^2$$

and therefore

$$Fx = Fu^2/2a = \tfrac{1}{2}mu^2,$$

where we have used the fact that $-F = m \times(-a)$. So the kinetic energy of a body equals ($\frac{1}{2}$ mass $\times$ velocity^2).

Kinetic energy $= \frac{1}{2}mv^2$

It is important to note that though this formula is very useful, it has its limitations: for example, it is not applicable to particles travelling at speeds approaching the speed of light.

Exercise 8.12

How much kinetic energy is lost by a trolley of mass 2 kg in slowing from $3\,\mathrm{m\,s^{-1}}$ to $2\,\mathrm{m\,s^{-1}}$?

In the above treatment we have considered only constant forces and accelerations, but we can extend the argument to a varying force without too much trouble. We have to split the distance moved by the force into a large number of subdivisions, each one so small that the force can be considered constant from one end of it to the other. The work done in getting through the first tiny section of length δx_1, for which the magnitude of the force is F_1, is $F_1\delta x_1$, and if the trolley has speed u at the start and v_1 at the end of the section, we can write

$$F_1 \delta x_1 = \tfrac{1}{2}mv_1{}^2 - \tfrac{1}{2}mu^2$$

Similarly for the next section

$$F_2 \delta x_2 = \tfrac{1}{2}mv_2{}^2 - \tfrac{1}{2}mv_1{}^2$$

and then

$$F_3 \delta x_3 = \tfrac{1}{2}mv_3{}^2 - \tfrac{1}{2}mv_2{}^2$$

and so on, until for the last section we shall get

$$F_n \delta x_n = \tfrac{1}{2}mv^2 - \tfrac{1}{2}mv_{(n-1)}^2$$

where v is the final velocity. The sum of all the terms on the left will

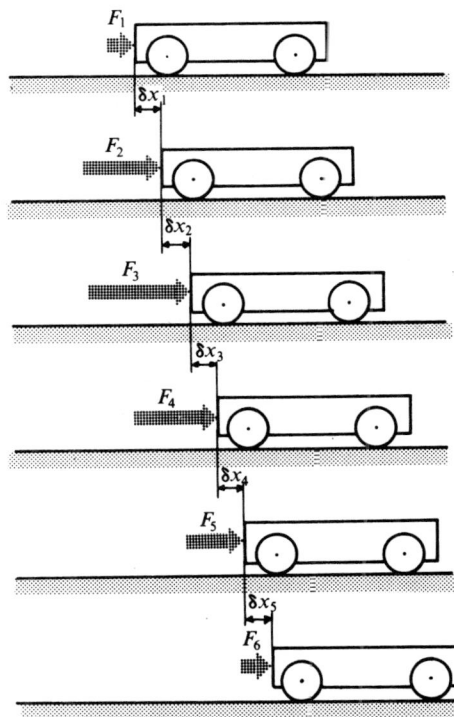

Work done by a varying force. The displacement of the point of application of the force is split into small sections, for each of which the force is considered constant. The smaller the sections, the smaller the error involved in this assumption.

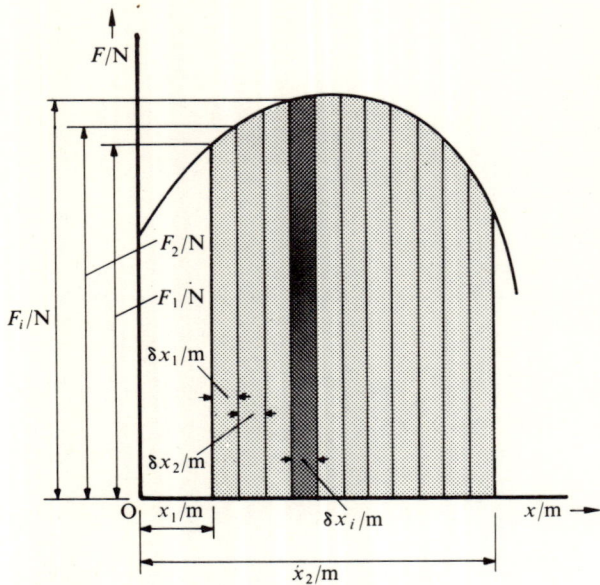

Figure 8.14

Graph of force F against displacement x. The area under the curve is a measure of the work done.

be the work done by the force over the entire distance, and when we add the right-hand terms the only contributions which do not cancel will be the terms involving the initial and final velocities, $\frac{1}{2}mv^2 - \frac{1}{2}mu^2$.

A graphical interpretation is shown in Figure 8.14 which indicates how the force F changes with x. Over the small displacement δx_i the area of the heavily shaded portion is not far from $(F_i/\text{N}) \times (\delta x_i/\text{m})$; and the smaller we make δx_i, the better the approximation becomes. The total work done between x_1 and x_2 is therefore represented by the sum of all such areas, that is, by the area under the graph betweeen (x_1/m) and (x_2/m).

Finally we can express the argument in calculus notation. This time we shall need to write

$$\text{work done} = \int_{x_1}^{x_2} F\,dx,$$

where x_1 and x_2 are the limits of the distance over which F acts. So

$$\text{work done} = \int_{x_1}^{x_2} ma\,dx$$

But the acceleration

$$a = \frac{dv}{dt} = \frac{dv}{dx} \times \frac{dx}{dt} = v\frac{dv}{dx}.$$

So

$$\text{work done} = \int_{x_1}^{x_2} mv\frac{dv}{dx}\,dx$$

$$= \int_u^v mv\,dv = \frac{1}{2}mv^2 - \frac{1}{2}mu^2.$$

Kinetic energy–potential energy exchanges

Figure 8.15 shows an experimental approach to the question of conversion from kinetic to potential energy. We start the trolley from rest at the top of the incline and we can find the velocity v it picks up by measuring how long it takes to cover one metre on the level – a wire W projecting from the trolley flips two small switches S, 1m apart, one to start an electric stop clock the other to stop it again. We can vary the height h from which the trolley starts and get a measure of v each time. A graph of v^2 against h turns out to be a straight line; h is proportional to v^2. This is an illustration of the energy conserva-

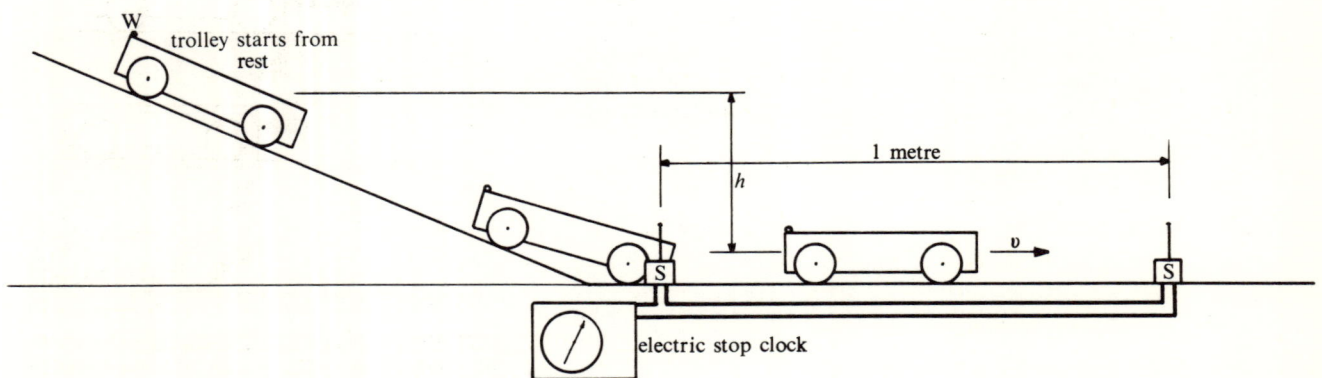

Figure 8.15

Conversion of potential energy of a trolley into kinetic energy.

tion law. At the top the trolley has potential energy mgh, and at the bottom it has kinetic energy $\frac{1}{2}mv^2$. The energy conservation law tells us that all of the potential energy mgh must be converted into another form, and the only form we can think of at the moment is kinetic energy, so

$$\text{p.e. loss} = \text{k.e. gain}$$
$$mgh = \tfrac{1}{2}mv^2$$
$$v^2 = 2gh.$$

Of course we could have worked that out without the law of energy conservation – we could have got it from Newton's laws, but it would have been more laborious.

Exercises 8.13, 14

13. Use $F = ma$ to show that a trolley falling freely from rest through a height h acquires a velocity v given by
$$gh = \tfrac{1}{2}mv^2.$$

14 Given that the trolley runs down a plane inclined at an angle α to the horizontal, write down (a) the component of the trolley's weight down the plane, (b) the acceleration this produces, (c) the speed v picked up by the trolley in a distance l down the plane, starting from rest. By writing down the connection between l and h, the magnitude of the vertical drop, show that $mgh = \frac{1}{2}mv^2$.

Now suppose we had to deal with a runway shaped as in Figure 8.16. We should find $F = ma$ very inconvenient to handle; at every stage of the run the accelerating force down the plane would be different, so the acceleration would vary in size and direction, making it very difficult to find the final velocity. But using energy conservation the problem is no more difficult than before; all we need to know is that the trolley has dropped through a height h, therefore it has lost potential energy mgh, therefore it has picked up kinetic energy $\frac{1}{2}mv^2$ equal to mgh. The final velocity is *independent* of the shape of the runway and depends only on the loss of height, and that too can be confirmed experimentally. Incidentally, though different paths give the same final velocity, the time taken to get to the bottom is different for each one. The path giving the shortest time is called the brachistochrone (*brachistos* shortest, *chronos* time), and it turns out to be a cycloid (see p. 271).

Pole vault. Kinetic energy gained in the run-up is converted into potential energy. The flexible pole stores some elastic potential energy during part of the ascent.

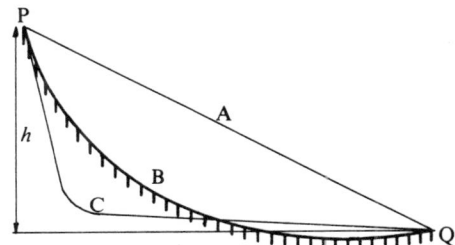

The brachistochrone. A, B and *C* are three different runways linking *P* and *Q*. A trolley starting from rest at *P* arrives at *Q* with a velocity independent of the runway used, but the time taken depends on the path. Path *B*, the brachistochrone, provides the shortest time.

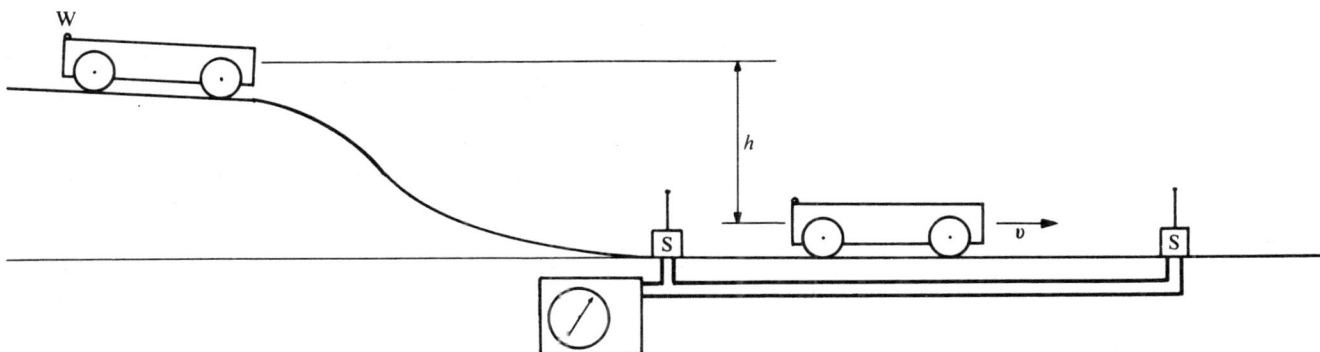

Figure 8.16

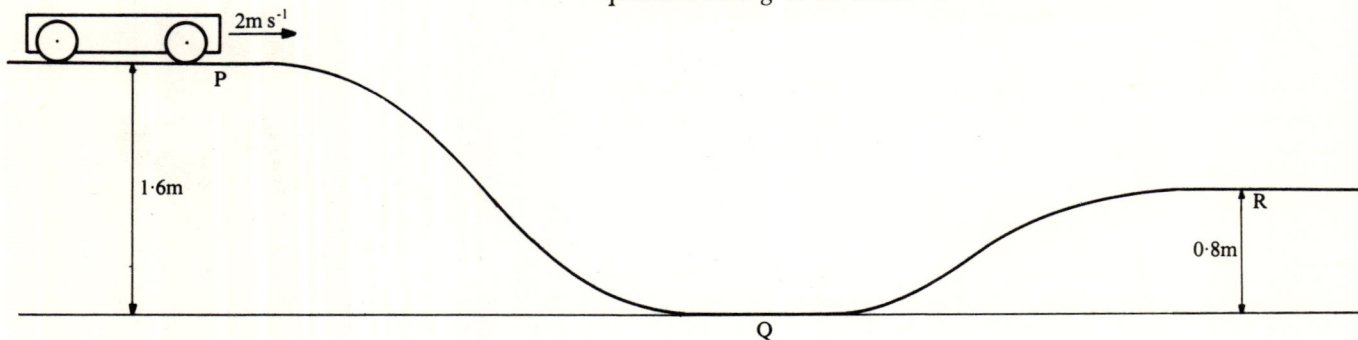

Exercise 8.15

A truck starting from the point P (Figure 8.17) with a velocity of $2\,\mathrm{m\,s^{-1}}$ coasts down the hill and up the other side. Assuming the sum of potential and kinetic energies remains constant, find the velocity at Q, $1\cdot6\,\mathrm{m}$ below P, and at R, $0\cdot8\,\mathrm{m}$ above Q. For this question take g to be $10\,\mathrm{m\,s^{-2}}$.

Figure 8.17 (Exercise 8.15)

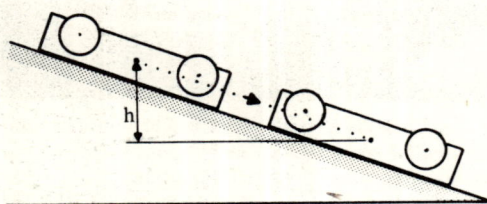

An apparent breakdown of energy conservation, $mgh \neq \frac{1}{2}mv^2$!

Energy conservation and friction

All our ideas to do with energy conservation can seemingly be demolished by one simple experiment: we turn the trolley upside-down and let it slide down the runway to the bottom, where it stops. Certainly potential energy, mgh, has been lost, but we do not need a clock to tell us that the kinetic energy $\frac{1}{2}mv^2$ is not equal to it!

Something has clearly gone wrong. Perhaps we might argue that now there is friction we cannot expect energy to be conserved; that just as Newton's first law applies only to an ideal situation so the law of conservation of energy also applies only when friction is absent. If this were the case the law would not be worth the paper it was written on; the fact is that *no* exception has yet been found to it in *any* situation, ideal or otherwise. We can see what has happened in this case by looking at the trolley and runway from a microscopic point of view. We have good reason for believing that the atoms making up the trolley and runway are not stationary–they cannot move about the trolley as a whole but they can and do vibrate about their mean positions. This means they have energy: kinetic energy because they are moving, and potential energy because, as their relative positions change in the vibration, work is done on the forces of repulsion and attraction between them. Now when the trolley slides down the incline, the atoms in the trolley experience forces from the runway. Neither surface will be smooth; there will be considerable irregularities on the atomic scale and a large number of atoms near the surface will get very hard knocks as the slide proceeds. This will start them vibrating randomly more energetically than before–their k.e. and p.e. will increase. And what the energy conservation principle says this time is that if we add up all the energy gained by all the atoms involved, it will come to mgh. Have we any justification for this belief? Will the trolley or runway be any different? They will be a little *hotter*–the temperature will rise, and we have learned to identify the macroscopic phenomenon of a temperature rise with the microscopic phenomenon of an increase of potential and kinetic energy of the atoms. This additional p.e. and k.e. is called 'internal energy', usually referred to in everyday language as heat. Whenever the principle of conservation of energy

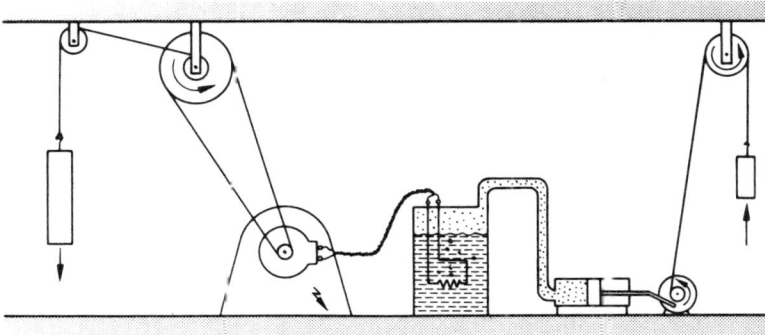

Energy can neither be created nor destroyed, but can be converted from one form to another....

appears to have failed we always find evidence that some of the energy has been converted into internal energy; something has got hotter. Before using the energy conservation principle to solve any problem therefore we must always check whether any internal energy is being produced. (As physicists, speaking carefully, we generally use the term 'internal energy' rather than 'heat', keeping the latter to refer to *energy in transit* between two bodies *by radiation or conduction*.)

We can also convert internal energy back into mechanical energy, with somewhat more trouble. We shall not develop this theme here; it takes us into the realm of thermodynamics.

Momentum or energy?

Although the example of the sliding trolley does not invalidate the law of energy conservation, it does mean that it is not very useful to us in working out the motion, unless we happen to know the size of the frictional force responsible or the amount of energy which has been used to cause rise of temperature. Certainly in cases where there are frictional losses we cannot simply convert potential energy loss into kinetic energy gain regardless. Exercise 8.16 shows how the method must be modified when friction is present.

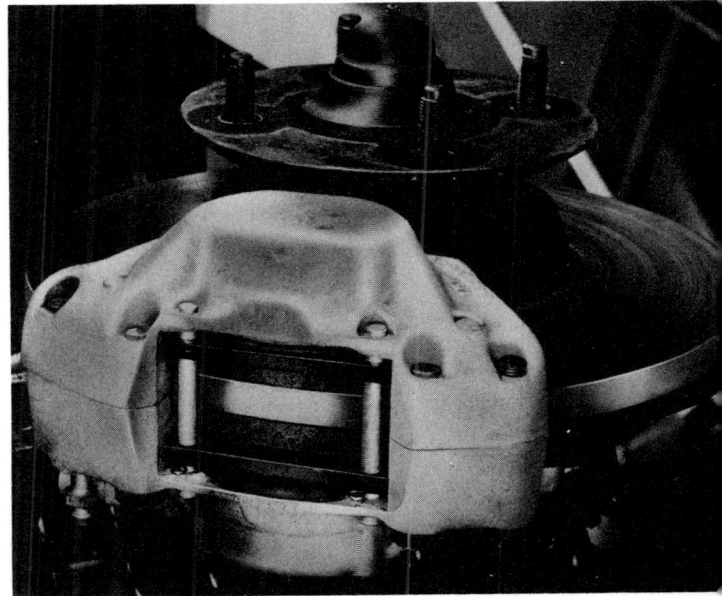

Disc brake. The k.e. of the braking vehicle must be dissipated as internal energy ('heat') in the region of contact between caliper and disc. The large surface of the disc encourages rapid cooling.

Exercise 8.16

A 1·5 kg block starts from rest and slides 2·0 m down a plane inclined at 30° to the horizontal. All the way down it experiences a 3 N frictional force. Find

a) the p.e. loss,
b) the work done against the friction,
c) the final k.e.,
d) the final velocity,
e) the gain in internal energy of the system.

Momentum is also conserved in the above situation, but it does not help us much in solving the problem, as this is a case where the

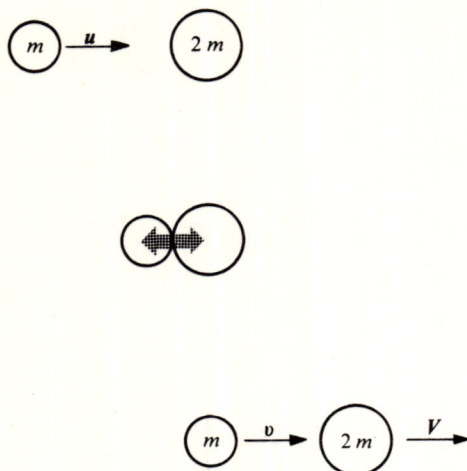

Figure 8.18

earth is involved in the 'closed system'. Momentum conservation really comes into its own in problems involving interactions between objects of comparable size–it is especially useful for collisions. The role of energy conservation in collisions is rather peculiar: sometimes it is useful, sometimes not, according to whether the collision is *elastic* or *inelastic*. These are terms which are discussed in Chapter 11, but the main point is that elastic collisions involve no permanent deformation of the bodies and no rise in temperature, whereas inelastic collisions deform the bodies, and during the deformation process some of the kinetic energy is converted to internal energy of vibration of the atoms. Unless we know how much energy is involved, the conservation of energy law is of no help.

Example

A particle of mass m and velocity u hits a stationary particle of mass $2m$ (Figure 8.18). Assuming that the velocities after impact have the same direction as u, describe the subsequent motion if (a) the collision is perfectly elastic, (b) 50% of the kinetic energy is lost in the collision, (c) the collision is so inelastic that the particles stick together after impact. What percentage of k.e. is lost in case (c)?

a) Suppose the final velocities are v and V for small and large particles respectively.

Momentum conservation gives

$$mu + m \times 0 = mv + 2mV$$
$$u = v + 2V \qquad \ldots(3)$$

and energy conservation gives

$$\tfrac{1}{2}mu^2 + \tfrac{1}{2}(2m) \times 0^2 = \tfrac{1}{2}mv^2 + \tfrac{1}{2}(2m)V^2$$

or

$$\tfrac{1}{2}u^2 = \tfrac{1}{2}v^2 + V^2$$

Eliminating v from the energy equation by means of equation (3), we have

$$\tfrac{1}{2}u^2 = \tfrac{1}{2}(u - 2V)^2 + V^2,$$

that is,

$$u^2 = u^2 + 6V^2 - 4uV$$
$$6V^2 = 4uV$$
$$V = 0 \quad \text{or} \quad 2u/3.$$

It is interesting that because the only facts we have put into the problem are that energy and momentum must be conserved, the analysis throws out the answer which would apply if the collision never took place, as well as the answer we want. Using our value of V in equation (3),

$$u = v + 4u/3$$

giving

$$v = -u/3,$$

so the smaller particle rebounds.

It is no accident that $(V - v) = 2u/3 + u/3 = u$. This always happens for perfectly elastic collisions and often provides a useful short cut: *the relative speed of separation equals the relative speed of approach.*

b) For this case also, momentum conservation gives

$$mu = mv + 2mV$$

but the energy equation yields

$$\tfrac{1}{2}mu^2 + \tfrac{1}{2}(2m)\times 0^2 = \tfrac{1}{2}mv^2 + \tfrac{1}{2}(2m)V^2 \quad + \quad mu^2/4$$

(k.e.)   (k.e.)   (k.e.) $\left(\begin{array}{c}\text{internal energy}\\ \text{gain}\end{array}\right)$

that is, $u^2/4 = v^2/2 + V^2.$

Eliminating as before, we get

$$V = u/6 \quad \text{or} \quad u/2.$$

Why should there be two answers this time? We have insisted that kinetic energy is lost, so there must be a collision; how can it be that there are two different possibilities? Things become clearer when we work out the respective values of v; they emerge as

$$\begin{array}{cc} V = u/6 & V = u/2 \\ v = 2u/3 & \text{and} \quad v = 0. \end{array}$$

The first pair of answers is very interesting; the smaller mass has emerged *faster* than the larger one – it has ploughed straight through.
c) Once again the momentum equation is

$$mu = mv + 2mV,$$

but this time we have no specific information about the energy; instead we are told that $v = V$. So, making use of the previous equation, we have,

$$v = V = u/3.$$

The k.e. after impact is $\tfrac{1}{2}mv^2 + mV^2$ or, in terms of u,

$$\text{k.e.} = mu^2/6,$$

that is, 66·7% of the original kinetic energy has been lost. This is the maximum amount of k.e. that can be lost in the collision we have been considering, and corresponds to the so-called *fully inelastic* case. (Notice that a fully inelastic collision does *not* imply *total* loss of k.e.)

Exercises 8.17, 18

17 Supply the missing steps in part (b) of the example.
18 What results would have emerged from the example if both particles had been of mass m?

Energy conservation in an oblique collision

For an oblique collision we may have to handle two momentum equations (for components of momentum in two perpendicular directions) and an energy equation, and the analysis that results can get a little more complex. We shall look at a rather interesting case, when the particles concerned have equal masses and there is no kinetic energy loss.

Example

A particle A of mass m moving with velocity $\boldsymbol{u}$ collides with a stationary particle B, also of mass m. Find the angle between the paths taken by the particles after the collision if it is perfectly elastic.

We shall take it that $\boldsymbol{u}$ is in the x-direction, and that after the collision A and B move off with velocities v and V making angles α and β respectively with the x-axis (Figure 8.19).

Momentum conservation in the x-direction gives

$$mu = mv\cos\alpha + mV\cos\beta$$

Series of photographs, taken at intervals of 0·1 s, of two billiard balls of equal mass involved in a direct collision. Is momentum conserved? Is kinetic energy conserved?

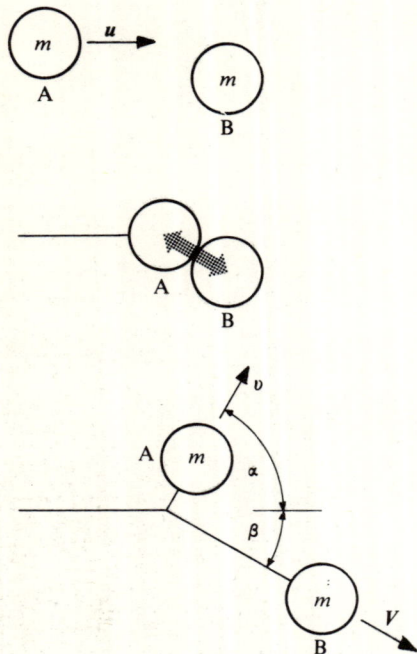

Figure 8.19
Oblique collision.

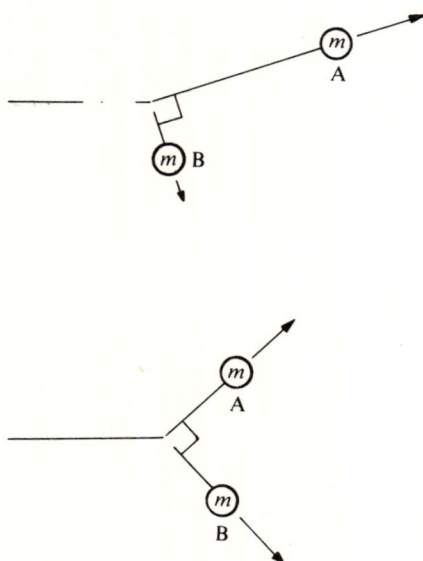

Figure 8.20
In any oblique perfectly elastic collision between particles of equal mass, one of which is initially stationary, the angle between the directions of the subsequent velocities is 90°.

and in the y-direction we get

$$0 = mv\sin\alpha - mV\sin\beta.$$

Finally, kinetic energy conservation gives us

$$\tfrac{1}{2}mu^2 = \tfrac{1}{2}mv^2 + \tfrac{1}{2}mV^2.$$

There is no question of taking components for this equation of course; energy is a scalar.

Reduced to essentials these equations become

$$u = v\cos\alpha + V\cos\beta \qquad \text{...(4)}$$
$$v\sin\alpha = V\sin\beta \qquad \text{...(5)}$$
$$u^2 = v^2 + V^2. \qquad \text{...(6)}$$

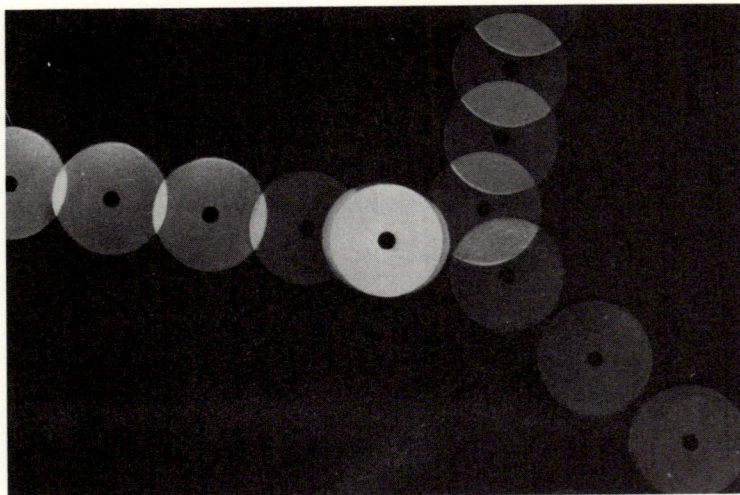

An almost perfectly elastic collision between two identical low-friction pucks, one initially at rest.

That finishes the physics; now we have to do some algebra. One route commences by eliminating u between equations (4) and (6) to give

$$v^2 + V^2 = v^2\cos^2\alpha + V^2\cos^2\beta + 2vV\cos\alpha\cos\beta,$$
$$v^2\sin^2\alpha + V^2\sin^2\beta = 2vV\cos\alpha\cos\beta.$$

Now we use equation (5) to substitute for V, getting

$$2v^2\sin^2\alpha = \frac{2v^2\sin\alpha\cos\alpha\cos\beta}{\sin\beta}$$

or

$$\cos\alpha\cos\beta = \sin\alpha\sin\beta$$
$$\cos\alpha\cos\beta - \sin\alpha\sin\beta = 0$$
$$\cos(\alpha + \beta) = 0$$
$$\alpha + \beta = 90°.$$

Notice that this result is independent of the individual values of α or β (see Figure 8.20).

Exercise 8.19

Given $\alpha = 30°$, find v and V in terms of u.

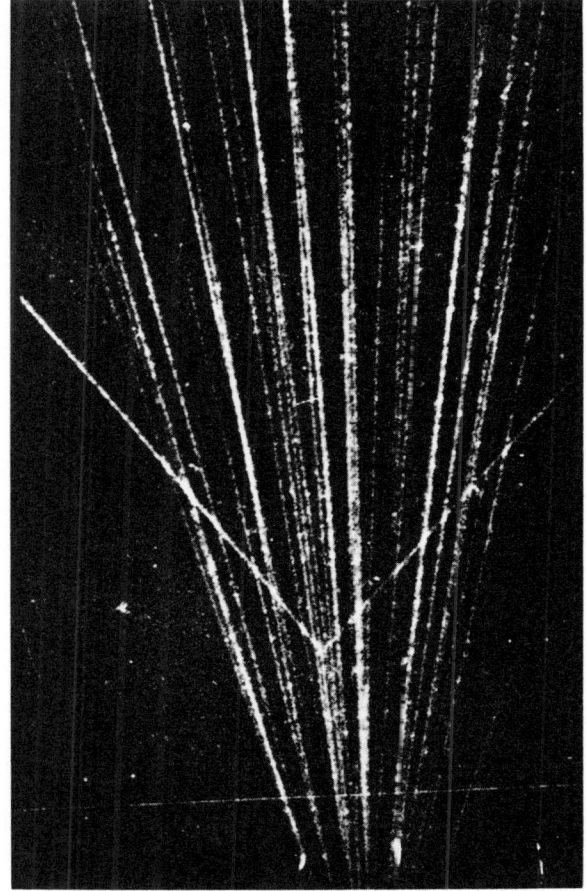

Left: Cloud chamber photograph of α-particle tracks in hydrogen. The forked track indicates a collision between the α-particle and a hydrogen nucleus. (α-particles are helium nuclei travelling at high speed; the mass of an α-particle is roughly four times that of a hydrogen nucleus.) *Right:* Collision of an α-particle with a helium nucleus (i.e. a collision between identical particles). The angle between the emerging particles does not appear to be quite 90° because the tracks are in a plane which is not parallel to the plane of the photographic plate.

Range of application of the conservation laws

There is a sense in which the laws of conservation of momentum and energy can never be proved by experiment, because we cannot try them in all possible situations. Nevertheless, when we have applied them properly and taken into account everything we can think of—the mass of the earth for example, or the kinetic and potential energy of atomic motion—we have never yet found them to fail. As an example of the trust we place in them, and the extent to which they justify that trust, we can mention a case where it seemed that both laws were seriously violated. The atoms of radioactive substances have nuclei which are *unstable*; an unstable nucleus is liable at any instant to eject a small particle, thereby turning into a different nucleus, or at least a modified form. Many types of unstable nuclei emit a β-particle, an electron travelling at high speed. In many cases it proved possible to measure the velocity of the electron and also the recoil velocity of the decayed nucleus, and when all the sums were done it was found that both momentum and energy conservation laws seemed to be violated in the process. This was so sensational that in 1930 Pauli suggested that not one but *two* particles were emitted, one being of such a character that it was not detected. He worked out the properties which this new particle would have to possess if the conservation laws were to hold; it was called a *neutrino*. For nearly thirty years nobody could say for sure whether this was really a case of non-conservation of energy and momentum or whether the neutrino really existed, though most people believed in the neutrino. Then in 1959 Reines and Cowan devised a way of detecting the presence of neutrinos if they were really there, and the test proved positive–the neutrino exists and the conservation laws still hold. Perhaps one day an exception will be found, but to date we know of no violation.

Bernoulli's principle

This is the name given to the formula we get if we apply the energy conservation law to a so-called *ideal* liquid. Bernoulli's principle is to a liquid what the equation

$$\tfrac{1}{2}mv^2 + mgh = \text{a constant}$$

is to a solid object. Think of a liquid flowing down a pipe. Generally speaking the motion is probably very complicated and untidy; there is swirling and eddying, and not only is the velocity different at every point of the pipe, it also changes from one instant to the next. That is far too difficult for us to handle, so we look instead at a simplified model: an ideal liquid, which possesses some, but not all, of the properties of real liquids. Our ideal liquid is rather like a flexible, slippery rubber cylinder flowing down the pipe, in all respects but one–the liquid is incompressible, and if the pipe narrows the tube gets correspondingly longer. The velocity is still different in different parts of the tube, but now at least the velocity at a given point does not change with time. The name for this situation is *streamline flow*, and the path taken by a small part of the liquid is then a *streamline*, though it should be noted that a streamline is *not* defined in this way when the flow is turbulent. In Chapter 16 we shall look at the conditions under which a real liquid flows along streamlines.

We have not yet finished simplifying our problem. We shall also take it that the walls of the pipe do not exert any drag forces on the ideal liquid, and that no part of the liquid has any tendency to slow down adjacent liquid flowing past it at a faster rate; in other words we are ignoring all *viscous* forces. These are also discussed in Chapter 16; they are analogous to frictional forces between solids and result in the conversion of the kinetic energy associated with the liquid flow into energy of random motion and vibration of the molecules, constituting a temperature rise.

These simplifications are fairly drastic, but in fact they lead to predictions which are quite close to observed behaviour in many practical situations; it must be that viscous drag and turbulence are relatively minor effects in such circumstances.

To derive Bernoulli's principle we look at a section of tube full of liquid density ρ (Figure 8.21); there is a vertical drop in height h over the section considered, and we take it that the pipe thickness is small compared with h. If the pipe is a thick one, the derivation can still go through if we consider instead a thin tube of liquid flowing along a streamline. The pressure, velocity and cross-sectional area all change as liquid flows through the section, from P_1, v_1 and A_1 at one end to P_2, v_2 and A_2 at the other. Perhaps the first point to emphasize is that $P_1 - P_2$ is *not* equal to ρgh; that was fine for a stationary column but is no good when the liquid is moving–in its derivation we assumed an equilibrium condition. We are going to

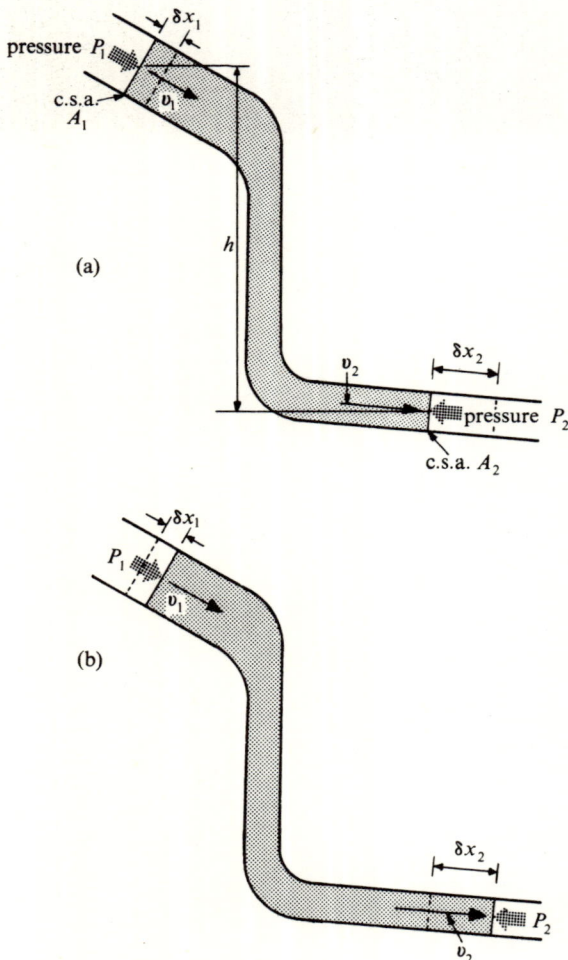

Figure 8.21

consider the energy changes which occur when the liquid occupying this section moves forward a small amount δx_1 at one end and a corresponding amount δx_2 at the other. Since the cross-sectional area changes over the section, δx_1 cannot equal δx_2; instead we can write

$$A_1\,\delta x_1 = A_2\,\delta x_2, \qquad\qquad ...(7)$$

because the volume of that part of the liquid we are concentrating on does not change as it moves down the pipe. An alternative way of arriving at equation (7) is to realise that every litre of liquid passing A_1 causes another litre to pass A_2; the *flow rate* (volume passing per second) is the same through any cross section of the pipe.

Exercise 8.20

Show that the flow rate can be written as $v_1 A_1$ or $v_2 A_2$ (see example on p. 59) and hence that $A_1\,\delta x_1/\delta t = A_2\,\delta x_2/\delta t$, where δt is the short time interval between the situations shown in Figures 21a and b.

Now although all the liquid has moved forward, the net result is just as though we had taken the volume $A_1\,\delta x_1$ of liquid from one end of the section and transferred it to the other. This helps us to write down both the gravitational potential energy change and the kinetic energy change very easily.

$$\text{p.e. loss} = (A_1\,\delta x_1\rho)gh$$
$$\text{k.e. gain} = \tfrac{1}{2}(A_1\,\delta x_1\rho)v_2^2 - \tfrac{1}{2}(A_1\,\delta x_1\rho)v_1^2.$$

Now comes another surprise; these two expressions are not equal. They would be equal if the liquid were falling freely, or even if it were sliding down the pipe in isolation, but it is not. There is liquid in front of our section pushing it back, and there is liquid behind pushing it on, and as a result work has to be done. The work done on our section by the liquid behind it is $P_1 A_1\,\delta x_1$ and the work done by it on the liquid in front is $P_2 A_2\,\delta x_2$, so the net work done on the section is $P_1 A_1\,\delta x_1 - P_2 A_2\,\delta x_2$, which we can write as

$$(P_1 - P_2)A_1\,\delta x_1,$$

using equation (7).

Now we can form the energy conservation equation; energy losses due to drag we have already decided to leave out. We can write p.e. loss + work done by the rest of the liquid = k.e. gain. That is,

$$A_1\,\delta x_1\rho gh + A_1\,\delta x_1(P_1 - P_2) = \tfrac{1}{2}A_1\,\delta x_1\rho(v_2^2 - v_1^2)$$

or

$$\rho gh + P_1 - P_2 = \tfrac{1}{2}\rho(v_2^2 - v_1^2).$$

Finally, if we write $h = h_1 - h_2$, we can put this in the form

$$gh_1 + P_1/\rho + \tfrac{1}{2}v_1^2 = gh_2 + P_2/\rho + \tfrac{1}{2}v_2^2. \qquad ...(8)$$

That is,

The quantity $gh + P/\rho + \tfrac{1}{2}v^2$ remains constant along a streamline.

This is Bernoulli's principle.

In the absence of forces from the surrounding liquid, that is, in a situation where P_1 and P_2 are zero, the equation reduces to

$$gh_1 + \tfrac{1}{2}v_1^2 = gh_2 + \tfrac{1}{2}v_2^2$$

or
$$gh = \tfrac{1}{2}v_2{}^2 - \tfrac{1}{2}v_1{}^2$$
where h is the height through which the liquid falls. This is the familiar equation for free fall.

Exercise 8.21

Write down the units for each term in the expression $gh + P/\rho + \tfrac{1}{2}v^2$ in terms of m, kg and s and confirm that all three terms have the same units.

Venturi flow meter

Figure 8.22 shows a simple application of the principle, a *venturi meter*, which can be used to measure the rate of flow of liquid in a pipe. The liquid is made to flow through a narrower portion, of cross-sectional area A_2, fitted with a vertical tube as shown. The

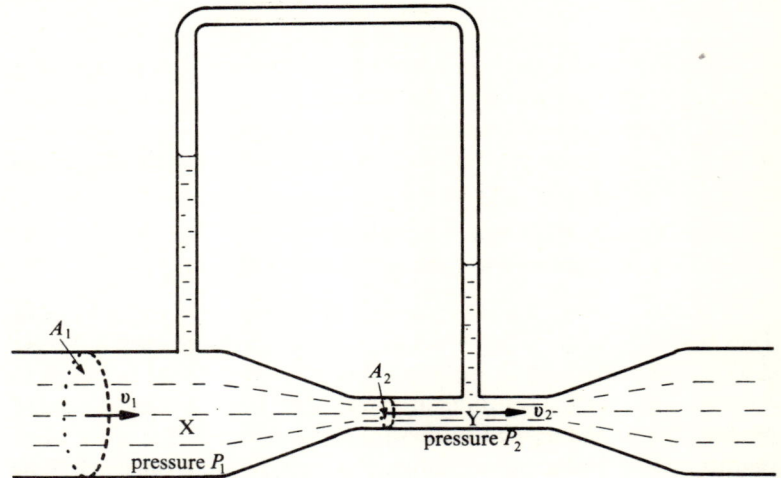

Figure 8.22
Venturi flow meter.

wider part with cross-sectional area A_1 has a similar junction, and the difference between the heights to which the liquid rises in the vertical tubes is a measure of the pressure difference $(P_1 - P_2)$ between points X and Y of the flow tube. As X and Y are on the same horizontal level, equation (8) gives
$$P_1/\rho + \tfrac{1}{2}v_1{}^2 = P_2/\rho + \tfrac{1}{2}v_2{}^2$$
and as $v_2/v_1 = A_1/A_2$ if the liquid is incompressible,
$$P_1 - P_2 = \tfrac{1}{2}\rho v_1{}^2[(A_1{}^2/A_2{}^2) - 1]$$

Exercise 8.22

Show that the flow rate Q (volume flowing per second) is given by
$$Q = v_1 A_1 = A_1 A_2 \left[\frac{2(P_1 - P_2)}{\rho(A_1{}^2 - A_2{}^2)}\right]^{1/2}$$
and that this is approximately $A_2 \left[\dfrac{2(P_1 - P_2)}{\rho}\right]^{1/2}$ if $A_2 \ll A_1$.

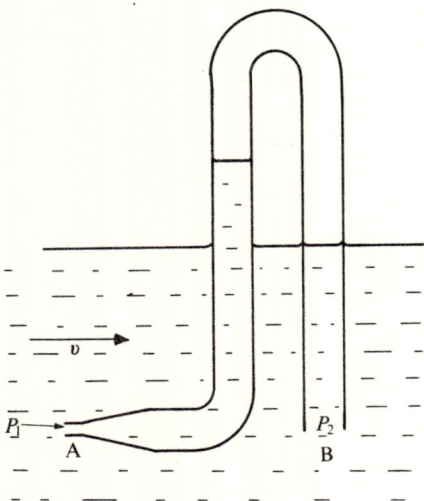

Pitot tube

This is a different type of flow meter useful when the liquid is not enclosed in a pipe; it can be used to measure rates of flow in a river

Figure 8.23
Pitot tube.

for example. The ends A and B (Figure 8.23) of the inverted U-tube are respectively facing upstream and at right angles to the flow. There is no tendency for liquid to enter B (it has no velocity component in the required direction) so the column of liquid in B has a height corresponding to the pressure P_2 at B. But the other column is higher, because it has to provide sufficient force to stop the flow at A. This time equation (8) gives

$$\frac{P_1}{\rho} = \frac{P_2}{\rho} + \tfrac{1}{2}v^2,$$

where v is the flow velocity and P_1 the pressure at A; $P_1 - P_2$ is indicated by the difference in levels in the U-tube.

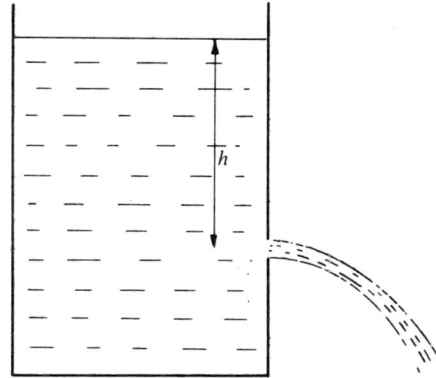

Exercise 8.23

A tank containing liquid of density ρ has a small hole of area A in one wall, a depth h below the surface. Show that the liquid flows from the hole with velocity given by $(2gh)^{1/2}$. The fact that the liquid emerges at the speed it would have acquired as a result of falling freely from the surface is known as Torricelli's theorem. (Surprisingly, it is *not* correct to assume that the volume of liquid leaving per second is Av. This is because the stream of liquid leaving the hole is *converging*; there are velocity components pointing towards the centre of the stream. A short distance out from the hole the stream diverges again, and if we want to work out the flow rate from the expression [$v \times$ cross-sectional area] we have to take the *minimum* cross-section of the stream, called the *vena contracta*, where the streamlines are parallel. If the hole is circular with a sharp edge, the *vena contracta* has an area of about $2A/3$.)

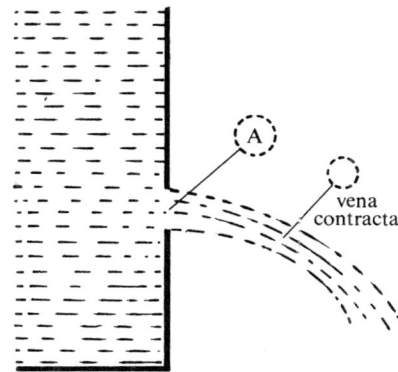

Vena contracta (see Exercise 8.23)

Bernoulli's principle and gases

Equation (8) is never strictly true for a real liquid, and it is even less suitable for a gas. Nevertheless the qualitative, and rather surprising fact that the pressure is lower in the region of faster flow is just as true for a gas as for a liquid, and has many applications. Atomisers and carburettors rely on this principle; in both cases the idea is to form a fine spray by squirting liquid from a jet into a fast-moving air stream. The air is made to flow through a constriction where the consequent increase in speed produces a region of reduced pressure. It is at this point that the liquid is introduced (Figure 8.24); it is forced into the low pressure region merely by the pressure of the atmosphere on the reservoir surface. Surface tension forces cause the liquid to form tiny droplets, and these are effectively mixed with the air on account of the turbulent conditions in the flow tube.

Aircraft wings owe much of their effectiveness to the same principle. The wing cross section is shaped in such a way that the velocity of air flow over the upper surface is greater than that over the lower one. This produces a region of reduced pressure above the trailing edge of the wing, and a consequent net upward force.

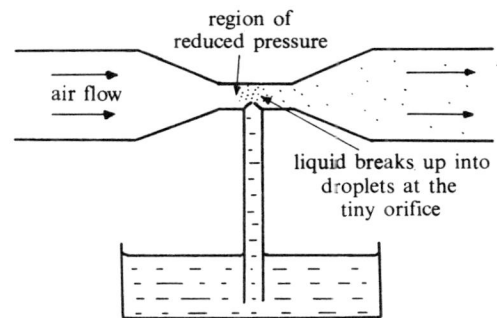

Figure 8.24

Power

Our civilisation runs on energy–large amounts of it. So far we have been able to find large stocks of chemical energy which we are able to convert speedily into internal energy (see p. 100), and thence to other forms such as kinetic energy of translation or rotation of

Conversion of chemical energy in leg muscle to gravitational potential energy.

machinery. On a molecular level we can picture it like this. Various substances we find in the earth's crust – oil, coal, natural gas – have molecules with the atoms arranged in such a way that they are able to release quite a lot of potential energy in chemical reactions. When these molecules are mixed with oxygen molecules and excited sufficiently, that is, when the temperature is raised, a rearrangement of atoms takes place into different, usually smaller molecules with less potential energy. This process results in a release of kinetic energy of random high-speed motion, sufficient to excite more molecules and constituting a very high temperature; the fuel is burning. Then we are able to devise mechanical arrangements which can convert some of this random kinetic energy into a more useful form; the gas molecules can be made to bombard a piston for example. Here we have the idea behind a heat engine, whether petrol-driven, diesel, steam or jet. Every engine makes energy available in this way; it becomes a point of importance at what rate energy can be made available, how much energy can be made available each second. The quantity which measures this is called *power*, and the unit is the *watt*: *a power of one watt is a release of one joule of energy per second*. A powerful car is one which can work quickly; it can acquire the potential energy needed to get to the top of a hill quicker. Of course it uses fuel at a faster rate! Engines are power limited, not energy limited – a humble two-stroke engine of small capacity can get a barge from one end of the country to the other, but it will not do it very quickly. [A rocket motor is an exception to the rule: in this case the power available is directly proportional to the speed attained, because the force is constant.]

The following example shows how the power of an engine can limit the top speed attainable by a vehicle.

Example

At its top speed of 100 km per hour on the level a certain 12 000 kg lorry experiences a total drag force of 3600 N.

a) What power does the engine develop?

b) What power would be needed to climb a gradient of 1 in 100 at the same speed if the drag force remained the same?

c) If the power calculated in part (b) were in fact available, what acceleration could be achieved at 100km per hour on the level? (Take g to be $10\,\mathrm{m\,s^{-2}}$.)

(a) $100\,\mathrm{km}$ per hour $= 100 \times 1000/3600\,\mathrm{m\,s^{-1}}$

Work done per second $=$ force $\times$ distance moved per second
$$= 3600 \times (100 \times 1000/3600)\,\mathrm{W}$$
$$= 10^5\,\mathrm{W}.$$

(b) Extra work done per second

$$= \text{mass} \times g \times \text{height gained per second}$$

$$= 12000 \times 10 \times \frac{1}{100}(100 \times 1000/3600)\,\mathrm{W}$$

$$= 3\cdot3 \times 10^4\,\mathrm{W}.$$

So

total power needed $= 1\cdot33 \times 10^5\,\mathrm{W}.$

(c) Power available $=$ force available $\times$ distance moved per second

so

available force $=$ (available power)/speed
$$= 1\cdot33 \times 10^5 \div (100 \times 1000/3600)\,\mathrm{N}$$
$$= 4800\,\mathrm{N}$$

Net accelerating force at 100km per hour on the level
$$= (4800 - 3600)\,\mathrm{N} = 1200\,\mathrm{N}$$

Accelerating force $F = ma$
$$a = (1200/12000)\,\mathrm{m\,s^{-2}} = 0\cdot1\,\mathrm{m\,s^{-2}}$$

Exercises 8.24, 25

24 If a 1000kg car can climb a gradient of 1 in 40 at 80km per hour find what acceleration it can achieve at this speed (a) on the level, (b) down a gradient of 1 in 60. Assume the drag force to be the same in each case and take g to be $10\,\mathrm{m\,s^{-2}}$.

25 A vehicle of mass m has an engine which can develop a power P given by $P = bv$, v being the vehicle velocity and b a constant. The vehicle experiences a drag force F given by kv^2, where k is another constant. Find an expression for the acceleration which can be achieved at speed v and show that the maximum speed is given by $(b/k)^{1/2}$. How would the answers be modified if the vehicle were climbing a gradient of 1 in 20?

Further exercises

Exercises 8.26–43

26 Discuss the law of conservation of linear momentum, and describe an experiment which illustrates the law.

A car of mass 1000kg, travelling in a straight line at $15\,\mathrm{m\,s^{-1}}$, is brought to rest in a distance of 60m by its brakes, which exert a constant retarding force. Find the magnitude of this force and the time for which it acts, and the change in the momentum and kinetic energy of the car.   (o)

27 A mass of 8kg is drawn along a level surface where friction is negligible by a horizontal thread of negligible mass which passes

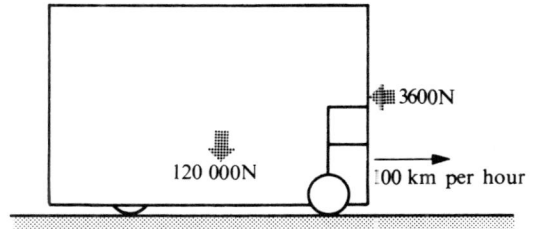

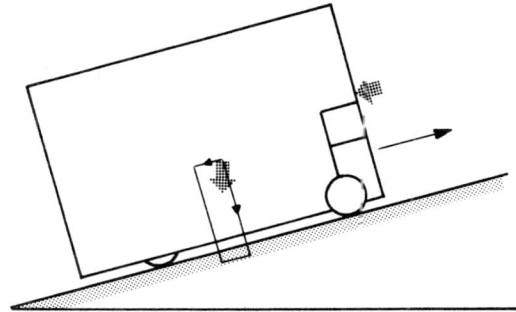

over a freely moving pulley and carries a vertically hanging load of 2 kg. Calculate (a) the distance travelled from rest by the large mass in 0·7 s; (b) the potential energy lost by the smaller mass in this time; (c) the kinetic energy gained by the larger mass in the same time. (For this question take the acceleration of free fall, due to gravity, as 10 m s^{-2}.) (SUJB)

28 Define momentum and state the law of conservation of linear momentum. Describe an experiment to verify this law.

A man stands on a trolley which can move without friction along a straight, horizontal track. The man weighs 100 kg, the trolley weighs 200 kg and both are initially at rest. Calculate what happens when the man

a) walks 5 m along the trolley parallel to the track and then stops,
b) fires a single bullet weighing 50 g at a muzzle velocity of 1000 m s^{-1} from a gun in a direction parallel to the track (neglect the mass of the gun),
c) fires a stream of 2 bullets per second each weighing 50 g at a muzzle velocity of 1000 m s^{-1} parallel to the track (neglect the mass of the gun and the loss of mass due to the escaping bullets). (O & C)

29 Sketch a graph of the relationship between the kinetic energy E (plotted on the vertical axis) and the distance travelled x (plotted on the horizontal axis) for a body of mass m sliding from rest with negligible friction down a uniform slope which makes an angle of 30° with the horizontal. What is the gradient of the graph equal to? (SUJB)

30 State the principle of the conservation of linear momentum and show how it follows from Newton's laws of motion.

A stationary radioactive nucleus of mass 210 units disintegrates into an α particle of mass 4 units and a residual nucleus of mass 206 units. If the kinetic energy of the α-particle is E, what is the kinetic energy of the residual nucleus? (JMB)

31 A space vehicle of mass M, moving towards the moon with velocity U, is split into two parts by an explosion. Immediately after the explosion the smaller portion (mass $M/4$) moves towards the moon with velocity V whilst the larger part (mass $3M/4$) has its velocity reduced to $U/3$.

Apply the principle of momentum conservation to find the value of V in terms of U. Write down expressions for the kinetic energy of the whole system (a) before, and (b) after, the break occurs. Given $M = 10^3$ kg and $U = 2 \times 10^3$ m s^{-1}, calculate how much energy will have to be supplied when the split occurs. (O & C)

32 State Newton's laws of motion and use them to derive the principle of conservation of momentum.

In nuclear reactors, fast neutrons are slowed down by elastic collisions with nuclei. Consider a neutron of mass m and velocity u making a direct collision with a stationary nucleus of mass M. Express the velocities of the neutron and nucleus after collision in terms of the initial velocity u.

Taking the mass of a neutron and of a proton to be exactly equal, what would be the effect, under the conditions of your calculation, of a collision between a neutron and (a) a hydrogen nucleus, (b) a deuterium nucleus which consists of a proton and a neutron?

How would the initial kinetic energy of the neutron be redistributed in the two cases?   (WJEC)

33 Distinguish between the momentum and the kinetic energy of a particle.

Calculate the recoil velocity of the uranium nucleus formed when a stationary plutonium nucleus (mass number 242) emits an α-particle with a velocity of $1\cdot54 \times 10^7 \mathrm{m s^{-1}}$.

What is the source of the kinetic energy of the final particles?
(JMB)

34 Calculate the percentage loss of kinetic energy of a neutron of mass m when it makes a perfectly elastic head-on collision with a stationary beryllium nucleus of mass $9m$. State the physical principles involved in your calculation.   (JMB)

35 State the law of conservation of linear momentum. A rocket of mass M travelling horizontally with a velocity V detaches its rear section of mass m and propels it horizontally backwards with velocity v *relative to the rocket* after the separation. With what *additional* velocity does the front section move forwards?
(SUJB)

36 Outline an experiment to demonstrate momentum conservation and discuss the accuracy which could be achieved.

Show that in a collision between two moving bodies in which no external forces act, the conservation of linear momentum may be deduced directly from Newton's laws of motion.

A small spherical body slides with velocity v and without rolling on a smooth horizontal table and collides with an identical sphere which is initially at rest on the table. After the collision the two spheres slide without rolling away from the point of impact, the velocity of the first sphere being in a direction at 30° to its previous velocity. Assuming that energy is conserved, and that there are no horizontal external forces acting, calculate the speed and direction of travel of the target sphere away from the point of impact.   (O & C)

37 a) A particle of mass m_1 makes a *perfectly elastic* head-on collision (i.e. there is no loss of kinetic energy) with a second particle of mass m_2 which is initially at rest. Under what conditions will m_1 (i) be brought to rest, (ii) have its velocity reversed in direction, (iii) continue moving in the same direction, as a result of the collision?

Show that if $m_1 = m_2$ (m_2 being initially at rest), and the collision is not head-on, the two particles will always move off at right angles to each other after the collision.

b) Two bodies A and B, each of mass m, approaching each other along the same straight line and each moving with velocity v, make a *perfectly inelastic* collision and are thus both brought to rest. Each body initially possessed kinetic energy $\frac{1}{2}mv^2$, so that a total amount of energy mv^2 is converted into heat in the collision.

An observer on A will see A stationary and B approaching with a velocity $2v$ and hence with kinetic energy $\frac{1}{2}m(2v)^2$ (or $2mv^2$). Why is the observer not correct in concluding that twice as much heat will be produced in the collision? Give the explanation of the apparent paradox.   (WJEC)

38 Discuss briefly laws which are applied to determine the motions of bodies after collision.

Show that if a large mass M, moving with a velocity V, makes a head-on elastic collision with a very small, stationary mass m, the energy lost by the large mass is approximately $2mV^2$.

The large mass passes through a gas of density ρ. If the mass presents a cross section of area A to the gas molecules (which may be treated as being stationary) show that the rate of loss of energy of the mass is $2A\rho V^3$. Hence, or otherwise, derive an expression for the time for the velocity of the mass to be reduced by a half. (c)

39 a) In the disintegration of a stationary radioactive nucleus, it is observed that the emitted electron and the residual nucleus travel in directions at 160° to each other. Discuss the implications of this observation.

b) In nuclear reactors it is frequently necessary to slow down the neutrons very quickly in order that they may be more effective in producing further fissions. This is done by allowing the neutron to suffer perfectly elastic collisions with stationary atoms in a moderator. Explain why heavy elements would be almost completely ineffective for this purpose. (WJEC)

40 Define linear momentum and state the principle of conservation of linear momentum. Explain briefly how you would attempt to verify this principle by experiment.

Sand is deposited at a uniform rate of $20\,kgs^{-1}$ and with negligible kinetic energy on to an empty conveyor belt moving horizontally at a constant speed of $10\,m$ per minute. Find
a) the force required to maintain constant velocity.
b) the power required to maintain constant velocity.
c) the rate of change of kinetic energy of the moving sand.
Why are the two latter quantities unequal? (o & c)

41 Define *linear momentum* and state the *principle of conservation of linear momentum*. Describe an experiment to demonstrate the conservation of linear momentum.

An open, water-tight railway wagon of mass $5 \times 10^3\,kg$ coasts at an initial velocity of $1 \cdot 2\,ms^{-1}$ without friction on a level track. Rain falls vertically downwards into the wagon. Calculate the velocity of the wagon after it has collected $10^3\,kg$ of water. What change has then occurred in the kinetic energy of the wagon and its contents if the initial kinetic energy of the falling rain is ignored?

How has this change occurred and what has happened to the lost kinetic energy? (o & c)

42 A straight pipe of uniform radius R is joined, in the same straight line, to a narrower pipe of uniform radius r. Water (which may be assumed to be incompressible) flows from the wider into the narrower pipe. The velocity of flow in the wider pipe is V and in the narrower pipe is v. By equating work done against fluid pressure with change of kinetic energy of the water, show that the hydrostatic pressure is lower where the velocity of flow is higher.

Describe and explain one practical consequence or application of this difference in pressures. (o & c)

43 State Bernoulli's theorem and use it to investigate the effect of a varying cross section in a horizontal tube on the hydrostatic pressure in an incompressible fluid flowing through the tube, neglecting the effects of viscosity. Explain the action of one of the following: (a) a filter pump, (b) a venturi meter, (c) an aerofoil.

A cylindrical vessel with its axis vertical is filled with liquid of negligible viscosity to a height z, and in its side is a small orifice at a height y, both z and y being measured from the base. Given that the jet of liquid strikes the horizontal plane through the base at a distance x from the base, find a relation between x, y, and z. For what value of y does x have a maximum and what is then the value of x?   (o & c)

9 Formulæ and dimensions

Models can be used to predict the behaviour of full-size systems, but we need to know the effect which the scaling down of one quantity has on all the other quantities involved. Dimensional analysis can sometimes assist with this problem.

In Chapter 2 we saw how the basic framework of physics consists of *quantities*, each of which is defined in terms of a unit and a counting process. A quantity is useful, that is to say, it is worth defining, if we can define relationships which link its size with the sizes of other quantities in various situations. These numerical relationships are the formulae and equations which provide the structure for the whole subject—they are the tools we work with. Sometimes a formula is used to define one of the quantities it involves, in which case it is an exact formula. 'Force equals rate of change of momentum' is like that; it is exactly right because we defined it to be so. More commonly the formula is likely to be an approximation, useful within certain limits. Boyle's law is a good example of this type; it is not a definition either of pressure or volume, nor of the quantity to which their product is equal, and it is only approximately true—and then only under certain circumstances for particular gases.

Dimensions

The equality between the left- and right-hand sides of an equation or formula is more than a *numerical* equality. Five apples plus three apples equal not just eight, but eight *apples*, and certainly do not equal eight pears. In the same way

$$x = ut + \tfrac{1}{2}at^2 \qquad \qquad \text{...(1)}$$

implies that both ut and $\tfrac{1}{2}at^2$ must indicate a certain number of *metres* if the equation is to make any sense. More generally, to step outside the International System and embrace all possible systems of units, we say ut and $\tfrac{1}{2}at^2$ must both have the *dimension* of length. This is the case of course; u is measured in $\mathrm{m\,s^{-1}}$ and therefore has dimensions (length/time), and a has dimensions (length/time^2), so that dimensionally the whole equation reads

$$\text{length} = \left(\frac{\text{length}}{\text{time}} \times \text{time}\right) + \left(\frac{\text{length}}{\text{time}^2} \times \text{time}^2\right)$$

or to use the approved symbolism,

$$[\mathrm{L}] = \left[\frac{\mathrm{L}}{\mathrm{T}}\right] \times [\mathrm{T}] + \left[\frac{\mathrm{L}}{\mathrm{T}^2}\right] \times [\mathrm{T}^2].$$

The preceding paragraph shows us what is meant by the word 'dimension' by implication; a formal definition is a little more difficult and we shall arrive at it by another example. The volume of a cube has dimensions $[\mathrm{L}^3]$, and we could express it as follows. A cube of side $1\,\mathrm{m}$ has a volume $1 \times 1 \times 1\,\mathrm{m}^3$; if one day we decided to introduce a 'new metre' exactly twice the length of the old one, then the volume of the cube in 'new cubic metres' would be $\tfrac{1}{2} \times \tfrac{1}{2} \times \tfrac{1}{2}$ or $(\tfrac{1}{2})^3$. In general if the units used to measure three base quantities P, Q and R are increased by factors a, b and c respectively and as a result the unit of some other, derived, quantity W is increased by a factor $a^\alpha b^\beta c^\gamma$, we say that W has the dimensions $[P]^\alpha$, $[Q]^\beta$ and $[R]^\gamma$. The base quantities used in this book are mass, length and time, but the complete list of SI base quantities includes current, temperature, luminous intensity and amount of substance.

Dimensional analysis

Going back to equation (1), we see that if the equation is to balance dimensionally then the term involving a *must* be of the form kat^2 where k is some pure number; we could not allow a term $\tfrac{1}{2}at^3$, for example. We can sometimes put this idea to good use; it can enable us to find the *form* of the equation linking a set of quantities without going through a full analysis or experimental investigation.

Example

Assuming the period T of small oscillations of a simple pendulum depends only on the mass m and weight mg of the bob and the length l of the pendulum, find the form of the equation which links these quantities.

What we need is a combination – or perhaps more than one – of m, mg and l which has dimensions $[\mathrm{T}]$; we hope to form this by choosing appropriate powers of the quantities concerned. Since our procedure is concerned only with the dimensions we can hardly expect to get the equation *numerically* correct, and we shall introduce a pure (dimensionless) number k to indicate our ignorance in this respect. We make a tentative start by writing

$$T = km^\alpha(mg)^\beta l^\gamma \qquad \qquad \text{...(2)}$$

where α, β, γ are the powers to which the mass, weight and length are raised in the formula. Of course we have given no reason why the

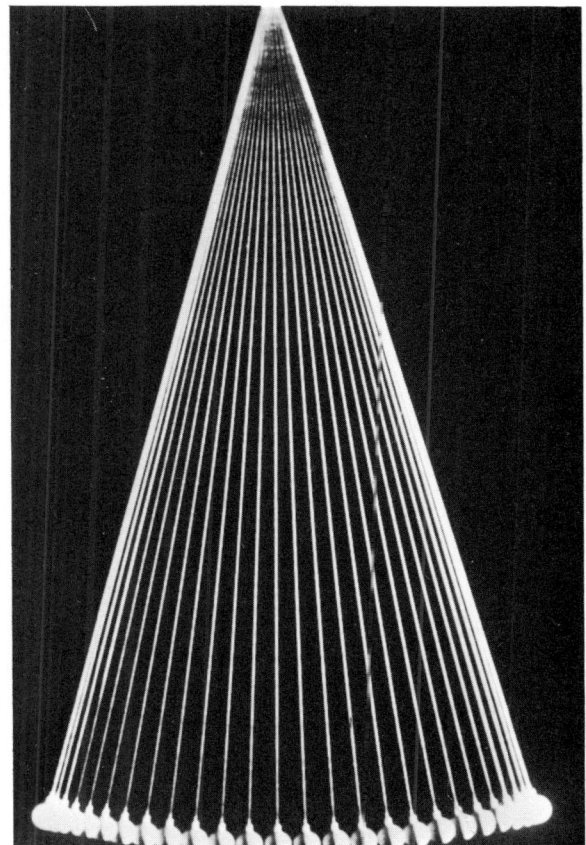

Pendulum illuminated by a stroboscopic lamp.

formula must be of this form; in fact we have no reason to suppose that there should be *any* simple mathematical relation between these quantities, but this is the sort of thing that we have come to expect. We now want to find α, β, and γ, and we start by writing down the dimensions; g is an acceleration with dimensions $[LT^{-2}]$, so we get

$$[T] = [M]^\alpha [MLT^{-2}]^\beta [L]^\gamma$$

We have to insist that the dimensions of the right-hand side must simplify and give us the dimensions on the left-hand side, and the problem is to choose α, β and γ accordingly. On the left-hand side for example there is no mention of any mass, whereas on the right-hand side we have $[M]^{(\alpha+\beta)}$; we can put this right if we insist that $0 = \alpha + \beta$. Balancing the other dimensions in the same way we get

$$\text{(Time)} \quad 1 = -2\beta$$
$$\text{(Length)} \quad 0 = \beta + \gamma$$

We now have three simultaneous equations.

Exercise 9.1

Solve the equations to find α, β and γ.

In this way we come to the conclusion that $T = k\sqrt{(l/g)}$ is the only form of equation (2) which is possible, given our initial assumptions. This *dimensional analysis* will not give any information about the size of k; in Chapter 15 we shall show that it is in fact equal to 2π (pp. 207 and 214).

Exercise 9.2

Assuming that the velocity v of a disturbance travelling along a stretched string depends on the tension T and *mass per unit length m* according to the formula

$$v = kT^\alpha m^\beta,$$

where k is a pure number, find α and β.

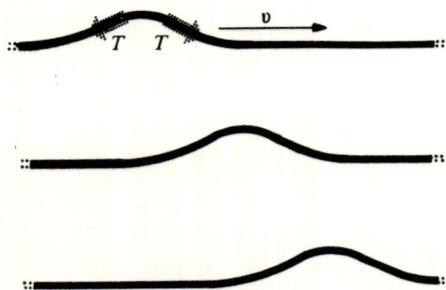

Disturbance travelling down a stretched string
(see Exercise 9.2).

Strengths and weaknesses of the method

Certainly dimensional analysis can sometimes be a very useful tool –in the right hands. In a few seconds it can give the form of an equation which may take considerable labour to derive by full analysis. It can be a useful aid to memory, but it can also perform a rather more vital role. Suppose we have an experiment involving measurement of two quantities a and b, and a quick look at the dimensions tells us that the quantity of real interest is a^4/b. We can then see immediately that any percentage error in a will be four times as important as a corresponding error in b, a very useful piece of information if we want to set up a system with maximum accuracy.

On the other hand, the method has its limitations. Leaving aside the obvious drawback that we get no information about numerical factors, there is the very real difficulty of deciding just which quantities should be considered in the analysis and which should be left out. Taking exercise (2) for example, although we may not find it hard to guess that the mass of some part of the string must somewhere appear in the formula, since presumably the inertia of bits of the string will influence how quickly one displaced section disturbs the next, we could well have tried the *density* of the string in place of the mass per unit length. It needs experience, not to say hindsight, to make the appropriate start.

In addition, there are some situations involving two different quantities with the same dimensions. Consider for example the case of a rigid pendulum–suppose we have a uniform bar of length l and mass m, having a hole a distance h from its centre by means of which we can pivot it and let it perform small oscillations. We might try dimensional analysis to find the period T; we could for example write

$$T = km^{\alpha}(mg)^{\beta}l^{\gamma}h^{\delta}. \qquad \ldots(3)$$

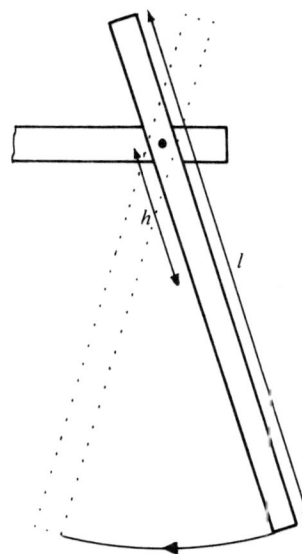

Rigid pendulum.

Exercise 9.3

Show that dimensional analysis gives
$$\alpha = \tfrac{1}{2}, \quad \beta = -\tfrac{1}{2} \quad \text{and} \quad (\gamma + \delta) = \tfrac{1}{2}.$$

This does not tie down the formula of course; many possibilities exist, like $T = k(lh)^{1/4}g^{-1/2}$ or $T = kl(gh)^{-1/2}$.

In fact it turns out that equation (3) does not even have the correct *form*; the right answer is

$$T = 2\pi\sqrt{\left(\frac{h^2 + \tfrac{1}{12}l^2}{gh}\right)}.$$

(See Chapter 19, p. 269.)

Nevertheless dimensional analysis can be very useful, and we shall make use of it several times in the chapters which follow. Further examples appear on pp. 227, 233 and 234.

Further exercises

Exercises 9.4–10

Some of the questions which follow contain references to quantities introduced in later chapters.

4 Imagine that you cannot recall whether (a) the acceleration of a body in uniform circular motion is v^2r, ω^2r, ωr^2, or v^2/r, and (b) the kinetic energy of a body in linear motion is $\tfrac{1}{2}m^2v$, $\tfrac{1}{2}(mv)^2$, $\tfrac{1}{2}mv$, or $\tfrac{1}{2}mv^2$, the symbols having the usual meaning. Use the dimensions of these quantities to identify the correct expression or expressions in each case. (WJEC)

5 The velocity v of waves of wavelength λ on the surface of a pool of liquid, whose surface tension and density are σ and ρ respectively, is given by

$$v^2 = \frac{\lambda g}{2\pi} + \frac{2\pi\sigma}{\lambda\rho},$$

where g is the acceleration due to gravity.

Show that the equation is dimensionally correct.

A vibrator of frequency $480 \pm 1\,\text{Hz}$ produces on the surface of water waves whose wavelength is $0{\cdot}125 \pm 0{\cdot}001\,\text{cm}$. Assuming that for this wavelength the first term on the right-hand side of the equation is negligible, calculate the value which these results give for the surface tension of water.

Discuss whether the assumption is justified. (O & C)

6 A laboratory is equipped with a stop clock which is known to have a rate error no larger than $0{\cdot}01\%$, and a ruler which is

known to have an error no greater than 0·05%. What is the smallest uncertainty that could be achieved in a determination of the acceleration due to gravity if the clock and the ruler were used? (The dimensional expression for acceleration is $[LT^{-2}]$.)

(o & c)

7 The dimensions of *energy*, and also those of *moment of a force*, are found to be 1 in *mass*, 2 in *length* and -2 in *time*. Explain and justify this statement.

A sphere of radius a moving through a fluid of density ρ with *high* velocity V experiences a retarding force F given by

$$F = ka^x\rho^yV^z,$$

where k is a non-dimensional coefficient. Use the method of dimensions to find the values of x, y and z. (o & c)

8 Explain the meaning of *dimensions of a physical quantity*.

Given that the velocity of propagation, v, of longitudinal waves in a thin rod depends only on the Young modulus, E, of the material and its density, ρ, use the method of dimensions to derive a relationship between v, E and ρ. (o & c)

9 Explain what is meant by the *dimensions* of a physical quantity.

The period of vibration T of a tuning fork may be expected to depend on the density ρ, and the Young modulus, E, of the material of which it is made and the length l of its prongs. Which of the following equations could represent the relation between T and the other quantities?

(a) $T = \dfrac{A\rho}{E}\sqrt{(gl^3)}$ (b) $T = Al\sqrt{(\rho/E)}$ (c) $T = \dfrac{AE}{\rho}\sqrt{\left(\dfrac{l}{g}\right)}$.

(A is a dimensionless constant and g is the acceleration due to gravity.)

The figures in Table 9.1 were obtained for a set of geometrically similar steel tuning forks:

Table 9.1

Frequency/Hz	256	288	320	384	480
Length of prong/cm	12·0	10·6	9·6	8·0	6·4

Use these figures to confirm (or otherwise) your choice of equations. (c)

10 A small drop of liquid, freely suspended in air, can vibrate about its spherical form. Assuming that the period of oscillation T depends only upon the radius of the drop a, the surface tension γ, and the density of the liquid ρ, obtain an expression for T.

(WJEC)

10 Forces in equilibrium

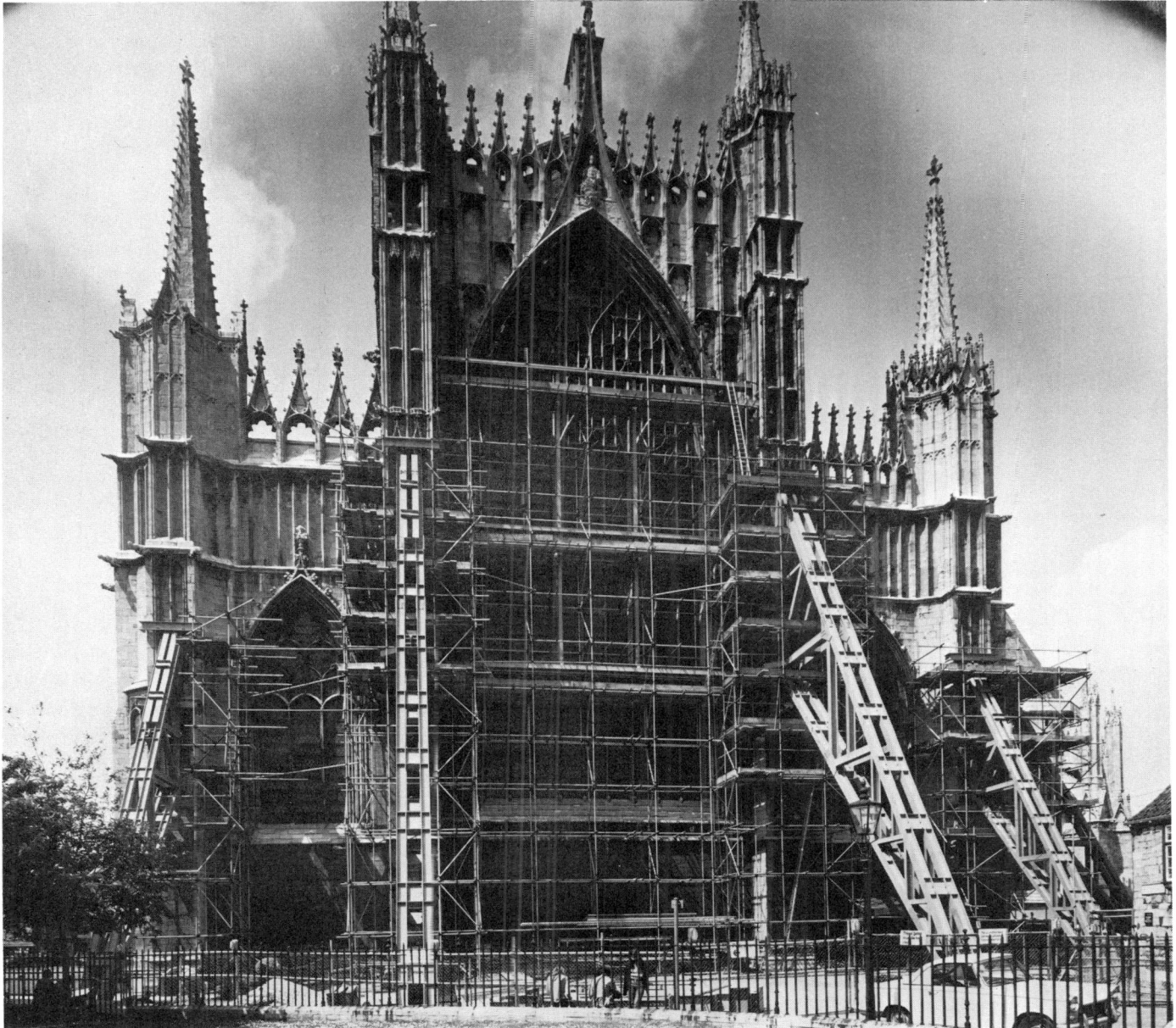

Maintaining equilibrium at York Minster. (Photograph courtesy of Shepherd Construction Co. Ltd.)

In Chapter 6 we looked at ways of working out the net effect of a number of forces acting together on an object; now we are going to consider the special situation when all these forces combine in such a way that the object, which we shall take to be *stationary*, has no tendency to move: it is in equilibrium. This means there must be no net force producing translational acceleration and it also means there must be no net couple producing rotational acceleration. The forces may still have a tendency to distort the object; we shall look into that possibility in Chapter 11.

Equilibrium and stability

If we succeeded in balancing a brick on one edge on a table we could argue that it would be in equilibrium, but of course it would not be likely to stay there very long; the least disturbance would upset it. There is more to a stationary body than a fine balance of forces; we have to look into the question of *stability*. Figure 10.1 shows very simple examples of three types of equilibrium. In each case there are

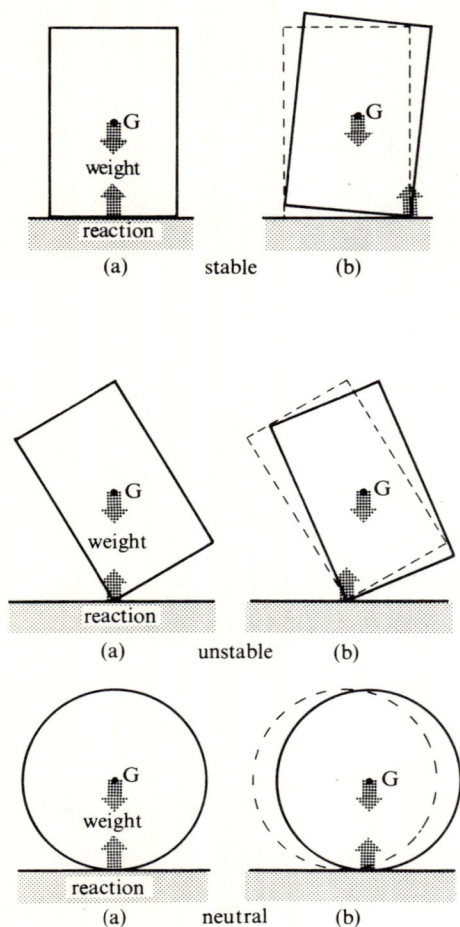

(a)   stable   (b)

(a)   unstable   (b)

(a)   neutral   (b)

Figure 10.1

Three types of equilibrium.

weight

reaction

table

extra
external force

weight

adjusted
reaction

table

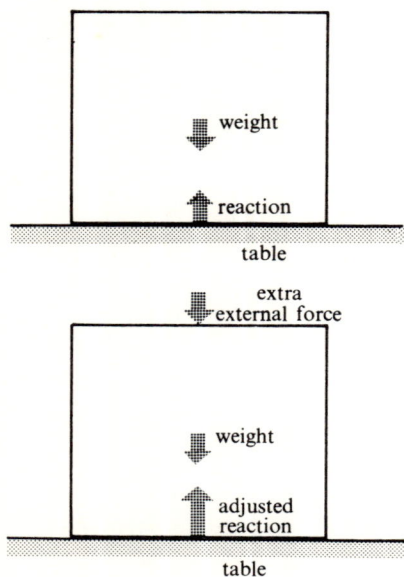

The reaction provided by the table is a self-adjusting force.

two forces involved, the weight and the reaction from the table. The brick resting on one face is not affected by small disturbances; it is stable. The brick balanced on its edge is highly unstable, and a very slight vibration will cause it to topple and take up a stable position. The cylinder represents an intermediate case; it is susceptible to small disturbances in the sense that it has no tendency to return to the original position, but on the other hand there is no avalanche effect; it is perfectly happy in some different position. We call this a case of neutral equilibrium.

Figure 10.1b shows one way in which we can distinguish the three cases; each object is shown slightly displaced. For the stable case the force system now consists of a couple acting so as to oppose the change and restore the cube to its original position. There is a couple in the unstable case too, but it is a couple which is encouraging the displacement. Displacement of the cylinder produces no couple; the forces are still in equilibrium.

Self-adjusting forces

The case of most practical significance and interest is the one we take for granted, the stable case. Most of the large-scale objects around us are more or less stationary with respect to their surroundings and almost all of them are stable. The implications are that the forces must all balance exactly, and they must respond to any small external disturbance in such a way as to maintain the *status quo*. The key to such stable conditions is to be found in the existence of some very common *self-adjusting* forces, which change size immediately in response to a disturbance. We can perhaps imagine what life would be like without them by thinking about the problem of keeping a rocket stationary in mid-air a few metres above the ground by suitable control of the engine thrust. We should need a number of small control rockets, acting in different directions, with thrust which could be varied in response to accelerometers and gyroscopes registering the first hint of change in attitude or height. Yet such an exact balance of all the forces on an object is achieved without any thought or labour on our part whenever we put a book down on a table. The reaction which the book experiences is a self-adjusting force; the table provides no more and no less than that which is exactly necessary to balance the weight, and if we press down on the book the reaction automatically increases to compensate.

If we push horizontally against the book another self-adjusting force comes into play–friction–and it also increases to keep pace exactly as we push harder, until an upper limit is reached beyond which there is not enough friction to prevent motion.

Microscopically it is the very rapid variation of interatomic forces with atomic separation that is responsible for the way objects can adapt to new external forces so easily. As we mentioned on p. 2, the net force between atoms changes from an attraction to a very strong repulsion over a small range of separations; a collection of billiard balls linked by springs makes a useful macroscopic analogy for our present purpose. Just as in such an arrangement the springs would contract or extend in response to any external attempt to change the overall shape, so in the case of the book and table the average interatomic separation adjusts to any external force, by an amount just sufficient to alter the reaction between the surfaces and restore equilibrium. As we press the book into the table we can imagine the average separation between the topmost layer of atoms in the table and the adjacent atoms in the book decreasing by a tiny amount, thereby providing the extra reaction needed to keep the book and table from passing through each other.

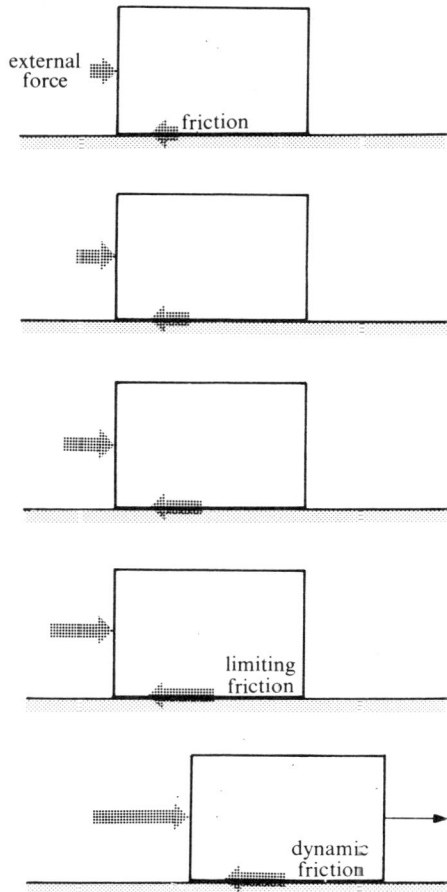

Friction is a self-adjusting force, up to a definite limit.

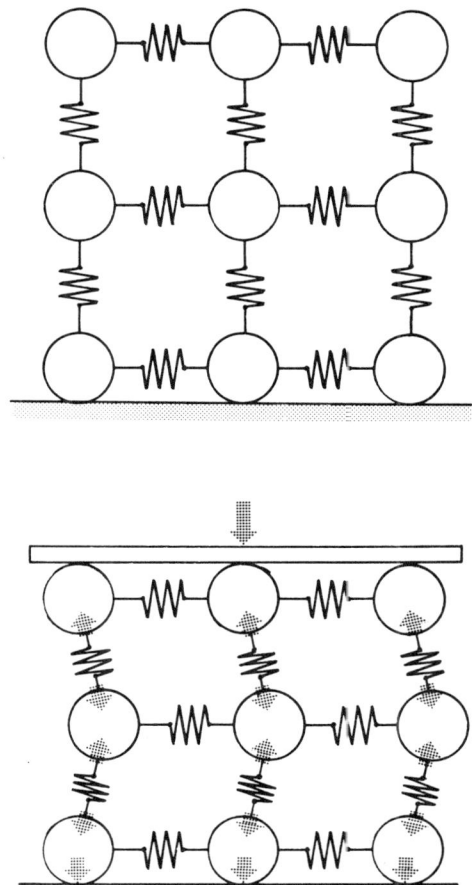

Simple model of the way in which forces between atoms adjust to new external forces.

Microscopic description of friction

Two surfaces in contact may look quite smooth, but microscopically they will be very uneven indeed, with hills, valleys, projections and ridges hundreds or even thousands of atoms deep. This is one reason why two lumps of metal do not normally stick together and form a single piece; the attractive forces are available, certainly, but only a tiny fraction of the surface atoms are sufficiently close to their opposite numbers to provide a worthwhile attractive force, and this is not enough to bind the two surfaces together. Measurements of contact resistance between metal surfaces suggest that the actual contact area may be as little as 1/10000 of the apparent area. Surface impurity is another important factor; a coating of oxide or layer of adhering molecules from the air will reduce the overall attractive force which can be brought into play to counteract any attempt to separate the surfaces. The methods employed to achieve a hot weld are designed to overcome both these obstacles. High temperature in the region of the required join produces local melting; the hills and valleys are filled in and impurities flow to one side. Two *very* smooth clean metal surfaces can adhere without a weld, though the attraction between them is not very great.

When one surface slides on another, we can imagine the projections getting in each other's way; it must be rather like rubbing two sheets of sandpaper together. Some of the friction between two metal surfaces is caused by the fact that the projections in one surface bite into the other, but there is another important effect which occurs as follows. As two metal surfaces are brought together the highest projections collapse, particularly in the surface of the softer metal, due to the very high pressures at these minute points of contact. This will continue until there is sufficient area of contact to provide the required normal reaction. Mechanical work is done in the process, and the extra internal energy made available ('heat'–see p. 100) is shared by relatively few atoms near the points of contact. Conduction of heat to the rest of the metal occurs very quickly, but temperatures as high as 1000°C, lasting for perhaps 10^{-4}s, have been recorded at the 'hot spots' by suitable thermocouples. Such temperatures are often sufficient to cause local melting, and the softer metal is welded to the harder metal at the points of contact. If we want to move one surface across the other we have to tear these welds apart, or *shear* them (see p. 141), and if we attempt to do this with insufficient force the interatomic separations in the welded regions change just enough to provide a self-adjusting force balancing the external force exactly.

The frictional force differs from the normal reaction in one important respect; it cannot keep up with the external force beyond a relatively small limit. Once there is sufficient external force to tear the welds apart, sliding commences. The size of this limiting force depends on the number and size of the welds (that is, on the *real* contact area) and on the shear strength of the metal. Hard metals do not yield a great deal when placed in contact, and they therefore produce relatively few, small welds, but these put up just as much resistance to shear as the larger and more numerous welds formed on the surface of a soft metal. Measurements of frictional forces for various surfaces show that there are no great differences between hard and soft metals, though friction can be reduced between two hard metals by using a very thin film of soft metal as a lubricant.

Exercise 10.1

A 1 kg lump of metal with a flat base 5 cm × 5 cm is placed on a metal surface. Assume that the projections in the surfaces collapse by about 10^{-6} mm (about five atomic diameters) in order to make order of magnitude estimates of the following.

a) The mechanical work done by the gravitational force during the 'sinking in' of the block

b) The number of atoms in a horizontal layer of the block one atom thick

c) The number of atoms in contact with the bottom surface if the real contact area is 1/10000 of the apparent contact area

d) The mass of the number of atoms given by (c) if the block contains 10^{25} atoms

e) The temperature rise which an amount of energy (a) would produce in a mass (d) of block if 400 J are required to raise the temperature of the block through 1 K.

The laws of friction

The most convenient macroscopic description of friction is in terms of a pair of equal and opposite forces; each acts on one of the two surfaces concerned along a line in their plane of contact. We can make one or two experimentally based generalisations about the magnitude of these forces, some exact and some very approximate; they are usually known as the laws of friction. In what follows we shall refer to one of the frictional forces, with reference to its action on one of the surfaces concerned.

1 *Friction always acts in a direction such that it opposes any tendency towards relative motion of the surfaces; the size of the frictional force is self-adjusting, and is just sufficient to prevent motion, up to a certain maximum value known as limiting friction.* This is an exact statement as long as we do not take it to mean that limiting friction has a reliable value in a given case; it is approximately the same in similar circumstances but is highly dependent on the precise condition of the surfaces, which can vary from one instant to the next.

2 *The value of limiting friction F_L is proportional to the normal reaction R between the surfaces.* Here we use the word 'normal' to emphasise that we are talking about the repulsive force *perpendicular* to the surfaces at the point of contact. We have to make this clear because it is common practice to combine the normal reaction with the frictional force to form what is called the *total reaction* between the surfaces.

In symbols,

$$F_L \propto R$$

and so we write

$$F_L = \mu R$$

where μ is called the coefficient of static friction; it depends on the nature and condition of the surfaces in contact.

3 *The value of limiting friction is independent of the area of contact of the surfaces.*

Statements 2 and 3 are at best approximations, and it is unrealistic to make more than a rough estimate of μ in any given situation.

Exercise 10.2

Figure 10.2 shows a simple arrangement for finding the coefficient of static friction between two surfaces. If the weight of the block is 14 N and the spring balance reads 3·5 N just before the block starts to move, write down (a) the normal reaction, (b) the frictional force, (c) the coefficient of friction, (d) the new force needed to move the block if a 2 kg mass is placed on it, assuming that the laws of friction are obeyed.

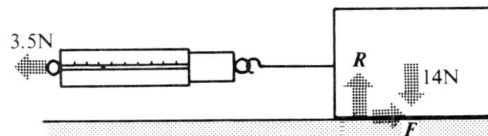

Figure 10.2 (Exercise 10.2)

We can find a microscopic interpretation of the laws of friction in terms of our notion of tiny surface projections. As the surfaces are brought into contact the more prominent projections will give way, until sufficient points of contact exist to support the upper surface. Whether these points of contact are spread over a large or small macroscopic surface is immaterial: hence the unimportance of apparent contact area. On the other hand, the greater the normal reaction required, the more points of contact are needed, and each

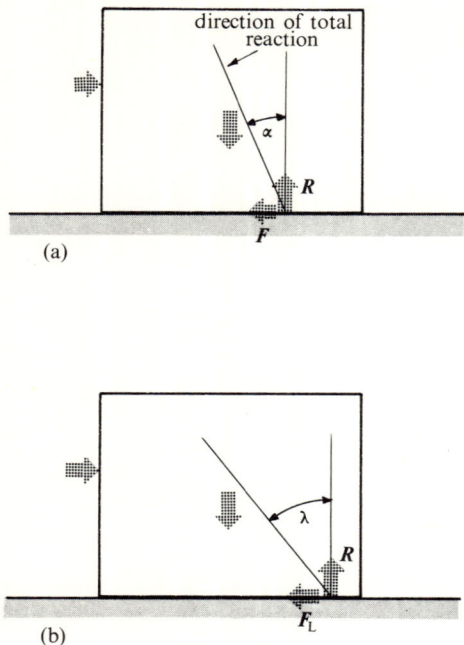

(a)

(b)

Figure 10.3

Direction of total reaction: (a) general case, when $F < F_L$, (b) limiting case when $F = F_L$.

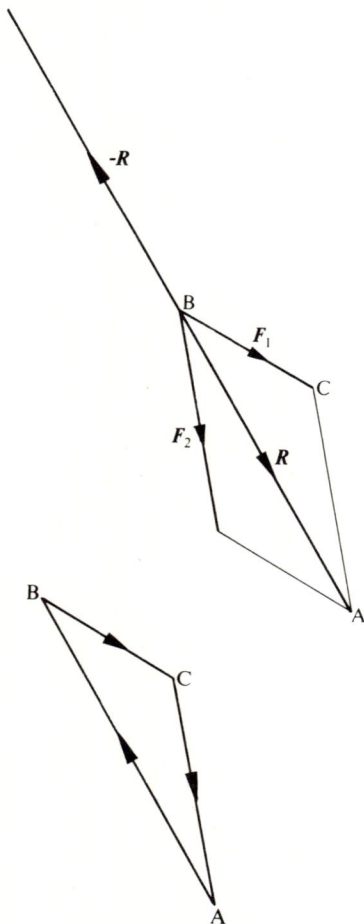

Figure 10.4

Triangle of forces.

of these contribute to the frictional force when any attempt is made to slide one surface over the other.

Friction does not disappear when sliding commences of course; we can add some laws of *dynamic* friction between moving surfaces as follows:

4 *Dynamic friction acts in such a direction that it opposes the relative motion of the surfaces, and has a size F_D which is proportional to the normal reaction.* We can use this statement to define a coefficient of dynamic friction μ_D analogous to that of static friction.

$$F_D = \mu_D R$$

5 *The frictional force F_D is independent of the area of contact and is less than the value of limiting friction.* Very roughly speaking it is also independent of the relative speed of the surfaces.

Angle of friction

We have already mentioned that it is sometimes convenient, especially in the static situation, to combine the normal reaction R and frictional force F, both arising from atomic repulsions and attractions, into a single force called the *total reaction*. The total reaction acts at an angle α to the normal given by

$$\tan \alpha = \frac{F}{R}$$

(see Figure 10.3). If the size of the frictional force happens to be the limiting value F_L (and it is a trap for the unwary to assume this too hastily) then the corresponding value of α is called the *angle of friction* and is usually given the symbol λ. Since in this case

$$F = F_L = \mu R,$$

we see that

$$\tan \lambda = \mu,$$

though it needs to be stressed that $\tan \alpha$ is *not* equal to μ in general; μ is merely the maximum value it can take. The value of $\tan \lambda$, that is, of μ, can exceed unity in some circumstances, particularly when surface impurities are removed.

Necessary and sufficient conditions for equilibrium of a system of forces

Now that we have looked at the character of some of the forces which are so vital in maintaining many common equilibrium situations, we want to tackle the problem of the precise conditions needed for equilibrium. We know the general requirements: no net force, and no net couple. What we want now is a set of rules to establish whether or not any given combination of forces meets these requirements. In Chapter 6 we looked at two methods by which we could reduce a system of forces to a simpler form, and in fact we could use either method to test whether or not we have an equilibrium situation. The first method involves combination of the forces stage by stage by the parallelogram rule until all the forces are included (p. 70); if we end up with no net force and no couple then we have equilibrium. This method is quite useful when only three forces are involved: in fact it is the basis of the *triangle of forces theorem*. This states that *if three forces are in equilibrium, they can be represented in size and direction by the sides of a triangle taken in order*. Figure 10.4 shows how the theorem follows immediately from the parallelogram law; if R is the resultant of forces F_1 and F_2 then $-R$ is the force which will *balance* F_1 and F_2, and $-R$, F_1 and F_2 are adequately represented in size and direction (though not in line of action) by the sides AB, BC, CA respectively. If three forces are to be in equilibrium they must obey the triangle of forces theorem, and that means they must all lie in the same plane. It is also necessary that they meet in a point.

If we have more than three forces to deal with, stage-by-stage application of the parallelogram law can rapidly become cumbersome and we shall concentrate on the alternative technique developed in Chapter 6, relying on the processes of resolution and taking moments. Now there is a difference between the set of conditions which always apply in any equilibrium situation, and the set of conditions we need to test before we can be sure that equilibrium in fact exists. We call the first set the *necessary* conditions and the second set the *sufficient* conditions. It is necessary for equilibrium that the sum of the resolved components of the forces in any chosen direction is zero, otherwise there is a net force in that direction. It is also necessary that the sum of the moments about any axis perpendicular to the plane of the forces must be zero. Fortunately, to satisfy ourselves that equilibrium exists we do not have to resolve in every conceivable direction, nor do we have to take moments about every possible axis. We can work out a set of sufficient conditions as follows; we shall deal only with cases where the forces all lie in one plane. First, we resolve in some chosen direction, let us say parallel

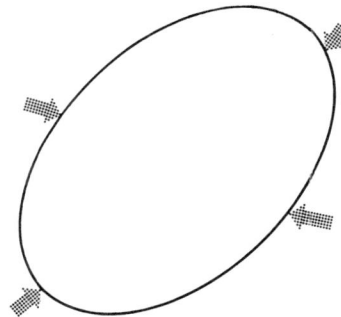

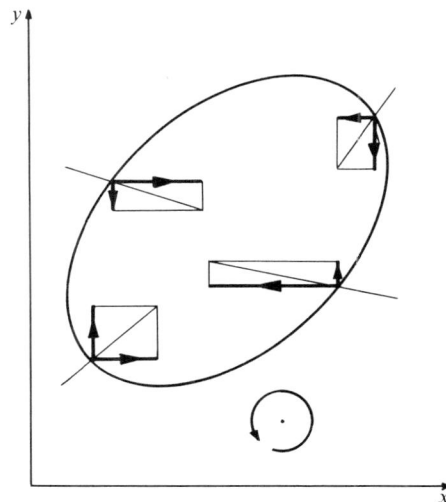

Testing for equilibrium of a system of coplanar forces. One set of sufficient conditions: sum of x-components = 0, sum of y-components = 0, sum of moments about any axis perpendicular to the plane = 0.

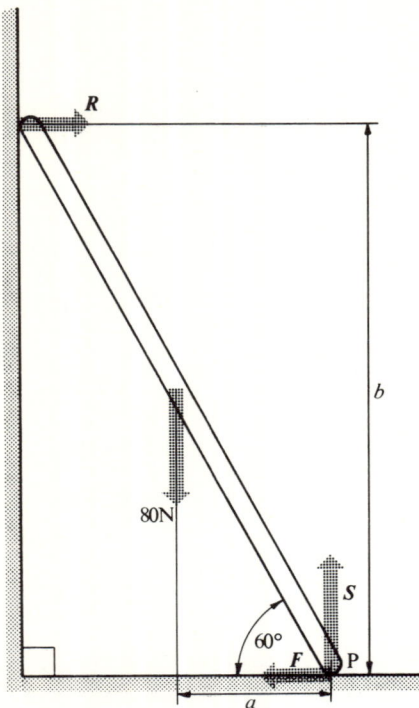

Figure 10.5

to the x-axis. If the sum of the resolved components is zero, well and good–there is no net force in the x-direction. In fact, the only direction in which there may still be a net force is perpendicular to the one we have chosen. Any other net force would have a component in the x-direction and would have shown up already. The next step therefore is to resolve in the y-direction, and if that gives us a zero result too we can be sure there is no net force at all. There could still be a couple, but a couple has the same moment about *any* axis perpendicular to its plane, so if we take moments about just one such axis–any one–and get zero net moment, we can be sure there is no net couple either.

Usually what happens is that we know an object is in equilibrium but we do not know the size and direction of all the forces acting; this is particularly true of normal reactions and frictional forces. In such cases, resolving and taking moments can give us three different pieces of information, three independent equations which we can use to find up to three unknown quantities.

Example

A uniform ladder of weight 80 N rests against a wall so as to make an angle of 60° with the horizontal. The wall exerts negligible frictional force on the ladder. Find the normal reactions R and S at the wall and ground, and the frictional force F at the ground.

Four forces act on the ladder (Figure 10.5) and clearly the most convenient directions in which to resolve are horizontal and vertical.

Resolving horizontally,

$$R - F = 0 \qquad \ldots(1)$$

Resolving vertically,

$$S - 80\,\text{N} = 0 \qquad \ldots(2)$$

This type of problem is greatly simplified if we take moments about the most convenient axis. It is useful to remember that no force which passes through the chosen axis will appear in the moment equation, and it is usually a good idea to have as few unknown forces as possible in one equation. So we take moments about an axis through P, the base of the ladder, which we take to be of length l. Referring to Figure 10.5, we have

$$Rb - a \times 80\,\text{N} = 0.$$

That is,

$$Rl\sin 60° - (l/2)\cos 60° \times 80\,\text{N} = 0$$
$$R = (40/\sqrt{3})\,\text{N}.$$

Equations (1) and (2) then give

$$F = (40/\sqrt{3})\,\text{N} \qquad \text{and} \qquad S = 80\text{N}.$$

Exercise 10.3

Write down the minimum coefficient of friction between ladder and ground which will maintain equilibrium.

We can deal with up to three unknown quantities, but no more. A casual glance might lead us to think otherwise; there seems to be no limit to the number of equations we can write down. Why not resolve along the direction of the ladder and get another equation–surely four equations will deal with four unknowns? Well, they will not, because the four equations are not all *independent*–our fourth equation will simply be a heavily disguised combination of two of the others. The fact is that once we have resolved in two directions we have wrung out *all* the information we can get from the resolving process.

Exercise 10.4

Write down the equation obtained from resolving along the direction of the ladder and show that it consists of a combination of equation (1) multiplied by $\frac{1}{2}$ and equation (2) multiplied by $\sqrt{3}/2$.

An alternative set of sufficient conditions

The system we have been describing is not unique; we can devise other sets of conditions sufficient to ensure equilibrium. Suppose for example we start by taking moments about some axis perpendicular to the plane of the forces, and make sure the sum is zero. Already that excludes the possibility of a couple; all we can allow now is a single net force, and at that it must happen to pass through the axis we have chosen. So, we take moments about another axis; that must give zero result also. But the net force may still be eluding us; it could conceivably have a line of action passing through *both* the axes we have picked. Finally, then, we take moments about a *third* axis, making sure it is not in line with the other two (Figure 10.6). This completes another set of sufficient conditions.

Sliding and toppling

The block shown in Figure 10.7, a cube of side a having weight W, is resting on a horizontal surface but experiences a steadily increasing horizontal force P applied to an upper edge. Equilibrium cannot be maintained indefinitely; when it breaks down, will the cube slide along the table or will it fall over?

To answer this question we have to realise that the resultant of all the vertical repulsions between the atoms in the table and those in the block, that is, the normal reaction R, does not act through the middle of the block. There is a tendency for edge A to be pushed into the table, and this will move the line of action of R further and further towards A as P increases. When it is a distance b from X, the mid-point of the bottom face of the cube, we get the following set of equations, assuming the block is still in equilibrium.

Resolving horizontally,

$$F - P = 0 \qquad \text{...(3)}$$

and vertically,

$$R - W = 0 \qquad \text{...(4)}$$

and taking moments about X

$$Rb - Pa = 0 \qquad \text{...(5)}$$

Looking first at equations (3) and (4), we see that for all the different equilibrium situations which occur as P increases, R is always equal to W but F must increase to keep pace with P. Of course F cannot become greater than the limiting value μR.

Exercise 10.5

Show that this means P must not become greater than μW.

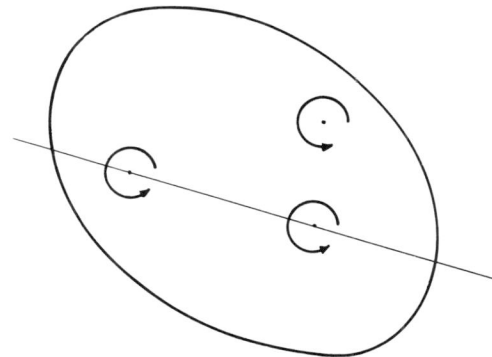

Figure 10.6

An alternative set of sufficient conditions for equilibrium of coplanar forces.

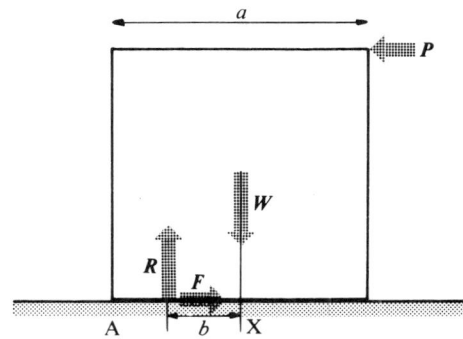

Figure 10.7

But equation (5) gives us another constraint; it is not physically possible for b to be greater than $\frac{1}{2}a$; we cannot expect the atomic repulsions to combine to form a force acting outside the base of the block. Once R has been driven to the edge, any further increase in P will mean that the moment equation can no longer be balanced – in other words the block will fall over.

Exercise 10.6

Show as a result that P must not be greater than $\frac{1}{2}W$.

Now the question is, which will happen, sliding or toppling? If P reaches a value μW before it reaches the value $\frac{1}{2}W$, the block will slide, otherwise it will topple. We say it will slide or topple according to whether μ is greater or less than $\frac{1}{2}$. Another example follows.

Example

Figure 10.8a shows a rectangular block of weight W resting on a plane inclined at an angle α to the horizontal; α is gradually increasing. The coefficient of friction between block and plane is μ. Find the minimum value of μ which will ensure toppling rather than sliding.

We have drawn the normal reaction R a distance s from X, the mid-point of the base. Because this problem involves only three forces we can be sure that they all meet in a point if equilibrium is to be maintained, so R has been drawn through the point of intersection of W and the frictional force F.

Resolving along the plane, we have

$$F - W\sin\alpha = 0$$

and perpendicular to the plane,

$$R - W\cos\alpha = 0$$

Dividing these gives us

$$F = R\tan\alpha$$

and as F cannot be greater than μR this means that

$$\tan\alpha_{max} = \mu$$

if sliding is to be prevented.

We can find the toppling criterion by allowing R to go right to the edge of the cube. Normally we would have to take moments to extract the information we need, but because the moment equation deals with the lines of action of the forces and we have already realised that they meet in a point, we have no real need to go to the trouble of taking moments in this case. Instead, referring to Figure 10.8b, we can see at once that if toppling is to be avoided

$$\tan\alpha_{max} = a/b.$$

So if the block is to fall over rather than slide, we must have $\mu > a/b$.

Exercise 10.7

Examine the conditions for sliding and toppling of the block in the last example if it is on a horizontal surface and acted on by a steadily

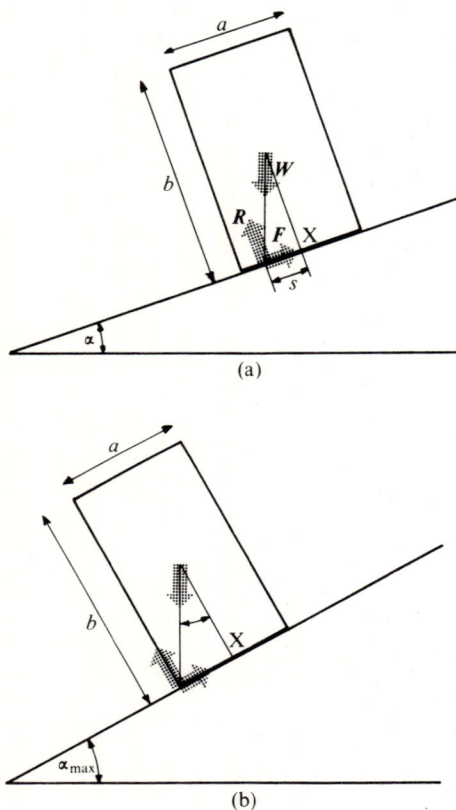

(a)

(b)

Figure 10.8

increasing force **P** applied to an upper edge, where
a) **P** is horizontal
b) **P** is pointing up at an angle of 30° to the horizontal.

Tensions and compressions, strings and struts

Tensions and compressions are two further types of self-adjusting force systems very useful in achieving and maintaining equilibrium. If we say that a string or strut has a tension T at a particular point, we mean that if it is broken at that point, T is the magnitude of the force which must be applied to each section to hold the two parts together. Referring to Figure 10.9, the tension T in the string very near the bottom end is equal to mg, since it is sufficient to support the weight. If we cut the string near the top we shall have to supply a force a little greater than mg to hold the ends together, on account of the weight of the string itself, but if this can be neglected we can safely say that *the tension in a light string or strut is the same at all points along its length* and equal to the force which each end exerts on the object to which it is attached. Similar considerations apply to struts under compression (Figure 10.10).

Equilibrium of composite bodies

If we have to deal with a composite body, such as the step-ladder in the example which follows, we have extra decisions to make about the way we go ahead with our processes of resolution and taking moments. Should we consider the body as a whole, or should we deal with the equilibrium of one part in isolation? Often the answer depends on whether we are interested in the forces which the different parts of the body exert on each other. Newton's third law tells us that whatever force part A exerts on part B, part B exerts an equal and opposite force on part A. If we look at the body as a whole we shall not see such 'third law pairs' appearing at all in our equations—they will cancel each other exactly, both when we resolve and when we take moments. On the other hand, if we look at the equilibrium of part A alone, we are bound to involve the force which part B exerts on it; without this force part A would not be in equilibrium.

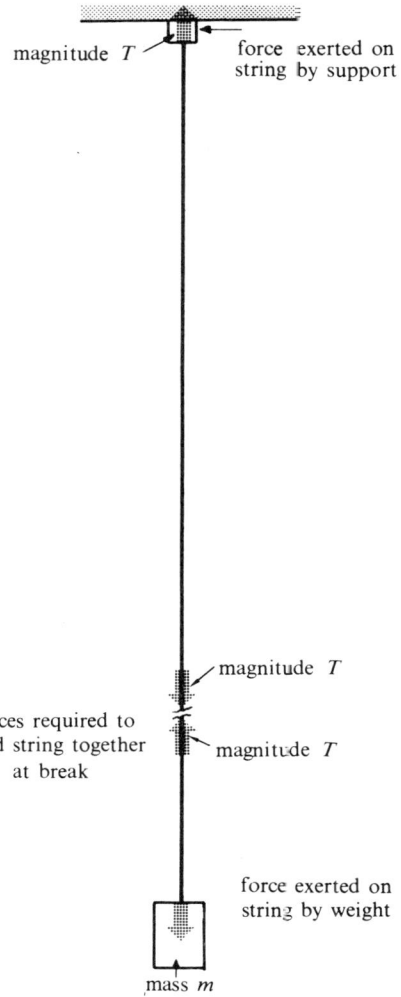

Looking down the suspension cable, Forth Road Bridge.

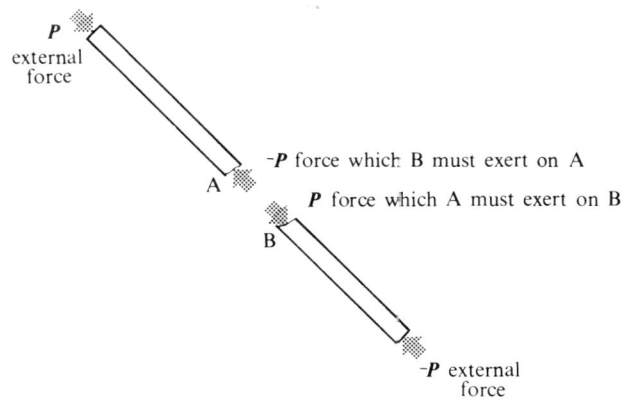

magnitude T — force exerted on string by support

magnitude T

forces required to hold string together at break — magnitude T

force exerted on string by weight

mass m

Figure 10.9

P external force

$-P$ force which B must exert on A

A

P force which A must exert on B

B

$-P$ external force

Figure 10.10
Compression in a light strut.

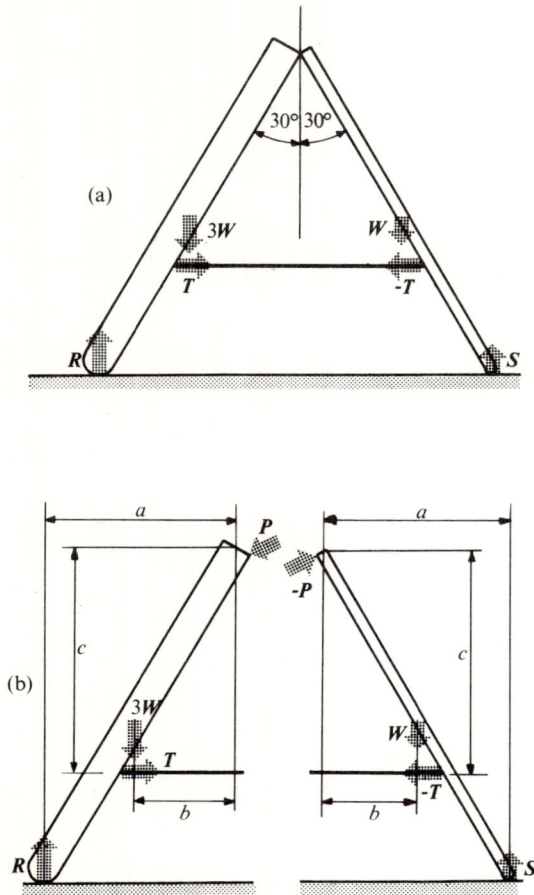

(a)

(b)

Figure 10.11

Example

Figure 10.11a shows a step-ladder with two sections hinged at the top, each making an angle of 30° with the vertical. One section has weight $3W$, the other W; both weights act through the centres of their respective sections. The sections are joined by a light horizontal string fixed one third of the way up each section, and the ladder rests on a floor providing negligible friction. Find the tension T in the string, assuming that the influence exerted by each section on the other at the hinge can be represented by a single force.

If we simply look at the ladder as a whole we shall get no information about T, nor about the pair of forces involved at the hinge. As far as the ladder is concerned these are internal forces, 'third law pairs'. The equation we get by resolving vertically for the whole ladder,

$$R + S - 4W = 0 \qquad \qquad \text{...(6)}$$

will be very useful, but in order to get T into the picture we must look at each section of the ladder in isolation (Figure 10.11b). Of course this brings in another unknown quantity, the force P at the hinge; in fact it brings in two unknown quantities because we do not know the size or the direction of P. However, by picking our way carefully we can keep these out of our equations; we take moments about the hinge for each section in turn. This gives us

$$Ra - 3Wb - Tc = 0$$

and

$$Wb + Tc - Sa = 0,$$

where a, b and c are indicated in the figure. All three can be written in terms of the length l of either section, so we get

$$Rl\sin 30° - \frac{3Wl}{2}\sin 30° - \frac{2Tl}{3}\cos 30° = 0$$

and

$$\frac{Wl}{2}\sin 30° + \frac{2Tl}{3}\cos 30° - Sl\sin 30° = 0,$$

which simplify to

$$R = \frac{3W}{2} + \frac{2T}{\sqrt{3}}$$

and

$$S = \frac{W}{2} + \frac{2T}{\sqrt{3}}.$$

Using these values in equation (6) we arrive at

$$\frac{3W}{2} + \frac{2T}{\sqrt{3}} + \frac{W}{2} + \frac{2T}{\sqrt{3}} = 4W,$$

that is

$$T = \frac{\sqrt{3}W}{2}.$$

Exercises 10.8, 9

8 Find the horizontal and vertical components of the force P which the lighter section exerts on the heavier one at the hinge and

hence show that the line of action of P is inclined at 30° to the horizontal.

9 What would be the tension in the string if the two sections of the ladder were identical and each of weight W?

Stability; the energy criterion

We now go back to the three types of equilibrium indicated in Figure 10.1 and think about them from a different standpoint; in each case we shall look at the potential energy change involved in a small displacement from equilibrium.

A glance at Figure 10.12 shows that if we displace a stable object from its equilibrium position the centre of gravity moves up, which means that the potential energy increases. On the other hand, displacement from a position of unstable equilibrium involves a decrease of potential energy; the centre of gravity moves down. In the neutral case the centre of gravity does not change height and the p.e. remains the same. We have here examples of quite general principles, namely that *an object in stable equilibrium is in a position of minimum potential energy, and an object in unstable equilibrium is in a position of maximum potential energy*. If the potential energy decreases for just *one* type of displacement the body is unstable; a cube balanced on one edge will not topple over a corner, but it will nevertheless topple.

These principles apply to cases involving all forms of potential energy; they are not restricted to changes in gravitational p.e. alone. It is a matter of common experience that all objects tend to decrease their potential energy whenever they are free to do so. There is always a possibility that dishes on a shelf will find their way to the floor, stretched springs which possess elastic potential energy will return to normal length if given a chance, precipices are inherently dangerous. More subtle examples occur when two different forms of potential energy, with interdependent values, occur together; we shall see in Chapter 12 that the shape of a drop of liquid on a polished surface is consistent with minimal potential energy. The connections between stability and potential energy can be understood if we remember that an object possessing p.e. can *do work*; this means there is a force waiting to do the work, and it is this force which accelerates the body if it is allowed to do so. The work is done at the expense of the p.e., which continues to decrease until the force has reduced to zero and can therefore do no more work. The p.e. has then reached a value as low as it can get. (The work done is not necessarily *useful*–it usually leads to an increase in kinetic energy, and frictional forces eventually convert this into extra internal energy, with its associated temperature rise.)

We can derive a useful formula which summarises this idea. Suppose we have an object, not in equilibrium, but possessing potential energy E, and the net force on the object, in let us say the x-direction, is F_x. Now we let the object move a short distance δx in the x-direction–so short that F_x does not change appreciably. The work done by the force system is approximately $\bar{F}_x \delta x$ and the corresponding potential energy change is $-F_x \delta x$. Writing this change as δE we get

$$\delta E \approx -F_x \delta x \quad \text{or} \quad F_x \approx -\frac{\delta E}{\delta x},$$

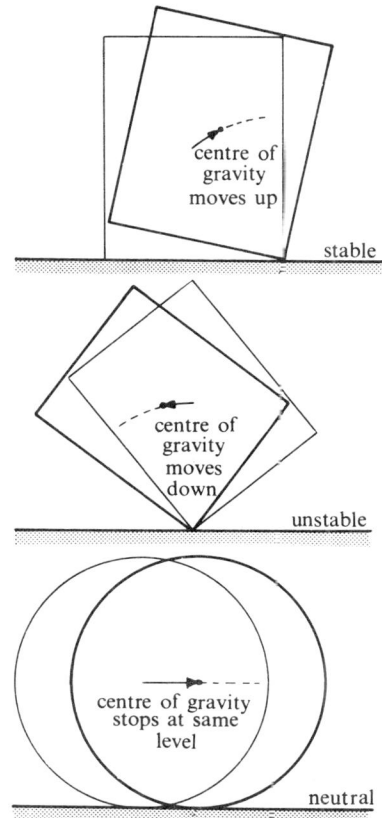

Figure 10.12
Potential energy criterion for equilibrium.

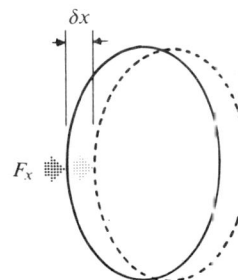

Potential energy change $\delta E = -F_x \delta x$.

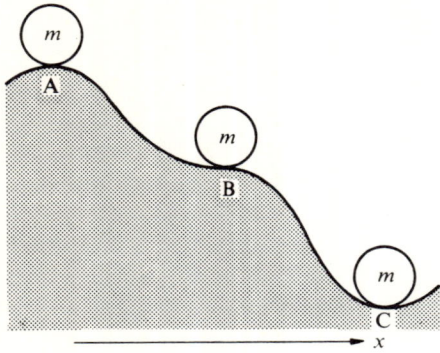

Figure 10.13

Three situations for which $\mathrm{d}E/\mathrm{d}x$ is zero, where E is the potential energy of the mass m. E is maximum at A (unstable equilibrium), minimum at C (stable equilibrium); point B corresponds to a point of inflexion for the graph of E against x, giving unstable equilibrium.

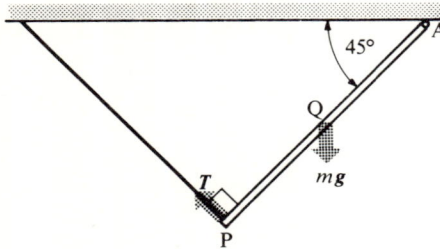

Figure 10.14

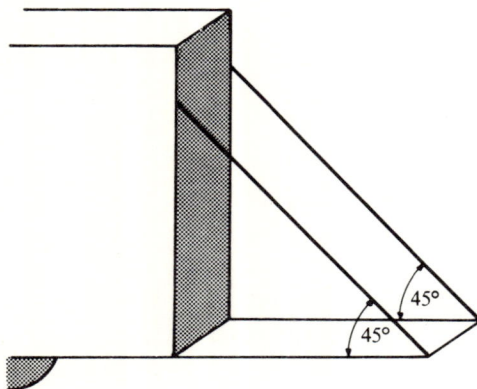

Figure 10.15

and as δx tends to zero this expression moves to the limit

$$F_x = -\frac{\mathrm{d}E}{\mathrm{d}x}.$$

We can expect equilibrium when $F_x = 0$, and elementary calculus tells us that if $\mathrm{d}E/\mathrm{d}x = 0$, E usually has a maximum or minimum value, though $\mathrm{d}E/\mathrm{d}x$ can also be zero at a point of inflexion (Figure 10.13).

Virtual work

The fact that a stable equilibrium position involves minimum p.e. can provide us with an alternative way of tackling a large number of problems. The idea is as follows. If we have an equilibrium situation, there is no net force, and therefore if the object concerned experiences a very small displacement, the net work done is zero. So we imagine the small displacement taking place, and we work out the consequent tiny amounts of work done on or by all the external forces involved and put their sum equal to zero. The operation of this *principle of virtual work* is best illustrated by a simple example.

Example

Figure 10.14 shows a uniform rod of mass m freely pivoted at A but kept at an angle of $45°$ to the vertical by a string attached to the end of the rod and perpendicular to it. Find the tension T in the string.

We imagine that the string suffers a tiny increase in length δl. The work done on the string in producing this increase is $T\delta l$. Now, if the point P swings through a small distance δl, the point Q swings through a distance $\delta l/2$ and drops through a vertical height $(\delta l/2)\cos 45°$. The potential energy lost by the rod is therefore $mg\,\delta l/2\sqrt{2}$. The force at A does not move its point of application, so no work is involved. We have to arrange the total work done on the system to be zero, so we write

$$T\delta l = mg\frac{\delta l}{2\sqrt{2}} \qquad \text{or} \qquad T = \frac{mg}{2\sqrt{2}}.$$

Exercise 10.10

Figure 10.15 shows the tailboard of a lorry, supported on each side by a light chain. The tailboard has a weight of 120 N, which can be considered to act at its centre, and when it is horizontal the chains make angles of $45°$ with it. Use the principle of virtual work to find the tension in each chain.

Further exercises

Exercises 10.11–16

11 A string of negligible mass is fixed horizontally between two rigid supports. A mass of 2·0 kg is suspended from the mid-point of the string. If each segment of the string now makes an angle of $10°$ with the horizontal, calculate the tension in each segment. (AEB)

12 State the conditions which must be satisfied if a number of non-parallel coplanar forces acting on a body maintain it in equilibrium. How would you demonstrate experimentally the truth of your statements? Explain why the lines of action of three such forces must be concurrent for equilibrium.

A uniform ladder 20·0 m long and of mass 15·0 kg leans with one end resting against a smooth vertical wall at a point 16·0 m above the ground and with the lower end on the rough horizontal ground. The force of the wall on the ladder acts in a direction normal to the wall. Draw a diagram to show the forces acting on the ladder.

A man of mass 60·0 kg now stands at a point one-quarter of the way up the ladder. Determine the angle which the force of the ground on the foot of the ladder makes with the horizontal.

(AEB)

13 Under what conditions is a body said to be in equilibrium? What is meant by (a) *stable equilibrium* and (b) *unstable equilibrium*? Give one example of each.

A pair of railway carriage wheels, each of radius r, is joined by a thin axle; the mass of the whole is m. A light arm of length l ($<r$) is attached perpendicularly to the axle and the free end of the arm carries a point mass M. The wheels rest, with the axle horizontal, on rails which are laid down a slope inclined at an angle ϕ to the horizontal. Show that, provided that ϕ is not too large and that the wheels do not slip on the rails, there are two values of the angle θ that the arm makes with the horizontal when the system is in equilibrium, and find these values of θ. Discuss whether, in each case, the equilibrium is stable or unstable.

(O & C)

14 It is often suggested that a car may be stopped most safely on an icy road by making repeated short applications of the brakes. Discuss whether this procedure is likely to bring the car to rest in a shorter distance than when the brakes are applied (a) hard, so that the wheels are locked, (b) just hard enough to prevent skidding. Explain whether any advantage is to be gained in icy conditions by (i) reducing the tyre pressures, (ii) adding an extra load over the driving wheels.   (C)

15 In what circumstances is a physical system in equilibrium? Distinguish between stable, unstable and neutral equilibria.

Discuss the stability of the equilibrium of a uniform rough plank of thickness t, balanced horizontally on a rough cylindrical fixed log of radius r, it being assumed that the axes of plank and log lie in perpendicular directions.   (JMB)

16 A straight rod of length l, small cross-sectional area a and of material density ρ is supported by a thread attached to its upper end. Initially the rod hangs in a vertical position over a liquid of density σ and then is lowered until it is partially submerged. Derive and discuss the equilibrium conditions of the rod, neglecting surface tension.   (JMB)

11 Forces between atoms in solids: elastic and plastic behaviour

A set of forces keeping an object in equilibrium causes neither translational nor rotational acceleration, but that does not mean the forces are having no effect at all. They may well be compressing it, stretching it, bending or twisting it, and if they get too big they will break it. In this chapter we are going to take a look at these deformations, both from macroscopic and microscopic viewpoints. Macroscopically, we simply measure the size of a deformation, measure the forces causing it and search for a formula to link the two sets of results. Then we look at our microscopic model to see if it can account for the observed behaviour, and perhaps refine or modify the model in the process. We shall deal here only with *isotropic* materials, that is, materials with properties which are the same in all directions.

Elastic behaviour

Push a block of rubber out of shape and it will spring back when released; push a block of metal out of shape and it will behave in the same way as long as the deformation is quite small. Figure 11.1 is a sketch graph showing how the amount of distortion of an object – extension for example, or angle of twist – depends on the size of the forces responsible. If the object is given a deformation corresponding to point *A* on the graph it regains its shape completely when the forces are removed; we say that point *A* is in the *elastic* region. Point *B* corresponds to a *plastic* distortion, that is, one which does *not* disappear completely when the object is released; instead the object ends up with a permanent deformation indicated by point *C*. Point *D* divides the elastic region from the plastic region; it is called the *elastic limit*. If we stretch a wire, for example, the elastic limit is reached when the fractional change in length is about one part in a thousand.

Hooke's law

Not only do most solids behave elastically when deformed slightly, they also obey the general law that *the amount of elastic deformation is proportional to the size of the forces responsible*. This rule is a little unreliable near the elastic limit but is followed closely for smaller deformations. A special case was investigated by Robert Hooke (1645–1703), who found that the extension of a metal wire in the elastic region is directly proportional to its tension. This is known as Hooke's law; one way of checking it is by means of the apparatus described on p. 138.

Microscopic view of elastic behaviour

We can use our observations of elastic behaviour to make a guess at the way the force between two atoms or molecules varies with their separation. In Figure 11.2 the distance *d* is the equilibrium separation of the centres of the atoms, when the net force between them is zero. We can imagine what happens to the atomic separations if we stretch a wire. The external forces are transmitted atom by atom down the length of the wire and all the interatomic separations along its length are increased slightly to an average separation *x* let us say, until there is sufficient attractive force between each section of wire to hold it together. When the external forces are removed, the attractive forces pull the atoms back to the original separation. That is something of an over-simplification, because of a very important aspect of our model, namely the atomic vibrations; each atom must be thought of as vibrating about its mean position. The vibrations are

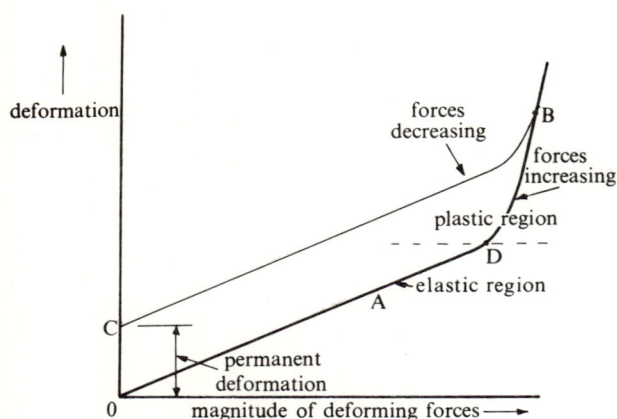

Figure 11.1

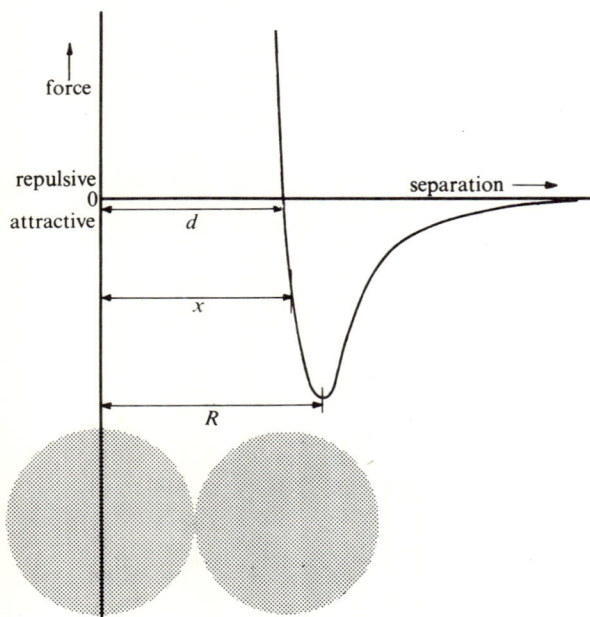

Figure 11.2

Graph showing variation of the magnitude of forces between atoms with their separation, according to our simple model.

in all directions and at all sorts of frequencies and some atoms vibrate more vigorously than others at any given time. So when we stretch the wire we must talk about a shift of the *mean* interatomic spacing, and if the forces on the wire are abruptly removed the atoms do not simply snap back into place. At first the wire vibrates as a whole, as a loaded spring oscillates when its equilibrium is disturbed, but in a very short time the ordered, large-scale vibrations give way to random atomic vibrations of increased energy, constituting a temperature rise.

As the tension in the wire is increased, the forces involved may become big enough to produce a reorganisation in the siting of the atoms, and the wire acquires a permanent deformation. We shall take a closer look at the behaviour of metals in the plastic region later in the chapter. Eventually there must come a point at which the mean interatomic separation at some part of the wire gets dangerously close to R, which is the separation for which the attractive force between atoms is as large as it can get, and any further demands on it cannot be met – the wire breaks.

Figure 11.2 gives us Hooke's law too. Any curve is a good approximation to a straight line over a sufficiently small region, and that part of the curve in the vicinity of the equilibrium separation d is no exception, so small departures from d in either direction produce *proportional* amounts of restoring force. To achieve equilibrium the restoring forces must exactly balance the external forces causing the deformation, and so we infer that there is direct proportionality between the size of the external forces and the change in length they produce.

Longitudinal stress and strain

We now want to write down the connection between the extension produced in a wire and the size of the forces responsible. We can expect it to depend on the shape, size and material of the wire.

Exercise 11.1

Given that a certain type of spring is stretched 1 cm when subjected to equal and opposite forces of 1 N, write down the total extension produced by these forces when applied to

a) the ends of a chain of ten such springs joined end to end.
b) ten springs all bunched in parallel.
c) the ends of a chain consisting of five parallel pairs.

(See Figure. 11.3.)

Every spring in a series chain experiences the full effect of the forces applied to the free ends, and the total extension is proportional to the length of the chain. On the other hand, springs bunched in parallel *share* the force applied, and the extension is inversely proportional to the number of springs in parallel. Now we can argue that atoms in a wire will behave in the same way within the elastic region, since each interatomic attraction acts rather like the tension in a spring. We can expect to find that the extension of the wire is proportional to the number of atoms 'in series' – that is, to the length of the wire l – and inversely proportional to the number of atoms 'in parallel', that is, to the cross-sectional area A, and both these ideas are con-

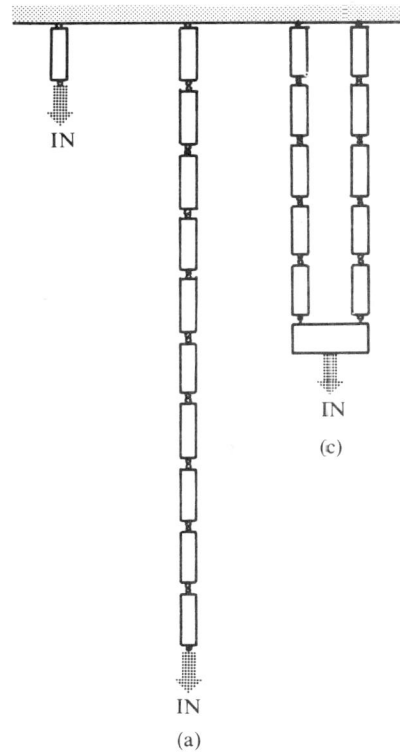

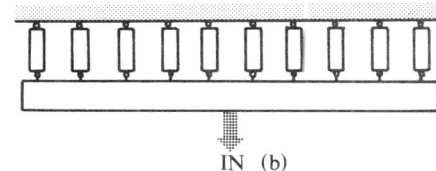

Figure 11.3 (Exercise 11.1)

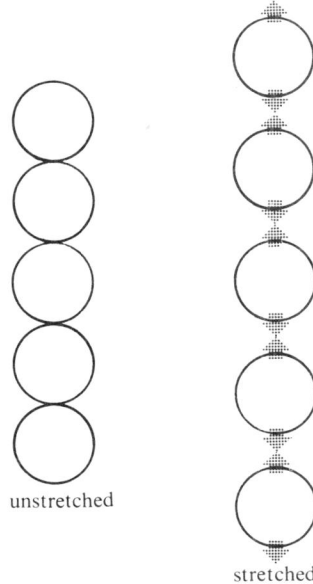

Simple model of a chain of atoms in a stretched wire.

firmed by experiment. This means we can write

$$\text{extension } x \propto Fl/A, \qquad \qquad ...(1)$$

where F is the size of either of the forces needed to produce the extension.

We find it convenient at this point to introduce two new quantities, called *stress* and *strain* respectively. Qualitatively, a stress is a system of forces tending to distort an object, and the resulting distortion is called the strain, but quantitatively these terms have a rather more restricted use; their exact definitions depend on the type of distortion considered. We shall define three types of stress and strain, corresponding to longitudinal extension, uniform compression and shear, and in each case we shall introduce a *modulus of elasticity*, which is a measure of the ability of the material to resist distortion, and which we shall define by the ratio stress/strain.

The longitudinal stress is defined by the quotient F/A in equation (1); it is the size of one of the extending forces divided by the area over which it acts. The longitudinal strain is defined as the fractional increase in length, x/l, and if we write equation (1) in the form

$$\frac{F}{A} = E \times \frac{x}{l}$$

we get

$$\text{stress} = E \times \text{strain}.$$

E, defined by this relation, is the modulus for longitudinal extensions; it is called the *Young modulus* and has units Nm^{-2}. The Young modulus is a constant for any object obeying Hooke's law, providing the elastic limit is not exceeded, but its significance does not end there. Because we have been careful to write into equation (1) the dependence of extension on the shape and size of the object, E does *not* depend on these factors; it is a constant for a given *material*.

Determination of the Young modulus

Figure 11.4 shows a standard piece of equipment called Searle's apparatus, which can be used to measure the Young modulus. It is best to work with a long thin wire–long because we are limited to an extension of about 0.1%, so that even for a two metre wire we shall have to be content with an increase of a millimetre or two, and thin because thicker specimens need bigger forces, which are difficult to handle. Searle's apparatus uses two loaded wires suspended from the same support, and as the load on one of the wires is increased the pivoted bar A dips down on one side. A spirit level indicates whether or not it is horizontal, and the micrometer B can be used to return it to its original position. This means that the extra extension of the wire is indicated by the micrometer, to within about $0.01\,mm$.

The point of having two wires is that a number of systematic errors are eliminated; any temperature fluctuation affects both wires, and the same is likely to be true of any yielding of the support. It is advisable to measure extensions produced by a number of loads and check that a linear graph is obtained, and it is essential to take readings as the load is decreased again to confirm that the elastic limit has not been passed. (The readings taken while the load is decreased may not be *exactly* equal to the corresponding readings taken during increase of load; this phenomenon is known as elastic hysteresis. With most metals the effect is so slight that it will generally be masked by experimental uncertainty.)

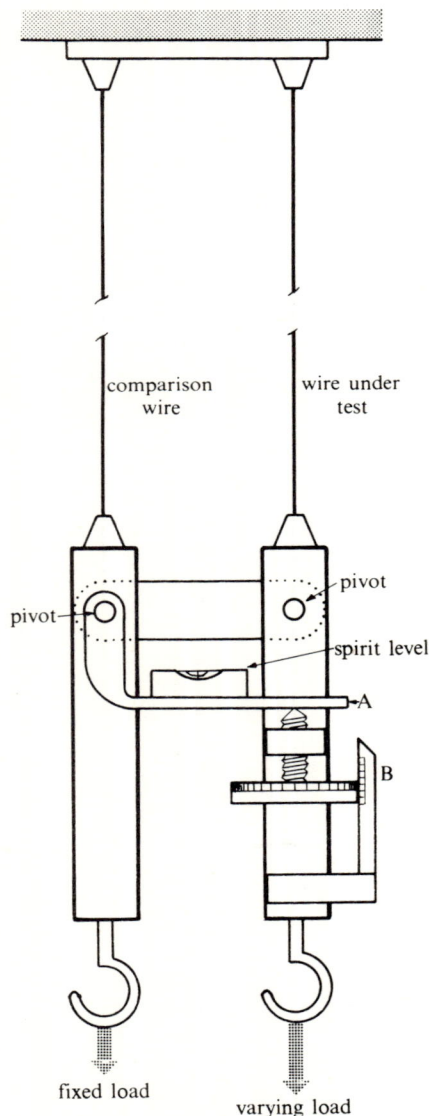

comparison wire

wire under test

pivot

pivot

spirit level

A

B

fixed load

varying load

Figure 11.4

The extension is not the only difficult measurement; we also need the cross-sectional area of the wire, which means we have to measure its diameter in several places and at different orientations to check for taper and non-circular cross section. It does not pay to use a wire with diameter less than a millimetre or so, otherwise the error arising from this part of the experiment exceeds the one involved in the measurement of the extension.

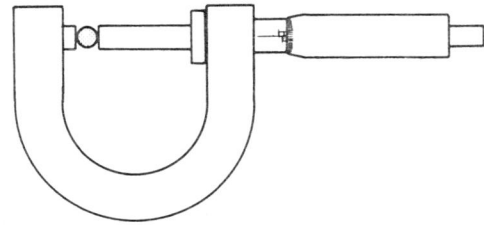

Exercise 11.2

The following results are obtained for a circular steel wire.
   Load provided by mass of (10.00 ± 0.01) kg.

$$\text{extension} = (1.55 \pm 0.01)\,\text{mm.}$$
$$\text{length} = (2.235 \pm 0.001)\,\text{m.}$$
$$\text{mean diameter} = (1.48 \pm 0.01)\,\text{mm.}$$

Find the value of the Young modulus for steel together with the estimated percentage error, taking g to be $9.80\,\text{m s}^{-2}$.
   Which measurement contributes most to this estimated error?

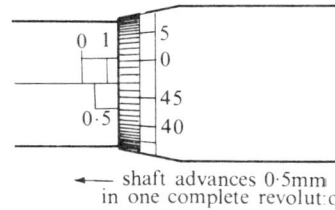

The Poisson ratio

When a wire is stretched it gets thinner; we can think of the atoms 'filling in the spaces' and moving closer together laterally as the longitudinal separations are increased. We might guess that the reduction in cross section occurs in such a way that the volume of the wire stays the same as before, but this is not so; there is some increase. The ratio of the fractional decrease in width of a slab of material to the corresponding fractional extension caused by stretching it turns out to be a constant for a given material obeying Hooke's law and is called the *Poisson ratio*, μ. We can easily show that if the volume were to remain constant the Poisson ratio would be just $\frac{1}{2}$. Taking a slab with length l and square cross section of side b, we can write the modified dimensions caused by a longitudinal stress as $l + \delta l$, $b - \delta b$ and $b - \delta b$.

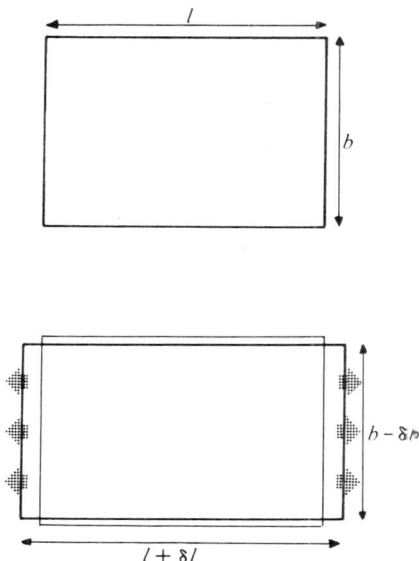

shaft advances 0·5mm
in one complete revolution

Micrometer. Reading 1·47 mm.

Lateral contraction associated with longitudinal extension.

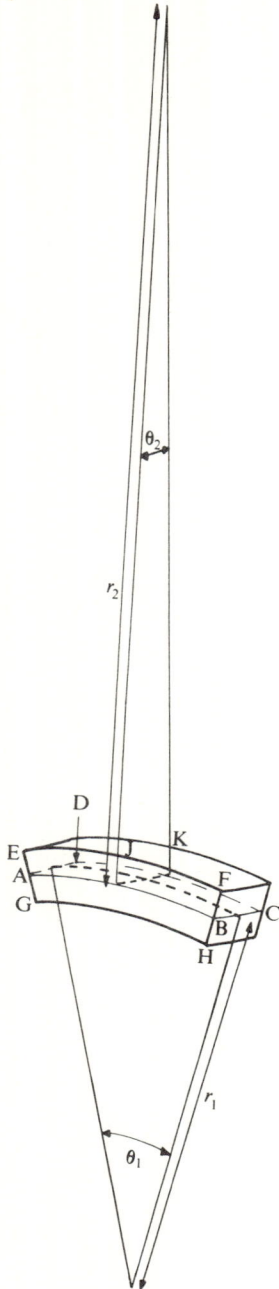

Figure 11.5

Exercise 11.3

Show that the volume change can be written as $b^2\,\delta l - 2lb\,\delta b$ if terms involving $\delta l\,\delta b$, δb^2 or smaller quantities are ignored. Hence show that for the volume change to be zero we would need $\delta l/l = 2\,\delta b/b$ and that the Poisson ratio would therefore be $\frac{1}{2}$ in this case.

The Poisson ratio is not easy to measure by direct means. We have to work with a reasonably thick rod to have a chance of measuring the decrease in width, remembering that the fractional change will be no more than about 0·05% before the elastic limit is exceeded. A thick rod requires a large tension to produce a reasonable extension, and this makes the question of support much more difficult. Figure 11.5 shows the basis of a more convenient method. If you bend a flat slab of indiarubber, or steel for that matter, the top face forms what is called an *anticlastic* surface; it will be convex in one plane, the plane of the bend, with radius of curvature r_1, but concave in the perpendicular plane, with radius of curvature r_2. We can show that the Poisson ratio is given simply by $\mu = r_1/r_2$, and if we polish up the surface until it acts as a mirror we can measure r_1 and r_2 by suitable optical methods.

Exercise 11.4

We take it that the original centre plane of the slab $ABCD$, though distorted, has the same face area as before, so that the length of AB is unchanged, the line GH is compressed, and EF is extended. Show that if the slab depth is d, the new length of EF measured along the curve can be written as $(r_1 + d/2)\theta_1$ so long as θ_1 is small, and hence that the fractional extension of EF is $d/2r_1$. Show by a similar argument that the fractional compression of JK is $d/2r_2$ and that $\mu = r_1/r_2$ follows.

The Poisson ratio for a metal is typically around 0·3.

Bulk modulus

We find it helpful to define two more elastic constants corresponding to two further simple forms of deformation. The first is the change in volume caused by a uniform compression from all sides, in other words the sort of deformation an object suffers to an imperceptible degree due to atmospheric pressure. Once again for convenience we want a number which depends only on the material and not on the geometry of the object, and the right quantities to work with this time turn out to be δP, the pressure change on the surface, and $-\delta V/V$, the corresponding fractional change in volume. The expression

$$\frac{\text{pressure change}}{\text{corresponding fractional change in volume}},$$

or $-V\delta P/\delta V$, is a constant for any material obeying Hooke's law, and is called the *bulk modulus K*:

$$K = -V\,\delta P/\delta V$$

As δP is a force per unit area, or stress, and $-\delta V/V$ is a fractional deformation, or strain, the bulk modulus, like the Young modulus, is the ratio of a stress to the corresponding strain, and has units Nm^{-2}.

Isothermal bulk modulus for a gas

Gases behave elastically when compressed, and we can derive a simple expression for the bulk modulus of a gas which obeys Boyle's law. The size of the bulk modulus turns out to depend on the conditions under which the gas is compressed, because there is usually an associated temperature rise. The microscopic explanation is that the gas molecules collide very many times per second with the piston responsible for the compression, and because the piston is necessarily moving, the molecules rebound from it with increased speed at each collision. So the mean speed of the molecules increases during the compression; speaking in macroscopic terms, the temperature rises. However, if the compression occurs at such a slow rate that heat can flow out relatively quickly to the surroundings, keeping the gas temperature substantially constant, we can use Boyle's law

$$P = k/V, \qquad ...(2)$$

where P is the pressure and V the volume, k being constant, and therefore write

$$\frac{dP}{dV} = \frac{-k}{V^2}. \qquad ...(3)$$

Putting the expression for the bulk modulus in the differential form

$$K = -V\frac{dP}{dV},$$

we can make use of equation (2) and (3) to get $K = P$. The isothermal bulk modulus of a gas is equal to its pressure.

Exercise 11.5

Complete the missing steps in the derivation.

Shear modulus

The third type of deformation sufficiently important to warrant its own elastic constant is a *shear*, a displacement of parallel planes of a block of material as indicated in Figure 11.6. We look at a very simple case, a rectangular block deformed into a parallelepiped; the figure shows a set of forces which will achieve this. A single force would clearly not do–it would cause acceleration. A pair of equal and opposite forces F and $-F$ acting tangentially along opposite faces of the block looks more hopeful, but such a system acting alone would produce rotational acceleration. There must be an equal and opposite couple acting on the block to keep it in equilibrium; in our case the lower face of the block is welded to a rigid support.

The deforming force F is shared out amongst all the atoms in the top face, and the appropriate quantity to work with is F/A, the force per unit area of the face to which it is tangential; this is called the *shear stress*. Having in mind the way we defined strain in the case of longitudinal extension, we define shear strain as the ratio x/l (Figure 11.6); this is equal to the angle γ measured in radians so long as it is small. Then we write

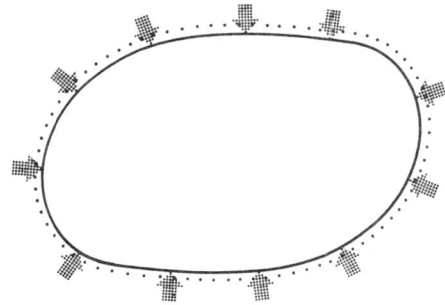
Deformation caused by uniform compression.

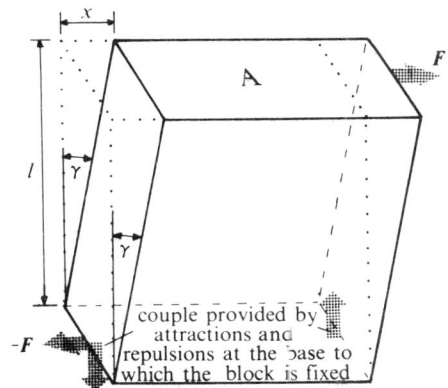
Figure 11.6
Shear stress and strain.

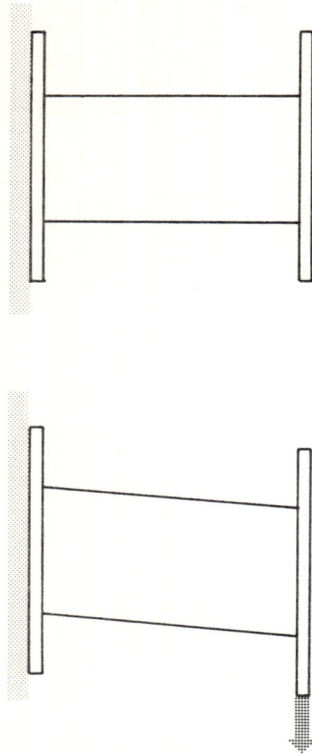

See Exercise 11.6.

$$\frac{\text{shear stress}}{\text{shear strain}} = \frac{F}{A\gamma} = G \qquad \dots(4)$$

where G is the *shear modulus*. Like the other elastic moduli, it depends only on the material used, is constant if Hooke's law is obeyed, and has units $N\,m^{-2}$.

Exercise 11.6

A cylindrical steel shaft of diameter 1·0cm is firmly embedded at each end in a flat metal plate; the two plates are parallel and separated by a 2·0cm gap. If one plate is securely clamped, what force applied to the other plate in its plane will cause it to shift 0·01mm? The shear modulus for the steel is $8\cdot0 \times 10^{10}\,N\,m^{-2}$.

Elastic energy

To produce a deformation we have to do work on the object concerned, and if the deformation lies within the elastic region this work can be made available again as the object regains its original shape.

To calculate the work done in a longitudinal extension, we need of course to write down the product of the force and the displacement of its point of application, but there is the complication that the force changes as the extension proceeds. If the total extension produced by a force F_0 is x_0, then at some intermediate extension x we are working against a force F given by

$$\frac{F}{F_0} = \frac{x}{x_0},$$

assuming that Hooke's law is obeyed. The work done in a very small additional extension δx is $F\,\delta x$ or $F_0 x\,\delta x / x_0$, and the total work done for the complete extension is therefore given by

$$\int_0^{x_0} \frac{F_0 x\,dx}{x_0} \quad \text{or} \quad \tfrac{1}{2}F_0 x_0$$

In words,

$$\text{energy stored in a stretched wire} = \tfrac{1}{2} \text{ final tension} \times \text{extension}$$

We can see that the factor $\tfrac{1}{2}$ arises because we really need some sort of *average* force over the whole extension, and the tension in the wire passes through all values between zero and the final full amount F_0, increasing steadily with the extension if Hooke's law is obeyed. The stored energy is represented by the area under the graph of F/N against x/m (Figure 11.7).

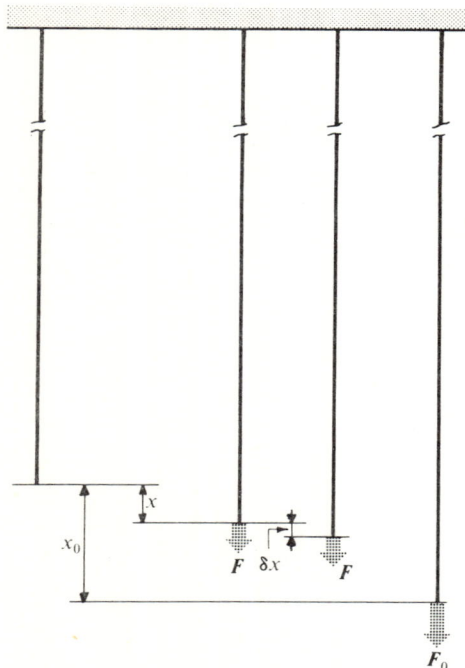

Figure 11.7

Graph of load F against extension x for a stretched wire. The area under the curve is numerically equal to the work done and has the value $\tfrac{1}{2}F_0 x_0/N\,m$.

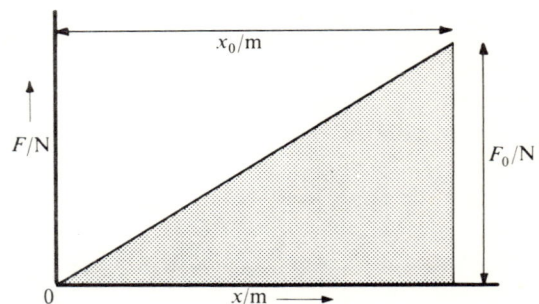

Stretching a wire; a varying force is involved.

Exercise 11.7

By writing F_0 and x_0 in terms of the longitudinal stress and strain, show that the elastic energy is equal to $\frac{1}{2}$ stress × strain × volume of the wire.

Similar considerations apply for the other deformations we have considered; the factor $\frac{1}{2}$ comes in each time, and in fact the statement

elastic energy per unit volume $= \frac{1}{2}$ **stress × strain**

turns out to be true for uniform compressions and shear strains also.

Exercises 11.8, 9

8 Show that if the radius r of a sphere decreases by a very small amount δr the consequent decrease in volume δV is $4\pi r^2 \delta r$, and that if this reduction is achieved by a pressure P on all sides the total work done is $\frac{1}{2} \times 4\pi r^2 P \times \delta r$. Use these results to demonstrate that the work done is equal to $\frac{1}{2} P \delta V$ and that this is equal to $\frac{1}{2}$ stress × strain × original volume.

9 Given that $\frac{1}{2} Fx$ is the work done in a distance x against a force which increases uniformly from zero to F over the distance, show that the work done in distorting the block in Figure 11.6 is $\frac{1}{2} Fl\gamma$, where l is the height of the block. Show that this is also equal to

$\frac{1}{2}$ shear stress × shear strain × volume of block.

Interdependence of the elastic moduli

Longitudinal extensions, uniform compressions and shears all involve changes of mean interatomic separation on the atomic level, that is to say they are all controlled by attractions and repulsions of the atoms, so it is reasonable to suppose that the values of the different moduli are connected. In fact once the Young modulus and Poisson ratio are known for an isotropic material the values of the bulk modulus and shear modulus can be calculated. We can get at the connections by macroscopic arguments. In Figure 11.8 we have a solid cube of material of side a acted on by a uniform pressure P.

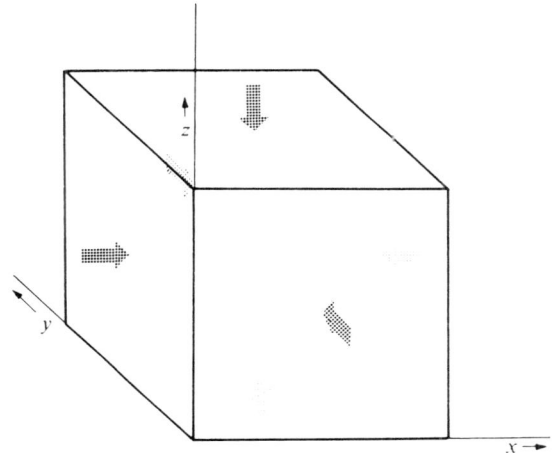

Figure 11.8
Cube under uniform compression.

Exercise 11.10

Show that the decrease in volume is $a^3 P/K$, where K is the bulk modulus.

Now we can find this volume decrease by a different approach. The uniform pressure can be thought of in terms of a force Pa^2 pushing in on each face, and we can write down the deformations such forces produce in terms of the Young modulus E and Poisson ratio μ. We define the x-, y-, and z-directions to correspond to those of the cube edges and consider the effect of the pair of forces acting in the x-direction.

$$\text{Longitudinal extension} = \frac{1}{E} \times \frac{\text{force}}{\text{area}} \times \text{original length.}$$

We can take it that the same formula applies to a *longitudinal compression* (as opposed to a uniform compression), so that the

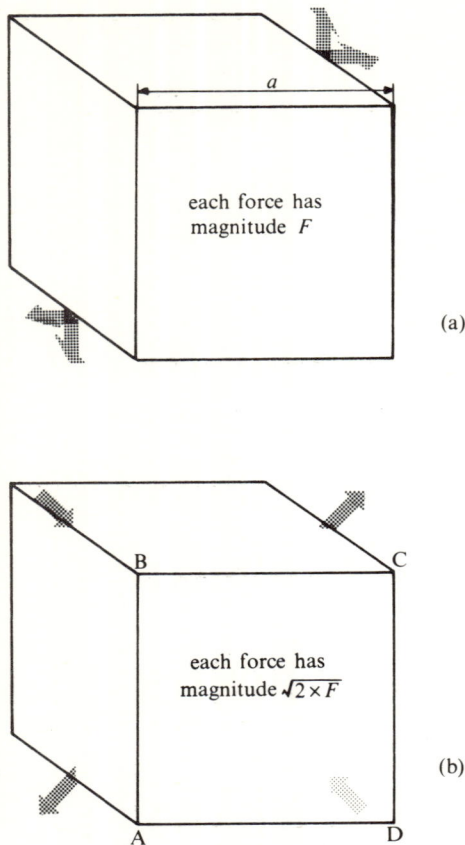

each force has magnitude F

(a)

each force has magnitude $\sqrt{2} \times F$

(b)

Figure 11.9

Two ways of looking at the same force system. (Though for simplicity each force is drawn as though it is applied at a point, it must in fact be applied over the whole surface.)

reduction in length in the x-direction due to this pair of forces is

$$\frac{1}{E} \times \frac{Pa^2}{a^2} \times a \quad \text{or} \quad \frac{Pa}{E}.$$

But that is not the end of the story. The forces in the x-direction also produce a *widening* of the cube in the y- and z-directions, of magnitude $\mu Pa/E$ in each case.

The other two pairs of forces produce corresponding effects, and so each edge of the cube experiences one compression and two extensions. If we can take it that these changes are superimposed on each other, the change δa in the length of each side is $-Pa(1 - 2\mu)/E$. Now the new volume is $(a + \delta a)^3$, which is approximately $a^3 + 3a^2\,\delta a$ if we are prepared to ignore the terms containing δa^2 and δa^3, which are both very small indeed.

Exercise 11.11

Write down the volume change in terms of δa and a and show that it equals $-3Pa^3(1 - 2\mu)/E$.

Hence use the result of Exercise 11.10 to show that $K = E/3(1 - 2\mu)$.

We can also use the solid cube to find the shear modulus G in terms of E and μ. Figure 11.9a shows two equal and opposite couples of moment Fa applied to the cube, and equation (4) enables us to write

$$G = \frac{F\gamma}{a^2} \quad \text{or} \quad \frac{F\,\delta x}{a^2 a},$$

where δx is the horizontal shift produced in the top surface. But again we can look at the forces in a different way (Figure 11.9b): we can resolve them into components which act in the directions of the *diagonals* AC and DB. In each case they have magnitude $2F/\sqrt{2}$ and they act over an effective area $a \times \sqrt{2}a$. The longitudinal stress is therefore of magnitude F/a^2 along both diagonals, tending to extend AC and shorten DB. We want to write down the new length of AC, and we have to remember that it increases on two counts, firstly because of the stretching forces, and secondly because the reduction of DB will automatically produce an increase in width in the plane perpendicular to it.

Exercise 11.12

Show that the forces acting in the direction AC will produce in it an extension equal to $\sqrt{2}F/aE$, and that the forces in the direction BD will compress BD by the same extent.

The extra extension of AC due to the compressive forces will therefore be $\mu \times \sqrt{2}F/aE$, giving a total extension e equal to $\sqrt{2}F(1 + \mu)/aE$. Figure 11.10 tells us how to connect this extension with x; we see that so long as the deformation is small $x = \sqrt{2}e$.

Exercises 11.13, 14

13 Show that as a result $G = \dfrac{E}{2(1 + \mu)}$.

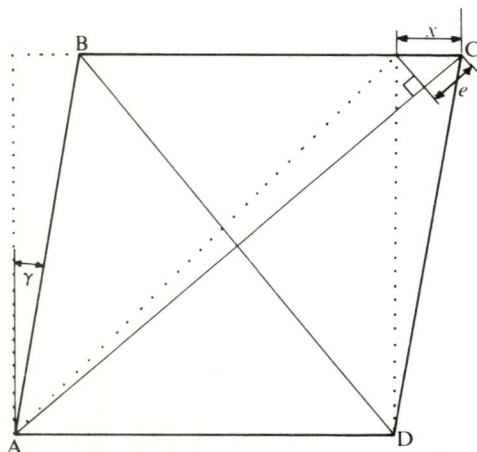

Figure 11.10

14 The values of E and G for copper are $13 \cdot 0 \times 10^{10} \, \mathrm{N \, m^{-2}}$ and $4 \cdot 83 \times 10^{10} \, \mathrm{N \, m^{-2}}$ respectively. Find K and μ.

Twisting a rod

We can make use of the elastic constants we have defined to describe more complex deformations, for which different parts of the object concerned experience different amounts of stress. We have already indicated, for example, that the curvature of a rod bent under stress can be worked out in terms of the Young modulus, and now we shall show that the shear modulus can be used to account for the opposing couples needed to twist a wire of circular cross section. In Figure 11.11 we can see that when such a wire is twisted, the almost rectangular section $ABCD$ has been distorted into the shape of $A'B'CD$, and we can imagine the whole wire split into similar sections. $ABCD$ is a rather special example, a more typical section being $PQRS$, parallel to it but lying within the body of the wire; it experiences the same type of distortion but to a smaller extent. If we say that the section $PQRS$ has a small thickness so that it constitutes an area δa of the cross section of the wire, and that it is experiencing a shear stress corresponding to a force of magnitude δF along $P'Q'$, then we can write

$$G = \frac{\delta F}{\gamma \, \delta a}$$

where G is the shear modulus and γ the angle indicated. Of course γ is not an angle we find easy to measure; a more convenient indication of the deformation is θ, the angle of twist. Fortunately there is a simple relationship between them, namely

$$\theta x = \gamma l$$

where $PQRS$ is distance x from the centre of the wire. The link in this equation is the length of arc PP' and it is worth noting that this length is given *exactly* by θx (θ in radians) but *approximately* by γl so long as γ is small. This means that providing the wire is long we shall not be restricted to small values of the angle of twist θ in our theory, since γ can safely be assumed small even when θ is quite large.

So putting $\theta x / l$ in place of γ we get

$$G = \frac{l \, \delta F}{\theta x \, \delta a} \quad \text{or} \quad \delta F = \frac{G \theta x \, \delta a}{l}.$$

Now we want to carry out an addition process. The idea is to try and find out how θ depends on the total torque T twisting the entire wire about its axis—of course we need two couples, equal and opposite, one at each end of the wire in order to produce distortion rather than rotation. T will be equal to the sum of the moments of all the tiny forces like δF about the axis of the wire. The moment δT of δF about this axis is given by

$$\delta T = x \delta F = \frac{G \theta x^2 \, \delta a}{l}.$$

Now G, l and θ are the same for every small section; if one part δa of the cross section rotates through $30°$, *every* part of that cross section rotates through $30°$. So we can write

$$T = \sum \delta T = \frac{G \theta}{l} \sum x^2 \, \delta a,$$

but now we have to write δa in terms of x. The only thing we must

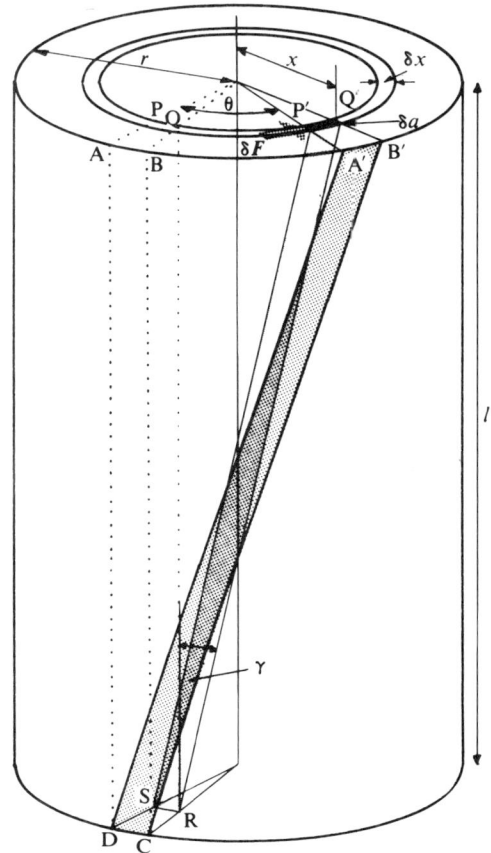

Figure 11.11

Twist of a cylindrical wire.

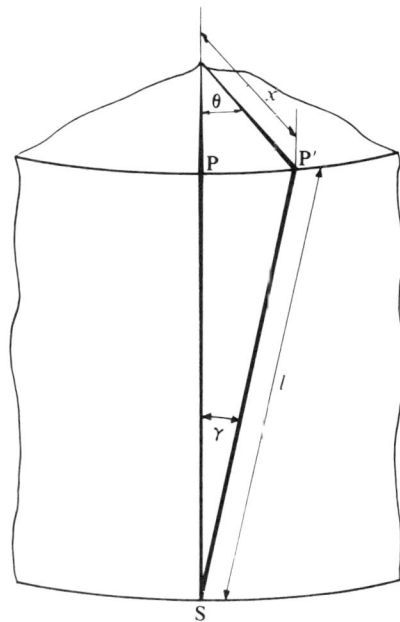

$\theta x = \gamma l$ as long as γ is small (the lengths of PS and $P'S$ can each be taken to be l).

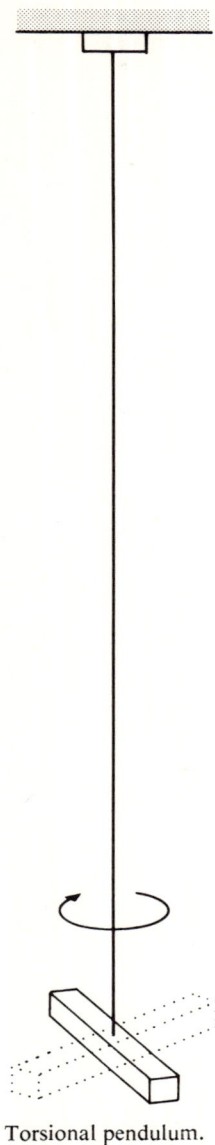

Torsional pendulum.

watch when we are selecting how big δa should be is that every part of it must have the same x, so the best element of area to choose is a *ring* of radius x and thickness δx, δx being very small. The area of such a ring is very nearly equal to the ring width times the circumference, that is, $2\pi x\,\delta x$, which gives

$$T \approx \frac{G\theta}{l} \sum 2\pi x^3\,\delta x,$$

or in integral form,

$$T = \frac{G\theta}{l} \int_0^r 2\pi x^3\,dx$$

where r is the wire radius. So

$$T = \frac{\pi G\theta r^4}{2l}. \qquad \qquad ...(5)$$

The quantity $\pi G r^4/2l$ is constant for a given wire and is called the *torsional constant c*, enabling us to write

$$T = c\theta \qquad \qquad ...(6)$$

The angle of twist of a wire in the elastic region is therefore directly proportional to the applied torque. The constant c has units N m rad^{-1}.

Equation (5) is the key to the most convenient way of measuring G for metals. The apparatus used is a *torsional pendulum*, a thin vertical wire of the metal clamped securely at its upper end and fixed into the centre of a uniform block of simple shape at the bottom. When the wire is given a twist it will then execute torsional oscillations as the block rotates back to and through the equilibrium position. In Chapter 19 we shall derive the formula for the period of such oscillations in terms of c (p. 269); this enables us to calculate G.

Exercises 11.15–17

15 If a wire of diameter 0·5 mm is twisted through 30° by a torsional stress, what twist occurs in a cylindrical rod of the same material, having the same length and subjected to the same forces, but with a diameter of 1 cm?

16 Show that the work done in twisting a wire through an angle θ is $\frac{1}{2}T\theta$ where T is the torque required.

17 Compare the torques needed to give equal angles of twist to each of the following, all made from the same material and having equal lengths.
   a) a solid rod of circular cross section, radius 5 mm.
   b) a hollow tube, inside radius 4 mm, outside radius 5 mm.
   c) a solid rod of circular cross section with radius such that its mass is equal to that of the tube.

Energy in a stretched spring
The extension of a coiled spring under tension is due partly to the bending and partly to the twisting of each section of wire making up the coils. It is possible to work out what extension can be expected from a given spring in terms of its shape and size, together with the elastic moduli of the material used, but the dependence is not a simple one, and it is convenient to introduce a single quantity to describe the stiffness of the spring, namely the *tension needed to produce unit extension*. This quantity, often called the *force constant*

for the spring, is given the symbol λ. The extension x produced by a pair of equal and opposite forces of size W is therefore W/λ.

Exercises 11.18, 19

18 Show that the energy stored in a spring stretched an amount x by a pair of equal and opposite forces each of magnitude F is $\frac{1}{2}Fx$ or $\frac{1}{2}x^2\lambda$.

19 The mass m is hung on the end of the light spring shown in Figure 11.12, and is gently lowered a distance x. Show that the energy increase of the system (spring plus mass) is given by $-mgx + \frac{1}{2}\lambda x^2$, where λ is the force constant for the spring, and that this expression has a minimum at $x = mg/\lambda$. This is the extension at which the tension in the spring balances the weight mg: another example of the principle that the p.e. is a minimum when the system is in equilibrium.

Beyond the elastic region

Figure 11.13 shows how the extension of a metal wire varies with the load beyond the elastic region. P is the elastic limit; it lies very close to, though not necessarily at, the end of the straight section of the curve, and beyond it the wire extends more than it would do if Hooke's law were still valid. After the point Q has been reached, large extensions, sometimes as great as one third of the original length, are produced by quite small increases in load; Q is called the *yield point*. From this point on almost all of the extension remains when the stress is removed; if we were to take away the load after reaching point R the wire would end up with a permanent extension indicated by point S.

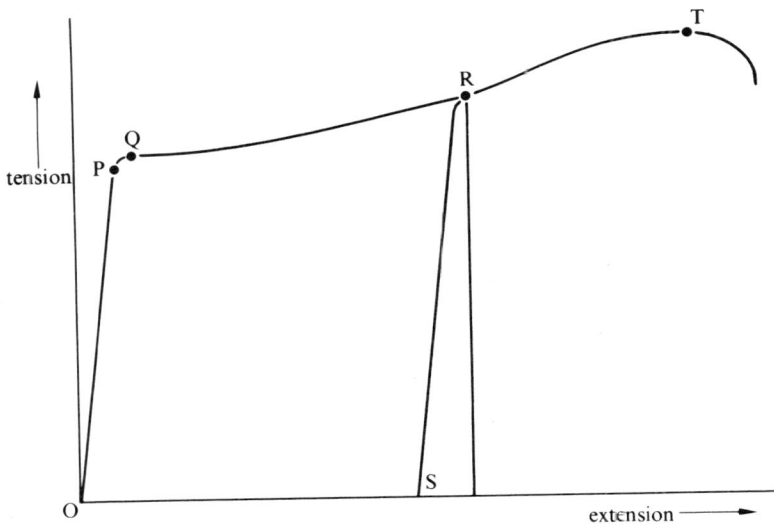

Figure 11.12 (Exercise 11.19)

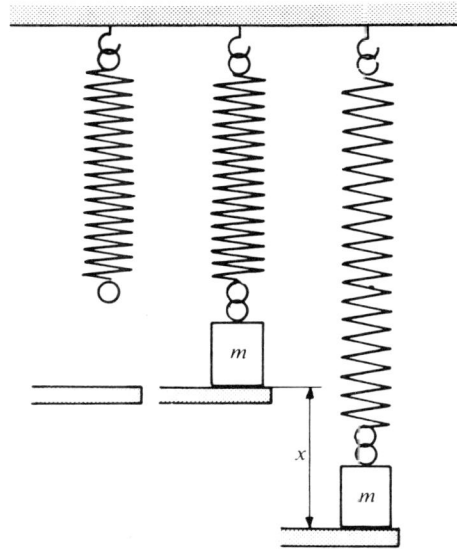

Figure 11.13
Graph of tension against extension for a wire of ductile metal.

As the tension is increased still further, the wire toughens somewhat, that is, the rate of change of tension with respect to extension increases, but when point T on the graph is reached some part of the wire gets thinner than the rest; a 'neck' forms. The stress (force

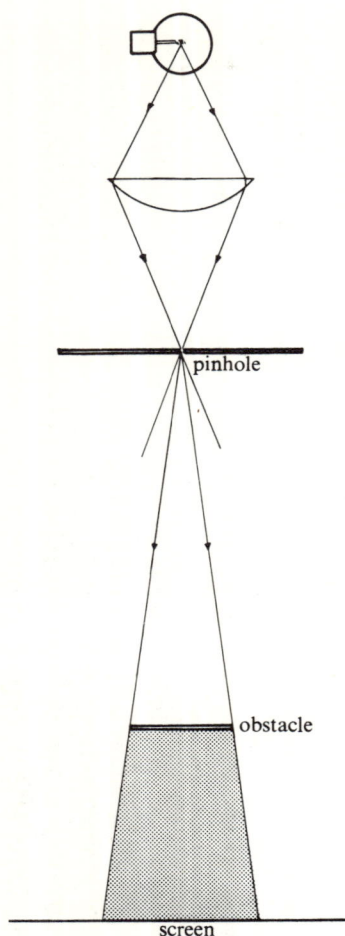

Figure 11.14

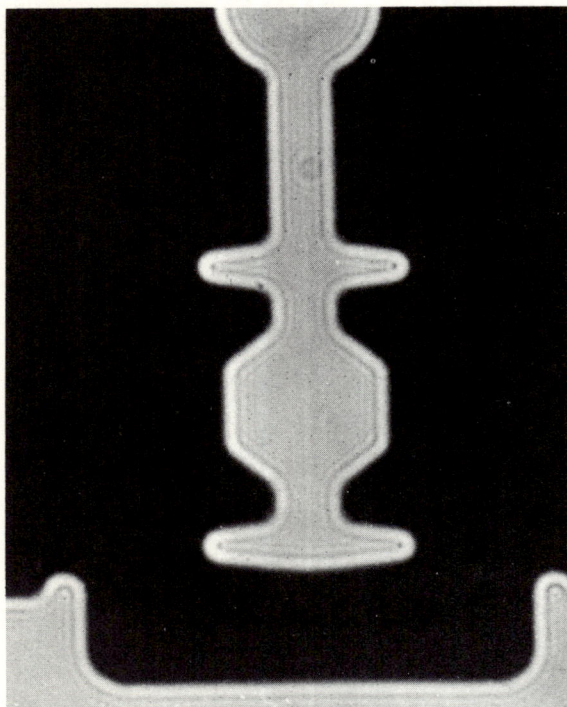

Figure 11.15

Diffraction of light from a point source by a razor blade.

per unit area) across the neck is greater than anywhere else because the cross-sectional area is smaller, and the neck rapidly gets thinner, soon resulting in a break–it is rather like trying to stretch a roll of putty. The maximum extending force per unit area which the wire can stand, corresponding to point T, is called the *breaking stress*.

In our graph we have shown the load *decreasing* beyond T. This is because the experiment is often carried out on a machine which increases the *strain* at a steady rate–usually by rotating a screw of large diameter–and records the corresponding stress. Beyond T, and sometimes just beyond the yield point also, the wire 'gives' at such a rate that the machine requires some time to catch up with it, and therefore cannot maintain, let alone increase, the stress on the wire until the extra extension is taken up. If we simply hang weights on a wire until it breaks, the graph will stop abruptly at T.

Work done in plastic deformation

Since the work done by the extending force is given by $\int F \, dx$ we can see that the area under the load–extension graph is a measure of the energy needed during a deformation. Most of this energy ends up as additional vibrational kinetic energy of the atoms, in other words a plastic deformation produces a rise in temperature; this is what happens to the kinetic energy of vehicles involved in a collision. If a car is designed so that it will deform plastically at a chosen impact force, much of the energy can be dissipated in the process, tending to mitigate the effect on the motorist. A very rigid frame member or steering column will not absorb energy in this way; hence the move towards front ends that crumple and steering columns that collapse in a controlled manner on impact.

Structure of metals

To understand the behaviour of a metal beyond the elastic limit we have to look a little more closely at the way the atoms are arranged. Experimentally this is easier said than done because atoms are too small to see. It is not a question of building more powerful microscopes; the point is that light can be used to view an object only when it is several wavelengths across, and optical wavelengths are about a thousand times larger than atomic sizes. Instead we have to look at the way X-rays are diffracted by the material. This is a very big subject and we shall merely sketch out the broadest outline.

A casual glance at the arrangement illustrated in Figure 11.14 shows that the obstacle in the path of the light produces a sharp shadow on the screen. More careful scrutiny of the edge of the shadow reveals that it is not quite sharp; there are light and dark bands, quite close together (Figure 11.15). These have nothing to do with the familiar penumbra produced by an extended light source; they are caused instead by *diffraction* of light round the edge of the obstacle in accordance with its wave nature. (Diffraction is a phenomenon dealt with fully in textbooks on light.) If the obstacle is very small, comparable with the wavelength of the light, the diffraction pattern is very wide in comparison, and all trace of a sharp shadow disappears. The precise nature of the pattern depends on the shape and size of the obstacle and the nature of the light source. Figure 11.16 shows the pattern produced by a narrow parallel beam incident on a small circular obstacle. In Figure 11.17 we have a new development–this pattern is produced by an obstacle consisting of regularly spaced lines forming a rectangular grid; a similar effect can be seen by looking at a street lamp through an umbrella. The light waves coming from various parts of the grid interfere constructively in some directions and destructively in others. It is possible to work out what the diffraction pattern looks like, given all relevant details about the source and the obstacle, and it is also possible, though a little more tricky, to work out what the *obstacle* looks like from the diffraction pattern. Now, if we can find some radiation having a wavelength comparable with atomic dimensions, we can use the atoms in a tiny slab of material as our obstacles and work out how they are arranged from the resulting diffraction pattern. Light waves are far too long for this job, but X-rays, having wavelengths in the region of 10^{-10} m, are just right. The patterns that emerge can be quite complicated, remembering that the obstacle is now a three-dimensional one; Figure 11.18 shows an example.

Figure 11.16
Diffraction of light by a circular obstacle (laser source).

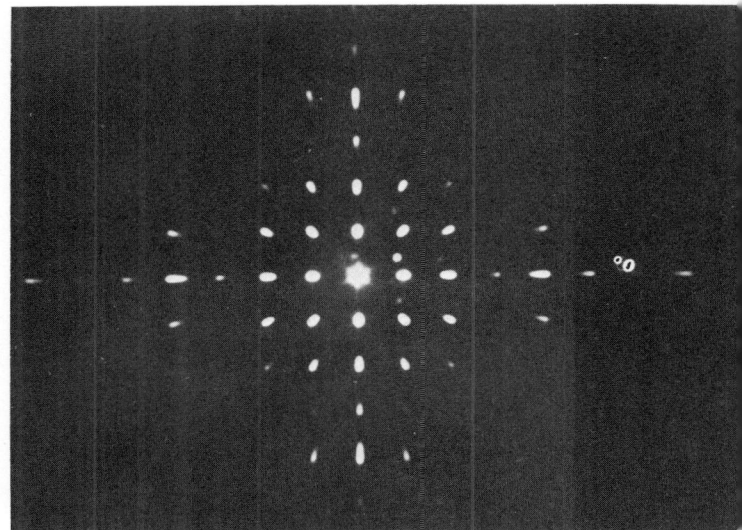

Figure 11.18
X-ray diffraction pattern

Figure 11.17
Diffraction of light from a point source by a rectangular grid.

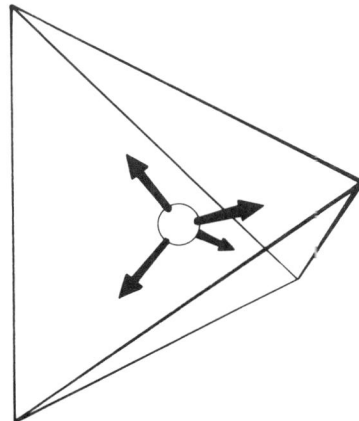

Directions of bonds for a carbon atom in diamond.

The first thing these patterns tell us is that in many materials the atoms are spaced in a regular fashion, so that the view from any point in the bulk of the solid is exactly repeated a little further on. Any substance having a regular structure like this is called a *crystal*, and metals form just one class of crystalline materials. There *are* solids which do not have a regular structure, glasses for example, but they form the exception rather than the rule.

Our simple billiard-ball picture of atoms is particularly successful in accounting for the behaviour of metals, for two reasons. The first is that only one type of atom is present, apart from any impurities, and the second is that the forces between atoms in a metal do not have any preferred directions, unlike the forces between carbon atoms for example, which act in the directions of lines drawn from the centre of a regular tetrahedron to its four corners. Metals act very much as though they are made up of sticky spheres. We cannot

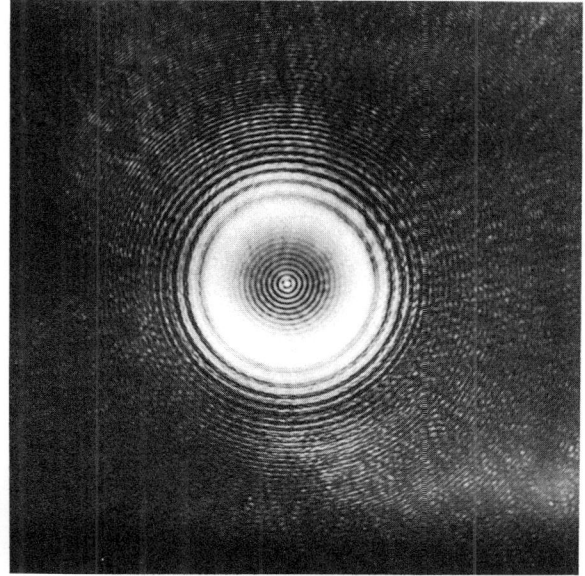

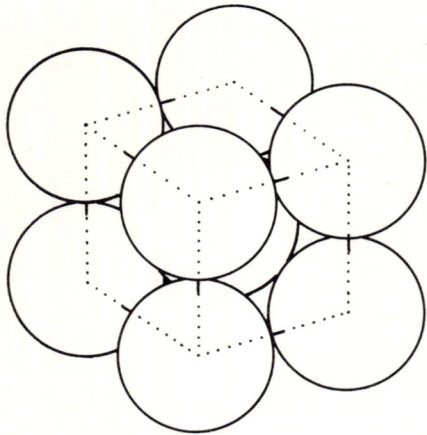

Figure 11.20
B.C.C. structure.

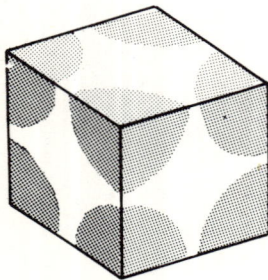

Figure 11.21
B.C.C. unit cell.

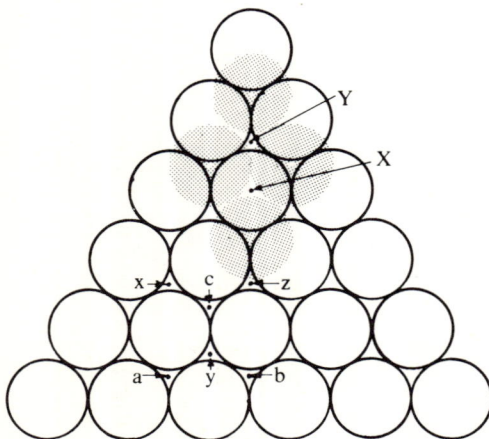

Figure 11.22

give an adequate explanation for the attraction between the atoms in terms of our simple model, but it is linked with the fact that one or two of the outer electrons of each atom are bound quite loosely. They are tied to the atom if it is isolated, but if it is surrounded by close neighbours, all exerting strong electrical forces, these outer electrons can no longer be said to belong to an individual atom but are free to wander about throughout the entire crystal, though they are discouraged from leaving it. This cloud of 'free electrons' is responsible for many of the properties peculiar to metals, such as high thermal and electrical conductivity.

How are the atoms themselves arranged? We get an extraordinarily good picture if we imagine them like billiard balls stacked in a box. There is more than one way of stacking, in fact there are three important ways as far as metals are concerned. If we imagine we have a box with a rectangular base, we can form one layer of atoms in a square array as shown in Figure 11.19. We are then left with a large

Figure 11.19

number of little valleys in which we can fit the atoms for the next layer. The atoms in the third layer will as a result fit directly on top of those in the first layer. Atoms of chromium and tungsten are stacked like this, and so are iron atoms at low temperature; the arrangement is called *body-centred cubic* (b.c.c.). Figure 11.20 explains why; all the details of the structure can be neatly summarised by talking about just nine atoms. Four of them are from the bottom layer, forming the corners of a square, four from the third layer complete a cube, and one from the second layer is in the middle. We can think of the entire structure as being made up of cubes like the one in Figure 11.21, all piled neatly in line with each other in rows, columns and layers. Of course the eight atoms at the corners belong to more than one cube, in fact each one belongs to *four* cubes, and only one quarter of each of them is inside the cube drawn. This cube is called the *unit cell* of the arrangement.

It may be a surprise to find that the b.c.c. array is not the one which will pack most atoms into a given volume. If we start instead by forming the first layer in rows making up equilateral triangles we can do a little better (Figure 11.22). There are two ways of starting the second layer: we can fill in the valleys either at *a*, *b*, *c*, or at *x*, *y*, *z*. We need only turn the picture round to make the first set look like the second, so it is immaterial which we choose, but the *third* layer can likewise be started in two different ways which are related differently to the *first* layer. We can start with an atom in valley *X*, thus getting the third layer immediately above the first, or we can start in valley *Y*, and although we get the same number in a given volume either way, the two possibilities give us different structures. The best way to describe the first is in terms of a hexagon in the first layer, made up of six outer atoms plus one in the middle, a similar hexagon directly above it in the third layer, and three atoms in the second layer (Figure 11.23). This arrangement is known as *hexagonal close packed*

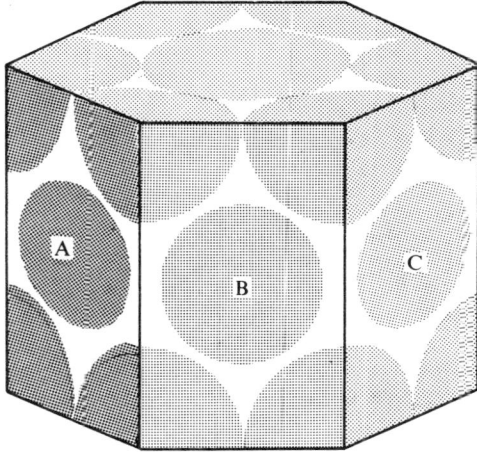

Figure 11.24

H.C.P. unit cell. If atoms *A* and *C* have their centres inside the cell, atom *B* has its centre outside.

(h.c.p.), and the hexagonal cylinder in Figure 11.24 forms the unit cell. Magnesium, zinc and titanium atoms stack this way.

Exercises 11.20, 21

20 To how many such hexagons do the atoms labelled *P*, *Q* and *R* respectively belong?
21 Show that although we can tile a floor so as to leave no gaps if we use tiles shaped as equilateral triangles, squares or regular hexagons, we cannot use regular pentagons, or in fact any other regular polygons.

If the third layer is started at *Y* instead (Figure 11.22), something rather interesting happens. Looking at the atoms labelled *A* to *F* in Figure 11.25, we may think we have regular hexagons again, *but the atoms are not all in the same plane.* A cube standing on one point and viewed from above also looks like a regular hexagon (Figure 11.26),

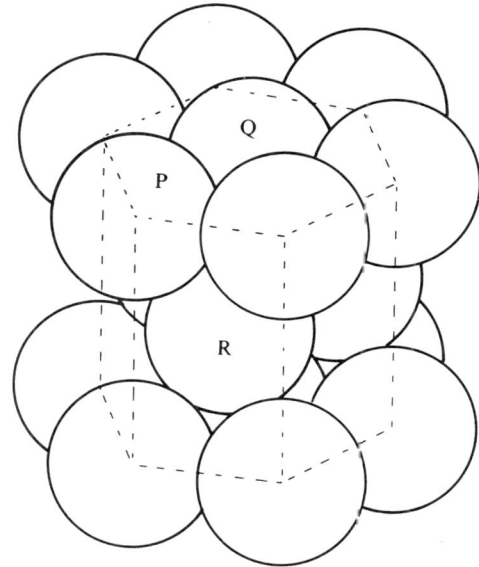

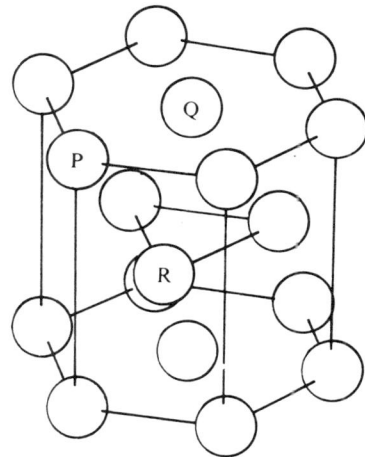

Figure 11.23
Two representations of the h.c.p. structure.

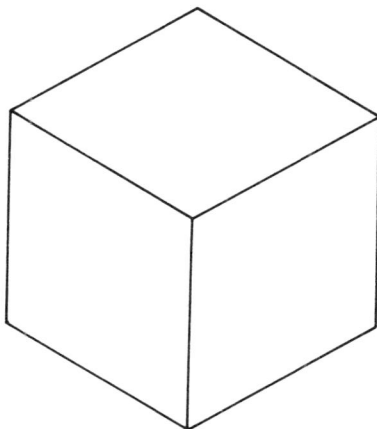

Figure 11.26
Regular hexagon or cube on one edge?

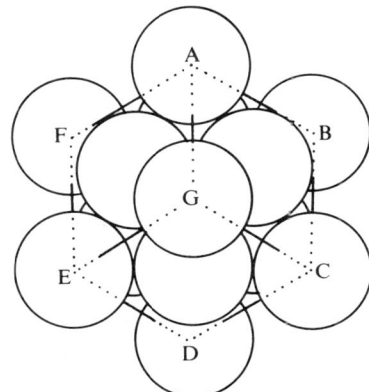

Figure 11.25

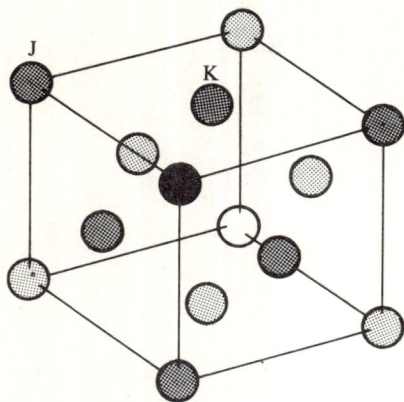

Figure 11.27

F.C.C. structure. The arrangement consists of one atom from the top layer (black), six atoms from the second layer (darker shading), six from the third layer (lighter shading) and one from the fourth layer (unshaded).

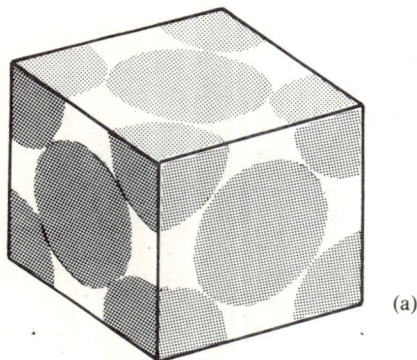

(a)

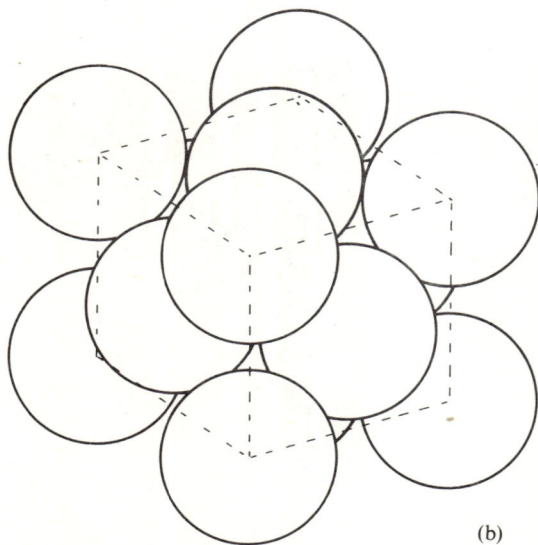

(b)

Figure 11.28

Further representations of the f.c.c. structure; (a) shows the unit cell.

and the best way to describe the new structure is by regarding atoms *A* to *F* as forming six corners of a cube. Atom *G* in the third layer makes the seventh corner, and the eighth must be supplied by an atom stacked underneath the first layer. This cube is larger than the body-centred cube; it does not have an atom in the middle but it does have an extra one in the middle of each face, and the structure is called *face-centred cubic* (f.c.c.). The unit cell is shown in Figure 11.28a. Copper and silver are two metals having f.c.c. structures.

Exercise 11.22

To how many unit cells do the atoms labelled *J* and *K* respectively belong? (See Figure 11.27.)

The bubble raft model

Although billiard balls are convenient to talk about they would not make a very good *practical* model with which to try out the ideas we have about the way atoms behave. The chief drawbacks are that we need a tremendous number of them to represent the tiniest fragment of metal, and, more serious, we need some close-range interatomic attractions. A much better basis for a working model was proposed by Sir Lawrence Bragg; this was a collection of tiny soap bubbles. By blowing gas from a steady supply through a small hole in the end of a pipette dipping into soap solution, a very large number of tiny identical bubbles are formed, a millimetre or less in diameter, and these arrange themselves in a raft on the surface. They have all the simple properties we have credited to atoms; they attract at short range due to surface tension effects (Chapter 12) and repel relatively strongly at even closer range due to the compression of the gas inside them.

Grain boundaries

The raft of bubbles formed on the surface by this process has the structure corresponding to the first layer of a hexagonal close packed or face-centred cubic crystal; the bubbles line up as shown in Figure 11.29. Indeed it is possible to form rafts several layers thick and get

Figure 11.29

Perfect bubble raft.

either type of structure emerging. But there is one important development: the raft is seldom perfect. Very often it is broken up into regions – we call them crystal grains – within which the bubbles are arranged correctly, but the lines in one region are at some odd angle with respect to those of the adjacent one. The regions are separated by irregular lines, two or three bubbles thick, within which there is an untidy arrangement as the order of one grain gives way to the order of another (Figure 11.30). These lines are called *grain boundaries*. We have good reason to believe that *the same phenomenon occurs with atoms in a metal*: indeed, the effects of it can be seen with the naked eye in some cases. If a metal surface is carefully etched, that is, attacked with acid so as to reveal a clean surface, the grains can be made visible, and are sometimes as much as a millimetre across. The acid attacks the metal in such a way that each grain becomes a reflecting surface with an orientation slightly different from the next one, and these surfaces catch the light in different ways. An old brass door-knob shows the effect very well; countless hand-grasps can produce an excellent etch.

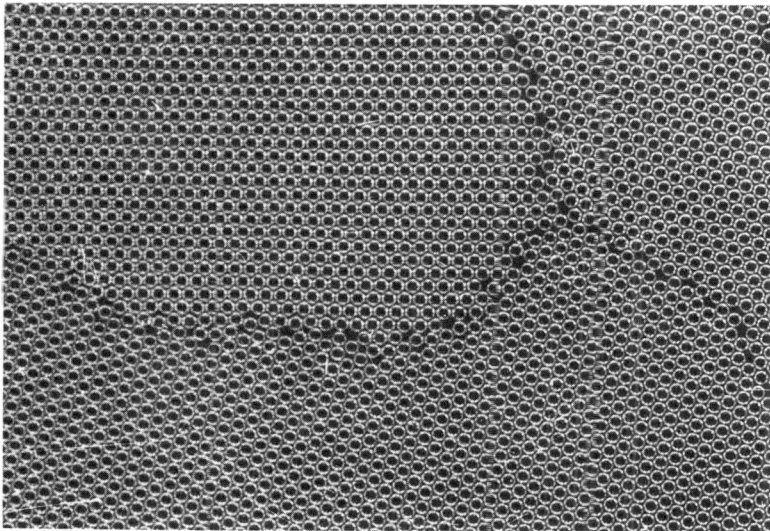

Field ion micrograph of the tip of a fine tungsten wire, showing a grain boundary. The small white dots show the positions of individual atoms.

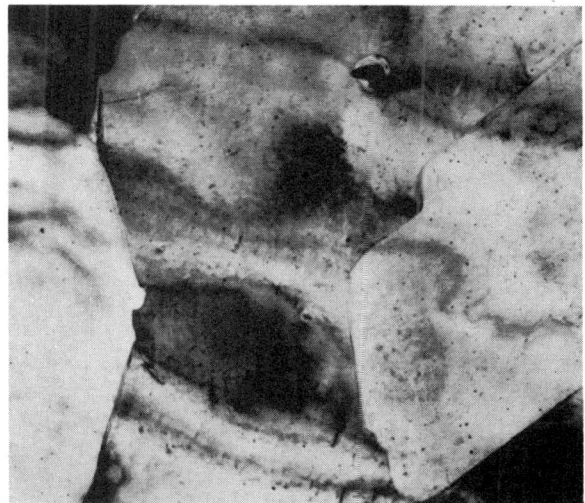

Figure 11.30
Grain boundaries in a bubble raft.

The grains in a bubble raft get larger as time goes on; of course they cannot *all* get larger, but some grow at the expense of others as bubble after bubble snaps into place. Over a *very* long period the same thing can happen with a metal, and if the metal is heated it can happen a lot quicker. A copper wire heated in a flame and then allowed to cool will contain relatively few grain boundaries. We can tell it is different by its lack of rigidity; it is much easier to bend than before, and we can find a very satisfactory explanation for this state of affairs.

Dislocations

Even within one crystal grain in a bubble raft the bubbles are seldom arranged quite perfectly. It often happens that at some point one or two bubbles are missing, and rather than simply leave a hole the bubbles nearby fill it in and 'share out' the fault, so that the effect is spread over perhaps twenty or thirty bubbles in a line (Figure 11.31). Such a region is called a *dislocation*, and dislocations turn

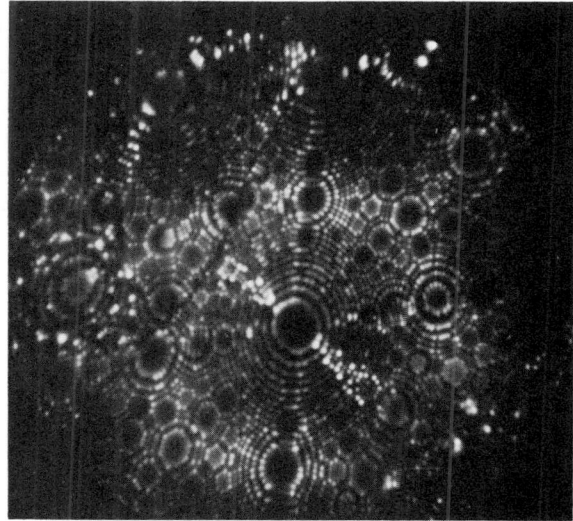

Grain boundaries in mild steel (using an electron microscope) Magnification: 16 000×.

Figure 11.31

A dislocation in a bubble raft.

out to be very important. If any attempt is made to stretch or compress the raft with suitable booms immersed in the soap solution, there is first some evidence of elastic behaviour, whereby each inter-bubble separation along the line of stress changes a little to compensate, but then plastic deformation occurs, and lines of bubbles slip with respect to each other. The way they slip is very interesting.

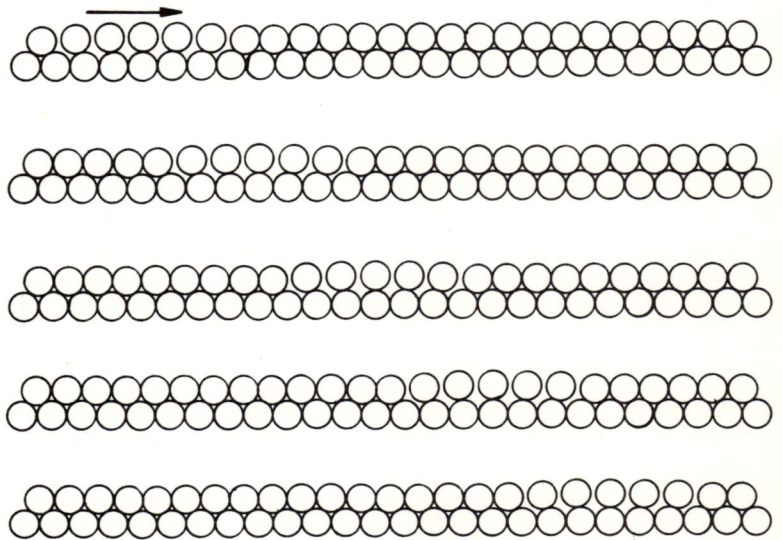

Five stages in the movement of a dislocation from left to right (view from side with book nearly at eye level).

What happens is that bubbles at one end of a dislocation plug more firmly into their proper places, as a result of which bubbles near the other end get pulled out of place slightly, and in this manner the dislocation moves very rapidly down the line until it hits a grain boundary, where it stops. The force needed to do this is far less than the one which would be required to lift all the bubbles in one line simultaneously over the hump and deposit them in the next site. It is rather like adjusting a heavy carpet; it is much easier to start a pucker at one end and work it down the length of the carpet than it is to drag the entire carpet a few centimetres.

Now we can understand why copper which has been *annealed* (heated and then cooled slowly) is so soft. The large crystals in the copper allow the dislocations to travel freely; they generally start at a grain boundary and move across the grain to the other side, and it is very easy to slide layers of atoms over each other. In the very act of bending the copper however, these large crystals are broken up into much smaller ones, and so the dislocations, which are halted at grain boundaries, have very much less freedom. What is more, too many dislocations get in each other's way, like traffic at a cross-roads, and as a result of all this the metal becomes harder. Easy to bend once, it is more difficult to bend again or to straighten the wire, which is said to be *work hardened*. One way of performing the strong man act with a poker is to anneal the poker beforehand to make it easier to bend, and then invite any member of the audience to straighten it again.

Impurities

A grain boundary stops a dislocation, and so does an *impurity* atom – anything to break up the regular pattern. The presence of small amounts of impurity in metals can drastically alter their behaviour under stress: witness the effect of adding proportions of carbon and other elements to soft iron to form steel.

Slip planes

The more densely the atoms are packed in a given layer, the easier it is for slip to occur between that layer and the next one because, as Figure 11.32 illustrates, there is less of a 'hump' to climb. The face-centred cubic structure has several such close-packed planes in various directions and dislocations can move freely along each one of these, so that metals with this structure are usually quite malleable and ductile. The body-centred cubic structure is not quite so closely packed, and the hexagonal close packed structure is quite asymmetric, the plane containing the hexagons being the only one with really close packing, as a result of which metals with this structure tend to be more brittle. Another simple example of the principle is provided by two well-known structures for carbon. The diamond structure is shown in Figure 11.33; the tetrahedral arrangement comes about because of the directional character of the bonds formed between the carbon atoms. The atoms in any given plane are quite well separated, resulting in a very rigid structure. In perfect graphite crystals on the other hand the atoms are packed close in parallel planes separated by an appreciable distance, and these planes can slip on each other very easily. This accounts for the fact that graphite is very effective as a lubricant.

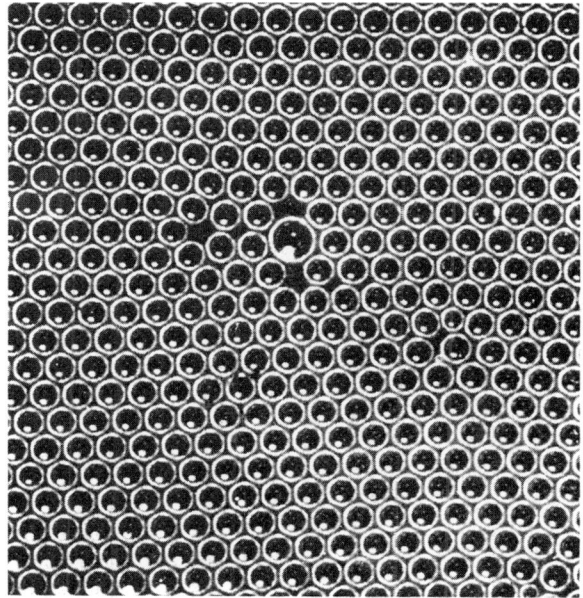

Bubble raft model of an impurity atom.

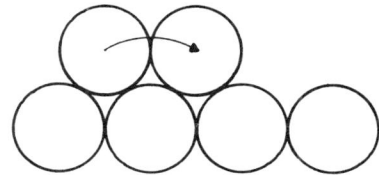

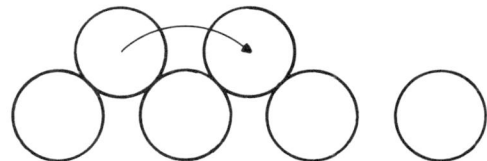

Figure 11.32

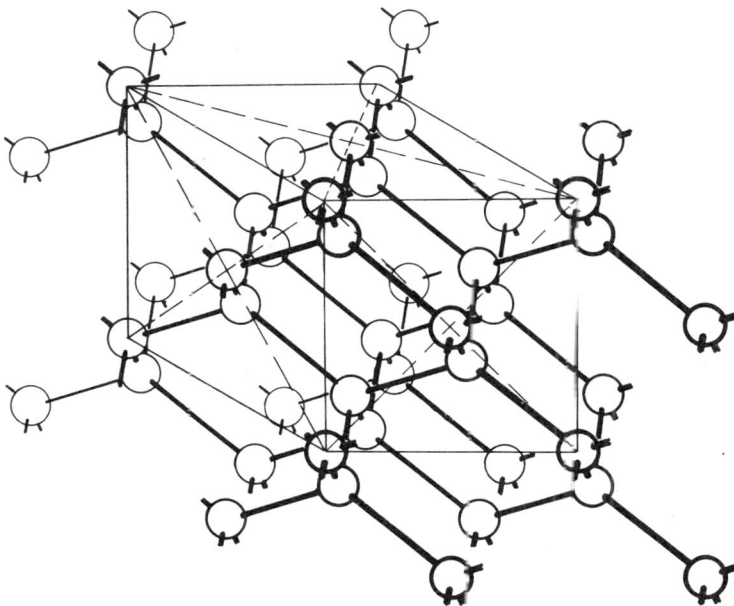

Figure 11.33
Structure of diamond.

Diamond

Whether a material is brittle or malleable depends a great deal on whether it is more resistant to shear or to tensile stress. Suppose by hitting an object we open up a tiny crack near the surface. We can think of that crack as a rather severe dislocation which sets out at high velocity; will it split the object in two, rather like opening a zip-fastener, or will it send out other dislocations in all directions like branches from a trunk, and spread the effect over the whole object? The answer to that depends on whether it is easier to pull atoms apart or slide them over each other. For the face-centred cubic structure sliding is always simpler, and metals having this structure are not brittle.

At the other extreme the diamond structure is extremely resistant to slip, and of course diamonds are brittle. What is more they are easier to break apart in some directions than others; a sharp tap on a fine blade held at just the right angle will cleave a diamond into two pieces having corresponding flat faces. The directions of the *cleavage planes* are decided by the atomic arrangement; they are planes across which there is less available attractive force than usual because of the interatomic spacing.

Further exercises

Exercises 11.23–37

23 In an experiment to measure Young's modulus for steel a wire is suspended vertically and loaded at the free end. In such an experiment (a) why is the wire long and thin, (b) why is a second steel wire suspended adjacent to the first?

Sketch the graph you would expect to obtain in such an experiment showing the relation between the applied load and the extension of the wire. Show how it is possible to use the graph to determine (a) Young's modulus for the wire, (b) the work done in stretching the wire.

Given that Young's modulus for steel is $2.00 \times 10^{11} \, \mathrm{N m^{-2}}$ calculate the work done in stretching a steel wire $1.00 \, \mathrm{m}$ in length and of cross-sectional area $3.0 \, \mathrm{mm^2}$ when a load of $100 \, \mathrm{N}$ is slowly applied without the elastic limit being reached.   (JMB)

24 State *Hooke's law* and explain the terms *elastic limit*, *Young's modulus*.

Draw a labelled diagram of an apparatus which could be used to determine accurately the extension of a long wire suspended vertically and loaded at the free end.

When the load on the wire, which is of length $2.00 \, \mathrm{m}$ and diameter $0.60 \, \mathrm{mm}$, is gradually increased from $17.5 \, \mathrm{N}$ to $42.5 \, \mathrm{N}$, the scale reading on the apparatus changes from $1.78 \, \mathrm{mm}$ to $2.61 \, \mathrm{mm}$. Calculate (a) Young's modulus for the material of the wire, (b) the additional potential energy stored in the wire.

(JMB)

25 In an experiment to measure Young's modulus by stretching a metal wire, the diameter of the wire was measured by a micrometer as $0.52 \, \mathrm{mm}$ with an error of $0.01 \, \mathrm{mm}$. The extension produced was $0.73 \, \mathrm{mm}$ with an error of $0.01 \, \mathrm{mm}$. The length of the wire, measured with a ruler was $2 \, \mathrm{m}$ with an error of $0.5 \, \mathrm{cm}$. The load measurement was without error. Calculate to 1 significant figure the maximum percentage error in the value of Young's modulus calculated from these results, explaining your calculation.

(SUJB)

26 Two wires, each one metre long and of $1\,mm^2$ cross-section, one of steel and the other of brass, are connected end to end. What tensile force would be required to extend the whole wire by 1 mm? (Young's modulus for steel is $2 \times 10^{11}\,Nm^{-2}$ and for brass is $10^{11}\,Nm^{-2}$.)   (O & C)

27 State Hooke's law. If a wire, of diameter d and length L, is subjected to a tensile force F, it increases in length by e. Write down the values of the stress and the strain in this case. Describe an experiment with a thin metal wire to investigate the limitations of Hooke's law and discuss the results obtained.

A steel wire, whose diameter is 0·52 mm and length 2·50 m extends by 2·88 mm for a load of 50 N. Find the value of Young's modulus for the steel of this wire. If the diameter was measured as 0·51 mm what would be the percentage error in the value of Young's modulus? What error in measuring the length of the wire would produce the same percentage error in the result?

(WJEC)

28 Define *modulus of elasticity*. What are its dimensions in terms of mass, length and time?

When waves are travelling along a thin rod made of an elastic solid of density ρ and Young's modulus E the velocity v of the waves is given by the formula $v = \sqrt{(E/\rho)}$. By dimensional considerations show that this is a *possible* formula.

The breaking strain of steel is 10^{-3}. Assuming that Hooke's law is obeyed until the rod breaks, find the tensile force required to break a steel rod 2 mm in diameter. Why is the work which must be done in order to break the rod dependent on its length although the force required to break it is not?

(Take Young's modulus for steel to be $2 \times 10^{11}\,Nm^{-2}$.)   (O)

29 A uniform steel bar of cross-sectional area $1·5\,cm^2$ and length 1·00 m at 10°C is heated to 60°C. At this temperature the ends of the bar are fixed to rigid supports. The bar is then allowed to cool to 10°C. Calculate (a) the tension in the bar, (b) the additional potential energy which is now stored in it.

(Coefficient of linear expansion of steel $= 1·2 \times 10^{-5}$ per °C. Young's modulus for steel $= 2·0 \times 10^{11}\,Nm^{-2}$)   (JMB)

30 A uniform rigid disc is suspended by means of four uniform parallel vertical wires which are clamped so that the plane of the disc is horizontal and each wire is under tension. These wires are each 2·00 m long and of cross-sectional area $0·5\,mm^2$. Three of the wires are made of steel and are attached to points equispaced on the periphery of the disc. The fourth wire is made of brass and is attached to the centre of the disc. When an additional mass of 40 kg is hung from the centre of the disc the wires are each extended by the same amount and the disc remains with its plane horizontal. Calculate the extension produced and the increase in tension in each wire when the 40 kg mass is added.

(Young's moduli: for steel $= 2·1 \times 10^{11}\,Nm^{-2}$ and
for brass $= 9·8 \times 10^{10}\,Nm^{-2}$.)   (O & C)

31 A copper wire and a steel wire, each 1·0 m long and $1.0\,mm^2$ in cross-sectional area, are laid side by side and are joined together at the ends. The composite wire is then placed vertically, with the upper end clamped and the lower end supporting a mass of 0·8 kg. Calculate the tension in each wire and the elastic potential energy of the system, stating any assumptions made.

(Values of Young's modulus: copper $= 1\cdot2 \times 10^{11}\,\mathrm{N\,m^{-2}}$
steel $= 2\cdot0 \times 10^{11}\,\mathrm{N\,m^{-2}}$.)   (C)

32 A helical spring of negligible mass is allowed to hang freely from one end, which is fixed. The other end is then attached to an object with a mass of 800 g which is placed on a support so that the spring is initially neither extended nor compressed. The support is then lowered very slowly, and when it has moved down by 5 cm the object is left hanging on the spring. Find (a) the loss of gravitational potential energy, (b) the work done on the support, and (c) the energy stored in the spring, assuming that Hooke's law is obeyed.

Describe the motion of the object if the support were suddenly completely removed instead of being lowered slowly.   (O & C)

33 Show that the energy stored per unit volume in a stretched wire is equal to half the product of the stress and the strain.

A catapult consists of two rubber cords of unstretched length 10·0 cm and area of cross section 0·40 cm². Assuming that all the energy stored in the stretched cords is converted into kinetic energy of the missile, calculate the maximum height to which a stone of mass 100 g could be projected if the cords were stretched by 5·0 cm.

(Assume Young's modulus for rubber to be $1\cdot00 \times 10^{7}\,\mathrm{N\,m^{-2}}$.)

(AEB)

34 Define *stress* and *strain*.

Describe the behaviour of a copper wire when it is subjected to an increasing longitudinal stress. Draw a stress–strain diagram and mark on it the elastic region, yield point and breaking stress.

A wire of length 5 m, of uniform circular cross section of radius 1 mm is extended by 1·5 mm when subjected to a uniform tension of 100 N. Calculate from first principles the strain energy per unit volume assuming the deformation follows Hooke's law.

Show how the stress–strain diagram may be used to calculate the work done in producing a given strain, when the material is stretched beyond the Hooke's law region.   (O & C)

35 Discuss the meaning of the term *elastic* in the following statements:

a) The molecules of a gas are assumed to be perfectly elastic;
b) A steel wire, subjected to an increasing load, is elastic up to a certain limit.

Describe an experiment to investigate the truth of the second statement.

A hollow cylindrical steel pillar of external diameter 0·10 m and length 3·00 m is required to support a compression load of $10^{5}\,\mathrm{N}$ without showing a reduction in length of more than 1·0 mm. What is the maximum internal diameter of the pillar?

(Young's modulus for steel $= 2\cdot0 \times 10^{11}\,\mathrm{N\,m^{-2}}$.)   (C)

36 Show that the energy stored in a rod of length L when it is extended by a length l is $\frac{1}{2}El^{2}/L^{2}$ per unit volume, where E is Young's modulus of the material.

A railway track uses long welded steel rails which are prevented from expanding by friction in the clamps. If the cross-sectional area of each rail is 75 cm², what is the elastic energy stored per kilometre of track when its temperature is raised by 10°C?

(Coefficient of thermal expansion of steel $= 1\cdot2 \times 10^{-5}$ per °C; Young's modulus for steel $= 2 \times 10^{11}\,\mathrm{N\,m^{-2}}$.)   (O & C)

37 Define Young's modulus and explain how it might be measured for a steel rod 1 m long, having a square cross section area of 25 mm².

The same rod is bent to form a circle and the two ends are welded together. The circle is heated until its temperature is raised by 180°C, when it is slipped over a wheel which it then fits exactly. Estimate the pressure exerted on the wheel when the circle cools to its original temperature.

(Young's modulus for steel $= 1\cdot95 \times 10^{11}\,\mathrm{N\,m^{-2}}$.
Coefficient of linear expansion $= 1\cdot10 \times 10^{-5}$ per °C.) (C)

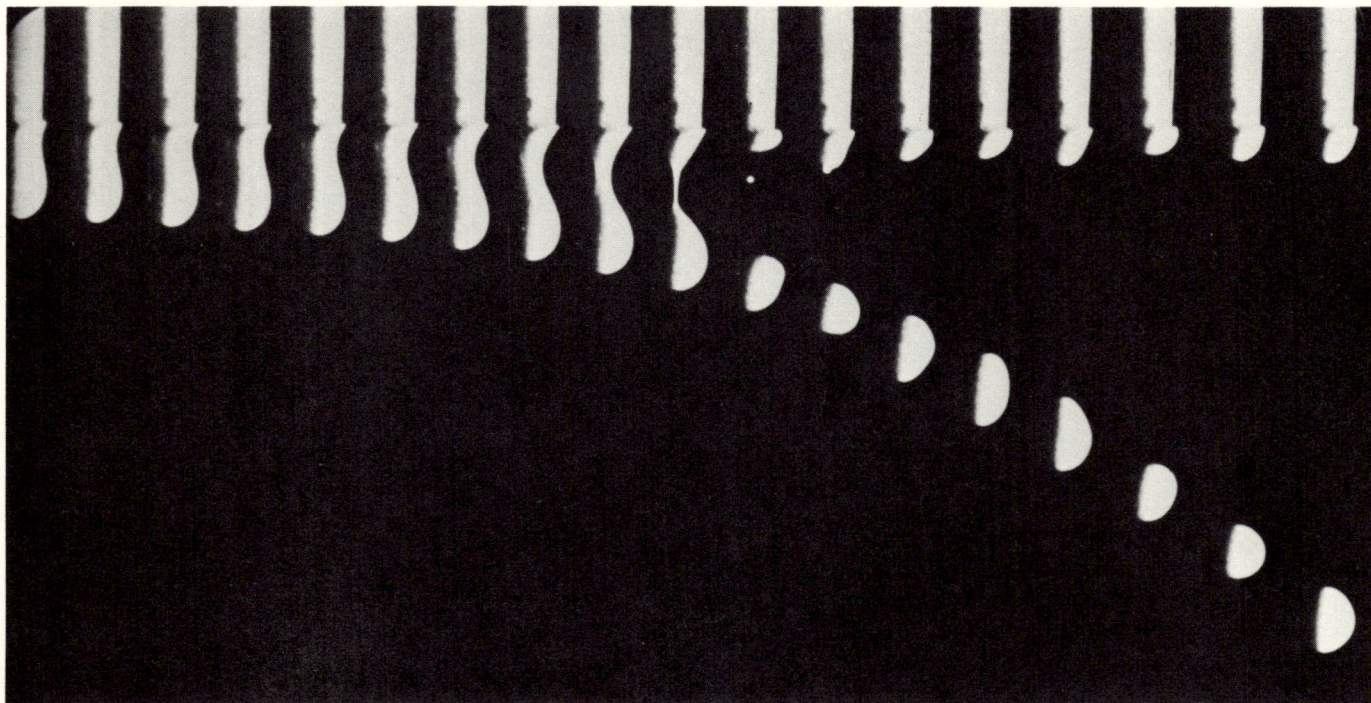

Stages in the fall of a drop of liquid.

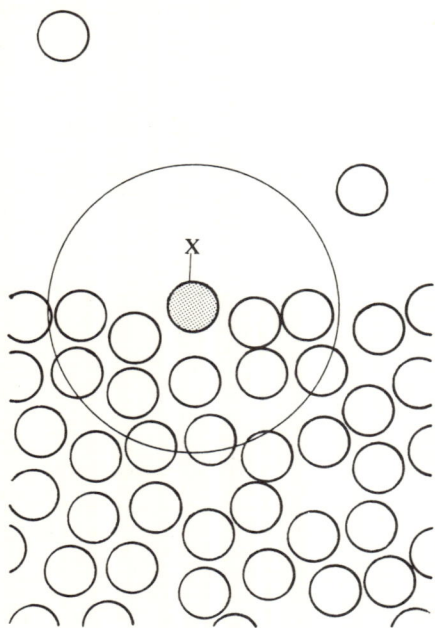

Figure 12.1
Sphere of influence for molecule X.

No doubt nearly everyone has seen the trick of supporting a slightly greasy steel needle or razor blade on the surface of water. We are very used to the fact that the surface has special properties not shared by the body of the liquid, and the smaller the quantity of liquid we are dealing with, the more important the surface effects become. A drop of water on a wax polished surface behaves very differently from a bucketful; it shows every tendency to draw into a sphere, and the smaller it is the more nearly it succeeds. In these and many similar situations the indications are that the surface is in a state of tension, and has a tendency to assume the minimum area possible, consistent with other factors such as gravity which may be present.

Can we find any explanation for such surface phenomena in terms of our microscopic model of a liquid? We talked in Chapter 1 in terms of a huge collection of molecules in random motion, vibrating and changing places continuously. They are held together as a body by short-range intermolecular attractive forces (see p. 302), but repulsive forces of even shorter range prevent any overlapping, so that liquids, like most solids, have large bulk moduli, and for most purposes can be considered virtually incompressible.

A molecule in the body of the liquid experiences pushes and pulls from all sides and on average these cancel each other out, leaving the molecule tolerably free to jostle its way through the crowd in any direction. We can see that the situation near the surface of the liquid will be a little different however. In Figure 12.1 we have drawn the picture our model gives of a molecule X in the interface between the liquid and the air above it. Any molecule outside the sphere drawn round X is too far away to attract it appreciably; all the attractions and repulsions it experiences come from other molecules within the sphere, which is just a few molecular diameters across. The obvious point to note is that only half of the sphere is filled with molecules—the gas molecules above the surface are too few and far between to count for anything. This means that the attractive forces on molecule X, which has wandered into the surface layer, pull it in towards the body of the liquid. A similar argument applies to every molecule within two or three molecular diameters of the surface. If we were dealing with a solid, we could say that X would be drawn in a little closer to its neighbours, just enough to produce a corresponding increase in repulsive forces, maintaining equilibrium. To that picture, however, we have to add the idea that all the molecules are continually jostling and changing places, so that all the time molecules return from the surface to the body of the liquid, their places being taken by other fairly energetic molecules near the surface which happen to be moving out. This can give a satisfactory explanation of the tendency towards minimum surface area; every molecule is encouraged to leave the surface and discouraged from joining it. Can we get a direct picture of lateral tension in the surface? This is a somewhat debatable point; the suggestion has been made (R. C. Brown, *Proc. Physical Soc.*, Vol. LIX, p. 429, 1947) that the surface is somewhat deficient in molecules, because by and large it is only the more energetic molecules that get to it. This would mean that the average intermolecular spacing in the surface layer is greater than in the body of the liquid. If this is so – and it would be difficult to find an independent test – then the surface layer acts in some respects like a skin under tension, since there is no longer exact balance of attractive and repulsive forces between adjacent molecules. Just as in the case of a stretched skin, such a situation can exist only if there are some external forces to maintain the tension; these, as we shall see, are provided at the edge of the surface, where it meets a solid wall for example. Whatever the explanation, experiment leaves no doubt that there is lateral tension in the surface.

Attraction between liquids and solids

Just as there are attractive forces between *identical* molecules when they are pulled apart (we call these *cohesive* forces), we find that there can also be attractive forces between molecules of different substances – *adhesive* forces. What happens at a liquid–solid interface depends on whether the adhesive forces between liquid and solid are stronger or weaker than the cohesive forces between the liquid molecules. A slightly greasy razor blade sits on the top of a clean water surface because the cohesive forces between the water molecules are stronger than any adhesive forces between water and grease, and the attraction which can be called into play between the surface molecules is sufficient to enable them to hold the line, like police linking arms to keep back a crowd. We say the water does not *wet* the blade. On the other hand if we dip the edge of a clean glass slide into water a very different picture emerges (Figure 12.2a); this time the adhesive forces are stronger and they are able to pull the water surface up as shown in order that as many water molecules as possible may be in contact with the glass. We say that water *wets* glass. If the glass is slightly greasy or dusty we may well get the shape in Figure 12.2b, because the adhesive forces between water and the surface impurity may be smaller than the cohesive forces between the water molecules. It is a general principle for all surface tension investigations that apparatus must be scrupulously clean; any impurities present seem to have a habit of congregating at the liquid surface and drastically altering its behaviour. (It should be said that adequate treatment of the interaction between liquids and solids is best attempted in terms of the *energies* involved, but it is not proposed to develop this approach here.)

Lower end of vertical glass plate immersed in water.

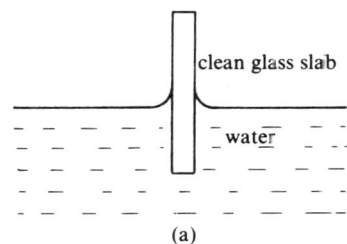

(a)

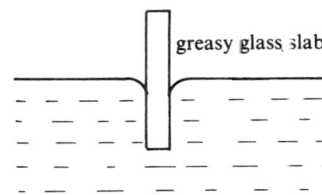

(b)

Figure 12.2

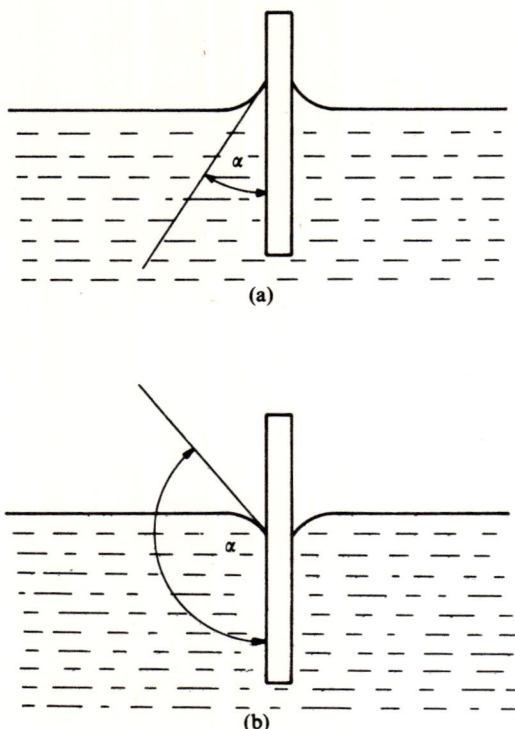

Figure 12.3

Angle of contact.

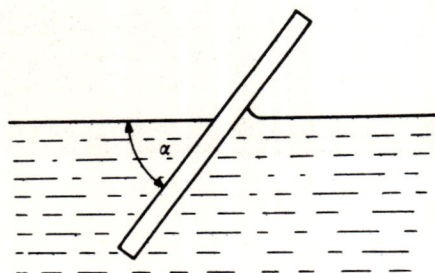

Figure 12.4

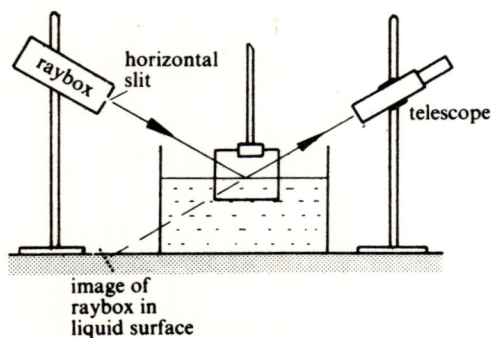

Figure 12.5

Angle of contact

The water surface in Figure 12.2a will in fact rise to meet the glass surface tangentially; this is not the case for all liquids meeting solid surfaces. A more general situation is shown in Figure 12.3a, and we define the *angle of contact* α as *the angle between the solid surface and the tangent to the liquid surface at the point of contact with the solid, measured in the liquid*. Figure 12.3b shows the angle of contact between a solid and a liquid which does not wet it; in accordance with the definition we measure this time an obtuse angle. For water and alcohol in contact with glass, α is usually taken to be zero; for paraffin in contact with glass it is about 25°, and for mercury in contact with glass it is 140°. The angle of contact is a rather uncertain quantity in a given situation, rather like a coefficient of friction, and like a coefficient of friction it is highly dependent on the state of the surface. It also depends on whether the surface is advancing or receding and how *quickly* it is advancing or receding. The standard conditions for measurement are for a stationary surface which has recently receded. It can be measured by dipping a plane slab of the solid into the liquid as shown in Figure 12.4. As the slab is rotated, a point is reached at which the liquid surface is horizontal right up to the slab; the angle α is then the angle of contact. Figure 12.5 shows a convenient way of recognising the correct setting; the telescope is focused on the image of an illuminated horizontal slit formed by reflection at the liquid surface, and α is adjusted until the image appears horizontal right up to the slab.

Macroscopic definition of surface tension

We now want to put the idea of tension in a liquid surface on a quantitative basis; to do that we look at the physical situation indicated in Figure 12.6. We have here a rectangular wire frame on which rests a wire which is free to move, and the whole arrangement has been dipped in soap solution and removed so as to form a rectangular film–a thin layer of liquid, perhaps one micrometre thick. If we break the left-hand portion of the film, the wire will be pulled to the right unless we provide a force to prevent it. The force required turns out to be proportional to the width of the frame but independent of the film thickness (we can vary the thickness simply by moving the wire to a different place on the frame after breaking the left-hand portion of the film). This confirms that we are dealing with a *surface* phenomenon. In symbols,

$$F \propto l,$$

where l is the length of the line of contact between the wire and the film; l is equal to *twice* the frame width because the film has two plane surfaces, upper and lower, with liquid between; both contribute to the force on the wire. Introducing a constant γ, we write

$$F = \gamma l,$$

where γ depends on the nature of the film but is independent of its thickness; it is called the *surface tension* (or the *coefficient* of surface tension, to distinguish it from the general phenomenon) of the soap solution and has units Nm^{-1}. An operational definition of surface tension can therefore be composed in the following terms: *surface tension is the force per unit length exerted by a liquid surface perpendicular to its edge and tangential to the surface, where it meets a solid to which it adheres.*

We must deal with an immediate objection; surely γ is a quantity which depends on the liquid *and* solid surface, since it is a measure of the force which the liquid exerts on the solid? Well, if we change the wire for another one made of a different metal, we find γ to be the same. The implication is that providing the liquid *wets* the wire (and if it does not we cannot form our film) the adhesive forces between film and wire are significantly stronger than the cohesive forces between liquid molecules. The molecules of solution actually in contact with the wire are held there firmly by adhesive forces, but if the film is to remain intact then molecules further out must hold on to their neighbours by cohesive forces. The strength of a chain is settled by its weakest link; it is the cohesive forces which decide the value of γ.

Direct measurement of surface tension in terms of the definition
Figure 12.7 shows a suitable arrangement for water, or indeed any liquid which wets glass; the glass slide is suspended from a balance.

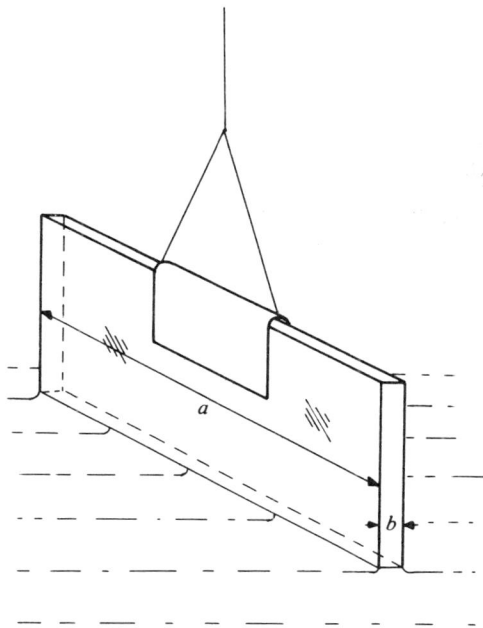

Figure 12.6

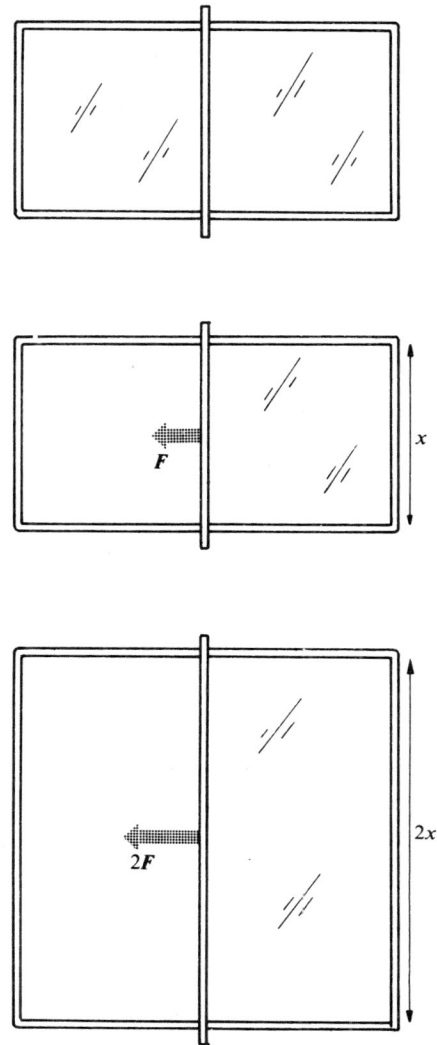

Figure 12.7
Direct measurement of surface tension.

We have to record the weight W_1 of the dry slide, and then the force W_2, indicated by the balance, which is just sufficient to support the slide with its base at the same height as the general level of the liquid. The arithmetic for this determination is very simple:

$$W_2 - W_1 = 2\gamma(a + b)\cos\alpha,$$

where a is the length of the glass slide and b is its thickness; α is the angle of contact. Perhaps it needs to be emphasised why it is the *perimeter* of the base of the slide and not its area which figures in the formula. Why do we not argue in terms of cohesive forces across the whole of the horizontal surface indicated by the line PQ (Figure 12.8) for example? Experiments with slides of different sizes can soon establish that it *is* the perimeter length $2(a + b)$ and not the area ab which controls the force required, but why should this be so? The answer lies in the fact that we are dealing with a liquid, not a solid. If the slide were adhering to a solid, then as we attempted to raise the slide above the general level, as in Figure 12.9, the average

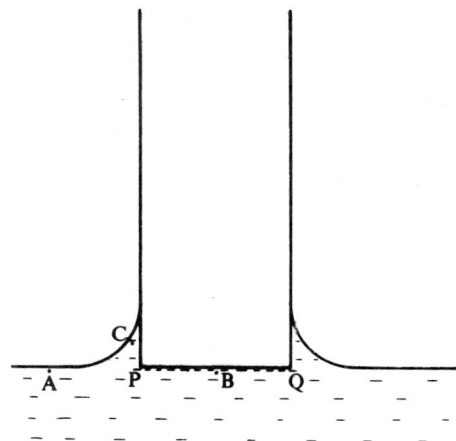

Figure 12.8

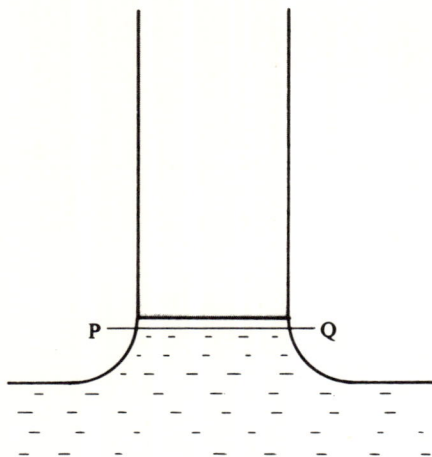

Figure 12.9

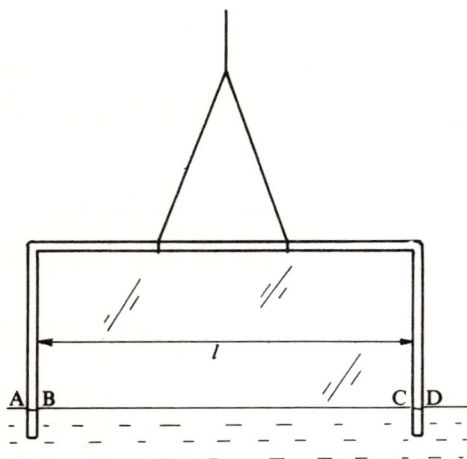

Figure 12.10

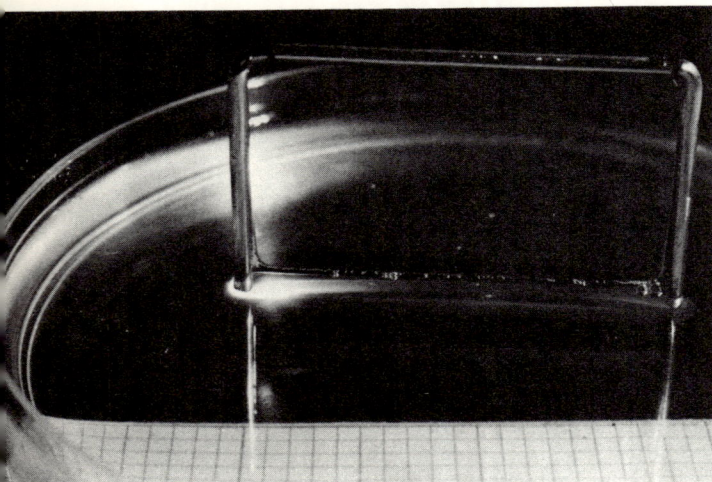

Soap film formed on a vertical wire frame.

vertical interatomic separation in the surrounding material would increase slightly to compensate, and every atom in the cross section PQ would contribute to the attractive force holding the solid together. In fact we would have the tensile-stress situation discussed in Chapter 11; the force would be proportional to the cross-sectional area. But a liquid cannot stand tensile stress in this way. As soon as there is a hint of an increase in mean intermolecular separation, molecules pile in from the sides to restore the situation and the volume stays constant. The only force available to lift the wedge of liquid is that provided by the surface molecules, and this is precisely the force we have written as $2(a + b)\gamma\cos\alpha$.

Surface tension and Archimedes' principle

In the experiment described above we need to make sure that the base of the glass slide is at the same height as the general level of the liquid when W_2 is measured, otherwise the usual hydrostatic upthrust produces a significant effect. It may be asked whether there is an upthrust even in this situation (Figure 12.8), since the sides are partially surrounded by liquid. That this is *not* so can be seen directly if we remember that any upthrust is due to a difference of pressures between top and bottom of the slide, and the pressure at B must be the same as that at A just below the general level of the liquid, that is, atmospheric pressure. This implies incidentally that the pressure at C must be *less* than atmospheric, a point which we shall return to later in the chapter.

Exercise 12.1

By considering the equilibrium of the liquid raised above the general level in Figure 12.8, show that its weight is equal to $2(a + b)\gamma\cos\alpha$.

Figure 12.10 shows a modification suitable for soap solution or liquids of similar character, able to form thin films. The wire frame is immersed in the solution and drawn out to form a film; we need the difference between the forces W_2 and W_1 needed to support the frame with and without the film respectively:

$$W_2 - W_1 = 2\gamma l,$$

where l is the length of the frame. We get the factor 2 because the film consists of two parallel surfaces, very close together. Note that l is the distance between the *inside* edges of the vertical wires; although there is extra length of surface perimeter around the two wires themselves (AB and CD in the figure) this will also affect W_1 to an equal extent.

We cannot use the wire frame for water because pure water cannot form a stable film in this situation. The film has to provide a little more tension near the top than near the bottom because it has to hold up the rest of the film. Soap solution can cope with this; a variation of soap concentration down the film produces the necessary gradation, which is self-adjusting–a rather remarkable phenomenon. A pure liquid cannot adjust in this way.

Exercise 12.2

A ring, 4 cm in diameter, made of thin wire and with mass 0·80 g, is suspended from a balance and placed horizontally on a clean water surface. Given that the balance reads 2·65 g before the ring is pulled clear, find the surface tension of water ($g = 9·8\,\mathrm{m\,s^{-2}}$, $\alpha = 0°$.)

Surface energy

If we think about the potential energy of molecules in a liquid we get an alternative way of looking at surface phenomena, very useful in some situations. Suppose we start with two isolated molecules a long way from each other and bring them together. They will lose potential energy as work is done by the attractive force between them. Figure 12.11 is a sketch of the way we can expect the force to vary with separation, and superimposed on it is the corresponding graph for the potential energy; note how the equilibrium separation, giving zero force, also gives minimum potential energy.

If a molecule is brought up to a *pair* of molecules in the same way we can expect an even greater potential energy loss, because the total attractive force is that much stronger, and if we extend this argument a little further we can say that the potential energy lost by the molecule is roughly proportional to the number of very near neighbours it acquires. A surface molecule has fewer near neighbours than a molecule in the body of the liquid, so it has more potential energy. Switching to a macroscopic statement, a large liquid surface area means an arrangement with high potential energy, and since an equilibrium situation occurs when the potential energy is a minimum, the surface has a tendency towards minimum area.

We can express this idea in quantitative terms. Suppose we have a surface which we can easily enlarge: the soap film on the wire frame in Figure 12.12, for example. The wire CD experiences a force $2\gamma l$ towards AB, where l is the width of the frame. If we now pull CD out a distance x, the mechanical work done is $2\gamma lx$, as a result of which an area $2lx$ of new surface is formed. This means that the potential energy per unit area of new surface is just γ; we can say that *the surface tension of a liquid is numerically equal to the mechanical work required to form unit area of new surface.* Notice how this calculation is more straightforward than the one we would have to carry out for a rubber skin, for which the tension would increase as we pulled. When we pull rubber it stretches–different parts of the surface have to move apart. The liquid surface does *not* stretch in this sense; instead more molecules go into the surface as it enlarges, so that the average intermolecular separation does not change, and there is no increase in tension.

In the above discussion we have left out an important effect. The extra p.e. needed by those molecules moving into the expanding surface must come from somewhere, and if the liquid is thermally insulated from its surroundings the average k.e. of all the molecules is reduced in consequence. In other words, the temperature drops, and this is likely to change the surface tension. Further consideration of the phenomenon lies beyond the scope of this book, but the expression we have derived for the surface energy is correct so long as the changes occur *isothermally*, that is, so long as we allow heat to flow in and out of the liquid to keep the temperature constant. Because internal energy is involved as well as mechanical energy, the mechanical contribution we have calculated ($2\gamma lx$) is called the *free* surface energy to differentiate it from the *total* surface energy, which includes the energy involved in the heat exchange as well.

We get exactly the same situation in the case of an expanding gas. Boyle's law holds only for an isothermal change, and if a gas is expanded *adiabatically*, that is, in such a way that no heat is allowed to enter or leave the system, the gas is cooled in the process and Boyle's law is not applicable.

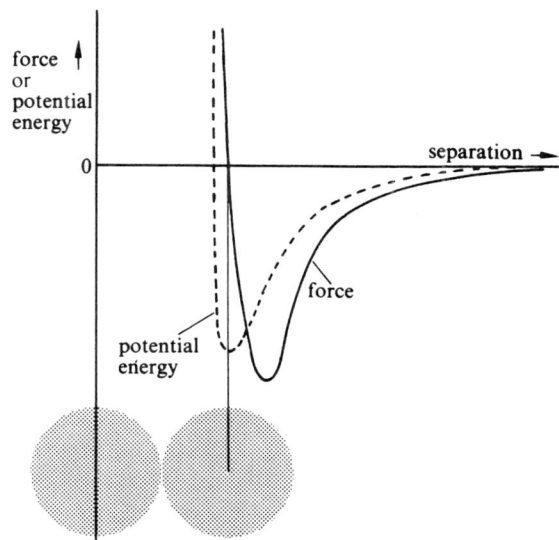

Figure 12.11

Sketch graphs showing variation of force and potential energy with separation of two molecules. The zero of potential energy corresponds to infinite separation.

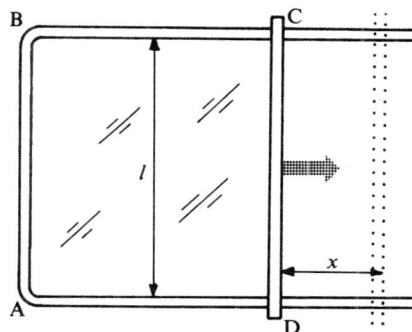

Figure 12.12

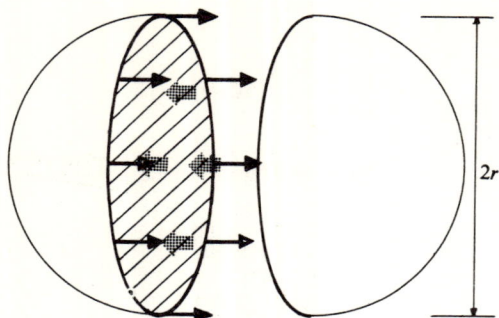

Figure 12.13

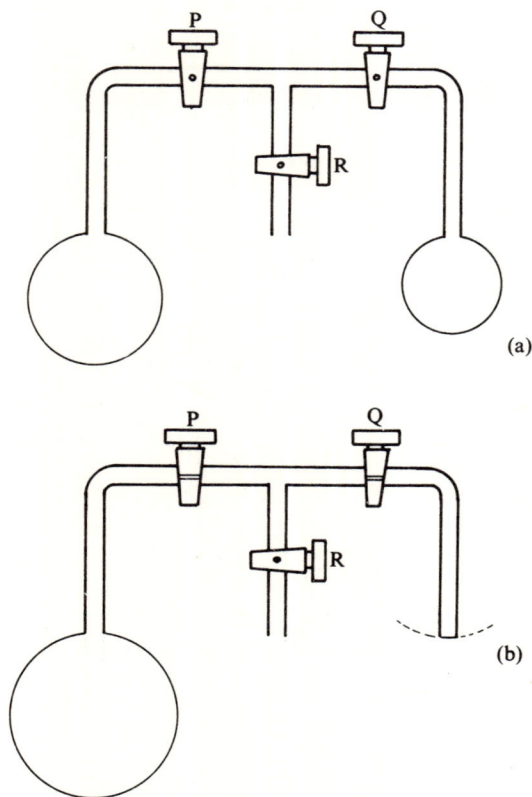

Figure 12.14

Equilibrium of a spherical drop or bubble

In the absence of any external forces to cause distortion, a drop of liquid draws into a sphere. This is consistent with the minimum-energy condition for equilibrium; with the help of a branch of mathematics known as the calculus of variations we can show that the solid with minimum surface area for a given volume is in fact a sphere. Because the surface of the drop has a tendency to draw in, the liquid suffers a certain amount of compression, and the pressure inside is a little in excess of that outside. We can calculate the size of the difference if we mentally divide the drop, of radius r, into two hemispheres and ask what force each half exerts on the other (Figure 12.13). Of course the molecules in the plane dividing the hemispheres exert considerable repulsion across the plane; most of this is in response to the atmospheric pressure acting in on all sides, but we are looking for the *extra* pressure P balancing the tension in the surface. The repulsive force between the two halves across the plane due to P is $\pi r^2 P$, and the force holding the two halves together around the perimeter of the plane is $2\pi r \gamma$.

So
$$\pi r^2 P = 2\pi r \gamma$$

$$P = \frac{2\gamma}{r}. \qquad \ldots(1)$$

This formula serves equally well for a 'drop' of gas in a liquid, in other words a bubble. It can also be modified to give the excess pressure inside a soap bubble; this time we have to deal with *two* perimeters in the diametric plane, for the inside and outside surfaces.

Exercise 12.3

Remembering that the soap bubble has a radius r considerably greater than its thickness, show that the excess pressure P is given by

$$P = \frac{4\gamma}{r}.$$

We can use the idea of free surface energy together with the principle of virtual work to provide an alternative derivation of formula (1).

Exercise 12.4

Show that if an air bubble of radius r with excess pressure P in a liquid with surface tension γ expands by a tiny amount δr, the increase in free surface area is approximately $4\pi \times 2r\,\delta r$ if terms in δr^2 are ignored, and that the energy needed is therefore $8\pi r \gamma\,\delta r$. Show that during the expansion the excess pressure P in the bubble is responsible for an amount of work equal to $4\pi r^2 P\,\delta r$ and hence that

$$P = \frac{2\gamma}{r}.$$

Figure 12.14 indicates a convincing demonstration that the excess pressure is greater for *smaller* bubbles. When the taps P and Q are opened, R being closed, the larger bubble grows at the expense of the smaller one. The arrangement does not settle into equilibrium until we reach the situation shown in Figure 12.14b; the criterion is that the radii of the two bubbles must be equal.

In general we can say that there is a pressure difference of $2\gamma/r$ across any spherical liquid surface, the greater pressure being on the concave side.

Exercise 12.5

Referring to Figure 12.15, by considering the excess pressure on the concave side of the film common to both bubbles show that

$$\frac{1}{R} = \frac{1}{R_1} - \frac{1}{R_2}.$$

Non-Spherical films

If the wide edges of two funnels are dipped in soap solution, placed face to face and pulled apart, the film formed has the shape shown in Figure 12.16. It can be turned into a cylinder by sealing the funnel ends and pushing them together a little to produce a small amount of excess pressure. The magnitude P of the pressure needed can be calculated by either of the methods used for the spherical film. Splitting the cylinder into two halves (Figure 12.17), we can write

repulsion across diametric plane due to excess pressure $=$ attractive force along AB and CD (two surfaces each)

$$2lrP = 4l\gamma$$

where r is the radius and l the length of the bubble. Hence

$$P = \frac{2\gamma}{r}.$$

Exercise 12.6

Use the energy method to obtain this result.

In fact we can include all these expressions for excess pressure into one general formula. If the surface is more complicated, having at each point different radii of curvature in different planes perpendicular to the surface, then one of these radii is smaller, and one greater, than any of the others. These maximum and minimum radii r_1 and r_2 lie in planes perpendicular to each other, and it can be shown that the excess pressure P across such a surface is given by

$$P = \gamma\left(\frac{1}{r_1} + \frac{1}{r_2}\right)$$

or, in the case of a film with two surfaces,

$$P = 2\gamma\left(\frac{1}{r_1} + \frac{1}{r_2}\right).$$

Figure 12.16 is an interesting case. Since the inside is open to the atmosphere, this surface has a shape such that at every point $r_1 = -r_2$.

Measurement of surface tension by means of excess pressure in a bubble: Jaeger's method

Figure 12.18 shows a way in which the excess pressure in a bubble can be used to find γ. Water drips slowly into the flask on the left and as a result the pressure in the apparatus builds up, causing a bubble to form at the end of a tube dipping into a beaker containing the liquid under test. The excess pressure is recorded by the manometer.

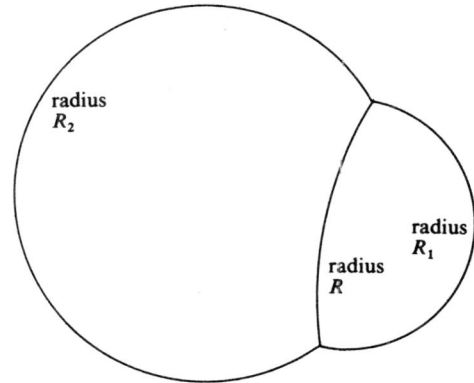

Figure 12.15

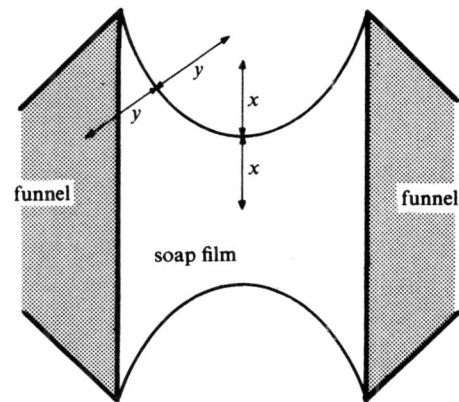

Figure 12.16

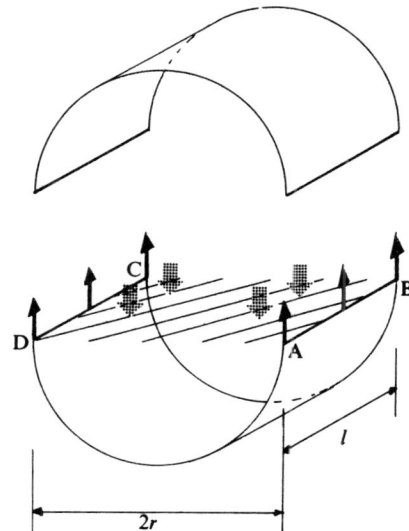

Figure 12.17
Excess pressure across a cylindrical surface.

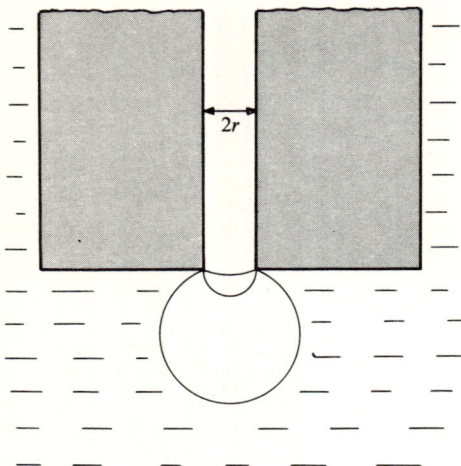

Figure 12.19

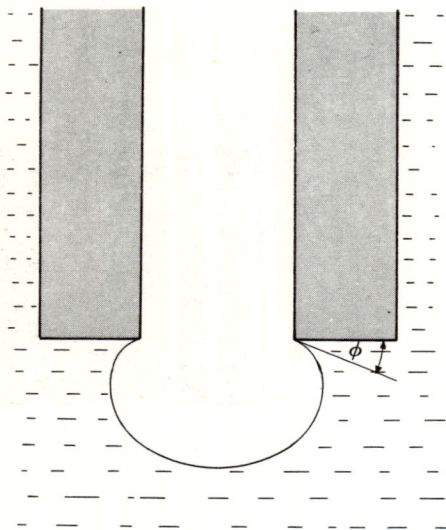

Figure 12.20

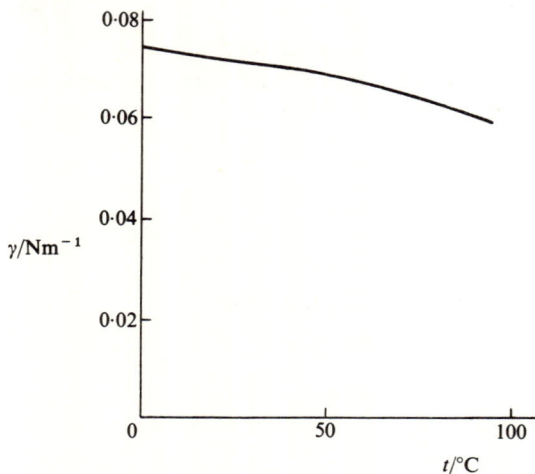

Graph of variation of surface tension γ of water with temperature t.

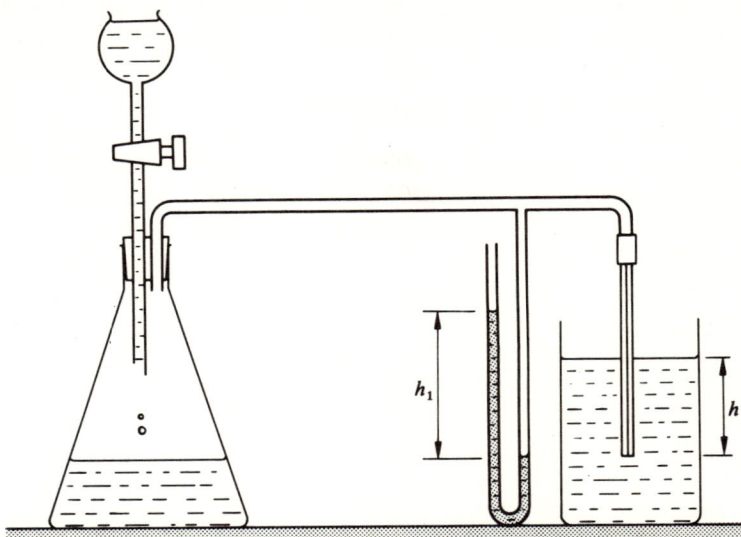

Figure 12.18
Jaeger's method for measurement of γ.

At some stage during its growth the bubble reaches unstable equilibrium, expands rapidly and breaks away, causing the manometer reading to drop abruptly. Another bubble begins to form immediately, breaking away in turn.

It is very natural to assume that the bubble becomes unstable when it is hemispherical, with diameter equal to that of the inside of the tube (Figure 12.19); this after all is the minimum diameter it can have, and further increase in size requires a *smaller* excess pressure for equilibrium, since excess pressure is inversely proportional to radius. If this were so we could write down a simple formula for the surface tension as follows. If the maximum difference in levels in the manometer is h_1, and the depth of the tube below the surface is h, then

$$\begin{array}{ccc} \text{Excess pressure} \\ \text{inside apparatus} \end{array} = \begin{array}{c} \text{pressure due to} \\ \text{head } h \end{array} + \begin{array}{c} \text{pressure change across} \\ \text{bubble surface.} \end{array}$$

that is,

$$\rho g h_1 = \sigma g h \qquad + 2\gamma/r$$

or

$$2\gamma/r = g(\rho h_1 - \sigma h), \qquad \qquad ...(2)$$

where ρ and σ are the densities of the liquids in the manometer and beaker respectively, and r is the internal radius of the tube.

Unfortunately equation (2) is not true except for tubes of very small bore; if the internal diameter of the tube is about 3 mm for example the bubble does not become unstable until it achieves a shape like that shown in Figure 12.20, with ϕ equal to about 20°. For this reason the equation cannot be used with any accuracy for *absolute* determinations of surface tension, but it can be used for comparisons, and is especially useful for studying the dependence of γ on temperature.

Jaeger's method has two advantages worth noting. First, the temperature can be controlled precisely because the bubble is formed in the body of the liquid, not at the surface, where there is likely to be a significant temperature gradient. Secondly, any impurities present tend to migrate to the surface of the bubbles and get carried away, so that later bubbles will be formed in pure liquid.

Equilibrium of a drop on a solid surface

The shape of a drop of liquid resting on a solid surface depends on whether or not the liquid wets the solid. We shall look at only the simplest possible case, that of a very large circular drop resting on a surface for which the adhesive forces are negligible (Figure 12.21) this means we can treat the surface of the drop in contact with the solid as though it were simply a liquid–air surface. There are two conflicting tendencies. The gravitational attraction flattens the drop into a sheet and the surface tension draws it into a sphere, and a compromise has to be arrived at. For a very small drop the surface tension wins and the drop is nearly spherical. A very large drop has an almost plane top; we shall take it to be plane right out to the edge and attempt to work out the thickness from the minimum energy condition. We take the drop volume as V with density ρ, and suppose it to have a height h and radius r.

Exercise 12.7

Show that the gravitational potential energy of such a drop with respect to the solid surface is $\frac{1}{2}\rho Vgh$.

Now we have to work out the surface area. We shall consider the drop to be so large that the area of the curved edges can be ignored compared with the top and bottom surfaces. As a result we can write

$$\pi r^2 h = V \quad \text{or} \quad r^2 = \frac{V}{\pi h}$$

Exercise 12.8

Show that as a result the surface area can be written as $2V/h$ approximately.

This means that the free surface energy is $2V\gamma/h$, so that the sum E of free surface energy and gravitational energy is given by

$$E = \tfrac{1}{2}\rho Vgh + \frac{2V\gamma}{h}.$$

Now we want the value of h which gives minimum energy, that is, the value for which dE/dh is zero and d^2E/dh^2 is positive. We find

$$\frac{dE}{dh} = \tfrac{1}{2}\rho Vg - \frac{2V\gamma}{h^2}$$

and this is zero when $h^2 = 4\gamma/\rho g$.

Exercise 12.9

Show that d^2E/dh^2 is positive for this value of h.

If there is adhesion with the solid surface we are not entitled to write the free surface energy of the bottom surface as γ per unit area. A more thorough approach to the problem gives the formula

$$h^2 = \frac{2\gamma(1 - \cos\alpha)}{g\rho},$$

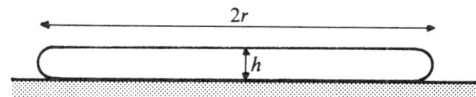

Figure 12.21

Shape of a large drop of water on a polished surface (to which it does not adhere).

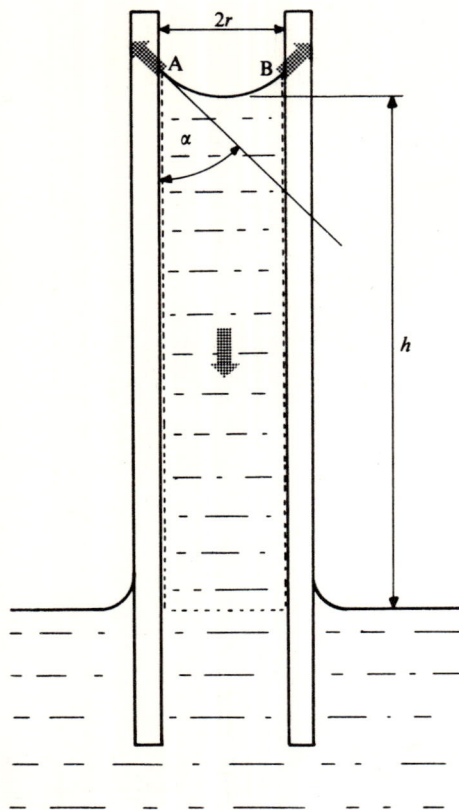

Figure 12.22

where α is the angle of contact. Notice that this agrees with our result $h^2 = 4\gamma/\rho g$ when $\alpha = 180°$, that is, when the liquid does not wet the solid surface.

Capillary rise

If we dip the lower end of a glass capillary tube into water, inside it a thread of the liquid rises a few centimetres above the general surface level. The adhesive forces are quite strong, and as many water molecules as possible are encouraged to cling to the glass walls.

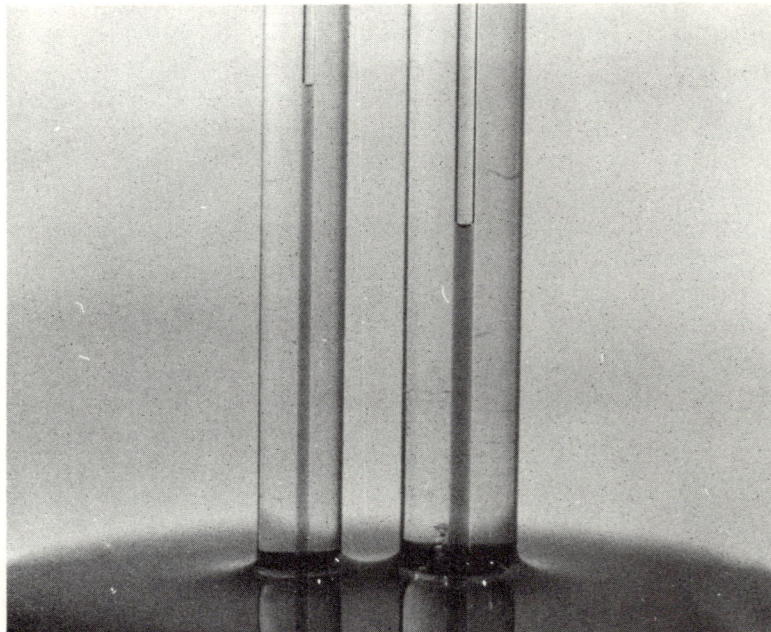

Capillary rise

However, the height to which the water rises is a measure not of the adhesive forces but of the *cohesive* ones, that is, it depends on the surface tension. We have here a situation similar to the one on p. 163; although the water molecules very close to the wall are held by adhesive forces, the molecules a little further out have to be supported by cohesive forces. It is fairly certain that a film of water one or two molecules thick adheres to the glass well above the level of the capillary rise, but the cohesive forces are not sufficient to pull the main column up any further. This idea gives us one way of working out the connection between the capillary rise and the surface tension, but it should be pointed out that the following derivation is only strictly applicable to capillary tubes with uniform bore.

Looking at the equilibrium of the column of liquid defined by the dotted line in Figure 12.22, that is, excluding only the molecules adhering strongly to the glass, and taking the general case with an angle of contact α, we can argue that the weight of the column is supported by the surface tension around the perimeter AB.

This gives $\rho g h \pi r^2 = 2\pi r \gamma \cos \alpha$, where h is the capillary rise, r the bore radius, ρ the density and γ the surface tension (we have ignored the weight of liquid above the bottom of the meniscus). So

$$\gamma = \frac{\rho g h r}{2 \cos \alpha}.$$

Now in fact this is the correct answer for *any* shape of bore, so long as the radius at the meniscus is *r*. If for example the tube is shaped as in Figure 12.23, the column of liquid is supported not only by the surface tension but also by the vertical components of the adhesive forces at the tapered walls, in fact the force $2\pi r\gamma \cos\alpha$ supports the liquid inside the dotted cylinder and the adhesive forces support the rest.

A more satisfactory approach to the whole problem is to consider the pressure change across the curved surface. Referring to Figure 12.24, *X* and *Y* are two points just outside the plane and curved surfaces respectively, *Z* is a point just under the plane surface and *W* is on the same level as *Z*. The pressure at all four points is almost identical, but the pressure at *V* just under the curved surface is significantly less. The pressure drop can be looked at in two ways; either we can say it is ρgh due to the head of liquid in the bore, or we can write it as $2\gamma/R$ because we get a pressure change of this magnitude across a surface of radius *R* (the surface is spherical as long as the bore radius is not more than a millimetre or so). The diagram shows that *R* is related to the bore radius *r* by $r = R\cos\alpha$. As a result we can write

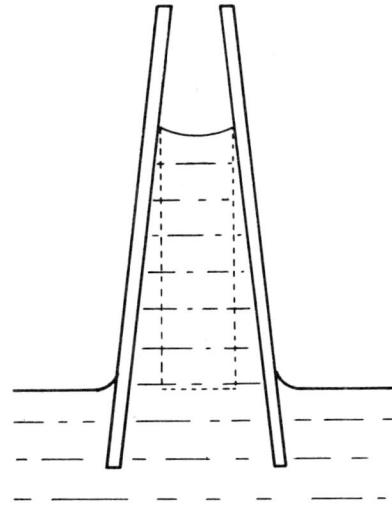

Figure 12.23

$$\rho gh = \frac{2\gamma \cos\alpha}{r},$$

giving

$$\gamma = \frac{\rho ghr}{2\cos\alpha}.$$

This can provide a convenient way of measuring γ; points to watch are that the tube must be dipped well in and withdrawn slightly to provide a receding surface, and *r* must be measured at the meniscus; this involves snapping the tube at the right level and using a travelling microscope. The method has its drawbacks; the temperature in the capillary tube may be significantly different from that in the beaker except at room temperature, and any impurities present are likely to migrate to the surface and stay there. In spite of these difficulties, the capillary rise method is the one most suitable for absolute determinations of γ.

Exercise 12.10

Find the height of the capillary rise of water up a vertical tube with internal diameter 0·75 mm if the surface tension is $7\cdot2 \times 10^{-2}\,\mathrm{N\,m^{-1}}$ and *g* is 9·8 m s^{-2}. Take the angle of contact to be zero.

If the capillary is very fine indeed the liquid can rise to appreciable heights; the absorbency of blotting paper depends on the spaces between the fibres acting as minute capillaries. In the same way water rises up a brick wall unless an impervious layer, the damp course, is included just above ground level.

Exercise 12.11

Figure 12.25 shows a film of water rising between two parallel vertical glass plates separated by a distance *d*.

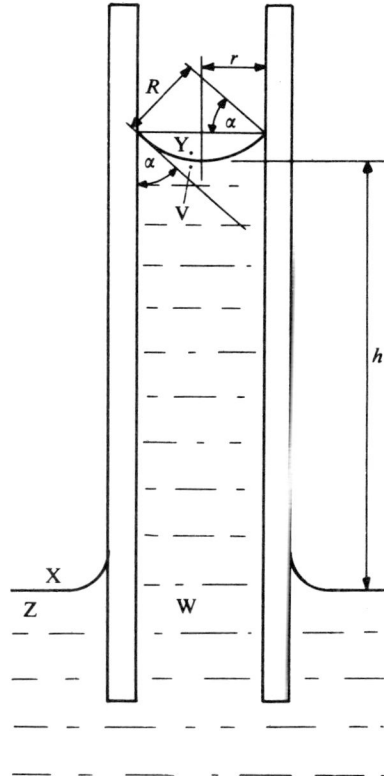

Figure 12.24

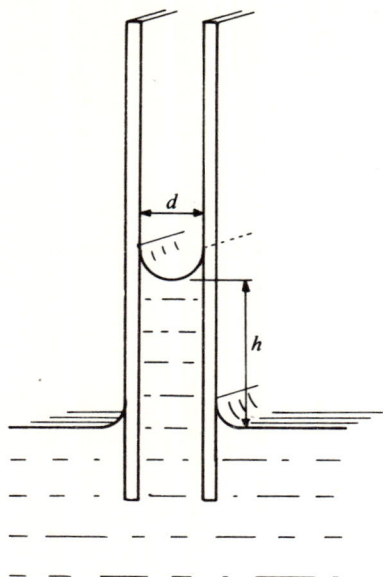

Figure 12.25 (Exercise 12.11)

Write down the pressure change across the cylindrical surface and hence show that the water rises a height h given by

$$h = \frac{2\rho g \gamma}{d},$$

where the symbols have the same meaning as previously.

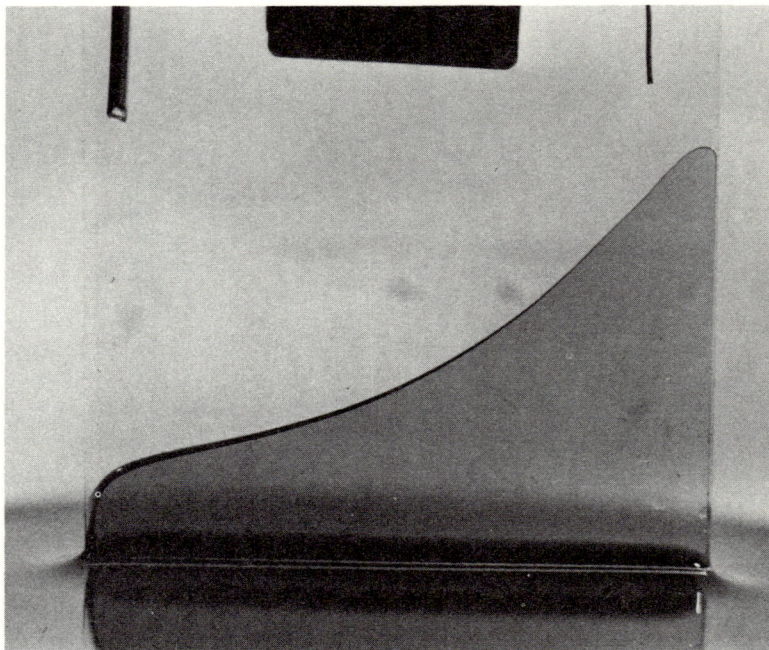

Liquid rising in the wedge-shaped gap between two vertical glass plates. The wire spacers are shown at top left and right.

Formation of small bubbles; boiling and bumping

What happens when a liquid is boiling in a beaker? Bubbles of the vapour form near the base of the beaker and as they rise to the surface they expand rapidly as the liquid evaporates into them. But how does a bubble start in the first place? That is not an idle question. Suppose we imagine that a vibration of the beaker or some other local fluctuation was able to produce a cavity in the liquid perhaps ten molecules across. We can calculate that the excess pressure needed to prevent a bubble of such a size from collapsing is very large indeed.

Exercise 12.12

Show that the excess pressure for a bubble as much as a hundred molecular diameters across in a liquid with surface tension $2 \cdot 5 \times 10^{-2}\,\mathrm{N\,m^{-1}}$ is about fifty atmospheres. Take the molecular size to be about $2 \times 10^{-10}\,\mathrm{m}$.

Now the vapour will simply not provide that sort of pressure. Vapour in equilibrium with its liquid can exert only one pressure, the *saturated vapour pressure* (s.v.p) at a given temperature (although the s.v.p. upon a curved surface is different from that upon a plane surface–see paragraph following) and if any attempt is made to compress it, it simply goes back into the liquid. What is more we find that a liquid boils at such a temperature that its vapour pressure is equal to the external pressure. We can imagine a bubble of visible size able to be maintained without collapsing, because the excess pressure needed is very small, and a tiny increase in liquid temperature is enough to provide sufficient vapour pressure. But the bubble in Exercise 12.8 requires far too much excess pressure-it would simply collapse. The fact is that bubbles usually form around centres or *nuclei* (not to be confused with atomic nuclei). Particles of dirt, projections in the beaker wall, bubbles of air trapped below the surface all form good nuclei; they all give quite an appreciable radius of curvature around which the bubble can make a start, and they form a very important part of the boiling process. If very pure water is heated carefully in a scrupulously clean beaker, avoiding vibrations and disturbances of all kinds, its temperature can be raised three or four degrees above the normal boiling point before boiling commences. This is called *superheating* the liquid; by removing as many nuclei as possible we have made it very difficult for bubbles to form at all. When they do at length form-and a slight vibration can start the process-they expand very rapidly indeed, often depositing a fair proportion of the liquid over the bench. The phenomenon is known as 'bumping'.

Saturated vapour pressure above a curved surface

We can use Figure 12.26 to argue that the vapour pressure upon a concave liquid surface of radius r must be less than that upon a plane surface by an amount approximately equal to $2\gamma\sigma/r\rho$, where σ and ρ are the densities of vapour and liquid respectively. The enclosed space above the liquid contains vapour from the liquid only, so the pressure at A is equal to the s.v.p. upon a plane surface. The pressure at B is less than this by an amount σgh due to the 'head' of vapour. If the s.v.p. at the curved surface were equal to that at the plane surface, continuous evaporation would result at B, with corresponding condensation at the plane surface, and there would be a flow of liquid up the capillary tube in consequence, in conflict with the law of energy conservation. We conclude that the s.v.p. upon the curved surface must be less than that upon the plane surface by an amount σgh, or $\sigma \times 2\gamma/r\rho$, making use of the formula for capillary rise.

Supercooled vapours

The s.v.p. upon a convex surface is *greater* than that above a plane surface, so that a small drop evaporates more readily than a large one. In microscopic terms we can say that once an adventurous molecule has jumped clear of the surface it is a little further away from many of the molecules attracting it back again, and is less likely to return (Figure 12.27). The result is that a very small drop is difficult to form; nuclei are needed, and if conditions are very clean a vapour can be cooled well below the temperature at which it would normally condense before droplets form. Such *supersaturated* vapours are common at high altitudes; town atmosphere by contrast generally contains more than its fair share of dust particles and impurities of all kinds, so that fogs are plentiful in industrial areas.

The size of the cohesive force

Before we leave the subject of surface tension we should like to see whether it can provide us with any clue as to the size of the attractive force between two molecules. Even an estimate of the order of

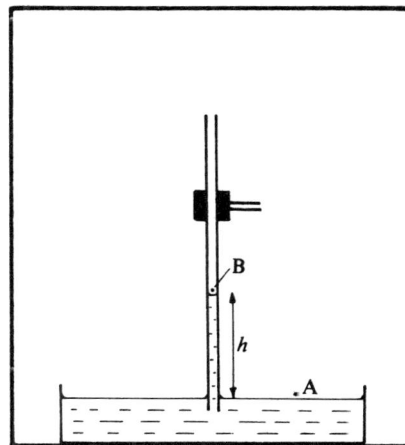

Figure 12.26

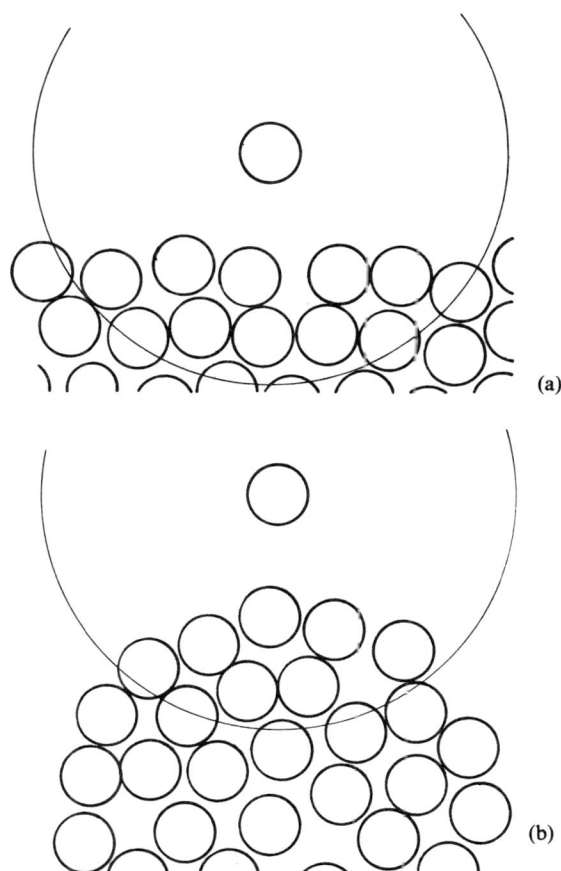

Figure 12.27

Sphere of influence for (a) a molecule above a plane surface, (b) a molecule above a convex surface. In case (b) there is less attraction back towards the surface.

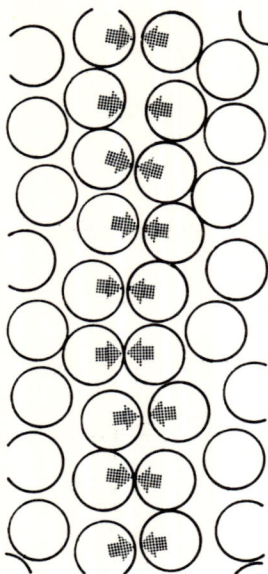

Figure 12.28

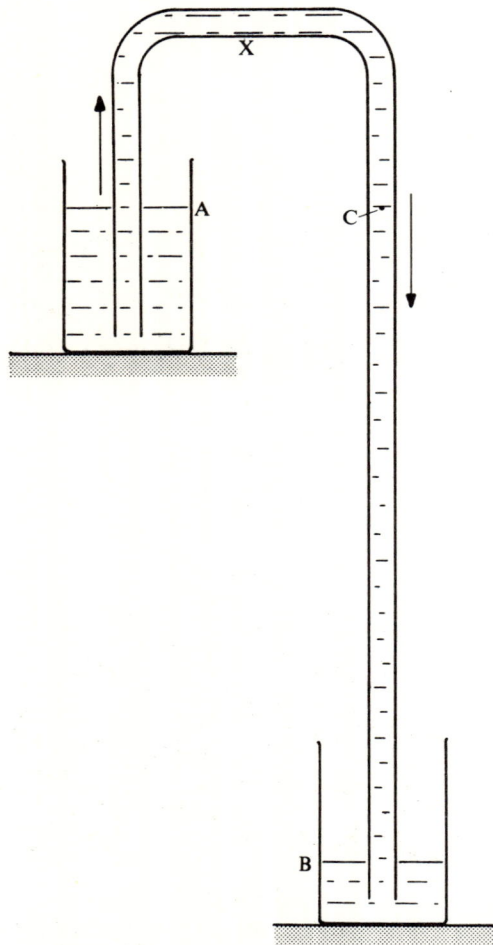

Figure 12.29

magnitude would be something, so we shall be prepared to make very bold simplifying assumptions to get an answer. We shall suppose that the depth of the surface layer responsible for surface tension is controlled by the effective range of intermolecular attractive forces, and assume that it is two or three molecules deep. If the molecular diameter is d, there are $1/d$ molecules along a line of unit length, and so the attractive force keeping the surface together along such a line is provided by $2/d$ or $3/d$ molecules on each side of it. The size of the attractive force across the line is γ per unit length, so that the attraction provided by each molecule is of the order of $\frac{1}{2}\gamma d$. If we imagined each molecule to attract only its opposite number across the line (as indicated in Figure 12.28, which shows the top layer only) then $\frac{1}{2}\gamma d$ would in fact be our estimate of the force between two molecules, but no doubt there are cross attractions as well. We need to guess how many molecules one side of the line are attracted by a molecule on the other side. It is certainly more than one and probably less than ten; we will settle arbitrarily on five. This gives the attractive force F which one molecule can exert on another as

$$F \approx \frac{\gamma d}{10}. \qquad \qquad ...(3)$$

Exercise 12.13

Taking γ for water as $7 \times 10^{-2}\,\mathrm{N\,m^{-1}}$ and d as $2 \times 10^{-10}\,\mathrm{m}$, estimate F for two water molecules.

Surface tension and the siphon

It is sometimes asserted that we should never speak of a liquid or gas being 'sucked' into a vacuum; instead we should say that the atmospheric pressure pushes it in. We now want to qualify that view in one important respect. Figure 12.29 shows a siphon transferring liquid from A to B; it works, of course, because the pressure at A equals the pressure at B and is therefore greater than the pressure at C on a level with A, and the pressure difference pushes the liquid over. The trouble is that a siphon works just as well if the height of X above A is greater than the barometer height for the liquid, and it can also work in a vacuum. We have to assume in such cases that the liquid really is being 'sucked up', each layer pulls up the next layer because of intermolecular forces. What is more, this pull does not occur simply over a perimeter, it occurs over the entire cross-sectional area, just as though the column were a flexible solid. How can we reconcile this with our ideas earlier in the chapter? We need to recall why a liquid cannot withstand tensile stress; the molecules come in from the sides so as to keep the volume constant. Now assuming there is good adhesion between liquid and wall, this *cannot happen* in the case we are looking at now, and so an unbroken column of liquid completely filling a tube which it wets is able to withstand tensile stress. It is thought that this is how water can rise right to the top of tall trees.

Exercise 12.14

In terms of the quantities appearing in equation (3), show that the attractive force across unit area due to the above effect is in the region of $5F/d^2$, assuming that each molecule on one side of the dividing plane effectively attracts about five molecules on the other side. Show that such a force is strong enough to support a column of water about 20 km high, and write down the pressure in atmospheres to which this corresponds.

Further exercises

Exercises 12.15–29

15 Explain, using a simple molecular theory, why the surface of a liquid behaves in a different manner from the bulk of the liquid.

Giving the necessary theory, explain how the rise of water in a capillary tube may be used to determine the surface tension of water.

A microscope slide measures $6 \cdot 0$ cm $\times 1 \cdot 5$ cm $\times 0 \cdot 20$ cm. It is suspended with its face vertical and with its longest side horizontal and is lowered into water until it is half immersed. Its apparent weight is then found to be the same as its weight in air. Calculate the surface tension of water assuming the angle of contact to be zero.   (AEB)

16 A hydrometer has a cylindrical glass stem of diameter $5 \cdot 0$ mm. It floats in water of density $1000 \, \text{kg m}^{-3}$ and surface tension $7 \cdot 2 \times 10^{-2} \, \text{N m}^{-1}$. A drop of liquid detergent added to the water reduces the surface tension to $5 \cdot 0 \times 10^{-2} \, \text{N m}^{-1}$. What will be the change in length of the exposed portion of the glass stem? Assume that the relevant angle of contact is always zero.   (JMB)

17 What is meant by *surface tension* and *capillarity* and what is the connection between them?

A vertical wire W (Figure 12.30) is suspended over a tall glass of water P from one end of a torsion balance ABC (the torsion wire being perpendicular to the arm at B, and C being the counterweight). What happens to A when P is raised slowly and vertically?

Using your answers devise a method of measuring the coefficient of surface tension of water, assuming that the water wets the wire.

If W is replaced by a thin-walled glass tube of the same external diameter, what would be the effect on the magnitudes of your observations?   (SUJB)

18 Explain briefly the meaning of *surface tension* and *angle of contact*.

Account for the following: (a) a small needle may be placed on the surface of water in a beaker so that it 'floats', and (b) if a small quantity of detergent is added to the water the needle sinks.

A solid glass cylinder of length l, radius r and density σ is suspended with its axis vertical from one arm of a balance so that it is partly immersed in a liquid of density ρ. The surface tension of the liquid is γ and its angle of contact with the glass is α. Given that W_1 is the weight required to achieve a balance when the cylinder is in air and W_2 is the weight required to balance the cylinder when it is partly immersed with a length $h \ (<l)$ below the free surface of the liquid, derive an expression for the value of $W_1 - W_2$. If this method were used to measure the surface tension of a liquid, why would the result probably be less accurate than that obtained from a similar experiment using a thin glass plate?   (O & C)

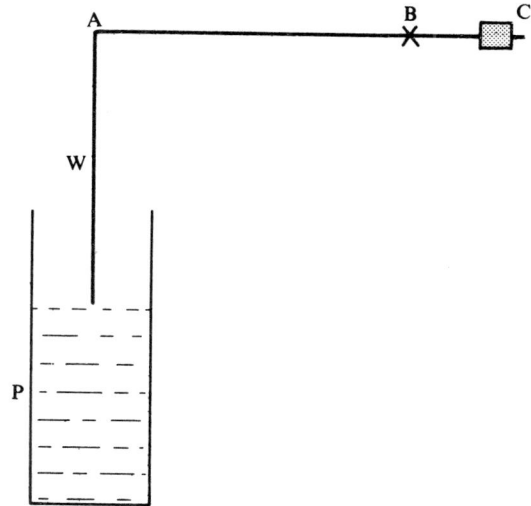

Figure 12.30

19 Explain (a) in terms of molecular forces why the water is drawn up above the horizontal liquid level round a steel needle which is held vertically and partly immersed in water, (b) why, in certain circumstances, a steel needle rests on a water surface. In each case show the relevant forces on a diagram. (JMB)

20 Derive, by the method of dimensions, the form of the expression for the excess pressure inside a bubble, assuming it to depend only on the surface tension of the liquid and the radius of the bubble.

Calculate the work required to break up a droplet of mercury of radius 2·0mm into drops each of radius 0·50mm. (Surface tension of mercury = 0·52 N m^{-1}.) (AEB)

21 A glass capillary tube of uniform internal diameter $3\cdot0 \times 10^{-2}$ cm, length 15·0cm and open at both ends is fixed vertically with one end 5·0cm below the surface of a liquid of density 1020 kg m^{-3}, surface tension $6\cdot0 \times 10^{-2}$ N m^{-1} and zero angle of contact with glass. Assuming the formula for the excess pressure inside a spherical liquid surface and that atmospheric pressure is $1\cdot0 \times 10^{5}$ N m^{-2}, draw a graph showing how the pressure at a point on the axis of the tube varies with its distance from the exposed end of the tube.

Calculate the minimum pressure in excess of atmospheric pressure which must be applied to the open end of the tube to cause a bubble of air to escape at the lower end.

Describe, with the aid of a diagram of the apparatus used, how you would confirm the result of this calculation experimentally. (JMB)

22 Explain what is meant by the following statements:

a) the coefficient of surface tension of mercury is 0·46 N m^{-1}
b) the angle of contact between mercury and glass is 137°

A clean open-ended glass U-tube has vertical limbs, one of which has a uniform internal diameter of 4·0mm and the other of 20mm. Mercury is introduced into the tube; it is observed that the height of the mercury meniscus is different for the two limbs. Explain this observation, stating in which tube the level is the higher, and calculate the difference in levels.

Discuss whether the measurement of the difference in levels in a tube of these dimensions provides a satisfactory method for determining the coefficient of surface tension of mercury.
(Density of mercury = $1\cdot36 \times 10^{4}$ kg m^{-3}.) (C)

23 Explain in terms of molecular forces why some liquids spread over a solid surface whilst others do not.

A glass capillary tube of uniform bore of diameter 0·50mm is held vertically with its lower end in water. Calculate the capillary rise. Describe and explain what happens if the tube is lowered so that 4·0cm protrudes above the water surface. Assume that the surface tension of water is $7\cdot0 \times 10^{-2}$ N m^{-1}. (JMB)

24 Why does a liquid form a meniscus where its surface is in contact with the containing vessel? Discuss the effect of the shape of the meniscus in determining whether or not the liquid rises in an open-ended glass capillary tube placed vertically in a vessel of the liquid.

Describe how you would determine the surface tension of a liquid by the capillary-rise method.

A glass capillary 0·6 mm in diameter, open at both ends, is held vertically with the lower end immersed in water, of surface tension $7.3 \times 10^{-2}\,\mathrm{N\,m^{-1}}$. Estimate the height the water rises up the tube.

What would you expect to happen if the capillary tube were fairly long, and sealed at the upper end? (o)

25 Show that the pressure inside a small droplet of liquid (such as a fog droplet suspended in the air) differs from the outside pressure by $2\gamma/r$, where γ is the surface tension of the liquid and r is the radius of the droplet.

Figure 12.31 (exaggerated for clarity) shows a liquid of surface tension γ which has risen to a height h in a capillary tube, radius R, which it wets. Points A and B are a negligible distance apart, but on opposite sides of the surface; so are D and E and G and F. The density of the liquid is ρ; ignore the density of air.

a) List all lettered points which have the same pressure as A.

b) Calculate the pressures at all other marked points, indicating whether they are greater or less than that at A.

c) Explain why and under what conditions the difference between the pressures at E and G may sometimes be neglected.

d) Explain carefully, in the light of your previous answers, why the surface of a liquid in a very narrow capillary tube must be almost perfectly spherical.

e) Use your answers to derive a formula connecting h, ρ, R and γ, and calculate γ in the case where $h = 5.0\,\mathrm{cm}$, $\rho = 800$ $\mathrm{kg\,m^{-3}}$ and $R = 0.20\,\mathrm{mm}$.

f) Briefly outline the method you would use to find R if you were performing a very exact determination of γ. (SUJB)

26 Define surface tension.

Give an account of a method for measuring the surface tension of a liquid, including the necessary theory.

If a drop of water is placed between two pieces of plate glass, it is very difficult to pull the plates apart normally but easy to slide one over the other. Explain this.

What would be the value of the normal force required to separate the plates if they are circular and each of 5 cm radius, and are separated by a water film $0.01\,\mathrm{mm}$ thick?

Assume that the angle of contact is zero and the surface tension of water is $7.0 \times 10^{-2}\,\mathrm{N\,m^{-1}}$. (WJEC)

27 The glass tube ABC (Figure 12.32) is made up of two capillary tubes AB and BC joined in a straight line at B. Their radii are uniform and equal to $0.2\,\mathrm{mm}$ and $0.1\,\mathrm{mm}$ respectively. The tube is held *horizontal* and a bead of mercury is blown along it until some is in AB and some in BC. Describe the behaviour of the bead after the pressure is removed.

Assuming that the angle of contact does not alter, describe and explain the initial movement of the mercury if, when the pressure is removed, the tube is *vertical* with (a) end A, (b) end C uppermost, the two mercury surfaces being initially about $4\,\mathrm{cm}$ apart. (Surface tension of mercury in air $= 0.50\,\mathrm{N\,m^{-1}}$; angle of contact with glass $= 135°$; density $= 13600\,\mathrm{kg\,m^{-3}}$.) (o & c)

28 A drop of liquid of surface tension $4.0 \times 10^{-2}\,\mathrm{N\,m^{-1}}$ and density $1200\,\mathrm{kg\,m^{-3}}$ is placed in the centre of a flat glass plate of diameter 10 cm and initially takes up the form shown in Figure 12.33 in which $d = 2\,\mathrm{cm}$ and $t = 0.4\,\mathrm{cm}$ and the diametral cross section may be assumed to have parallel, straight sides and semi-circular ends. An identical glass plate is placed on top of the drop with its centre vertically above the centre of the lower plate. Describe

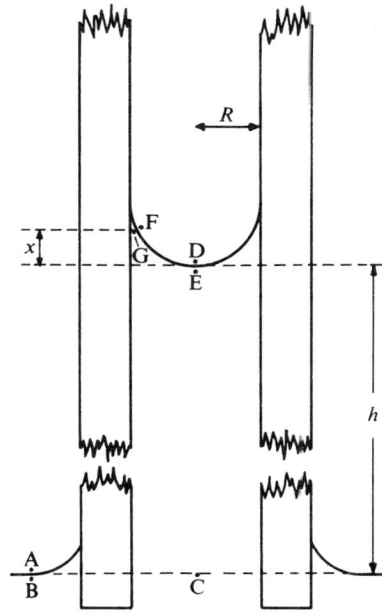

Figure 12.31

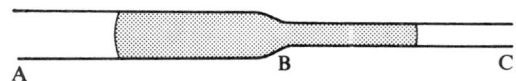

Figure 12.32

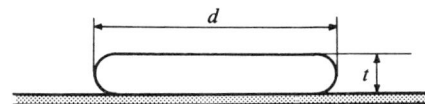

Figure 12.33

what happens as the upper plate is gradually lowered until it rests freely on the drop and is supported by it, assuming that the upper plate remains horizontal and central. Estimate the final separation of the plates if the upper plate has a mass of 10 g and the angle of contact between the liquid and glass is 180°. Make any approximations which you consider valid. What difference would it make if the angle of contact were zero?   (o & c)

29 One method of measuring the surface tension of a liquid is to suspend a microscope slide with its long edge horizontal and its largest face vertical from an arm of a balance, and find the additional mass m needed on the other arm of the balance to counter the downwards force of surface tension when a container of liquid is raised so that the liquid surface just touches the bottom edge of the slide. The equation used is $mg = 2\gamma(l + d)$, where γ is the surface tension and l and d are the length and thickness of the slide. The bottom edge of the slide should then be level with the plane surface of the liquid. Calculate the percentage error of the value which would be obtained for a liquid if γ is actually $0{\cdot}05\,\text{N}\,\text{m}^{-1}$, l is $8{\cdot}0\,\text{cm}$, d is $1{\cdot}0\,\text{mm}$, and if the bottom edge of the slide is $0{\cdot}2\,\text{mm}$ below the level of the plane liquid surface. The density of the glass is $3000\,\text{kg}\,\text{m}^{-3}$, and that of the liquid $1000\,\text{kg}\,\text{m}^{-3}$.

Referring to Figure 12.34a, discuss whether there is an error in the method when correctly performed owing to (a) upthrust on the shaded part of the slide; (b) the weight of the shaded part of the liquid.

Figure 12.34b shows two soap bubbles connected by a tube. Explain why the system is not in equilibrium and calculate the approximate radii of curvature of the soap films and the pressure within the system when equilibrium has been reached. Make clear any assumptions and approximations. For soap solution $\gamma = 0{\cdot}03\,\text{N}\,\text{m}^{-1}$.   (SUJB)

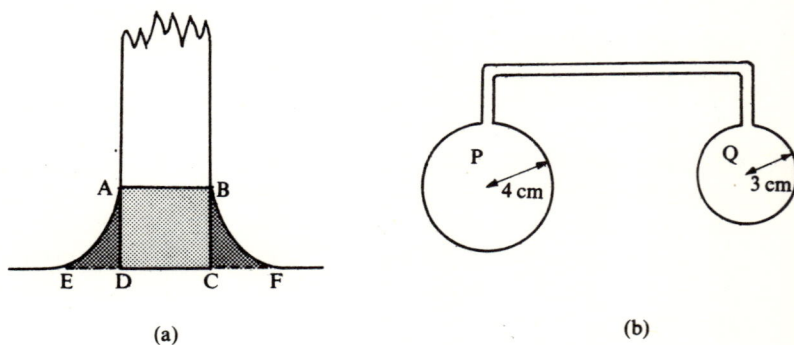

(a)   (b)

Figure 12.34

13 Unbalanced force, constant in size and direction; projectiles

We now want to turn our attention to some important applications of $F = ma$; we shall start by taking a further look at the simplest case when F is constant, both in size and direction. The equations for motion with constant acceleration are set out on p. 42; here we want to derive some of them again by a rather more general approach. Although what follows (pp. 179–80) is not essential to the understanding of the rest of the chapter, it will be useful later, particularly in Chapters 15 and 17. Writing the acceleration of a mass m produced by a constant force F acting in the y direction as d^2y/dt^2, we can put $F = ma$ in the form

$$\frac{d^2y}{dt^2} = \frac{F}{m}. \qquad \ldots(1)$$

The equality must still hold if we integrate both sides, so

$$\frac{dy}{dt} = \frac{Ft}{m} + c \qquad \ldots(2)$$

where c is some constant. This is an expression for the velocity v_y in the y-direction at any instant t after the start of the motion, but at first sight it does not look very useful, because apparently c can be any value we please. Now if equation (2) is to have any meaning it must describe a real situation, and the velocity in a real situation *cannot* be what we please; it has some definite value at any particular instant, controlled by what has been going on up to that instant. There must be some aspect of the situation which we have not yet fed into the mathematics, some piece of information which makes one situation different from another. That piece of information is the *initial velocity*. Two objects with the same acceleration can have different velocities at time t if they start with different velocities when $t = 0\,\text{s}$. Equation (1) says nothing about the initial velocity, it covers all possible cases, and that is why equation (2) gives a whole range of possible answers. We now have to pick out the answer which is correct in a given situation, and we shall do that by making sure it is correct when $t = 0\,\text{s}$.

Exercise 13.1

Write down the value of v which equation (2) gives when $t = 0\,\text{s}$ and hence show that $c = u$, the initial velocity.

Once we have fixed the value of the constant c at one particular instant it must continue with that value right through the motion; we can write $c = u$ for *every* value of t. Equation (2) therefore becomes

$$\frac{dy}{dt} = (F/m)t + u.$$

(Compare with equation (3) of Chapter 4, p. 42.) Now we integrate both sides again, and obtain

$$y = \frac{Ft^2}{2m} + ut + k, \qquad \ldots(3)$$

where k is another arbitrary constant, implying that there must be *another* piece of information about our object yet to be incorporated. This time we write in the position of the object at the start.

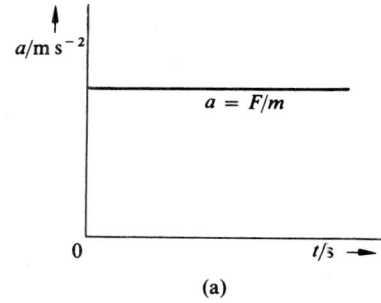

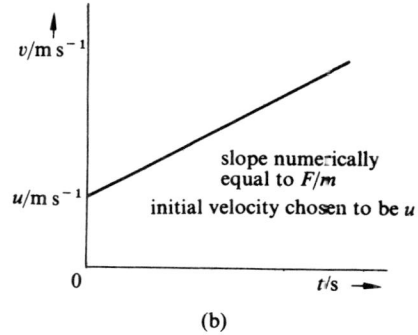

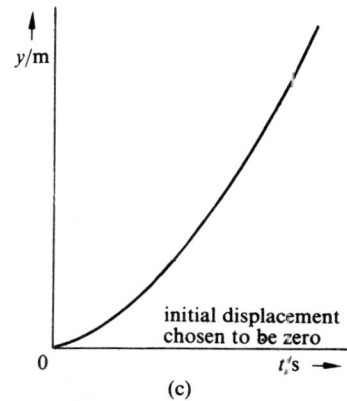

(a)

(b)

Graphs illustrating the integrations carried out in the text: (a) acceleration a, (b) velocity v, (c) displacement y, each plotted against time t.

Exercise 13.2

Write down the value that equation (3) gives for y when $t = 0$s. In most cases we can choose our axis so that the object is at the origin when $t = 0$s. Show that for this case $k = 0$m.

Equation (3) therefore becomes

$$y = \tfrac{1}{2}(F/m)t^2 + ut$$

(Compare with equation (6) of Chapter 4, p. 42.)

Differential equations

Equation (1) is a simple example of a *differential equation*, that is, an equation which includes differential coefficients, such as dx/dt and d^2x/dt^2, as well as x (just as quadratic equations include x^2 as well as x). Differential equations play an important part in more advanced physics and many of the features which characterise them have turned up even in this simple case. Very often the application of a very general physical law (in this case $F = ma$) results in a differential equation which must be solved. Sometimes it can be solved by direct integration, but more often that is not possible. When you are dealing with ordinary equations and your algebra is not very good, you can sometimes guess the answer and fit it in to see if it works. With some differential equations the *only* way to solve them is to guess an answer and fit it in to see if it works. We shall deal with a differential equation by that method in Chapter 15.

Another general feature is that the solutions are not constant numbers, they are *functions of a variable*; in this case the variable is the time t. We would hardly expect the solution of equation (1) to be $y = 5$m, or any other fixed distance; if we have applied $F = ma$, the law which we believe entirely controls the situation, then we expect to be able to get from it the complete details of the motion as time goes on – we expect a function of t.

Again, differential equations generally have too many solutions, and it is necessary to pick the right one for the job in hand by putting in information about the situation at the start, or at the origin, or at some other time or place that we know all about. This process is called *inserting boundary conditions*, and it removes the arbitrary constants from the solution and replaces them by definite values.

Exercise 13.3

A particle starts from the point $y = 3$m with a velocity $u = 4\,\mathrm{m\,s^{-1}}$ and a uniform acceleration of $-2\,\mathrm{m\,s^{-2}}$ in the y-direction. Using equation (3), find the appropriate value of k in this case and the largest value of y that results. Find also how long after the start the particle passes through $y = 0$m.

Motion in two dimensions; projectiles

So far, we have dealt only with the situation where the object, if it is moving at all at the start, is moving in the direction of the force. We are very familiar with what happens when the initial velocity is in some other direction; throw a ball out horizontally and it travels in a curve as shown in Figure 13.1. Here we have the motion of two similar objects illuminated by a lamp flashing at regular intervals;

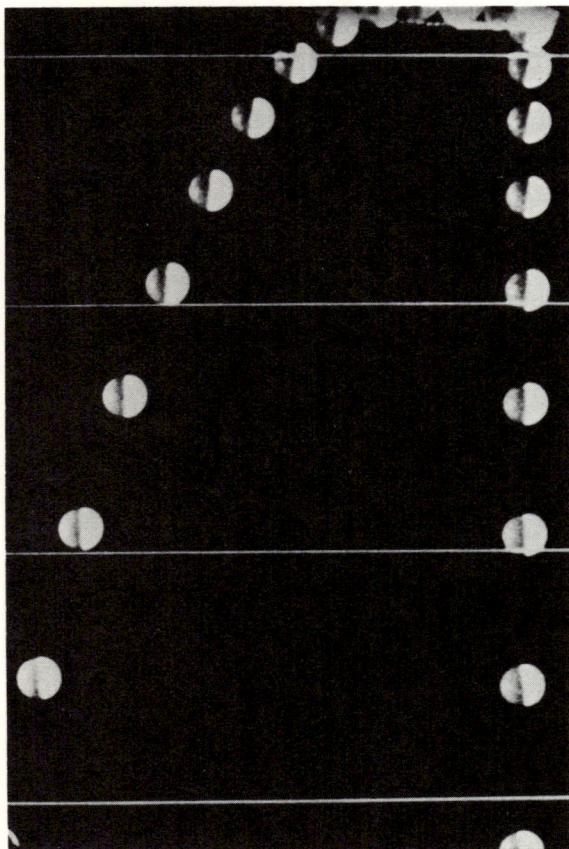

Figure 13.1

the object on the left was released from rest, and at the same instant the one on the right was projected horizontally from the same level. We can see that the vertical displacements in any given time interval are almost identical.

Exercise 13.4

Take measurements from the figure to test whether the object on the right travels through equal horizontal displacements between successive flashes, as would a billiard ball rolling on a smooth horizontal table.

These observations confirm our assumption that we can treat the components of motion in two mutually perpendicular directions separately.

Equations for a projectile

Ignoring air resistance, we can write the following equations for the horizontal and vertical displacements x, y and velocities v_x, v_y for a particle falling freely in the xy-plane:

$$v_x = u_x \qquad\qquad ...(4)$$

where u_x is the initial velocity in the x-direction;

$$x = u_x t, \qquad\qquad ...(5)$$
$$v_y = at + u_y \qquad\qquad ...(6)$$

where u_y is the initial velocity in the y-direction, and $a = -9.8\,\mathrm{m\,s^{-2}}$;

$$y = \tfrac{1}{2}at^2 + u_y t. \qquad\qquad ...(7)$$

These equations not only tell us where the particle is at various times, they also enable us to write down the equation describing the trajectory. Eliminating t between equations (5) and (7), we get

$$y = \tfrac{1}{2}ax^2/u_x^2 + (u_y/u_x)x,$$

which is the equation of a parabola. The way in which these equations can be put to use is illustrated in the following problem.

Example

A boy standing on the edge of a vertical cliff 40 m high throws a stone out to sea at an angle of 30° above the horizontal and at a speed of $20\,\mathrm{m\,s^{-1}}$. Find:

a) how long the stone is in the air
b) how far out the splash occurs from the base of the cliff
c) at what angle with respect to the horizontal the stone is moving just before it hits the water.

Adapting equations (4) to (7) to our immediate needs, and writing the initial velocity as u with components $u\cos\alpha$ and $u\sin\alpha$ (Figure 13.2) we get

$$v_x = u\cos\alpha \qquad\qquad ...(8)$$
$$x = ut\cos\alpha \qquad\qquad ...(9)$$
$$v_y = at + u\sin\alpha \qquad\qquad ...(10)$$
$$y = \tfrac{1}{2}at^2 + ut\sin\alpha \qquad\qquad ...(11)$$

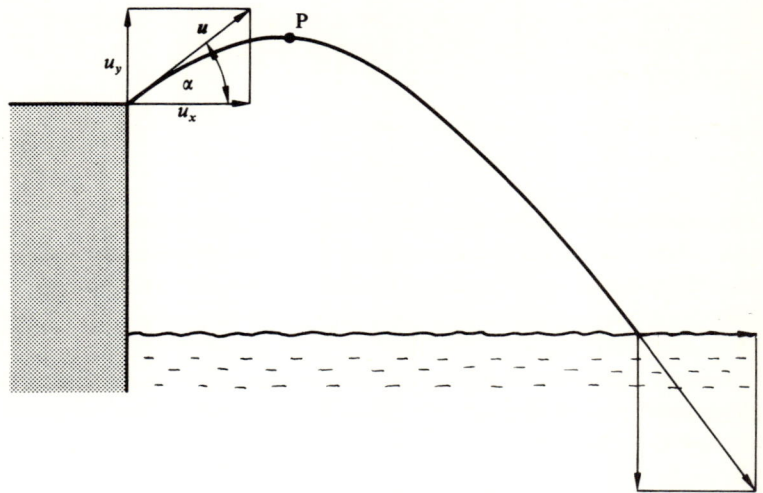

Figure 13.2
Trajectory of a projectile thrown from a cliff.

a) Equation (11) supplies us with the answer; the splash occurs when $y = -40\,$m, so substituting in the values given and putting $t = X$s we get

$$-40 = -4\!\cdot\!9X^2 + 20X/2,$$

giving

$$X = -2\!\cdot\!02 \quad\text{or}\quad 4\!\cdot\!05,$$

$$t = -2\!\cdot\!02\,\text{s} \quad\text{or}\quad 4\!\cdot\!05\,\text{s}.$$

Quadratic equations have a habit of giving us more answers than we want; the negative solution tells us how long before the start the stone should have left sea level if we wanted it to arrive in the boy's hand with the correct velocity to commence the motion we are discussing. The answer we really want is $t = 4\!\cdot\!05\,$s.

b) We can use this value in equation (9) to find how far out the stone travels in the x-direction before the splash. Writing this distance as Zm, we get

$$Z = 20 \times 4\!\cdot\!05 \times \sqrt{3}/2$$
$$= 70\!\cdot\!2,$$

so that the splash occurs $70\!\cdot\!2\,$m from the base of the cliff.

c) The angle β the trajectory makes with the horizontal at any instant is given by $\tan\beta = v_y/v_x$, or using equations (8) and (10),

$$\tan\beta = \frac{at + u\sin\alpha}{u\cos\alpha}.$$

Substituting in the value of t at which the splash occurs we get for the angle β_0 at which the stone enters the water,

$$\tan\beta_0 = \frac{2(-9\!\cdot\!8 \times 4\!\cdot\!05 + 10)}{20\sqrt{3}}$$

$$= -1\!\cdot\!71$$

$$\beta_0 = -59\!\cdot\!7°.$$

We get a negative value for β_0 because the trajectory dips down with respect to the horizontal; β starts positive but goes through zero as the stone passes point P in the figure.

Exercise 13.5

Compare how far out the splashes come if the stone is thrown out with the same speed but (a) horizontally, (b) 45° above the horizontal.

Maximum range on a horizontal plane

The angle of launch giving maximum range to the stone in Exercise 13.5 could be found analytically, but it would be a rather tough problem. We shall look at a simpler question: what is the best angle at which to throw a ball on level ground if it is to land as far away as possible? We can use equations (8) to (11) again, but this time we shall be interested in the value of x when y returns to zero; we shall be ignoring the height of the throwing arm above the ground. Since we want to know how this value of x varies with α, not with t, we shall need the equation for the trajectory. Combining equations (9) and (11), we have

$$y = \frac{ax^2}{2u^2 \cos^2 \alpha} + x \tan \alpha,$$

and when $y = 0\,\text{m}$ this gives $x = 0\,\text{m}$ or

$$- (2u^2 \cos^2 \alpha \tan \alpha)/a.$$

Remembering that

$$\tan \alpha = \frac{\sin \alpha}{\cos \alpha},$$

and that

$$2 \cos \alpha \sin \alpha = \sin 2\alpha,$$

we can write the range as $-(u^2 \sin 2\alpha)/a$ (remember a is negative). We want to pick α so that the range is as large as possible; the greatest value it can take occurs when $\sin 2\alpha = 1$, giving

$$2\alpha = 90°,$$
$$\alpha = 45°,$$

and for this angle the range is $-u^2/a$. If, for example, a cricket ball can be thrown at $30\,\text{ms}^{-1}$ and we take a to be $-10\,\text{ms}^{-2}$, the range is $90\,\text{m}$ ignoring air resistance.

Exercise 13.6

Show that there are *two* angles which give a range of $-u^2/2a$ and find them. Remember that since α is between $0°$ and $90°$, 2α will be between $0°$ and $180°$, and $\sin 2\alpha$ is still positive in the range $90°$ to $180°$.

Frames of reference

We now want to take an alternative view of the projectile problem, one that leads our thinking to a new realm. Imagine launching a projectile at some angle to the horizontal, say 50°, and then watching it from a vehicle moving along the ground with the same horizontal velocity. The projectile would appear to move straight up and down; in fact because of the independence of the two components of the motion it would appear to fall just as though it were thrown up vertically and viewed from a fixed point. Looking at the motion from a *moving frame of reference* like this is often a very useful device, and we shall use it to derive the projectile equations again.

The way to describe the projectile as it appears from the moving vehicle is to talk about its co-ordinates with respect to a set of axes fixed in the vehicle itself and moving with it; we shall call these co-ordinates x' and y' and refer to this system of co-ordinates as the

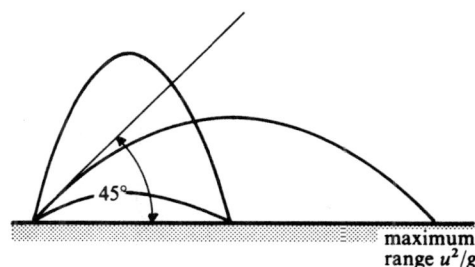

Range of a projectile on a horizontal plane.

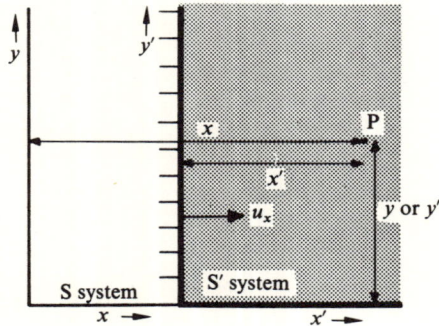

Figure 13.3
Point *P* referred to two different frames of reference.

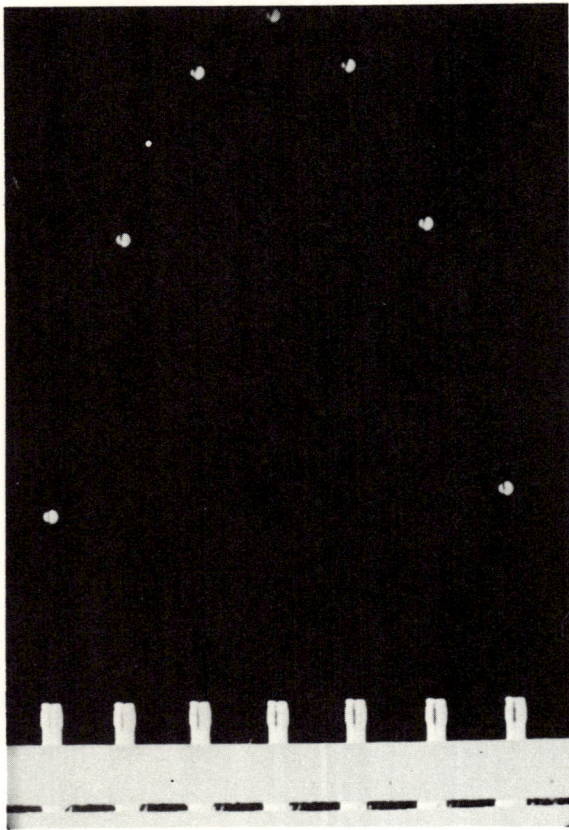

Multiflash picture of a ball projected from a vertical tube having a uniform horizontal velocity. The ball moves in a parabolic path but is vertically above the tube at every instant.

S'-system. We also need a clock in the vehicle, registering the time *t'*. In the *S'* system the motion of the projectile is described by the equations

$$y' = u_y\, t' + \tfrac{1}{2}at'^2 \qquad \ldots(12)$$

and
$$x' = 0 \qquad \ldots(13)$$

where $u_{y'}$ is the initial velocity.

Now every point (*x'*, *y'*) involved in these equations could equally well be labelled with respect to a *stationary* system of axes, the *S*-system; of course it has different labels (*x*, *y*) in the *S*-system, but we can write down the rules to get from one set of labels to the other. Referring to Figure 13.3, because the *x*- and *x'*-axes are collinear, the object has the same height when viewed from either system. So the *y* label is identical with the *y'* label,

$$y = y' \qquad \ldots(14)$$

(and if we have to deal with points outside the plane indicated,

$$z = z').$$

Similarly we take it that a stationary clock agrees with the clock in the moving vehicle, so

$$t = t'. \qquad \ldots(15)$$

However, *x* is not equal to *x'*. If the two systems are coincident at the start then *x* equals *x'* when *t* = 0 s, but if the *S'*-system moves off at speed u_x, then a point which is stationary in this system increases its *x*-value by u_x every second, so that

$$x = x' + u_x t.$$

These equations give the rules for changing from labels in the *S'*-system to the corresponding labels in the *S*-system; they constitute what is called the *Galilean transformation*. If we use them on our projectile equations (12) and (13) they tell us how the projectile moves when viewed from the *S*-system.

Exercise 13.7

By using the Galilean transformation $x' = x - u_x t$, $y = y'$ and $t = t'$ in equations (12) and (13), obtain the projectile equations (5) and (7).

Perhaps in this case the idea of transformation of axes is quite a large sledge hammer for cracking a rather small nut, but there are situations where it can be very useful indeed.

Inertial frames of reference

There is one very significant feature of the view we would get of the projectile from the moving vehicle—*it would look right*. It would look just like an object falling vertically. This means that the laws of motion—Newton's laws—are correct when they are used in the *S'*-frame of reference. We have to be clear what we mean here. Obviously we could take a set of observations of *x'*, *y'*, and *t'*, use the Galilean transformation to get the corresponding values of *x*, *y* and *t* and then show them to be consistent with Newton's laws. But we do not mean that. We mean that the raw observations *x'*, *y'*, *t'* *as they stand* are consistent with Newton's laws; the projectile looks as though it is falling vertically in the usual way. All such frames of reference in which Newton's laws hold are called *inertial* frames, and if we can find one inertial frame then all frames moving at uniform velocity with respect to it are also inertial frames. Because Newton's laws are valid in any inertial frame, no single inertial frame is 'special', in the sense that it is *stationary*. Rather we must say that all inertial frames are moving relative to each other. A frame stationary with respect to the earth's surface is approximately inertial, though we shall have more to say about that in Chapter 14.

Non-inertial frames

Can we view the projectile from a vehicle moving in such a way that the observed values of x', y' and t' do not tally with Newton's laws? We get this state of affairs as soon as our vehicle *accelerates*. Imagine a game of billiards played in a train pulling away from the station. It will be a little more difficult to play than usual – the balls will all congregate at one end of the table, and they will not travel in straight lines when they are struck. Someone viewing the situation from the trackside would understand immediately what is happening. He would see that the train is accelerating, that if the balls are to stay on the table they must be accelerating also, and that the only agent which can provide the necessary accelerating force is the cushion nearer to the rear of the train. But the player on board, if he did not look out of the window, might be forgiven for being puzzled. The balls he hits do not travel in straight lines, they curve towards the rear end of the table, and Newton's laws do not appear to be obeyed. If he insists that Newton's laws *must* be obeyed, he may begin to think the balls are experiencing a mysterious force, an attraction towards the rear coach. He is encouraged in this belief by the fact that he has to lean forwards to stay on his feet – all the indications are that *he* is experiencing a similar attraction. In fact, according to *Einstein's principle of equivalence*, the experiences he goes through at any given instant could equally well be accounted for *either* by thinking that the train is accelerating forwards *or* by thinking that a horizontal gravitational pull towards the rear of the train has been superimposed on the usual vertical one. The trackside observer, from his superior viewpoint in an inertial frame, says that the first answer is correct and that there is no extra gravitational pull. But because the experimental facts presented to the moving observer would fit either explanation, so that he could use either as a basis from which to predict the motion of the billiard balls as a result of a given disturbance, he may well find it useful to work with this idea of a 'switched on' force; it is called a pseudo force or inertial force. How real is a pseudo force? It is about as real as a virtual image. When we look into a mirror we are absolutely convinced that we see objects behind the mirror, because we are used to having our light rays travel in straight lines. Though we may know all the physics of reflection, and realise with our minds that the rays of light in this particular case do *not* travel in straight lines, but are suddenly bent at the mirror, our brain will not make use of this knowledge when interpreting information received from the retina. Instead it will tell us that those rays travelled in straight lines all the way from their source and therefore came from behind the mirror. Similarly it is no use telling an astronaut accelerating skywards in his rocket that he is not really heavier. His mind may be able to appreciate that his rather unpleasant experiences arise from the necessity of being accelerated with the vehicle he is riding in, but all his nervous system

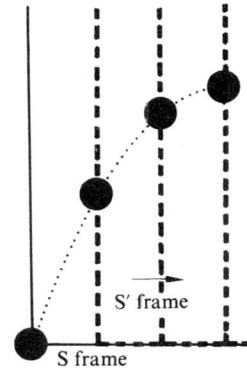

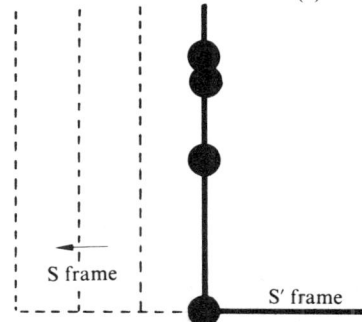

Motion of a projectile viewed from (a) a frame of reference stationary with respect to the ground, (b) a frame of reference having the same horizontal velocity as the projectile.

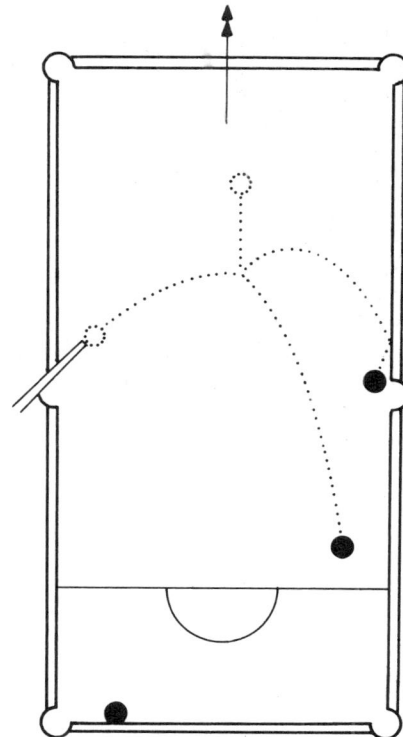

Billiards played in an accelerating vehicle.

will get across to him is that the earth is pulling at him with three or four times the normal force.

Now although it is useful to understand the idea of a pseudo force, and permissible to use it as a mathematical device which gives correct answers, a warning is not amiss. It is easy to misuse the idea, and it is certainly easy to give the wrong impression if words are not picked carefully. For this reason, although we shall give an example where the principle of equivalence is very helpful, you are recommended to work all problems from an *inertial* frame of reference, in which no pseudo forces are necessary for the successful application of $F = ma$. The problems on pp. 56–7 of Chapter 5 were worked in this way; the following example is by way of an exception to the rule. (It could also be solved by writing the equations appropriate to an inertial frame.)

Example

A pendulum of length l is suspended from the roof of a vehicle which is initially stationary on the earth's surface. Given that the vehicle then moves off with uniform horizontal acceleration a, find the angular amplitude of the swing of the pendulum.

According to the principle of equivalence, the effects of the acceleration inside the vehicle can be reproduced entirely by a horizontal gravitational pull towards the rear of the vehicle, of size great enough to produce an acceleration a on a free particle. If we combine this pull with the usual vertical one, able to produce an acceleration g, we get a resultant which would give a free particle an acceleration of size $(a^2 + g^2)^{1/2}$, in a direction making an angle α with the vertical given by

$$\tan \alpha = \frac{a}{g}$$

(see Figure 13.4.) This gives the direction of the new equilibrium line for the pendulum, and it swings about this line with angular amplitude α, or $\tan^{-1}(a/g)$.

Weightlessness

We have seen that a man's 'weight', as measured by instruments he carries with him in his vehicle, and as indicated by all his physiological sensations, can vary widely according to the acceleration of the vehicle; in fact if he is falling freely or in free orbit round the earth (see p. 304) his apparent weight is zero. This gives rise to unusual physiological experiences, for example, the customary difference of blood pressure between head and feet disappears, and tensions are altered in internal tissues which normally have to support the weight of organs. Readers are no doubt familiar with the 'stomach in mouth' feeling when travelling in a lift which suddenly accelerates downwards.

The general practice is to reserve the word 'weight' for 'pull of the earth' and use the term 'apparent weight' when we are referring to the reading on a spring balance which the man carries in the vehicle. If we refer all our measurements to an inertial frame there is no problem because the odd readings given by the instruments are all accounted for by the acceleration. It is only if we refer the measurements to a frame fixed in the accelerating vehicle, and expect

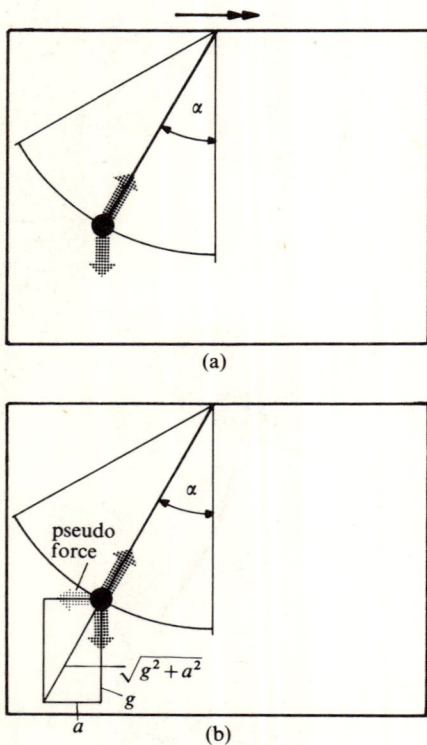

Figure 13.4

Pendulum in an accelerating vehicle, (a) from the point of view of an inertial frame, (b) from the point of view of a non-inertial frame moving with the vehicle, with the introduction of a pseudo force.

Newton's laws still to hold, that things start going wrong and the weight appears to change.

The breakdown of the Galilean transformation

Perhaps the most significant point about equations (15) and (16) is that they are not quite correct. They are very good indeed if the S'-system is moving slowly with respect to the S-system but beyond about $5 \times 10^7 \, \mathrm{ms^{-1}}$ they begin to break down to a noticeable extent. This is a healthy reminder that all our logic must be submitted to experimental test, because the reasoning leading to the Galilean transformation looks completely watertight. To get to the heart of the matter and see the paradox in its simplest and consequently most mysterious form, we need only make one short statement about the speed of light in a vacuum, c. Experiments have shown that c is the same when measured from any inertial frame of reference. The implications do not dawn until we think of a concrete example. Suppose you are travelling away from the earth at $2 \times 10^8 \, \mathrm{ms^{-1}}$. A communication is sent to you, by radio waves or a laser beam. It leaves the earth at $3 \times 10^8 \, \mathrm{ms^{-1}}$, so it will catch you up eventually. At what speed will you observe the radio wave or light beam to go past your window? At $1 \times 10^8 \, \mathrm{ms^{-1}}$? Wrong–it will pass at $3 \times 10^8 \, \mathrm{ms^{-1}}$. That is the paradox, and that is the strange way nature behaves. It was in 1887 that Michelson and Morley performed the classic experiment which established that the speed of light is independent of the speed of the observer, and their discovery made necessary a tremendous amount of rethinking. The Galilean transformation is wrong; it is an excellent approximation at low speeds, but for atomic particles travelling at high speed it is hopelessly wrong and must be replaced by the correct form, the *Lorentz* transformation. For the record, the equations are

$$y = y'$$
$$z = z'$$
$$x = (x' + u_x t)/(1 - u_x^2/c^2)^{1/2}$$
$$t = (t' + u_x x'/c^2)(1 - u_x^2/c^2)^{-1/2}$$

The Lorentz transformation puts quite a strain on our ideas of logic but it gives the right answer, and that is the only objective test we have. The last equation says for example that a clock stationary in the S'-system will appear to be running slow when read in the S-system. It is also true that a clock stationary with respect to the S-system will appear to be running slow when read from the S'-system. Oh yes, we might say, we can explain that; because the clock is going away from us very fast and the light signals have come from further and further away, obviously the ticks will appear further apart in time. No, that is not what is meant. The paradox is that *when the time taken for the signals to arrive is taken into account* the moving clock *still* appears to run slower than the stationary one. It is a real surprise, and it does not match up at all with our ideas of common sense.

Further exercises

Exercises 13.8–12

8 A boy, standing on a train travelling horizontally at $30 \, \mathrm{ms^{-1}}$, throws a ball vertically upwards at $9.8 \, \mathrm{ms^{-1}}$ relative to the train at the instant the train passes a crossing. How far is the boy from

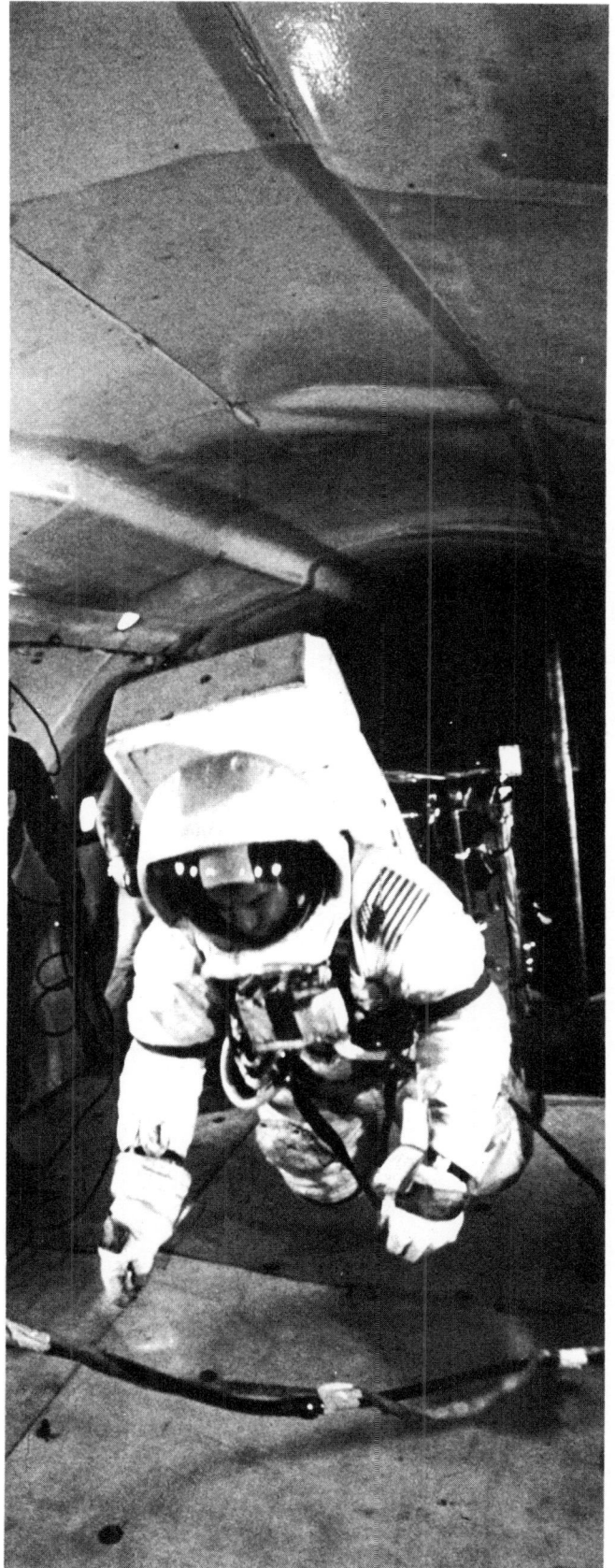

Zero-*g* simulation in an aircraft.

the crossing when he catches the ball? Draw a sketch to illustrate the path of the ball (a) when viewed by an observer travelling on the train, and (b) when viewed by an observer on the ground. ($g = 9.8\,\text{m s}^{-2}$)   (AEB)

9 A dart player stands 3·00 m from the wall on which the board hangs and throws a dart which leaves his hand with a horizontal velocity at a point 1·80 m above the ground. The dart strikes the board at a point 1·50 m from the ground. Assuming air resistance to be negligible, calculate (a) the time of flight of the dart, (b) the initial speed of the dart and (c) the speed of the dart when it hits the board.   (AEB)

10 A ball of mass 0·2 kg is projected upwards with a velocity of 40 m s^{-1} at 45° to the horizontal. Neglecting air resistance, taking time t to be measured from the instant of projection, and velocities to be measured relative to the ground, find
a) the velocities of the ball at $t = 1.5$ s and at $t = 3.5$ s
b) the change of momentum which takes place in this time interval
c) the rate of change of momentum during this time interval.

   Which, if any, of the above results would be altered if they were calculated relative to an observer in a car moving with uniform velocity? Explain.   (O)

11 Show that, when air resistance is neglected, the path of a projectile moving under gravity is a parabola.

   A particle projected from a fixed point on a horizontal plane at an angle of 30° to the plane strikes the plane 10 s later. Find its greatest height and its range on the plane.

   The particle is now projected at the same speed as before from a platform which is travelling along the horizontal plane in a straight line at 20 m s^{-1}; this straight line is in the same vertical plane as the line of projection of the particle, and the angle between the two lines is 30°. Show that the particle again describes a parabola. Find its greatest height and its range on the horizontal plane.   (O)

12 An object is dropped from a window 200 m high on the east-facing side of a skyscraper which is situated on the equator. Calculate the distance from the foot of the building to the point at which the object will reach the ground.   (WJEC)

14 Unbalanced force perpendicular to the velocity: circular motion

In Chapter 4 a distinction was drawn between *speed*, a scalar, and *velocity*, a vector. One of the consequences of that distinction is that a car going round a bend at constant speed is nevertheless changing its velocity; it is *accelerating*. At first sight it looks as if we are playing with words; simply because we choose to say that a different direction implies a different velocity we end up with the idea that an object travelling at constant speed can at the same time be accelerating. There is more to it than words however; not only can we work out the magnitude and direction of the acceleration, we can also show that it is perfectly consistent with $F = ma$.

For a start, there must be a force to produce the acceleration; in the example chosen it is a component of the frictional force exerted by the road on the tyres. Generally speaking, the frictional force has components parallel and perpendicular to the motion; for the purposes of this chapter we shall be considering only the perpendicular component, which we shall simply refer to as 'the frictional force on the tyre'. If that force were not present the car would keep going in a straight line, as an attempt to corner on an icy road will soon confirm. Of course if the car is to have constant speed the acceleration cannot be in the direction of motion; it cannot even have a

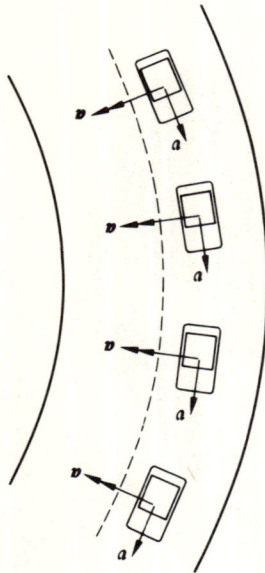

Figure 14.1

Showing how the directions of velocity and acceleration change as a car goes round a bend at constant speed.

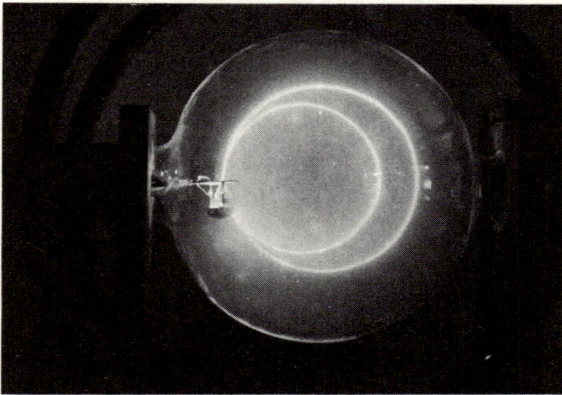

Figure 14.2

Fine beam tube. Electrons emitted from a hot cathode and accelerated by a positive voltage applied to the small conical anode cause ionisation in hydrogen gas at low pressure. The large circular coils provide a magnetic field perpendicular to the plane of the page.

component in the direction of motion–it must be perpendicular to the velocity *at every instant*. This means its direction must change continuously as the direction of motion changes (Figure 14.1).

Now, if the accelerating force is of constant magnitude we can take it that the car is deviating from its straight course at the same rate at every point of the bend, that is, it goes round in a curve of constant curvature–a circle. When any moving object is acted on by a single constant force always perpendicular to its velocity it moves in a circular path. A force like this is called a *centripetal force*. A stone whirled round in a horizontal circle on the end of a string is a familiar example. Here the centripetal force on the stone is provided by the string which is under tension, and it is always perpendicular to the stone's direction of motion. Naturally the stone moves in a circle: the centre of the string is fixed and the string cannot change length. A more convincing example is provided by an electron beam moving in a plane perpendicular to a uniform magnetic field. Here we do not have any string, nor is there an obvious centre for the circle, but we do have a centripetal force: its direction is perpendicular both to the magnetic field and to the velocity of the beam, and so the electrons obediently take a circular path. The electron beam is invisible in a vacuum but causes ionisation in a gas at low pressure, resulting in the emission of visible radiation. In Figure 14.2, two trajectories are shown, corresponding to different electron speeds (it is not possible to establish both trajectories simultaneously–the camera shutter was left open for two different accelerating potentials). The common point of the circles is the conical anode from which the electrons emerge.

The size of the acceleration

Figure 14.3 shows the circular path taken by a particle moving with constant speed v. To get its acceleration we look at the velocity change occurring between two points A and B on the circle–the velocities at A and B are v_1 and v_2 respectively, and are represented in the accompanying vector diagram. The velocity change between A and B is given by that velocity which must be added to v_1 (vectorially of course) to get v_2: it is called δv in the vector diagram. If P is halfway along the arc between A and B, δv is directed along PO. We can get the magnitude of δv from the diagram; it is just $2v\sin(\delta\theta/2)$, where $\delta\theta$ is the angle between OA and OB; v represents the magnitude both of v_1 and of v_2. If now we imagine A and B to be very close together we shall be able to use the approximation $\delta\theta \approx \sin\delta\theta$ and put

$$\delta v \approx v\,\delta\theta.$$

From this we get the useful expression

$$\frac{\delta v}{\delta t} \approx \frac{v\,\delta\theta}{\delta t}, \qquad \qquad ...(1)$$

where δt is the time taken by the particle to get from A to B. The ratio $\delta\theta/\delta t$ is a very useful quantity called the *angular velocity* ω; it is the rate at which the angle θ changes as the particle moves round the circle (regarding C as a fixed point), and it has units $\text{rad}\,\text{s}^{-1}$. We can connect ω with the speed along the arc as follows.

Length of arc travelled through in sweeping out angle $\delta\theta$ $= r\,\delta\theta$ (p. 14).

Speed along the arc = distance travelled/time taken
$$= r\,\delta\theta/\delta t$$
$$= r\omega$$

$$v = r\omega.$$

It is also useful to know the connection between ω and the time T required for one complete revolution. One radian of angle is swept out in a time $1/\omega$, so 2π radians require a time $2\pi/\omega$:

$$T = 2\pi/\omega.$$

We must now think carefully about the meaning of $\delta v/\delta t$ in equation (1). If A and B are very close together, then the direction of the acceleration, being always towards O, is substantially the same over the entire section considered, and its magnitude is equal to the magnitude of the velocity change divided by the time taken, that is, $\delta v/\delta t$. This is true *only* when we are looking at a *small* portion of the arc, so that the acceleration does not vary significantly in direction. So if we go to the limiting case where δt tends to zero and points A and B both merge into P, equation (1) gives us

acceleration a towards centre $= v\omega.$

Now we use the formula $v = r\omega$; we can use it to eliminate either v or ω. If we eliminate v we get

$$a = \omega^2 r$$

and if we eliminate ω we get

$$a = \frac{v^2}{r}.$$

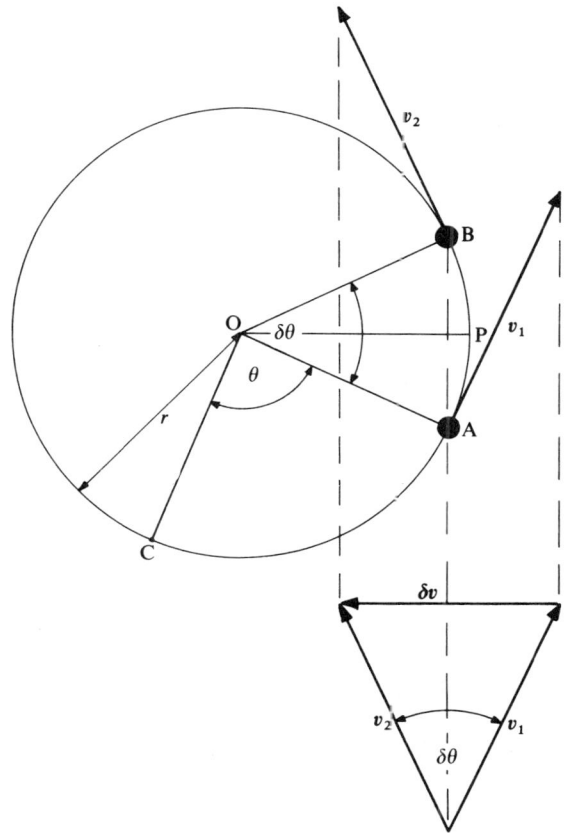

Figure 14.3

We shall now look at how these formulae can be used in conjunction with $F = ma$ to describe circular motion in various situations.

Object whirling round on the end of a horizontal string
We start with the simple case of an ice puck sliding around a fixed point O with speed v on the end of a string of length r (Figure 14.4). We are ignoring at present any frictional force between ice and puck; the ice simply acts as a support. The acceleration v^2/r towards the centre is provided by the tension P in the string, so Newton's second law gives us

$$P = \frac{mv^2}{r}.$$

Because the force P is always perpendicular to the motion it does no work; we do not have to provide any energy to get the puck round the circle, nor does it lose any kinetic energy in doing so. In practice of course friction would bring it to rest, but we are not attempting to describe this aspect here.

Exercise 14.1

A diesel locomotive of mass 6×10^4 kg takes a bend of radius 400 m at $20\,\text{ms}^{-1}$. Find the horizontal force the rails must exert on the wheels and their flanges.

Conical pendulum
Figure 14.5 shows a conical pendulum; a particle of mass m moves round at speed v in a horizontal circle on the end of a string which

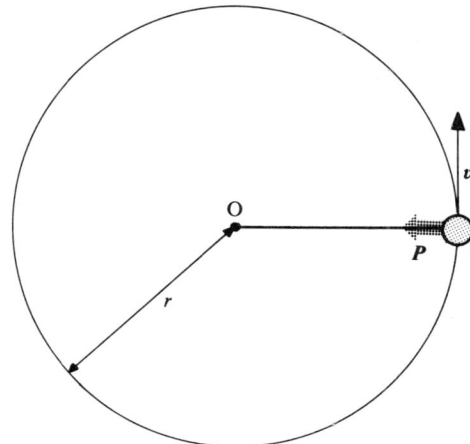

Figure 14.4

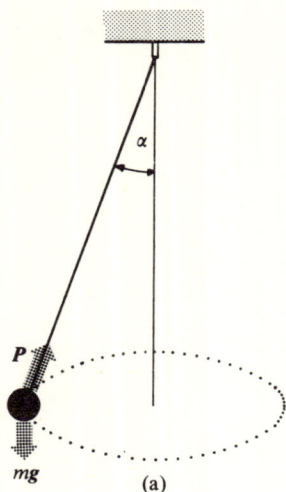

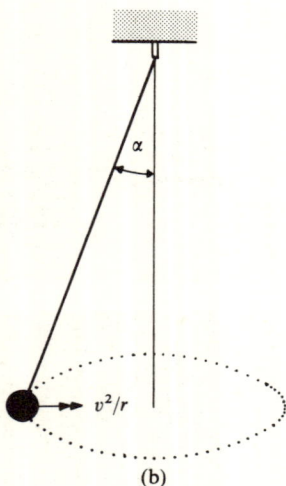

(a)

(b)

Figure 14.5

sweeps out a cone with semi-angle α. The forces are shown in Figure 14.5a, and the consequent acceleration in Figure 14.5b; it is often a good plan to use two diagrams in this way. There are two forces on the particle, its weight mg and the force P due to the tension in the string, and their resultant produces a horizontal acceleration v^2/r. Since there is no vertical acceleration, the vertical components of the forces must balance, so

$$mg = P\cos\alpha.$$

The weight obviously has no horizontal component, so it is the horizontal component of P which is left to provide the necessary acceleration. Newton's second law then gives us

$$P\sin\alpha = m\omega^2 r.$$

Exercise 14.2

Show that $\omega^2 = (g\tan\alpha)/r$ and hence that the period of revolution T is $2\pi\sqrt{(h/g)}$, where h is the vertical height of the point of support above the plane in which m moves.

Banking

The friction needed between tyres and road to get a car round a bend can be reduced by banking the road at an angle. We shall simplify the problem by considering the car to be a point mass, and represent the forces on all the tyres by just two forces, the normal reaction R and the frictional force F, acting perpendicular and parallel to the road surface respectively. The only other force on the car we need to consider is its weight, of magnitude mg. We can show that for any angle of bank on a bend of given radius there is one particular speed v_c for which the frictional force called upon vanishes almost entirely. If the car is travelling at a speed less than v_c it tends to slide down the bank, and the frictional force must act up the slope, whereas if the speed is greater than v_c, F must act down the plane, as in Figure 14.6, to keep the car in its circular path. The acceleration v^2/r must be directed towards the centre of the bend, of radius r;

Banked bend of an indoor running track.

although the road is banked we assume it stays at the same level all the way round the bend so that the vehicle is moving in a horizontal circle. There is no vertical acceleration, so there is no net vertical force; the sum of the vertical components is zero:

$$mg + F\sin\alpha - R\cos\alpha = 0,$$

where α is the angle of banking. The horizontal forces providing the acceleration are $R\sin\alpha$ and $F\cos\alpha$, so

$$R\sin\alpha + F\cos\alpha = mv^2/r.$$

Exercise 14.3

Show that if the force F is to be zero, v_c must be such that

$$\tan\alpha = v_c{}^2/rg.$$

Taking a bend on two wheels

This is a rather more tricky problem, because we have to take into account the fact that the various forces concerned do not act at a point. If we had considered the last problem in greater detail we should have run into the same difficulty, but we simplified the situation to avoid the implications; here they cannot be sidestepped. Figure 14.7 shows the forces on a motor-cycle as the rider leans over to take a bend of radius r. For a given speed there is only one correct angle of lean, α; if the rider attempts to set his machine at the wrong angle he will find himself having to compensate in some way to

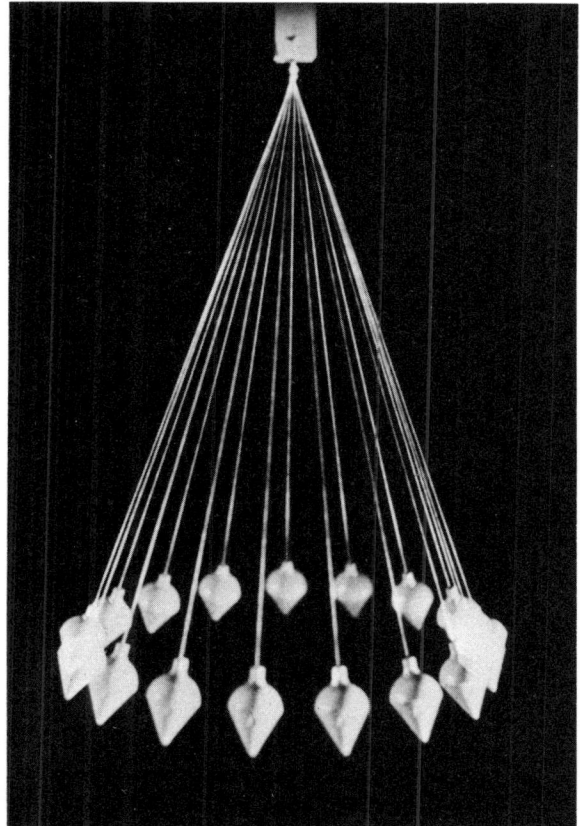

Conical pendulum illuminated by a stroboscopic lamp.

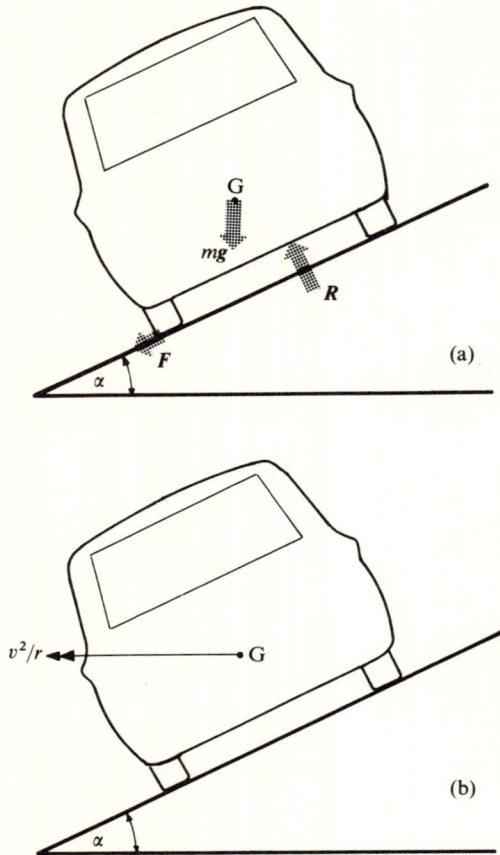

Figure 14.6

Car moving round a banked bend. The normal reactions provided by the road act on the tyres, but for simplicity we have combined them into the single force **R**.

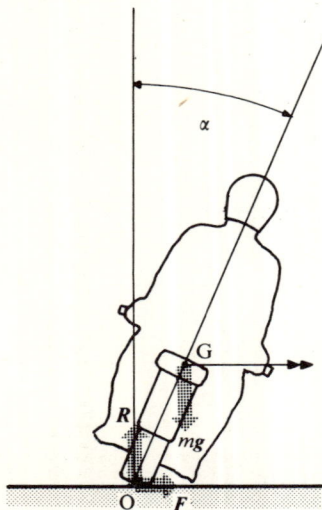

Figure 14.7

avoid falling off, either by altering the radius of his bend or throwing his body to one side.

There is no vertical acceleration, so

$$mg = R, \qquad \qquad \text{...(2)}$$

where m is the mass of machine plus rider and R is the normal reaction at the road surface. The frictional force F is the only force available to provide the acceleration towards the centre, so

$$F = \frac{mv^2}{r}. \qquad \qquad \text{...(3)}$$

Leaning over for the bend.

Unless there is enough grip from the tyres to provide a force of this magnitude a skid will result. High speeds and tight bends make the greatest demands on tyre grip, and there is a greater possibility of skidding in wet conditions, when the coefficient of friction between road and tyre is reduced.

So far our equations give no indication at all about the angle α. To find it, we have to use the fact that the net force on the motorcycle plus rider must be a horizontal force through their centre of gravity, otherwise it will provide not only an acceleration but also a turning effect which will tilt them to a different angle. Equations (2) and (3) provide us with the magnitudes of the forces and their directions, but they do not tell us anything about the *line of action* of the resultant. The most direct way to deal with the line of action here is to add up the forces parallelogram-fashion. **F** and **R** combine to form a single force **P** through O, and this force must combine with the weight mg to form a horizontal force going through their common point, because it has to produce a horizontal acceleration. We can easily see that the line of action of this horizontal force passes through the centre of gravity only if **P** also passes through the centre of gravity. This means we must insist that

$$\frac{F}{R} = \tan\alpha \qquad \qquad \text{...(4)}$$

Exercise 14.4

Show in the above case that $\tan\alpha = v^2/rg$, and find α for a bend of radius 50 m taken at 60 km h^{-1}.

We remember that in an equilibrium situation the matter of line of action is dealt with by applying the principle of moments. We can also tackle our present problem using moments, but we have to pick our way very carefully. For a start we cannot invoke the *principle* of moments – it applies only in equilibrium situations. What we *can* say is that because we require the net force, the centripetal force, to go through the centre of gravity, and therefore want its moment about the axis through the centre of gravity to be zero, we must insist that *the net moment of the constituent forces about the axis through the centre of gravity must be zero.* In an equilibrium situation we can take moments about any axis we choose; here it *has* to be an axis through the centre of gravity.

Exercise 14.5

Put the algebraic sum of the moments of F, R and mg about an axis through the centre of gravity equal to zero and show that this leads to $F/R = \tan\alpha$.

Sliding and toppling on a bend

We shall now take a closer look at a four-wheel vehicle on a bend, this time confining ourselves to a horizontal road surface (Figure 14.8). Each tyre experiences its own normal reaction and frictional force, but we have combined the forces on front and rear tyres so that we can work with just two normal reactions R_1 and R_2 and frictional forces F_1 and F_2 as shown (the centre of curvature of the bend is to the right of the figure). There is no vertical acceleration, so

$$R_1 + R_2 = mg \qquad \text{...(5)}$$

where m is the mass of the vehicle, and Newton's second law gives

$$F_1 + F_2 = \frac{mv^2}{r}, \qquad \text{...(6)}$$

where v is the speed and r the radius of the bend. There are two critical conditions we want to investigate; one concerns the possibility of sliding tangentially due to lack of tyre grip, and the other the possibility of turning over. We look at the sliding condition first and take it that the coefficient of friction between road and tyres is μ. At the critical speed when sliding is about to commence, friction must reach its limiting value for both sets of tyres, so

$$F_1 = \mu R_1$$
$$F_2 = \mu R_2.$$

(This does not mean that F_1/R_1 is *always* equal to F_2/R_2; it may well be that the friction reaches the limiting value for one set of tyres a little before it does for the other, but clearly slipping does not commence until the limit is reached for both sets of tyres.) Combining these equations, we have

$$F_1 + F_2 = \mu(R_1 + R_2) \qquad \text{...(7)}$$

Exercise 14.6

Show that equations (5), (6) and (7) give a critical velocity of $\sqrt{(\mu g r)}$ above which the car will skid. Note that this is independent of the mass of the car.

Now we shall look at the conditions under which the car is likely to turn over. Equation (7) is no longer relevant; the assumption is

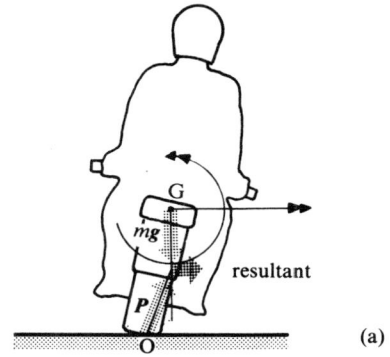

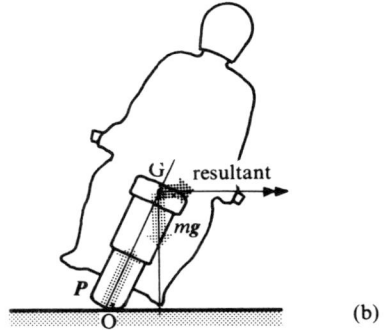

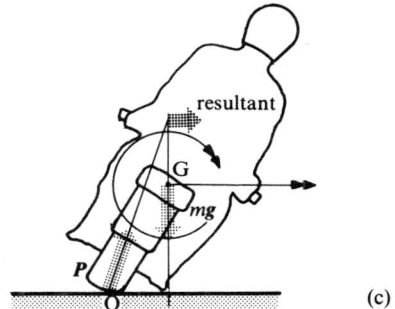

(a)

(b)

(c)

There is only one correct angle of tilt for a given speed and bend radius: (a) too small – the machine will fall outwards, (b) correct, (c) too large – the machine will fall inwards.

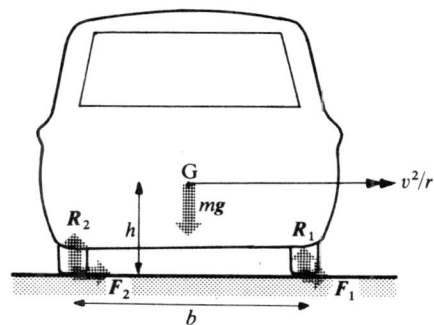

Figure 14.8

that the tyre grip is good enough to prevent sliding, and limiting friction is not reached. Equations (5) and (6) still hold; in fact they can be simplified a little because at the commencement of toppling the wheels nearer the centre of the bend leave the road, so F_1 and R_1 are both zero. This gives

$$R_2 = mg \quad \text{and} \quad F_2 = mv_c^2/r,$$

where v_c is the speed at which the car commences to overturn. These equations are not sufficient to find v_c; we need to know about the inherent stability of the vehicle, which is controlled by the position of its centre of gravity and the distance separating the wheels (called the 'track'). We are going to have to consider lines of action again, and we must insist that F_2, R_2 and the weight mg combine to form a horizontal force which acts through the centre of gravity if toppling is to be avoided. As for the case of the motorcycle therefore, we must have F_2 and R_2 combining to form a single force through the centre of gravity.

Exercise 14.7

Show that the critical speed v_c for turning over is given by

$$v_c^2/rg = b/2h$$

where h is the height of the centre of gravity above the road and b the distance between the tyres (assume that the centre of gravity is midway between the wheels). Note that this is also independent of the mass of the car. Find v_c for a car taking a bend of radius 50 m if $b = 1\cdot5$ m and $h = 0\cdot5$ m.

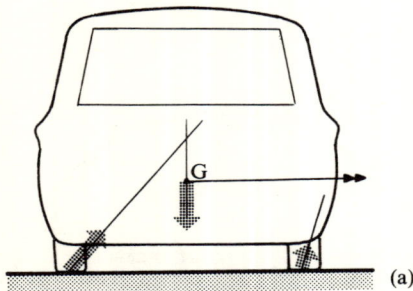

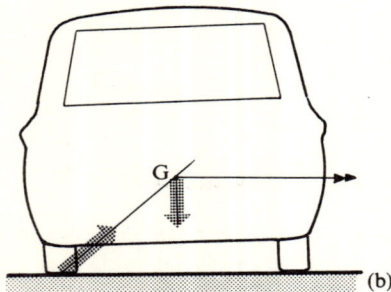

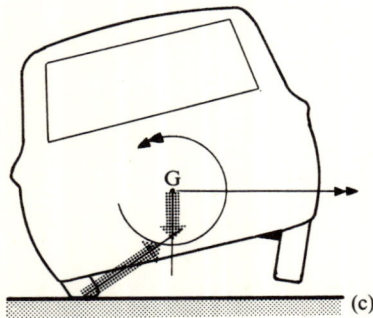

We should like to consider further why this represents the critical condition. Suppose we increase v; this means that we need more centripetal force to provide the acceleration v^2/r towards the centre, so more frictional force F_2 is called into play. As a result the force P intersects the line of action of mg at a point below the centre of gravity, tending to turn the car over in an anticlockwise sense. If on the other hand v is decreased, P intersects the line of action of mg above the centre of gravity and this tends to turn the car the other way, so the other two wheels make contact with the road and provide F_1 and F_1 to compensate.

Exercise 14.8

Work out the critical velocity for overturning by considering moments about the centre of gravity.

Motion in a vertical circle

Figure 14.9 illustrates an intriguing problem, if not a very practical one. If we have a particle moving around on the inside of a smooth vertical circular hoop, how fast must it be going at the bottom in order to avoid falling off at the top? We no longer have a constant speed round the circle: as the potential energy rises the kinetic energy falls, and we shall use the energy conservation principle to provide one of our equations. Taking the hoop radius to be r and the speeds at top and bottom to be v_T and v_B respectively, we can write

$$\tfrac{1}{2}mv_T^2 + mg(2r) = \tfrac{1}{2}mv_B^2 \qquad \ldots(8)$$

as long as all energy losses due to friction can be ignored.

Critical condition for overturning on a bend: (a) all four tyres experiencing reactions from the road, (b) critical condition, wheels on the right about to lift off the ground, (c) line of action of the net force is below G; car overturns. This simplified treatment has ignored the influence of the suspension, and the moment of inertia of the vehicle.

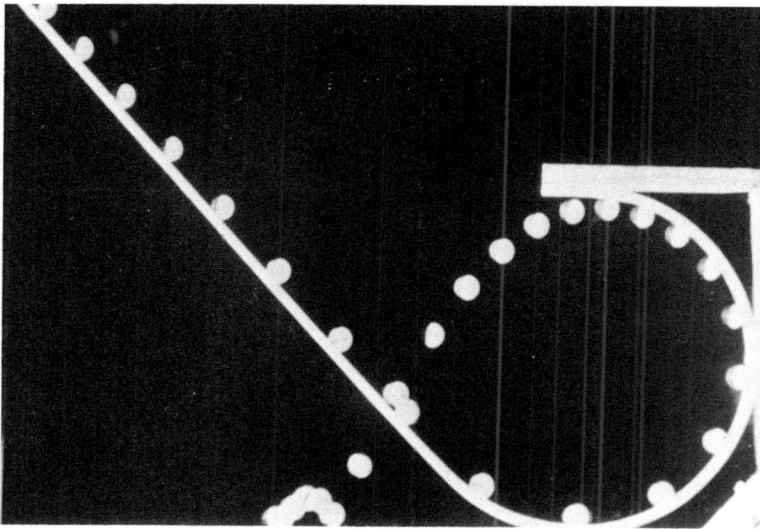

Figure 14.9

Though the particle speed is no longer constant we can still use the expression v^2/r for that component of the acceleration perpendicular to the direction of motion. The net force F on the particle has in general a component along the direction of motion, producing tangential acceleration, and a component directed towards the centre; it is the latter that we must put equal to mv^2/r. If the normal reactions provided by the hoop at top and bottom are R and S respectively, at the top we can write

$$R + mg = mv_T^2/r \qquad \qquad ...(9)$$

and at the bottom

$$S - mg = mv_B^2/r. \qquad \qquad ...(10)$$

We are looking for the speed at which the particle just fails to leave the hoop at the top; this will happen when R becomes zero, which means that equation (9) simplifies to

$$mg = mv_T^2/r.$$

Combining this with equation (8) we get

$$v_B^2 = 5gr.$$

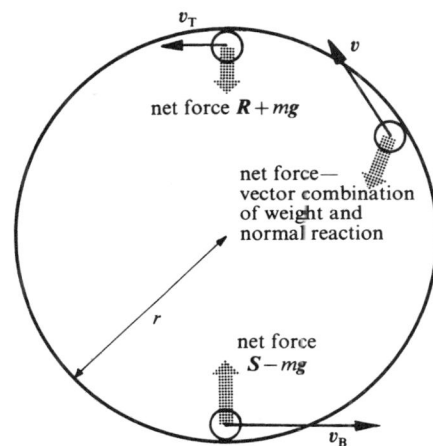

Motion of a particle constrained to move in a vertical circle.

Exercises 14.9, 10

9 If the particle reaches the required speed v_B by falling from rest down a smooth chute through a vertical height h, show that $h = 5r/2$.

10 Given that the chute is extended so that $h = 4r$, find the new values of v_B, v_T, R and S in terms of m, g and r.

Circular motion and pseudo forces

So far we have looked at the problem of circular motion from the point of view of an outside unaccelerated observer: we have been using an inertial frame of reference (see p. 184). What do things look like from the point of view of someone inside the vehicle? This involves looking at the situation from an accelerating frame of reference, so either we shall have to put up with the fact that we can no longer write $F = ma$ for the forces and accelerations we measure with respect to our chosen frame, or we shall have to introduce a

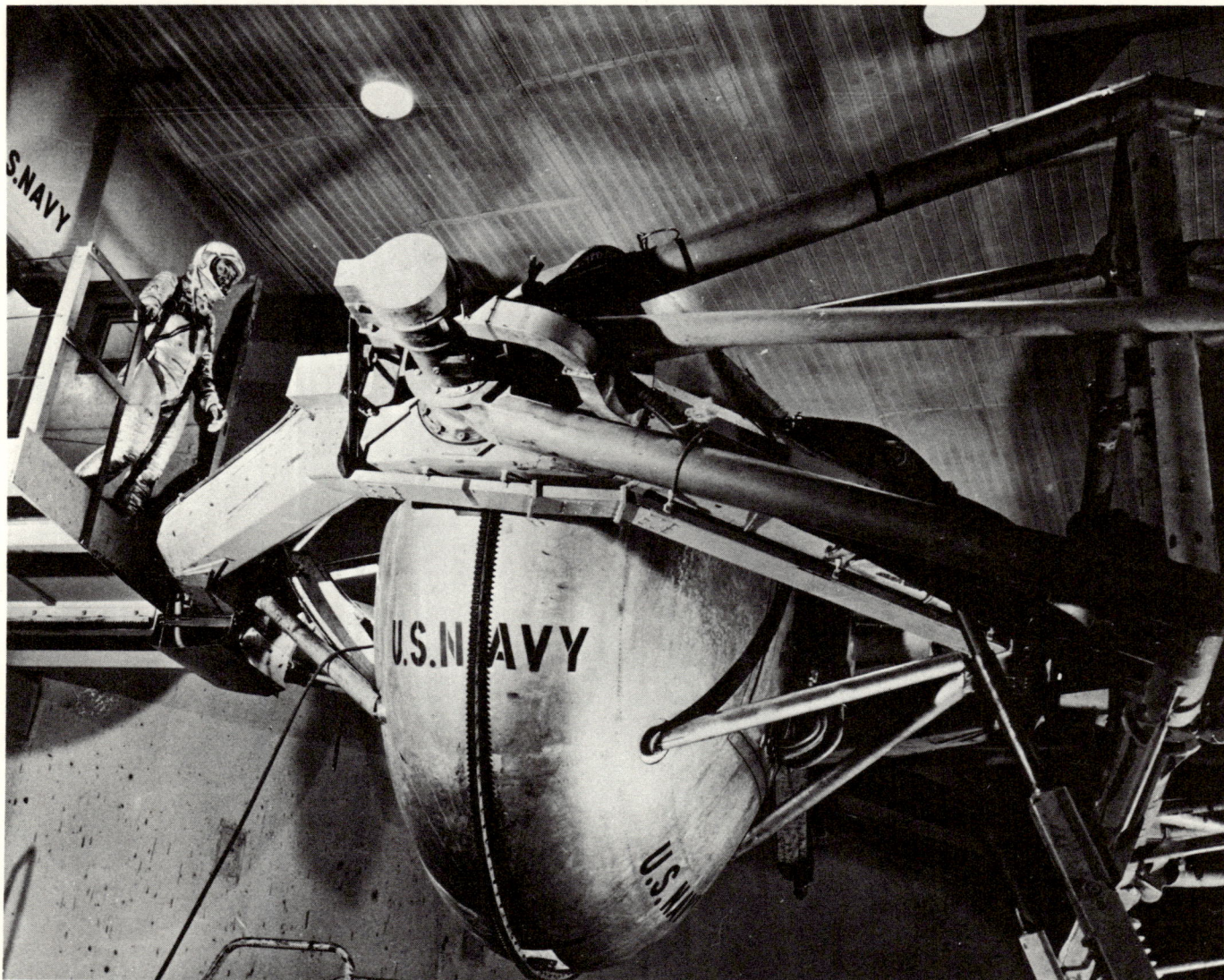

Centrifuge used in astronaut training programmes.

pseudo force to get things right. This particular pseudo force is called *centrifugal force*, and it needs to be emphasised that for an observer watching from the centre it does not exist. However it does have real significance for someone measuring things inside the vehicle; in fact if the vehicle is a rapidly rotating centrifuge of the type used in astronaut training programmes it can have enough significance to cause the observer to blame the pseudo force for his blackout! As far as he is concerned he has suddenly become several times heavier, and 'down' has become a direction nearer to 'away from the centre of rotation' than 'towards the centre of the earth'. It is no good the unaccelerated observer telling him that it is only a pseudo force, that if he *does* insist on riding an accelerating vehicle he must expect the seat to impart a force on him towards the centre; his physiology and any spring balances he has with him suggest to him that something peculiar has happened to gravity.

The earth's rotation

An object on the earth's surface is moving at about $30\,\mathrm{km\,s^{-1}}$ in an approximately circular orbit round the sun. For this reason alone a reference frame fixed on the earth is not an inertial frame. Even more important is the earth's rotation on its own axis; the fact is that a spring balance does not record the true weight of any object – the pull of the earth on it – because the spring balance is being used in an accelerating frame of reference. So we have to differentiate between the true weight of an object and its apparent weight; we have to distinguish between the acceleration of a falling object measured from an inertial frame of reference and the acceleration measured from an accelerating frame, fixed with respect to the earth's surface. We shall simply consider the more important effect, that of the earth's rotation. Suppose the true weight of an object at latitude α is mg', and that the apparent weight, as measured by a spring balance at that point, is mg. We shall use the fact that the apparent acceleration due to gravity, as measured by a rotating observer, added vectorially to the *observer's* acceleration, must give the true acceleration g'. The observer's acceleration $\omega^2 x$ (Figure 14.10) is small compared with g'; this means that we can put OP approximately equal to OQ in the vector diagram. The magnitude of g is therefore very close to $(g' - \omega^2 x \cos \alpha)$ or $(g' - \omega^2 r \cos^2 \alpha)$, where r is the radius of the earth. The inclination of the vertical, as measured by a plumbline, to the line of action of the true weight of the object, ϕ, is given by

$$\phi \approx \frac{PQ}{QO} \approx (\omega^2 x \sin \alpha)/g$$

or

$$\phi \approx (\omega^2 r \sin \alpha \cos \alpha)/g.$$

A standard trigonometric substitution converts this to

$$\phi \approx (\omega^2 r \sin 2\alpha)/2g,$$

and as $\sin 2\alpha$ is a maximum when $2\alpha = 90°$ it shows that the angle ϕ is largest at latitude 45°, neglecting a small error due to our approximations.

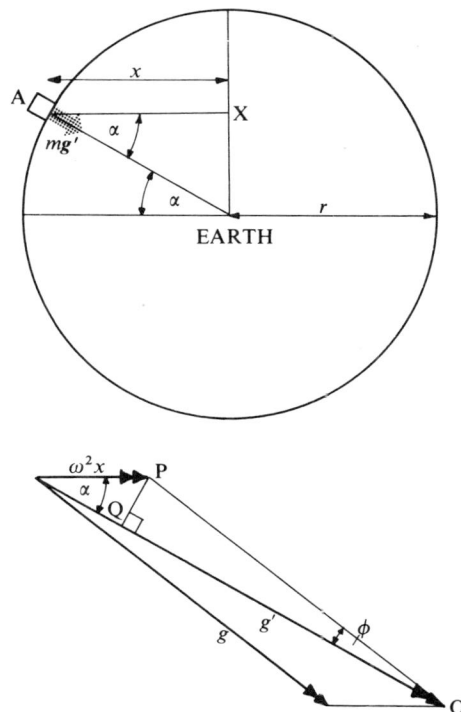

Figure 14.10
The acceleration of free fall of an object is g' with respect to the centre of the earth, but to an observer on the surface, himself accelerating towards X, the acceleration of the object appears to be g. A plumbline at A hangs along the direction of g.

Exercises 14.11, 12

11 Evaluate the values of ϕ and $(g' - g)/g$ caused by this effect at (a) latitude 60°, (b) the equator. Take the earth's radius as 6400 km and g as $10 \, \mathrm{m \, s^{-2}}$.

12 What would the period of rotation of the earth have to be in order that a mass resting on the surface at the equator should have negligible apparent weight?

Further exercises

Exercises 14.13–24

13 A light elastic string of natural length l is extended by an amount Fl/λ when subject to a tension F. A small body of mass m is attached to a point O on a smooth horizontal table by means of this elastic string. The body moves in a horizontal orbit of constant radius $5l/4$ and centre O with a tangential velocity v. Find the value of v and calculate the ratio of the kinetic energy to the elastic stored energy. Discuss qualitatively the motion of the body if the string were cut.   (o & c)

14 Use Newton's laws of motion to explain the following facts concerning a body moving with constant speed in a circle:

a) although the speed is constant, the body is accelerated;

b) although the radius is constant, the body has an acceleration towards the centre.

Explain exactly how this acceleration is caused in the following cases:

c) a train on a circular track;
d) a conical pendulum.

A small bob of mass 100g is suspended by an inelastic string of length 50cm and is caused to rotate in a horizontal circle of radius 40cm which has its centre vertically below the point of suspension. Show on a sketch the two forces acting on the bob as seen by an outside observer. Find (i) the resultant of these forces, (ii) the period of rotation of the bob.   (c)

15 A plumbline hangs freely in a railway carriage. Describe and explain its behaviour when the train is moving (a) along a straight track at increasing speed; (b) along a straight track with constant speed; (c) round a curve with constant speed.

If in case (c) the radius of the curve is 400m, and the plumbline makes an angle of 5° with the vertical, what is the speed of the train?   (o & c)

16 Prove that a body of mass m constrained to move in a circle of radius r with uniform angular velocity ω must be acted on by a force $m\omega^2r$ directed towards the centre of the circle.

A bob whose mass is 100g hangs by a light string from a fixed point and executes a circular orbit in a horizontal plane (*conical pendulum*). Given that the length of the string is 50cm, and the bob makes 10 revolutions in 14s, calculate the radius of the orbit and the tension in the string.   (o & c)

17 One end of a light inextensible string 1m long is attached to a fixed point. A small sphere of mass 0·5kg is attached to the string at its free end and describes a horizontal circle of radius 0·5m.

a) Find   (i) the angular velocity of the sphere in its path,
          (ii) the tension in the string
          (iii) the period of revolution
          (iv) the kinetic energy of the sphere.
b) Explain how the linear momentum of the sphere changes during the course of one revolution.   (o)

18 A fairground roundabout consists of a vertical pillar with a number of horizontal radial arms fixed at its upper end. Each arm supports a seat suspended from a cable of length 5m and negligible mass, the upper end of each cable being attached to a horizontal arm 4m from the axis of rotation. During a ride, one of the cables is inclined at 30° to the vertical when a passenger is seated in it. The centre of gravity of passenger plus seat is 1m from the end of the supporting cable in line with the cable. What is the rate of rotation of the roundabout?

Describe qualitatively the motion of the passenger as the roundabout is slowed down to rest.   (o & c)

19 Explain the following:

a) why a cyclist has to lean over when rounding a corner,
b) why the bonnet of a car dips downwards when the brakes are applied to retard forward motion,
c) why a car tends to overturn when going too fast round a corner.

Illustrate your answers by diagrams showing the forces involved.

A motor car, of which the breadth of the wheelbase is 1·69 m, has its centre of gravity at a point mid-way between the wheels and 0·81 m above the ground. What is the maximum speed, in $km h^{-1}$, at which the car can take a bend of radius 6·4 m without beginning to overturn? (AEB)

20 A small body of mass m is attached to one end of a light inelastic string of length l. The other end of the string is fixed. The string is initially held taut and horizontal, and the body is then released. Find the values of the following quantities when the string reaches the vertical position: (a) the kinetic energy of the body, (b) the velocity of the body, (c) the acceleration of the body, and (d) the tension in the string. (O & C)

21 A spirit level is placed lengthwise on the window sill of a railway carriage when it is stationary on a horizontal straight stretch of track and the bubble takes its position in the centre of the instrument. Describe and explain the behaviour of the bubble when the train moves with uniform acceleration.

$ABCD$ is a U-tube in which AB and CD are straight and vertical while BC is straight and horizontal. The U-tube is partly filled with liquid. Explain why, if the tube is rotated with the arm CD as axis, the liquid surfaces in AB and CD set at different heights. Calculate this difference if $BC = 30$ cm and the speed of rotation is 75 revolutions per minute. You may assume the bore of the tube to be small and surface tension effects negligible.
(JMB)

22 Assuming that the earth is a homogeneous sphere of radius 6400 km, find the value for the acceleration of a freely falling body at the equator, if its value at the pole is $9·82 m s^{-2}$.

Two small masses m_1 and m_2 are joined by a light thread. They are placed on a horizontal circular platform with the thread lying along a radius. The platform is set in rotation about a vertical central axis with gradually increasing speed. Find the speed at which the masses start to slip along the radius, the coefficient of friction being μ. Show that any mass, having the same coefficient of friction, placed at the centre of gravity of the original masses would slip at this same speed. (WJEC)

23 Small balls, each of mass 100 g, are attached at distances of 1 m, 2 m and 3 m from end D of a 3 m length of string. The string is then rotated with uniform angular velocity in a horizontal plane about D. If the outside ball is moving at a speed of $6 m s^{-1}$, what are the tensions in the three parts of the string? If the angular velocity of the string is increased, which part of the string will break first? (Neglect gravitational effects.) (JMB)

24 A ship is equipped with a very sensitive apparatus for the measurement of the acceleration due to gravity g and is steaming eastward along the equator at 30 km per hour. The value of g is measured as $9·79257 m s^{-2}$. What would you expect the measured value to be if the ship steamed west at the same speed? (JMB)

15 Unbalanced force dependent on position; simple harmonic motion

Sequence of photographs of the sun, taken at hourly intervals from within the Arctic circle. Because of the inclination of the axis of rotation of the earth to the line joining the centres of earth and sun, the motion appears to be oscillatory.

In Chapters 4 and 13 we derived equations to describe the motion of accelerated objects, equations such as

$$x = ut + \tfrac{1}{2}at^2 \qquad \text{and} \qquad v = u + at$$

(p. 42). Although these are very useful, they have one important restriction; they apply only to cases for which the acceleration is constant, that is, caused by a constant force. We now want to extend our ideas to deal with situations involving forces which are *not* constant, but which vary with the position of the mass which is being accelerated. Take the simple case of a mass on the end of a spring; the more the spring is stretched the greater the force it exerts, providing it is not pulled too far. A space vehicle experiences gravitational attraction which decreases as it gets further from the planet responsible. The force between two electric charges varies with their separation. How can we set about describing the motion in such situations? The *physics* will be unchanged; the key is still $F = ma$, and it will be true moment by moment throughout the motion. The mathematics will look very different however; F will be varying continuously, and so a will also vary continuously, and equations such as

$$x = ut + \tfrac{1}{2}at^2 \qquad \text{and} \qquad v = u + at$$

will just not do. In Chapter 13 we saw that these equations can be arrived at by solving the differential equation

$$F = m\frac{d^2y}{dt^2},$$

where F is a constant force in the y-direction. This equation continues to apply when F is varying, but now it is more difficult to solve. In the present chapter we shall be concerned mainly with motion confined to one direction, let us say this time the x-direction, and therefore we shall be dealing with a force F which is also in the x-direction. In addition we shall defer until Chapter 16 the possibility that the magnitude F may depend on the speed of the object concerned; in other words F will be defined entirely by the displacement x of the object. We can therefore write

$$F(x) = m\frac{d^2x}{dt^2} \qquad \text{...(1)}$$

where $F(x)$ means that F depends in some definite way on x. Later in the chapter we shall look in detail at how to solve equation (1) for one extremely important form of $F(x)$, but before that we can make some useful generalisations.

Energy considerations
Whatever form equation (1) may take, it is very likely that energy will be a useful quantity to deal with. Suppose the object moves from position x_1 to x_2. The work done by the force F is just force times displacement, but $F(x_2 - x_1)$ will not do because F changes; we need to use the full treatment.

$$\text{Work done on the object} = \int_{x_1}^{x_2} F dx.$$

If F is the only force involved then this work serves to increase the kinetic energy of the accelerated object, so that if the original speed was v_1 and the final speed v_2 we get

$$\int_{x_1}^{x_2} F dx = \tfrac{1}{2}mv_2^2 - \tfrac{1}{2}mv_1^2. \qquad \text{...(2)}$$

In fact we can show that this equation follows directly from equation (1). Writing v as the (variable) velocity in the x-direction, we can put

$$F(x) = m\frac{dv}{dt}$$

or

$$F(x) = m\frac{dv\,dx}{dx\,dt},$$

that is,

$$F(x) = mv\frac{dv}{dx},$$

giving

$$\int_{x_1}^{x_2} F(x)\,dx = \int_{v_1}^{v_2} mv\,dv = \tfrac{1}{2}m(v_2^2 - v_1^2).$$

An example involving the energy stored in a stretched spring will show how equation (2) can be put to good use.

Example

A spring with force constant λ (see p. 146) is fixed at its upper end. The free end is attached to an object of mass m but is initially supported so that the spring is just slack. The object is then allowed to

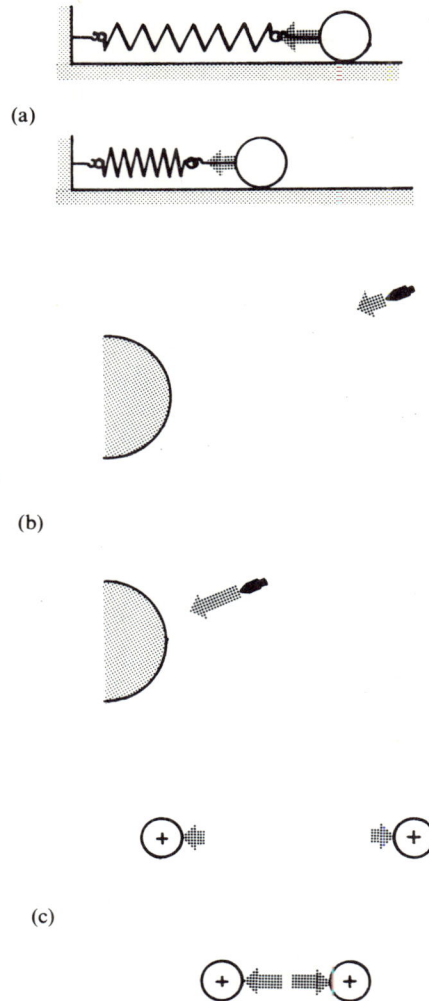

(a)

(b)

(c)

Three examples of forces dependent on position: (a) tension in a spring, (b) gravitational attraction on a space vehicle approaching a planet, (c) electrostatic forces.

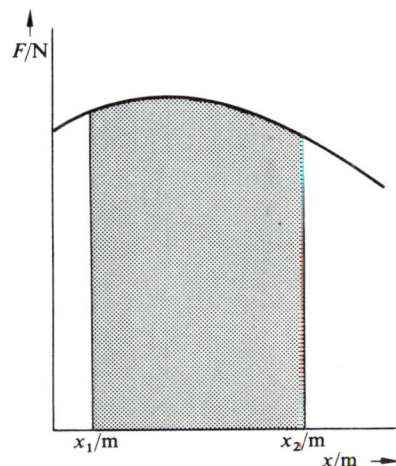

Work done $= \displaystyle\int_{x_1}^{x_2} F\,dx$ is numerically equal to the area under the graph of force F against displacement x, between x_1 and x_2.

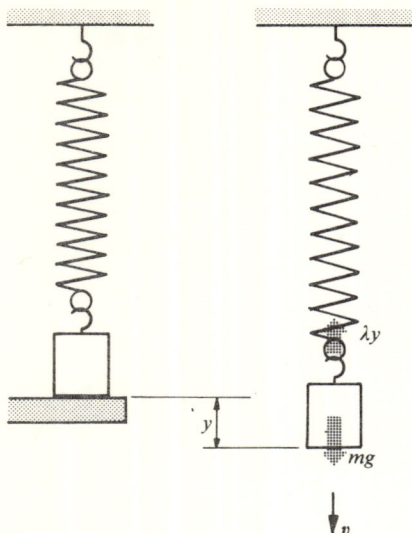

Figure 15.1

fall, and at no time during the subsequent motion does the spring become stretched beyond its elastic limit. Show that if all the loss in potential energy appears as kinetic energy of the mass m, its velocity v at a depth y below the starting point is given by

$$v^2 = 2gy - \frac{\lambda y^2}{m},$$

and find the values of y for which v is zero.

The net downward force F on the mass m at depth y (Figure 15.1) is given by

$$F = mg - \lambda y$$

and therefore the potential energy W lost by the system (mass plus spring) as a result of the displacement can be written as

$$W = \int_0^y F \, dy = \int_0^y (mg - \lambda y) \, dy$$
$$= [mgy - \lambda y^2/2]_0^y = mgy - \tfrac{1}{2}\lambda y^2.$$

Both these terms are familiar; the first is simply the loss in gravitational potential energy and the other is the energy stored in the stretched spring; in fact if we had used the result of exercise 11.18 we could have written this expression down at the start. The p.e. loss W appears as k.e. gain $\tfrac{1}{2}mv^2$, so

$$\tfrac{1}{2}mv^2 = mgy - \tfrac{1}{2}\lambda y^2$$
$$v^2 = 2gy - \frac{\lambda y^2}{m}.$$

The mass is therefore stationary when

$$2gy = \frac{\lambda y^2}{m},$$

that is, when $y = 0$ or $2mg/\lambda$.

Exercise 15.1

Show that the equilibrium position of the mass lies half-way between these values, and find the velocity as it passes through this point. Show that the kinetic energy is a maximum for this particular value of y.

The same approach can be used even when frictional forces are present as long as they do not depend on the velocity, as illustrated in the next example.

Example

Figure 15.2 shows a spring with force constant λ, lying on a horizontal table with one end fixed and with a mass m attached to the free end. The mass is pulled out a horizontal distance x_0, and when it is released it slides back but experiences a constant frictional force P during the motion. Find its velocity v when it is a distance x out from its original stable position.

At extension x_0 the stored elastic energy is $\tfrac{1}{2}\lambda x_0^2$, and at extension x it is $\tfrac{1}{2}\lambda x^2$, so that in going from x_0 to x an amount of energy $\tfrac{1}{2}\lambda(x_0^2 - x^2)$ is made available. The work done against the frictional

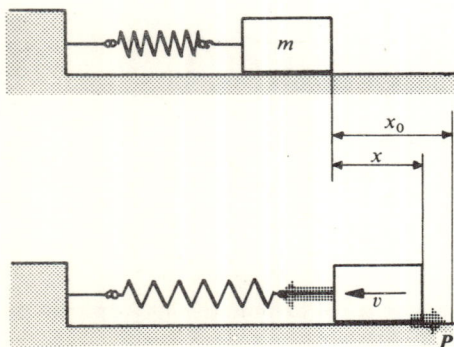

Figure 15.2

force is $P(x_0 - x)$, so the net kinetic energy gained is

$$\tfrac{1}{2}\lambda(x_0{}^2 - x^2) - P(x_0 - x)$$

giving

$$v^2 = \frac{\lambda}{m}(x_0{}^2 - x^2) - \frac{2P}{m}(x_0 - x).$$

Exercise 15.2

Show that the block just gets back to its original stable position if $P = \tfrac{1}{2}\lambda x_0$.

We shall be using equation (2) in situations involving gravitational forces in Chapter 20.

Simple harmonic motion

We shall now look in greater detail at the equation

$$F(x) = m\frac{d^2x}{dt^2}$$

when $F(x)$ takes the particularly important form

$$F(x) = -kx,$$

k being a positive constant. The equation takes this form whenever an object displaced from a position of stable equilibrium experiences *a restoring force proportional to the displacement*, and we find that many physical situations can be described at least approximately in such terms; for instance, as we saw in Chapter 11, elastic deformations of all types give rise to proportional amounts of restoring force. Stretched springs, twisted wires, plucked strings are just some examples, and they all share a particular type of *oscillatory* motion as a result.

The equation of interest to us is

$$-kx = m\frac{d^2x}{dt^2}, \qquad \text{...(3)}$$

where the negative sign takes account of the fact that the force is acting in a direction opposing the displacement x; it is a *restoring* force. We shall find it convenient to write k/m as ω^2; equation (3) therefore becomes

$$\frac{d^2x}{dt^2} = -\omega^2 x. \qquad \text{...(4)}$$

If ω is a real number ω^2 must be positive, ensuring that $-\omega^2$ is negative.

Exercise 15.3

What is the SI unit for ω?

Equation (4) says that the acceleration in the x-direction is proportional to the displacement and always directed back to the equilibrium position. We can get a qualitative picture of the resulting motion as follows. If the mass, initially at rest, is released from P (Figure 15.3a) it accelerates towards the equilibrium position Q. Although the acceleration decreases all the way from P to Q, it does not become zero until Q is reached; this means that the velocity builds up continuously as the object moves from P to Q, though the *rate* at which it builds up falls off as Q is approached. Figure 15.3b

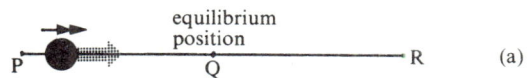

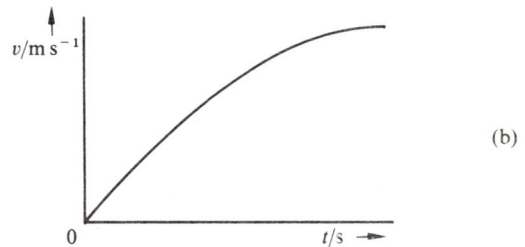

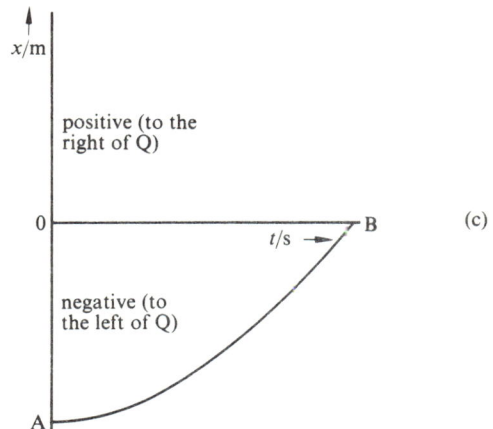

Figure 15.3

(a) Particle attracted towards Q with a force proportional to its displacement from Q. (b) The resulting velocity–time graph. (c) The corresponding displacement–time graph.

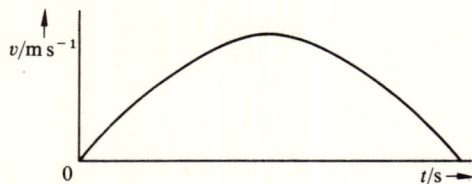

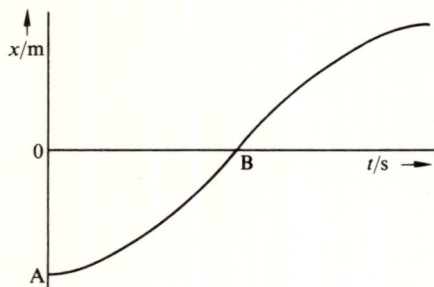

Figure 15.4

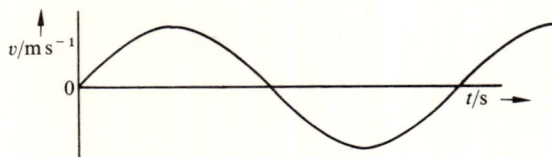

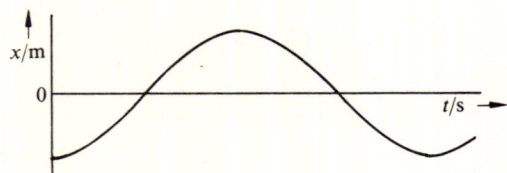

One complete cycle of the motion.

is a sketch of the velocity–time graph we can expect over this part of the motion; the slope of the line, being a measure of the acceleration, decreases continuously to zero.

Figure 15.3c is the corresponding displacement–time graph; we have adopted the usual convention that displacements to the left of Q are negative. The general shape of this graph can be worked out by remembering three things. First, the object starts at P and ends at Q; this defines the ordinates of the end-points A and B on the graph. Secondly, the object starts with zero velocity, so the initial slope is zero; the line must be horizontal at A. Thirdly, the velocity increases all the way to Q, so the line must curve up with ever-increasing slope.

Once the object has passed through Q, the acceleration is directed back towards Q, and the further the object gets from Q the larger the magnitude of this acceleration becomes. We can see that there is a definite symmetry in the situation; whatever force acts to the right between P and Q acts to the left at the corresponding point between Q and R. This must mean that the reduction in velocity to the right of Q is simply a reversal of the velocity increase between P and R. We can therefore extend the velocity–time graph as shown in Figure 15.4a; Figure 15.4b is the corresponding displacement–time graph, and we can argue from the symmetry that the particle does not stop until it reaches R, where $PQ = QR$.

The accelerating force towards Q does not disappear merely because the object comes momentarily to rest at R. The acceleration to the left continues, and the object commences to retrace its path. The resulting velocity to the left builds up and decays in just the same way as the velocity to the right, so that we can extend our graphs still further, merely reversing the signs of all the velocities. This brings the object back to a momentary rest at P, after which the entire process repeats itself again and again. So we see that the motion consists of a series of symmetrical oscillations about the equilibrium point Q, with a constant *amplitude* (maximum displacement) equal to PQ (or QR). The name we have for such behaviour is *simple harmonic motion* (s.h.m.); formally, *a particle performing s.h.m. has an acceleration which is directed at all times towards a fixed point, and has a magnitude proportional to its displacement from this point.* This definition applies at least approximately to a large number of physical situations, especially when the displacements involved are small. A few common examples follow.

Loaded spring

The spring in Figure 15.5 has force constant λ and when a mass m is hung on the end the extension l in the equilibrium condition is given by

$$\lambda l = mg. \qquad \qquad ...(5)$$

If the mass is now pulled down a further distance y, the total tension in the spring is $\lambda(l + y)$ and so the mass, when released, experiences a net upward force of value $\lambda(l + y) - mg$, or simply λy, using equation (5). Applying $F = ma$ to the mass,

$$-\lambda y = m\frac{d^2 y}{dt^2},$$

where the negative sign indicates that the force is tending to reduce y and restore equilibrium. This is therefore an example of equation (4), with x replaced by y, and ω^2 equal to λ/m.

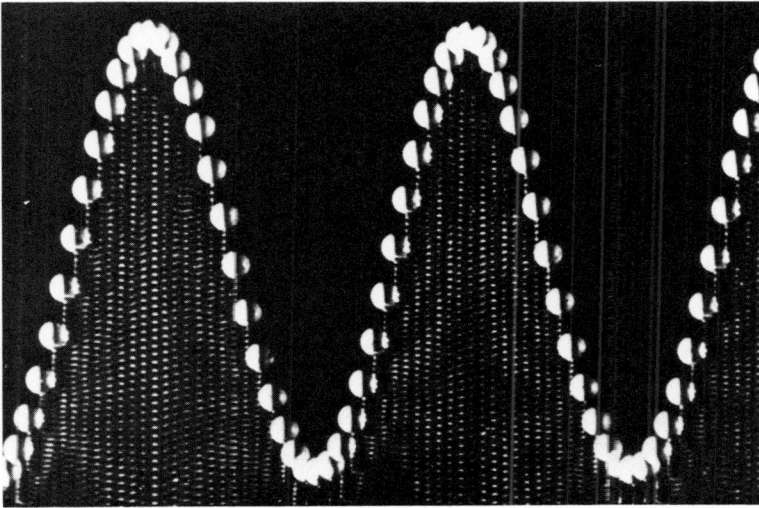

Vertical motion of a mass suspended from a spring, illuminated by a stroboscopic lamp. The film was given a uniform horizontal velocity to displace the images horizontally, so that the pattern corresponds to a displacement–time graph.

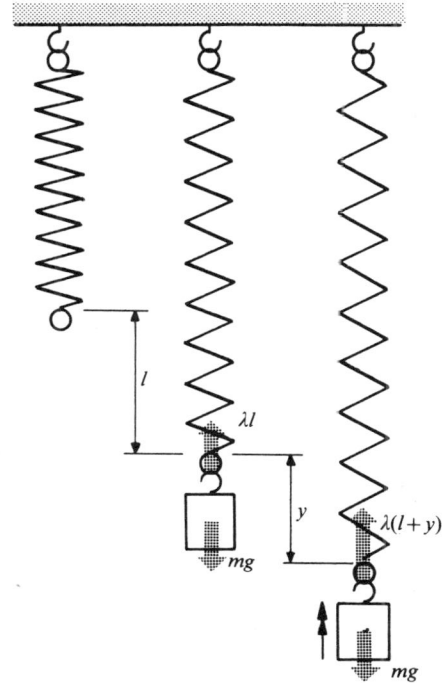

Figure 15.5

Floating body

Figure 15.6 shows an object floating in a liquid of density ρ; the mass of the object is m and its cross-sectional area in the region of the surface is A. The object is now pushed in a distance y; y must not be so large that the cross-sectional area at the liquid surface changes appreciably, and the object must be stable in the water; apart from this there is no restriction on size or shape. (We are ignoring the inertia of the liquid which must be moved in the process, also any change in the level of the liquid surface.) In equilibrium the upthrust exactly balances the weight, but as a result of the depression an extra volume of liquid Ay is displaced, so that the upthrust is increased by ρAyg. This net force produces acceleration upwards; Newton's second law gives

$$-\rho Agy = m\frac{\mathrm{d}^2 y}{\mathrm{d}t^2},$$

so we have equation (4) again, this time with ω^2 equal to $\rho Ag/m$.

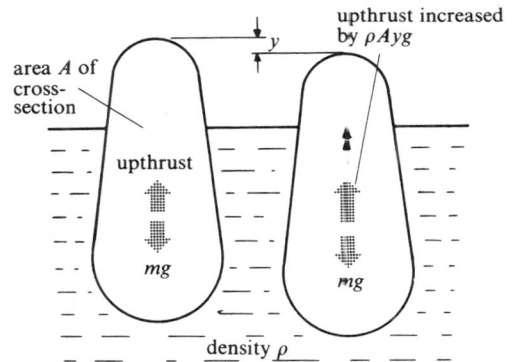

Figure 15.6

Simple pendulum

Figure 15.7 shows the bob of a simple pendulum drawn aside through an angle θ; the forces on the bob as it is released are its weight mg and the force T due to the tension in the string. The bob moves along the arc of the circle of radius l; there is no acceleration perpendicular to this arc so there can be no net force perpendicular to it either; we can write

$$T\cos\theta = mg.$$

The net force F along the arc is $mg\sin\theta$, or $mg\theta$ approximately if θ is small. Replacing θ by x/l, where x is the length of arc through which the bob has been displaced, we obtain

$$F = -mgx/l$$

where the negative sign again indicates a restoring force. If θ is

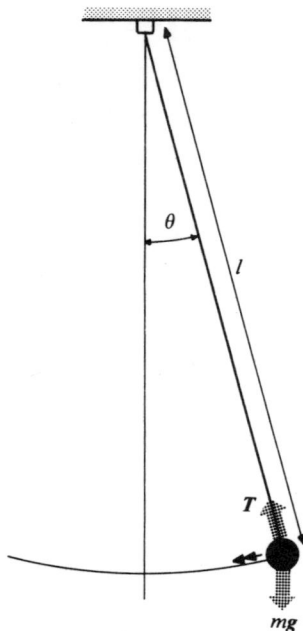

Figure 15.7

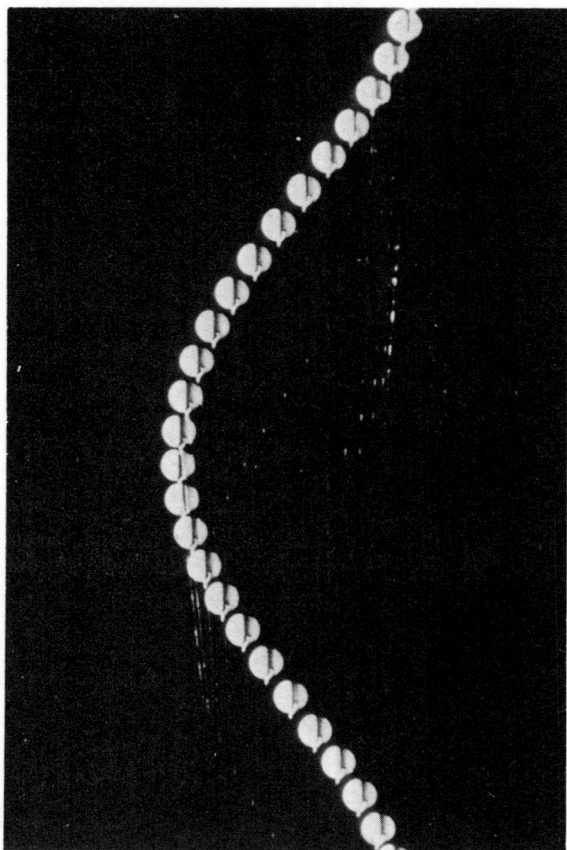

Motion of a simple pendulum illuminated by a stroboscopic lamp. The film was given a uniform vertical velocity to displace the images vertically.

small it also follows that we are dealing with a section of the arc which is very close to a straight line, so we can apply $F = ma$ to accelerations along the arc just as though it *were* a straight line. This gives

$$\frac{-mgx}{l} = m\frac{d^2x}{dt^2},$$

providing yet another example of equation (4), with ω^2 equal to g/l.

Exercise 15.4

The mass m in Figure 15.8 rests on a smooth horizontal surface between two light horizontal springs each stretched to a length $(l + a)$ but originally of length l. The tension in each spring is P. Find the restoring force on m if it is displaced a distance $x(<a)$ to the right. Hence write down the differential equation which defines its motion and show that $\omega^2 = 2P/ma$.

Solution of the equation; numerical method

Some differential equations can be solved with a certain amount of algebra to give a neat solution embodied in a formula; most cannot. Equation (4) happens to be one that can, but before we look for the formula it may help to see what is involved if we tackle it as if no neat answer were available. We shall work out how x changes as t changes by an approximate numerical method; this is the method to use when all else fails. The idea is as follows. Though no system behaving according to $d^2x/dt^2 = -\omega^2x$ has a constant acceleration, we *can* use

$$x = ut + \tfrac{1}{2}at^2 \quad \text{and} \quad v = u + at,$$

which strictly apply only to constant acceleration, *if we look at the motion during an interval which is very short compared with the time taken for one oscillation*, so short in fact that the acceleration does not have time to change appreciably. Suppose we decide to try a time interval of 0·1 s. We can write down the values of position, velocity and acceleration at the start and use $x = ut + \tfrac{1}{2}at^2$, $v = u + at$ and $F = ma$ to work out the values of position, velocity and acceleration at the end of 0·1 s. These new values, albeit approximate, are then used as the starting values of the *next* 0·1 s interval and the whole process is repeated. It is tedious by hand, but is a job for which a computer is ideally suited; a computer programmer is never more at home than when he can instruct the machine to work in a cycle, feeding the answer to a sum back into the start and going through the process again. Quite a modest computer can do that so quickly that it can carve the time into intervals of less than 0·001 s and still work the result out faster than it can happen in real time. Of course there is some error in each stage, and the error accumulates as the calculated starting conditions for the next stage get further and further from what happens to the real thing. Space vehicles have their trajectories worked out by numerical methods, and at various stages in the flight the motors are switched on for short periods to correct the velocity, compensating for the difference between observed and calculated position. The computation can then be resumed with updated information.

We shall not go into all the various refinements which make for better accuracy; the example which follows merely illustrates the principle in the simplest possible way. We consider a particle behaving according to $d^2x/dt^2 = -\omega^2 x$; let ω^2 be equal to 1 for simplicity and suppose that when $t = 0$s we allow the particle to start from rest at $x = 10$m. We can use this x-value to get the acceleration a by putting

$$a = -\omega^2 x$$

and then find our approximate value for the displacement at the end of 0·1 s by means of

$$v = u + a\delta t \qquad \text{and} \qquad \delta x = u\delta t + \tfrac{1}{2}a\delta t^2.$$

Here u is the velocity at the start of the section and v the velocity at the end of it; δx is the extra displacement in time δt, which we put equal to 0·1 s. The calculation for the first 0·1 s is set out in Table 15.1.

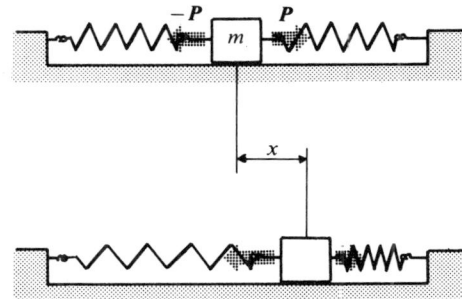

Figure 15.8 (Exercise 15.4)

Table 15.1

t/s	x/m	a/ms^{-2}	u/ms^{-1}	$a\delta t$/ms^{-1}	v/ms^{-1}	$u\delta t$/m	$\tfrac{1}{2}a\delta t^2$/m	δx/m
0	10	−10	0	−1	−1	0	−0·05	−0·05

Now we have to update the information ready for the second section. The new value of x will be $(10 - 0.05)$m and the new u will be the old v, $-1\,\text{ms}^{-1}$. The arithmetic will continue as indicated in Table 15.2.

Table 15.2

t/s	x/m	a/ms^{-2}	u/ms^{-1}	$a\delta t$/ms^{-1}	v/ms^{-1}	$u\delta t$/m	$\tfrac{1}{2}a\delta t^2$/m	δx/m
0	10	−10	0	−1	−1	0	−0·05	−0·05
0·1	9·95	−9·95	−1	−0·995	−1·995	−0·1	−0·04995	−0·14995
0·2	9·80	−9·80	−1·995					

Exercise 15.5

Find what value this process gives for x at the end of 0·4 s.

Figure 15.9 shows the graph of x against t which comes from following the calculation through the first eight seconds; it is very rough because the time intervals are rather large, but the oscillatory behaviour is plain. The time for one oscillation, the *period*, depends on the value of ω^2, and ω^2 is controlled by two independent factors, the *inertia* of the system and the *size of the restoring force per unit displacement*. Small inertia and large restoring force both make for rapid vibrations, witness the note from a tuning fork or the ring of a glass tumbler; large inertia and small restoring force make for slow oscillations, such as the pitch and roll of a big liner.

Algebraic solution of $d^2x/dt^2 = -\omega^2 x$

There are various ways of tackling this equation algebraically; we shall adopt here the hit-and-miss method of a trial solution (Appendix 6 provides an alternative approach). We know what to expect as the solution of a differential equation, we want x expressed as some function of t, but how will we recognise the right function when we

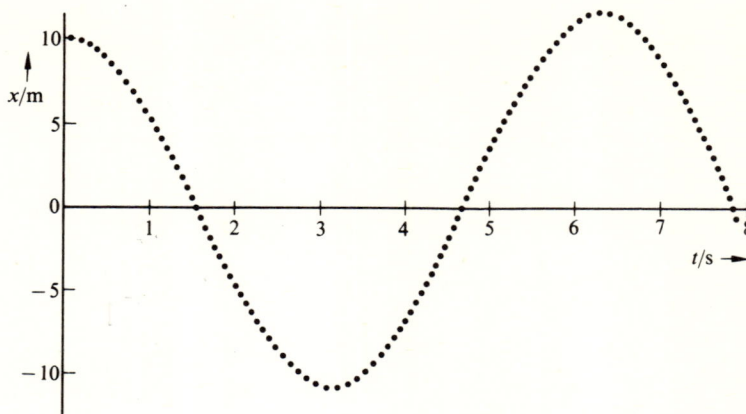

Figure 15.9
Displacement–time graph which results from the approximate numerical method.

see it? Let us start by trying a function which we can be certain will *not* work, to see what makes for an unsuccessful attempt; we shall try

$$x = bt^3$$

where b is some constant. By deciding to try this function we have committed ourselves to a lot more, for if

$$x = bt^3$$

then

$$\frac{\mathrm{d}x}{\mathrm{d}t} = 3bt^2 \quad \text{and} \quad \frac{\mathrm{d}^2x}{\mathrm{d}t^2} = 6bt.$$

It is a package deal; we cannot have the first without the other two. If these equations are to constitute a solution, then they must be capable of fitting into and obeying the equation

$$\frac{\mathrm{d}^2x}{\mathrm{d}t^2} = -\omega^2 x$$

for every moment of the entire motion. This means we shall need

$$6bt = -\omega^2 bt^3,$$

and this can only be so if $t = 0$ or $t^2 = -6/\omega^2$.

Quite apart from the fact that the second of these possibilities involves the square root of a negative quantity, this just will not do. The whole point about the differential equation is that it is true for every instant during the motion, and the correct solution, when we find it, must fit the equation at *every* value of t, not just at one or two special ones. It looks a very tall order–is there *any* function we can find which fits in precisely this way? In looking for clues we remember we are seeking a function that oscillates about $x = 0$; it is positive, then negative, then positive again. There are some functions we know which do that, only we have perhaps never thought of them as functions because we have been using them for a different job. One of them is $\sin\theta$. If we let θ increase steadily from 0 through $\pi/2$, π, $3\pi/2$, 2π rad and so on, $\sin\theta$ grows until it is just 1, then goes back to 0, out to -1 and back to 0 again, and then the whole cycle is repeated (Figure 15.10). We shall now borrow $\sin\theta$ for a new role; we shall see if we can let the size of θ stand for the magnitude of the time lapse from the start, and then $\sin\theta$ will be a number

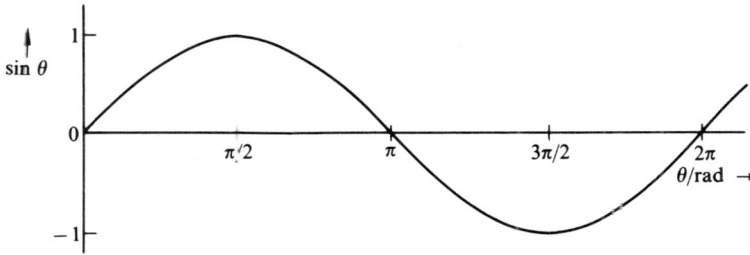

Figure 15.10
Graph of $\sin\theta$ against θ.

which is first positive, then negative, then positive again as time goes on. We want to see if this oscillating number can be made to represent the way the displacement of our particle varies with time.

Our first thought may be simply to put t instead of θ and try

$$x = \sin t$$

as a solution, but this will not do: the units do not balance. We cannot find the sine of a time interval, we can find the sine only of an angle, so to put that right we introduce a constant p with units $\text{rad}\,\text{s}^{-1}$ and work with $\sin pt$. Still we are not quite right; a sine is a ratio and has no units, and we must multiply this pure number by some constant length a if it is to represent a displacement. So our trial solution needs to read

$$x = a\sin pt.$$

This is another package deal; we are committed in the same breath to

$$\frac{dx}{dt} = ap\cos pt \qquad \text{or} \qquad \dot{x} = ap\cos pt$$

and

$$\frac{d^2x}{dt^2} = -ap^2\sin pt \qquad \text{or} \qquad \ddot{x} = -ap^2\sin pt,$$

introducing the more compact 'dot notation' for the rate of change of a quantity with respect to t. Now we have to see whether these expressions can be made to conform to

$$\ddot{x} = -\omega^2 x.$$

They do if

$$-ap^2\sin pt = -\omega^2 a\sin pt. \qquad\qquad \text{...(6)}$$

The left- and right-hand side of the equation look remarkably alike: in fact if we arrange it so that $\omega^2 = p^2$ they are *identical, no matter what the value of t may be*. This is what we have been aiming for. We have found that a solution of

$$\ddot{x} = -\omega^2 x$$

is

$$x = a\sin\omega t.$$

The quantity ωt is called the *phase angle* or simply the *phase*; it increases regularly as time goes on and takes x through all the values occurring in an oscillation. The natural question is, what is a? If we simply have to ensure that equation (6) looks the same on both sides we can have a any value we like, which seems rather unsatisfactory. What is more, $x = a\sin\omega t$ is not the only solution we can find.

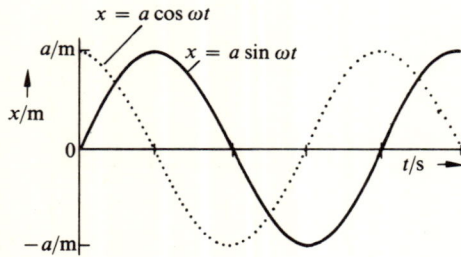

Figure 15.11

The sine solution is appropriate in situations for which $x = 0$ when $t = 0$, the cosine solution in situations for which $x = a$ when $t = 0$.

Exercise 15.6

Test whether any of the following expressions are solutions of $\ddot{x} = -\omega^2 x$:

a) $x = a\cos\omega t$

b) $x = b\sin\omega t + c\cos\omega t$

c) $x = a\sin(\omega t + \varepsilon)$

where a, b, c, ε are all constants.

Far from finding it difficult to find a solution therefore, we are now faced with an overwhelming number of solutions, but we recall that this was just the situation we had presented to us in Chapter 13 (p. 179). The selection of the one solution that applies to a particular case depends on putting in information which *applies* to the particular case, that is, the *boundary conditions*. Usually this means writing in the size of x and $\dot{x}$ at the commencement of the motion. We have a slightly more difficult problem here than in Chapter 13 however; we have to select not only the value of arbitrary constants like a and b, we have also to select the *form* of the solution; do we want $\sin\omega t$ or $\cos\omega t$ to be present, or both? A glance at the graphs of $a\sin\omega t$ and $a\cos\omega t$ shows how to decide this (Figure 15.11); when $t = 0$ the sine graph gives $x = 0$, but the *gradient* of this graph, numerically equal to $\dot{x}$, is positive. So this solution is tailor-made for a particle which is starting out from the origin with finite positive velocity when $t = 0$. On the other hand, the cosine graph has a value a/m when $t = 0$, and its gradient is zero at this point, so the solution

$$x = a\cos\omega t$$

is the one we shall need if our particle starts from rest some distance from the origin. If it has non-zero values of velocity *and* displacement we shall need to use the third or fourth solution, but we shall concentrate on the simpler cases.

Amplitude of the motion

What is the greatest value that x can take, assuming that our solution is $x = a\sin\omega t$ or $x = a\cos\omega t$? Usually to answer a question about maxima or minima we have to differentiate and put $\dot{x} = 0$, but we are so familiar with the graphs of $\sin\theta$ and $\cos\theta$ that we shall not need to resort to differentiation on this occasion. We know that neither $\sin\theta$ nor $\cos\theta$ can possibly be greater than 1 or less than -1, which means that for either solution x cannot become greater than a or less than $-a$. This sets the limits of the motion; a is the *amplitude*, the maximum value which x takes.

Example using the cosine solution

A particle initially at rest starts from $x = 0.2\,m$ when $t = 0\,s$ and moves according to the equation $\ddot{x} = -\omega^2 x$ where $\omega = 2\ s^{-1}$. Find where it will be when $t = 0.1\,s$ and $t = 5.0\,s$.

Because the particle has a displacement of $0.2\,m$ but no velocity at the start, the cosine solution is the appropriate one to choose, and so we can write not only

$$x = a\cos\omega t \qquad \ldots(7)$$

but also

$$\dot{x} = -a\omega\sin\omega t$$

nd

$$\ddot{x} = -a\omega^2\cos\omega t.$$

We have to fix the value of a; common experience may tell us that it will be 0·2 m but if we have any doubt we can use the following argument. Equation (7) must give the correct value of x at any value of t we choose. In particular if we put $t = 0$ s it gives us

$$x = a\cos 0 = a.$$

But we happen to know that when $t = 0$ s, $x = 0\cdot 2$ m. So $a = 0\cdot 2$ m. Now we can find x at any required instant.

When $t = 0\cdot 1$ s,

$$x = 0\cdot 2\cos(2 \times 0\cdot 1)\,\text{m}$$
$$\text{(angle in radians remember)}$$
$$= 0\cdot 2 \times 0\cdot 980\,\text{m}$$
$$= 0\cdot 196\,\text{m}.$$

When $t = 5$ s,

$$x = 0\cdot 2\cos(2 \times 5)\,\text{m}$$
$$= 0\cdot 2\cos(3\pi + 0\cdot 575)\,\text{m}$$
$$= -0\cdot 2 \times 0\cdot 839\,\text{m}$$
$$= -0\cdot 168\,\text{m}.$$

The fact that the angle is greater than 2π rad means that the particle has already completed one full oscillation.

There are a few delicate points to note in these calculations. Firstly, why *must* the angles be in radians? How can we be so sure that these are the units which will give the right answer? Are radians so special compared with degrees? In fact they are; if we write

$$x = a\sin\omega t$$

we are entitled to write

$$\dot x = a\omega\cos\omega t$$

only if ωt is in radians. This is because the differentiation of $\sin\theta$ from first principles involves the assumption that $\sin\delta\theta \approx \delta\theta$ when $\delta\theta$ is small, and this is true only if $\delta\theta$ is in radians. If $x = a\sin\phi$, where ϕ is in degrees, then

$$\frac{\mathrm{d}x}{\mathrm{d}\phi} = \frac{\pi a}{180}\cos\phi.$$

Since we have used differentiation in establishing our solution and have not used any factors like $\pi/180$, we have committed ourselves to angles in radians.

Secondly, we may well have to deal with angles greater than 2π rad. Of course the whole point about these trigonometric functions is that they repeat themselves every 2π rad; we merely have to subtract as many multiples of 2π rad as we can from the angle we

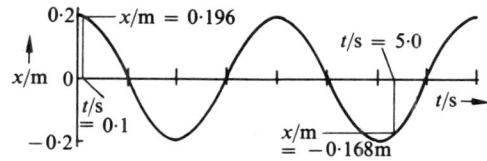

Graphical interpretation of the example involving the cosine solution: values of x corresponding to $t = 0\cdot 1$ s and $t = 5\cdot 0$ s.

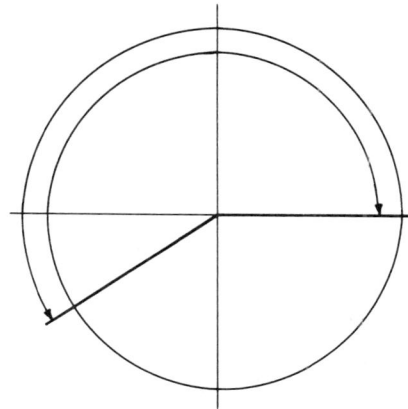

An angle of 10 rad, or $(3\pi + 0\cdot 575)$ rad.

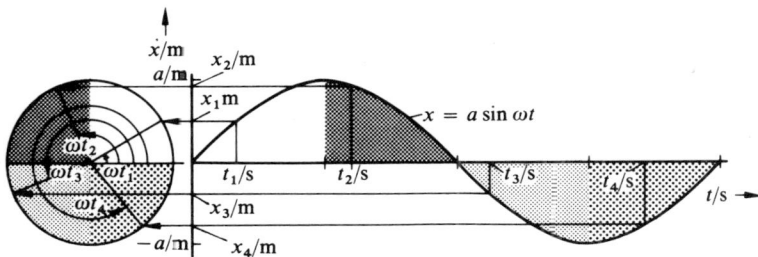

Figure 15.12

Examples of values of ωt occurring in all four quadrants, and the corresponding values of $x = a\sin\omega t$.

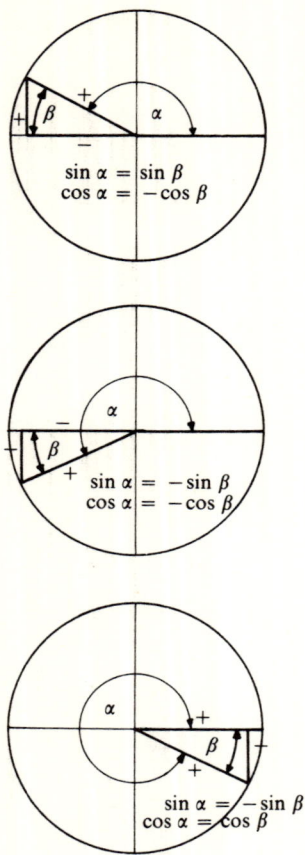

$$\sin \alpha = \sin \beta$$
$$\cos \alpha = -\cos \beta$$

$$\sin \alpha = -\sin \beta$$
$$\cos \alpha = -\cos \beta$$

$$\sin \alpha = -\sin \beta$$
$$\cos \alpha = \cos \beta$$

Figure 15.13

Trigonometric expressions involving angles between $\pi/2$ rad and 2π rad can be replaced by expressions involving corresponding acute angles.

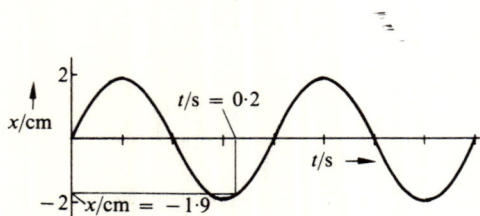

Graphical interpretation of the example involving the sine solution: the value of x corresponding to $t = 0.2$s.

are given and work with what is left. Then we may have to cope with angles extending to any one of the four quadrants of the circle shown in Figure 15.12; the rules for dealing with angles of any magnitude up to 2π rad are summarised in Figure 15.13.

Exercise 15.7

Referring to the last example, find where the particle will be when $t = 0.2$s and when $t = 2.0$s.

Example using the sine solution

A particle obeying the equation $\ddot{x} = -\omega^2 x$ starts from $x = 0$m when $t = 0$s with a velocity of 0.5ms^{-1}. Given that $\omega = 8\pi \text{ rad s}^{-1}$, find the displacement of the particle at 0.2s.

This time we need to put

$$x = a\sin \omega t \qquad \ldots(8)$$

which means that

$$\dot{x} = a\omega \cos \omega t \qquad \ldots(9)$$

and

$$\ddot{x} = -a\omega^2 \sin \omega t.$$

The information which enables us to find a is that $\dot{x} = 0.5 \text{ms}^{-1}$ when $t = 0$s. If we substitute these values into equation (9) we get

$$0.5 \text{ms}^{-1} = a\omega$$

giving

$$a = 0.5/8\pi \text{ m} \qquad \text{or} \qquad 2.0 \text{cm}.$$

Then when $t = 0.2$s, using equation (8) we have

$$x = 2.0 \sin 1.6\pi \text{ cm}$$
$$= 2.0 \sin(2\pi - 0.4\pi) \text{ cm}$$
$$= -2.0 \times 0.95 \text{ cm} = -1.9 \text{ cm}.$$

Relation between period of oscillation and ω

The expression $x = a\sin \omega t$ repeats itself every time ωt increases by 2π rad. For example, $x = a$ whenever

$$\sin \omega t = 1$$

which happens when $\omega t = \pi/2$, $(2\pi + \pi/2)$, $(4\pi + \pi/2)$, ..., $(2n\pi + \pi/2)$, ...

This means that one complete oscillation occurs every time ωt increases by 2π rad, that is every time t increases by $2\pi/\omega$. It follows that the period T of oscillations is given by

$$T = \frac{2\pi}{\omega}$$

A period of 0.2s implies that there are five complete oscillations per second; the *frequency f* of oscillations is 5s^{-1}. The unit of frequency is s^{-1}, but when we are dealing with regular recurrences, as in this situation, we speak of five *hertz* (Hz). In general,

$$f = 1/T = \omega/2\pi$$

Exercises 15.8, 9

8 Find the periods for the loaded spring, floating body and simple pendulum discussed on pp. 206–8.

9 A particle starts from $x = 0$ m with velocity $2\,\text{m\,s}^{-1}$ and performs
s.h.m. about its starting point with frequency 4 Hz. Find the
amplitude of the vibrations and write down equations giving its
displacement and velocity at any subsequent time t.

**Example of the calculation of times at which a given displacement
occurs**
A particle performs s.h.m. about the point $x = 0$ m with amplitude
0·5 m and frequency 4 Hz. Given that it is momentarily at rest at
$x = 0\cdot5$ m when $t = 0$ s, find the times at which it passes through
$x = 0\cdot3$ m.

The appropriate solution is $x = a\cos\omega t$ with $a = 0\cdot5$ m and
$\omega = 2\pi f = 8\pi\,\text{rad\,s}^{-1}$. So if the particle passes through $x = 0\cdot3$ m
after Z s,

$$0\cdot3 = 0\cdot5\cos(8\pi Z)$$
$$\cos(8\pi Z) = 0\cdot6$$
$$8\pi Z = 0\cdot927.$$

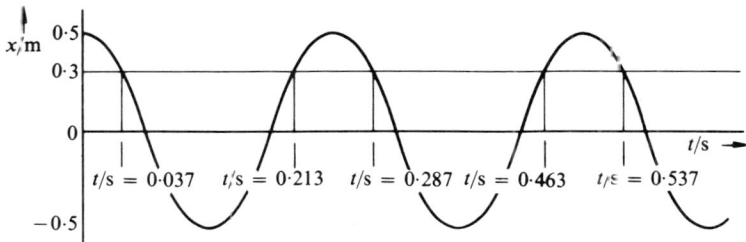

Graph indicating the times at which $x = 0\cdot3$ m in the example.

But 0·927 rad is by no means the only angle with a cosine of 0·6; not
only do the angles $(0\cdot927 + 2\pi)$ rad, $(0\cdot927 + 4\pi)$ rad, $(0\cdot927 + 6\pi)$
rad, . . . have the same cosine, but so also do angles $(2\pi - 0\cdot927)$ rad,
$(4\pi - 0\cdot927)$ rad, $(6\pi - 0\cdot927)$ rad, because angles in the fourth
quadrant have positive cosines. So we should write instead

$$8\pi Z = 0\cdot927,\ (2\pi - 0\cdot927),\ (2\pi + 0\cdot927),$$
$$(4\pi - 0\cdot927),\ (4\pi + 0\cdot927),\ \ldots$$

that is,

$$Z = 0\cdot927/8\pi,\ (0\cdot25 - 0\cdot927/8\pi),\ (0\cdot25 + 0\cdot927/8\pi),$$
$$(0\cdot5 - 0\cdot927/8\pi),\ (0\cdot5 + 0\cdot927/8\pi),\ \ldots$$
$$= 0\cdot037,\ 0\cdot213,\ 0\cdot287,\ 0\cdot463,\ 0\cdot537,\ \ldots,$$

giving

$$t = 0\cdot037\,\text{s},\ 0\cdot213\,\text{s},\ 0\cdot287\,\text{s},\ 0\cdot463\,\text{s},\ 0\cdot537\,\text{s},\ \ldots,$$

as indicated in Figure 15.14.

Exercise 15.10

Find the times for which x is (a) 0·25 m, (b) −0·25 m.

Example of the calculation of the velocity for a given displacement
What velocity does the particle in the last example have when
$x = 0\cdot2$ m?

The equations we need for this calculation are

$$x = a\cos\omega t \qquad\qquad \ldots(10)$$

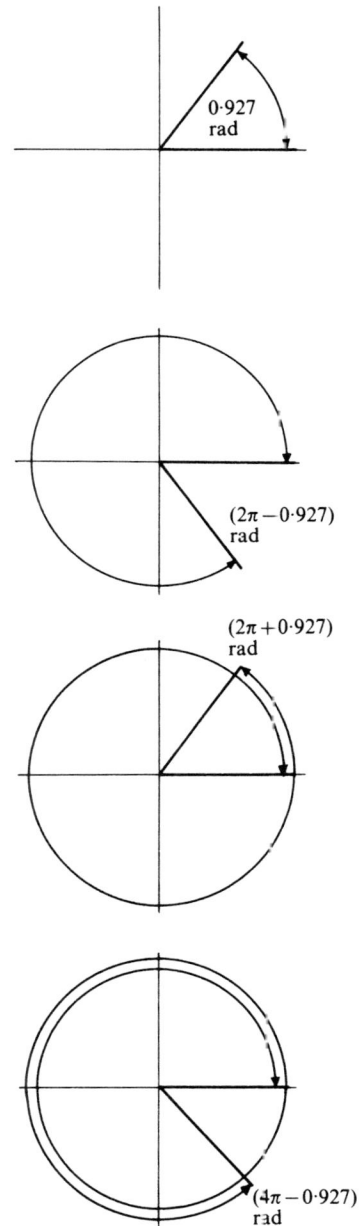

The first four angles with cosines equal to 0·6.

and

$$\dot{x} = -a\omega\sin\omega t. \qquad \ldots(11)$$

The problem could be tackled by using equation (10) to find values of t for which $x = 0.2\,\mathrm{m}$, and then substituting these values into equation (11). This is rather more laborious than is necessary, however a shorter route is as follows. We write equations (10) and (11) in the form

$$\cos\omega t = \frac{x}{a} \quad \text{and} \quad \sin\omega t = \frac{-v}{a\omega},$$

where we have written v for $\dot{x}$. Using $\cos^2\theta + \sin^2\theta = 1$, we obtain

$$(x^2/a^2) + (v^2/a^2\omega^2) = 1$$
$$v^2 = \omega^2(a^2 - x^2). \qquad \ldots(12)$$

This is a useful equation, worth remembering. Inserting numerical values, we have

$$v^2 = 64\pi^2(0.5^2 - 0.2^2)\,\mathrm{m^2\,s^{-2}}$$
$$v = \pm 11.5\,\mathrm{m\,s^{-1}},$$

and of course both answers are significant; the particle could be moving out or back depending on which stage of the oscillation is involved.

Exercises 15.11, 12

11 Find the velocity when $x = 0.4\,\mathrm{m}$.

12 By starting with the sine solution $x = a\sin\omega t$ and $v = a\omega\cos\omega t$ show that equation (12) is true in this case also.

The general solution $x = a\sin(\omega t + \epsilon)$

We now want to look at the implications of the rather more fearsome solutions mentioned in Exercise (6), p. 212. First we can show that solution (b) is in fact a rather complicated way of writing solution (c).

Exercise 15.13

By expanding $\sin(\omega t + \epsilon)$ in terms of $\sin\omega t$, $\cos\omega t$, $\sin\epsilon$ and $\cos\epsilon$, show that

$$b\sin\omega t + c\cos\omega t = a\sin(\omega t + \epsilon)$$

as long as $b = a\cos\epsilon$ and $c = a\sin\epsilon$. As a result show that

$$a^2 = b^2 + c^2 \quad \text{and} \quad \tan\epsilon = c/b.$$

Now the only difference between

$$x = a\sin\omega t \quad \text{and} \quad x = a\sin(\omega t + \epsilon)$$

is that *they start at different times*; the shapes of the graphs obtained are otherwise identical. For example, the first equation gives us $x = 0\,\mathrm{m}$ when

$$\sin\omega t = 0,$$

that is, when

$$\omega t = 0,\ \pi,\ 2\pi,\ 3\pi,\ 4\pi,\ \ldots,\mathrm{rad}.$$
$$t = 0,\ \pi/\omega,\ 2\pi/\omega,\ 3\pi/\omega,\ 4\pi/\omega,\ \ldots\mathrm{s}.$$

The other equation gives $x = 0\,\mathrm{m}$ when

$$\sin(\omega t + \epsilon) = 0,$$

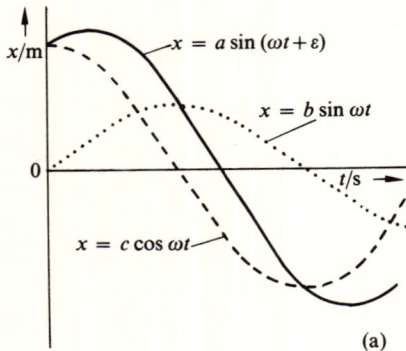

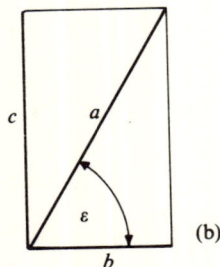

(a) Two sinusoidal disturbances, not in phase but having the same frequency, combine to form another sinusoidal disturbance. (b) Geometric representation of the rules for combining such disturbances.

that is, when

$$(\omega t + \varepsilon) = 0, \pi, 2\pi, 3\pi, 4\pi, \ldots, \text{rad}$$
$$t = -\varepsilon/\omega, (\pi - \varepsilon)/\omega, (2\pi - \varepsilon)/\omega,$$
$$(3\pi - \varepsilon)/\omega, (4\pi - \varepsilon)/\omega, \ldots \text{s},$$

that is to say, every one of these occasions is ε/ω earlier. The graph of $x = a\sin(\omega t + \varepsilon)$ is therefore the graph of $x = a\sin\omega t$ shifted to the left by ε/ω (Figure 15.14).

We have a term for this; we say that the motion described by $x = a\sin(\omega t + \varepsilon)$ is *not in phase* with that described by $x = a\sin\omega t$; in fact it is *out of phase* by ε (in radians) or ε/ω (in seconds) or a fraction $\varepsilon/2\pi$ of an oscillation or cycle (remember that the phase angle increases by 2π rad per oscillation).

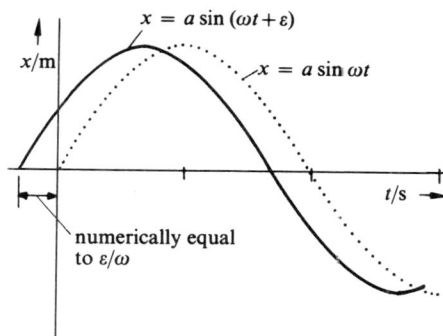

Figure 15.14
Comparison of $x = a\sin\omega t$ and $x = a\sin(\omega t + \varepsilon)$.

Exercise 15.14

Express the phase difference between $x = a\sin\omega t$ and $x = a\cos\omega t$ in radians, in seconds and as a fraction of a cycle.

For the general solution $x = a\sin(\omega t + \varepsilon)$, then, neither x nor dx/dt is zero when $t = 0$s, so this is the solution to choose if we are confronted with rather more involved initial conditions. Alternatively we could deliberately choose a different starting time so that either the sine or cosine solution is applicable.

Kinetic and potential energy of simple harmonic motion

Any particle performing s.h.m. has a continuously varying velocity and therefore a fluctuating kinetic energy. It does not matter which solution we look at; we will take $x = a\sin\omega t$ and therefore write the kinetic energy as

$$\text{k.e.} = \tfrac{1}{2}mv^2 = \tfrac{1}{2}ma^2\omega^2\cos^2\omega t.$$

This has a maximum value of $\tfrac{1}{2}ma^2\omega^2$ when $\cos^2\omega t = 1$, corresponding to the times at which the particle passes through the equilibrium position. The minimum value is zero, when $x = \pm a$.

We have a good deal of faith in the energy conservation principle and so we infer that the potential energy also fluctuates in such a way that the total energy remains constant moment by moment. We can in fact calculate the potential energy independently. If the restoring force distance x from equilibrium is $-F$, then the work done against this force in getting the particle to this point from the equilibrium position is $\displaystyle\int_0^x F\,dx$; we are dealing with a variable force of course so we must use this integral form. We can find an expression for F in terms of x by remembering that our differential equation came in the first place from Newton's second law,

$$F = -m\ddot{x}.$$

We have been writing $F = kx$ to get

$$kx = -m\ddot{x}$$

and then we replaced k/m by ω^2.

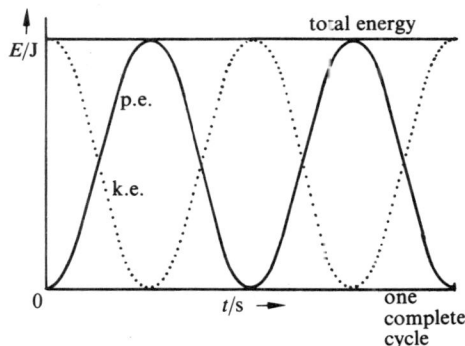

Variation of k.e., p.e. and total energy during one cycle of s.h.m. ($x = a\sin\omega t$).

Exercise 15.15

Show from these equations that $F = m\omega^2 x$.

So the p.e. at x, measured with respect to zero p.e. at the

equilibrium position, is

$$\int_0^x m\omega^2 x \, dx = \tfrac{1}{2}m\omega^2 x^2 \qquad \text{...(13)}$$

that is,

$$\text{p.e.} = \tfrac{1}{2}m\omega^2 a^2 \sin^2 \omega t$$

Exercise 15.16

Show that the total (k.e. + p.e.) is therefore constant and write down its magnitude.

We can see from this that the energy involved in s.h.m. is proportional to the square of the amplitude for a given frequency and to the square of the frequency for a given amplitude, that is, it is proportional to the product of these two squares.

S.H.M. and atomic vibrations

Equation (13) tells us how the potential energy varies with displacement from equilibrium, and the corresponding graph is sketched in Figure 15.15. It is not unlike the lower part of the graph showing variation of potential energy with separation for two atoms, discussed on p. 165 and repeated here (Figure 15.16), in fact for very small departures from the equilibrium atomic separation the two are identical, since both arise from a restoring force proportional to the displacement. As the atoms in a solid are in a state of continuous vibration we can think in terms of an interchange between potential and kinetic energy very like the energy exchanges we have been thinking about for a macroscopic case of simple harmonic motion. Let us imagine a pair of atoms in a solid and attempt to push the limits of our microscopic model a little further. According to our simple picture the atoms will be vibrating along the line joining their centres, and Figure 15.16 gives an indication of how the potential energy varies during the oscillations. At some temperature T_1 we can imagine the separation of two average atoms to vary between NA and NB; A and B are the points for which the energy E_1 of the vibration is all p.e., and as the separation passes through the value NP corresponding to equilibrium separation the kinetic energy will be E_1. At a higher temperature T_2 the average maximum kinetic energy E_2 will be greater, allowing the separation to vary between QC and QD, at which limits the p.e. is E_2 with respect to that of the equilibrium separation. But in this region the curve has departed appreciably from parabolic shape, and DR is greater than CR, which means that the average separation of the atoms is greater than NP. We cannot think of the motion as simple harmonic any more; the oscillations are *anharmonic*. As the temperature rises still further the atomic separation makes even greater departures from equilibrium, and if the vibrational energy rises to E_3 the atoms escape completely from each other.

This is a very idealised picture, but it can provide some qualitative explanation for several macroscopic phenomena. Firstly, we can associate the connection between mean interatomic separation and temperature with the *thermal expansion* of a solid. Secondly, as the temperature increases further and the vibrations cause larger fluctuations in interatomic separations we can visualise a situation where an atom, while not having energy corresponding to E_3, can

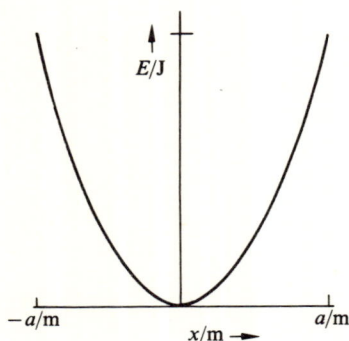

Figure 15.15

Graph of p.e. against displacement from equilibrium for a particle performing s.h.m.

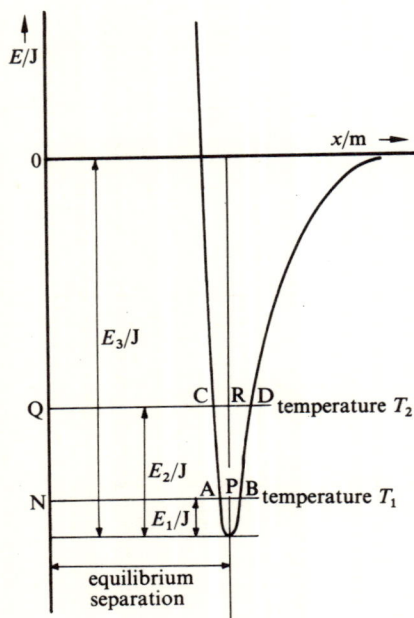

Figure 15.16

Graph of p.e. against separation for two atoms or molecules according to our simple microscopic model.

make a sufficiently large traverse from its usual place to become attracted to and associated with atoms in an adjacent part of the material; this would correspond to a liquid. Finally when the vibrations become very energetic indeed, atoms, or more usually molecules, can acquire sufficient vibrational energy to escape completely – we have a gas.

In the above discussion we have been talking about average effects. As in all populations individuals vary; some atoms may have enough energy to escape, causing evaporation even though the bulk of the liquid is not energetic enough for boiling. Similarly some atoms can escape from solids, giving rise to a measurable vapour pressure below the melting point.

If we are going to picture the vibrational energy of atoms in terms of s.h.m., what significance can we attach, if any, to the *frequency*? After all, a graph like Figure 15.16 presupposes some law of force between atoms, and this, together with their inertia, should define a definite frequency f_0 of vibration. The answer is that even if atoms behaved in all respects like billiard balls joined by springs we should not expect the vibrations to have a definite frequency, because if one billiard ball is disturbed, the other ends of the springs to which it is attached are not fixed; they will move too, and pass the disturbance on to their neighbours. What we have here is an example of *wave* propagation, and a disturbance with practically *any* frequency can be produced so long as it obeys the fundamental formula for a wave

$$c = f\lambda,$$

where c is the velocity of the wave, f the frequency and λ the wavelength. The vibrations of an atomic lattice can therefore be thought of as a tremendous jumble of wave-like disturbances with all sorts of frequencies and in all directions. The frequency f_c will have *some* significance; it will represent the *maximum* frequency present, corresponding to a wavelength of one or two atomic separations. The foregoing ideas form a basis for quite successful predictions of the variation of vibrational energy, and hence of specific heat capacity, with temperature for solids.

Combinations of simple harmonic motions in perpendicular directions
We get a further insight into s.h.m. if we look at a particle which is moving with s.h.m. in the x-direction and *at the same time* is moving with s.h.m. in the y-direction. Starting with the simplest case, we shall think about the motion in the xy-plane described by

$$x = a\sin\omega t \quad \text{and} \quad y = a\sin\omega t.$$

Clearly a particle behaving according to these equations starts at the origin and moves out along a 45° diagonal (Figure 15.17); its distance r along this diagonal at any instant is given by

$$r^2 = x^2 + y^2.$$

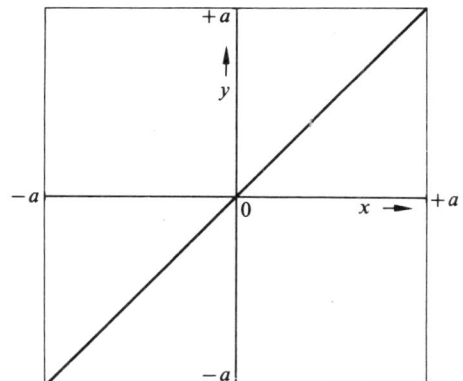

Figure 15.17

Exercises 15.17–19

17 Show that $r = \sqrt{2}a\sin\omega t$.
18 What path is taken by a particle moving according to the equations $x = a\sin\omega t$, $y = -a\sin\omega t$?
19 Draw the path taken by a particle moving according to the equations $x = a\sin\omega t$, $y = b\sin\omega t$, when $a = 5\,\text{cm}$ and $b = 10\,\text{cm}$.

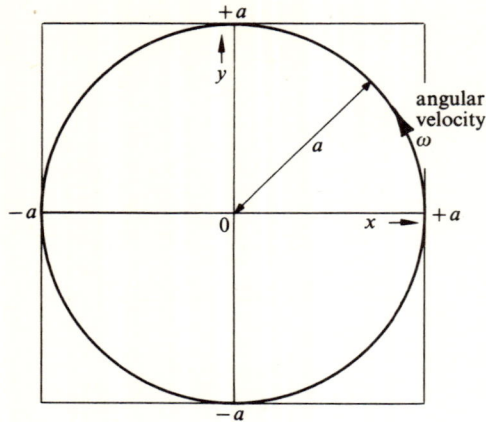

Figure 15.18

Now we come to a more significant situation, described by the equations

$$x = a\cos\omega t \quad \text{and} \quad y = a\sin\omega t.$$

This time the particle does *not* start from the origin, for when $x = 0$, $y = a$ and when $x = a$, $y = 0$. In fact the particle will *never* be at the origin because there is no value of t for which x and y are both zero. Instead, the particle goes round in a circle of radius a (Figure 15.18). We can tell it is a circle, because if we work out $x^2 + y^2$ we get an answer a^2 independent of t, and this tells us that the particle is always a constant distance a from the origin. What is more, the angle swept out in time t is ωt, so that the particle has an angular velocity of magnitude ω, and it therefore completes one revolution, or 2π radians, in a time $2\pi/\omega$, a familiar expression for the period of s.h.m.

Here then is a new view of simple harmonic motion. If we have a particle moving round in a circle at constant speed, in the xy-plane

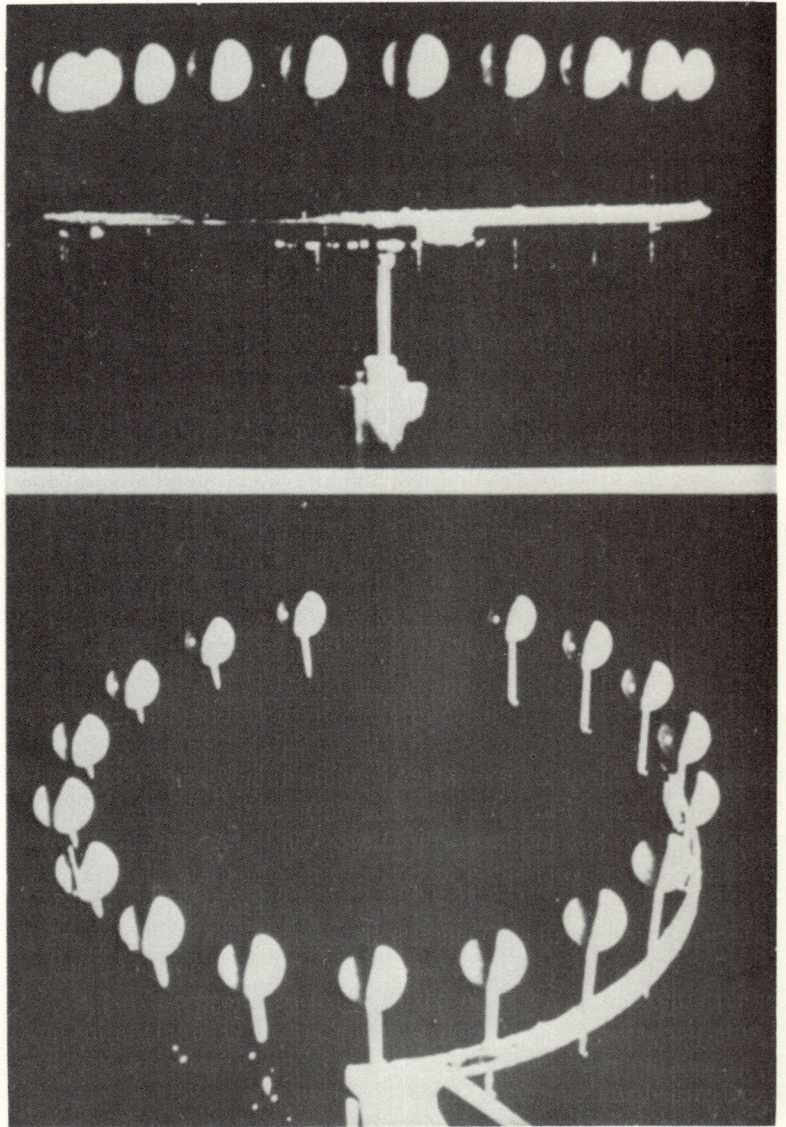

Uniform circular motion viewed from a distant point in the plane of the circle appears as s.h.m.

say, and we view the circle end on from a point a long way away on the x-axis, we simply see one *component* of the motion, and the particle appears to be performing s.h.m. *Uniform circular motion can be thought of as a combination of two perpendicular simple harmonic motions of equal amplitude and frequency, but differing in phase by $\pi/2$ rad or $1/4$ cycle.*

We can take this a step further if we look at

$$x = a\cos\omega t, \qquad y = b\sin\omega t.$$

The path traced by the particle behaving according to these equations is shown in Figure 15.19; it is a 'foreshortened circle'–an ellipse. Finally, if the equations instead read

$$x = a\cos\omega t, \qquad y = b\cos(\omega t + \varepsilon),$$

we get an ellipse with major and minor axes inclined to the x and y axes (Figure 15.20).

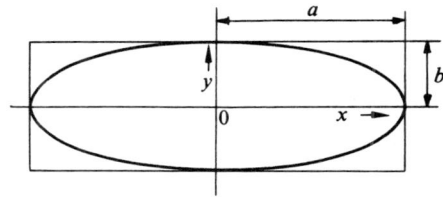

Figure 15.19

Exercise 15.20

Attempt a rough sketch of the path you would expect a particle to take if it moved according to the equations

$$x = a\sin\omega_1 t, \qquad y = a\sin\omega_2 t,$$

where

$$\omega_1 = 2\omega_2.$$

The curves obtained by such combinations of simple harmonic motions with different frequencies, each of which is a fairly simple multiple of a particular frequency, are called *Lissajous figures*.

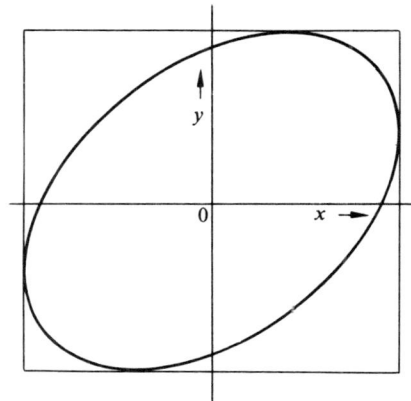

Figure 15.20

Further exercises

Exercises 15.21–34

21 The displacement x of a body of mass m moving with simple harmonic motion in a straight line varies with time according to the equation $x = a\sin\omega t$, the symbols having their usual meanings. Give equations to show how the following quantities vary with the displacement of the body from the centre of the motion: (a) the velocity, (b) the acceleration, (c) the potential energy, and (d) the total energy of the body. Draw sketch-graphs to show how the quantities (b), (c) and (d) vary with the displacement x.

Describe how you would use a helical spring, a series of known masses, a stopwatch and a metre rule (together with any other apparatus you might require) to determine the acceleration due to gravity.   (C)

22 A particle, mass 5g, executes simple harmonic motion of amplitude 2cm and frequency 2Hz. Draw a graph showing how the force acting on the particle varies with its displacement from the equilibrium position. What are the maximum and minimum values of the kinetic energy of the particle?   (JMB)

23 A 100g mass is suspended vertically from a light helical spring and the extension in equilibrium is found to be 10cm. The mass is now pulled down a further 0·5cm and is then released from rest. Stating any assumptions you make, show that the subsequent motion of the mass is simple harmonic motion.

Calculate (a) the period of oscillation, (b) the maximum kinetic energy of the mass.   (JMB)

24 Define simple harmonic motion and explain the meaning of the terms amplitude, period, and frequency.

A body of mass 100g hangs from a long spiral spring. When pulled down 10cm below its equilibrium position A and released, it vibrates with simple harmonic motion with a period of 2s.
a) What is its velocity as it passes through A?
b) What is its acceleration when it is 5cm above A?
c) When it is moving upwards, how long a time is taken for it to move from a point 5cm below A to a point 5cm above?
d) What are the maximum and minimum values of its kinetic energy, and at what points of the motion do they occur?
e) What is the value of the total energy of the system and does it vary with time? (wjec)

25 A body of mass 100g rests on a smooth horizontal surface and is attached to one end of a light spring, the other end of which is fixed so that the spring lies horizontally and just taut. How much work must be done to pull the mass out through 2cm, if the force constant of the spring, which obeys Hooke's law in extension and compression, is $0.5\,\mathrm{N\,m^{-1}}$?

If the body is then released, show how its acceleration and velocity vary with its displacement from the equilibrium position. How does the displacement vary with time? Will the motion proceed indefinitely under the conditions specified?

What can you say about the total energy of the system? Illustrate graphically how this is distributed between the spring and the mass as a function of the latter's displacement. (wjec)

26 The displacement x from its equilibrium position and the acceleration $\mathrm{d}^2x/\mathrm{d}t^2$ (both measured at time t) of a particle performing a simple harmonic motion are related by the equation
$$\mathrm{d}^2x/\mathrm{d}t^2 = -\omega^2x,$$
where ω is a constant. Other equations specifying the motion are
$$x = a\sin(\omega t + \varepsilon), \qquad T = 2\pi/\omega.$$
a) Define the symbols a, ε and T appearing in these formulae.
b) Draw rough graphs illustrating the displacement–time graph and the velocity–time graph for a particle performing simple harmonic motion. Insert on the graphs all the information that you consider necessary.
c) The amplitude of swing of the bob of a simple pendulum is $0.1\,\mathrm{m}$, and its maximum speed is $0.3\,\mathrm{m\,s^{-1}}$. What is the period of swing, and what is the length of the pendulum? (o)

27 A simple pendulum is found to make 500 complete swings in 1001s. How should the length of the pendulum be altered to make its period exactly 2s? (o & c)

28 A coin of mass 5.0g rests on a horizontal surface which vibrates vertically in a simple harmonic manner with an amplitude of 1.0mm. What is (a) the maximum frequency of the vibration that will permit the coin always to remain in contact with the surface, (b) the maximum force exerted by the surface on the coin at this frequency, and (c) the maximum speed of the coin? (aeb)

29 A particle rests on a horizontal platform which is moving vertically in simple harmonic motion with an amplitude of 10cm. Above a certain frequency, the thrust between the particle and the platform would become zero at some point in the motion. What is this frequency, and at what point in the motion does the thrust become zero at this frequency? (c)

30 A particle moving in simple harmonic motion, of period 2s, has an amplitude of 5cm. What is the value of the maximum acceleration and when does it occur?

A small circular cylinder is placed, with its axis vertical, on a horizontal surface which can be set in simple harmonic motion in its own plane with an amplitude of 5cm. If the frequency of oscillation is gradually increased, at what frequency would the cylinder slip if the coefficient of friction were 0·4? The diameter of the cylinder is one half of its height. What must then be the minimum value for the coefficient of friction if toppling is to occur before slipping?   (WJEC)

31 A particle P moves at a uniform speed v along the circumference of a circle, radius a, whose centre is at O. AOB is a diameter of the circle and Q is the foot of the perpendicular PQ on AOB. Write down expressions, in terms of the angle POA ($= \theta$), for (a) the distance OQ, (b) the velocity of Q along AOB, and (c) the acceleration of Q towards O. Show that the motion of Q is simple harmonic.

A *heavy* piston fits accurately into a cylinder which is closed at one end. The cylinder is stood on this end and the piston (compressing the air in the cylinder) comes to rest at a height h above the closed end. The piston is given a further *very small* depression and is released. Find the periodic time of its subsequent oscillations. (Neglect atmospheric pressure and assume that all the changes are isothermal.)   (O & C)

32 If a body is moving with simple harmonic motion what relationship must always exist between the force acting on it at a given instant and its position at that instant?

Explain briefly, with reference to the behaviour of springs, wires and rods when they are stretched or otherwise distorted, why it is that any oscillations which they may execute are usually simple harmonic.

A cylinder of cross-sectional area $3\,\text{cm}^2$, length 10cm and mean density $1\cdot00\,\text{gcm}^{-3}$, floats with its axis vertical in a beaker containing a liquid of density $1\cdot20\,\text{gcm}^{-3}$. The cross-sectional area of the beaker is $30\,\text{cm}^2$. What height is above the surface of the liquid? Calculate the restoring force if the floating cylinder is depressed through a small distance Xcm below its equilibrium position. Explain how this result suggests that, if released, the cylinder would then perform simple harmonic motion. What are the difficulties in calculating a period for the oscillations in this case?

How much work would have to be done to depress the cylinder through 0·5cm below its equilibrium position? Explain what happens to this energy from the instant the cylinder is released until oscillations have finally died away.   (SUJB)

33 The balance wheel of a watch vibrates with an angular amplitude of π radians and a period of 0·5s. Calculate (a) the maximum angular speed, (b) the angular speed when the displacement is $\frac{1}{2}\pi$ and (c) the angular acceleration when the displacement is $\pi/4$. Given that the radius of the wheel is r calculate the maximum radial force acting on a small dust particle of mass m situated on the rim of the wheel.   (O & C)

16 Unbalanced force dependent on velocity; viscosity

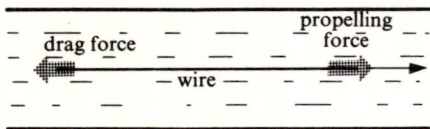

The next three chapters are chiefly concerned with the way in which an object moving through a medium is affected by drag forces; we shall start with the simplest situation, where the only force on the object, apart from the drag, is one with constant size and direction.

Energy loss due to drag forces

Any kinetic energy lost to the fluid through which the object moves will eventually result in local temperature rises, but it is helpful to think about two different ways in which the loss may occur. The first mechanism is straightforward; the object may have to exert forces to *accelerate* parts of the fluid, to make room for itself. The second effect is a little more subtle, and occurs even in the absence of such accelerations. Think of a very long thin wire moving longitudinally down a pipe full of liquid (Figure 16.1). The amount of water it has to displace is negligible, and yet it can experience an appreciable drag force due to the fact that different parts of the liquid are caused to move at different, though steady, relative velocities. If adhesive forces are strong compared with cohesive forces, the liquid actually in contact with the wire is pulled along with it, whereas the liquid in contact with the wall will remain stationary. Every other part of the liquid is at some intermediate velocity, depending on how far it is from the wire, and in this situation forces are brought into play which tend to reduce relative motion between adjacent layers—we call them *viscous* forces. They are important at low velocities, but at far faster rates of flow the drag due to acceleration of various parts of the liquid dominates. We shall look at some viscous forces first.

Figure 16.1

Microscopic description of viscous forces in a liquid

If we apply a shear stress to a block of metal there is a tendency for adjacent layers to move parallel to each other. If the stress is small, the resulting strain disappears again when the stress is removed, but a large stress produces inelastic deformation; there is slipping of parallel planes due to movement of dislocations, and much of the work done appears as an increase in vibrational kinetic energy of the atoms concerned, indicated by a temperature rise. We can think of the viscous forces between adjacent liquid planes as being a logical extension of such a process, with the big difference that *a liquid is unable to provide continual resistance to shear stress*. There may well be a *momentary* resistance; as soon as the stress is applied and molecules commence to move in response to it, intermolecular forces of attraction and repulsion are called into play to resist the change, but within the time that it takes two molecules to change places there is such a random resiting of molecules that all record of the situation which existed before the stress was applied is lost, and the molecules simply fall into the most convenient places consistent with the new conditions. This process continues as long as the stress is applied, and so a shear stress applied to a liquid does *not* produce a corresponding constant strain; it produces a constant *relative velocity* between adjacent layers. There is no question of elastic behaviour of course; the work done on the viscous force appears as extra vibrational kinetic energy of molecules, and the temperature rises a little in consequence.

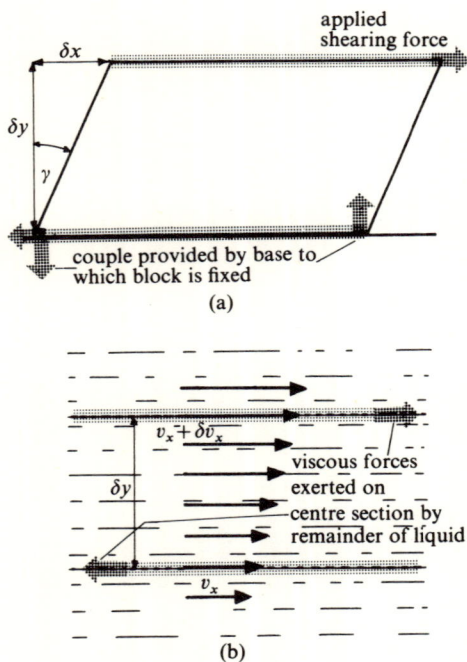

Figure 16.2

Comparison of shear stress in a solid and viscous drag in a liquid.

Macroscopic definition of viscosity

The analogy between viscous forces and shear strains described in

A glacier provides an example of very slow streamline flow. (Aletschgletscher–Jungfraujoch)

the preceding paragraph can be pursued further in quantitative terms. When dealing with shear stress in a solid we wrote

$$\frac{F}{A} = G\gamma,$$

(see p. 142) where F is the force one layer of area A exerts on another, γ is the angle defining the strain which results, and G is the shear modulus. An alternative expression for the strain is $\delta x/\delta y$ (Figure 16.2), that is,

$$\frac{\text{the displacement in the direction of the force } F.}{\text{the separation of the planes concerned}}$$

To describe the forces between adjacent layers of liquid we might try

$$\frac{F}{A} = \eta \frac{\delta v_x}{\delta y}$$

where δv_x is the *velocity* difference between planes separated by δy; we have written in the fact that the 'strain' is increasing at a constant rate. Just as the shear modulus G for a solid depends only on the material, so the quantity η is in many cases a constant for a given liquid at a given temperature; it is called the *coefficient of viscosity*, or simply the *viscosity*, of the liquid. (There are several important cases in which η is not constant. Many lubricating oils are *thixotropic*; as the velocity gradient increases in such liquids the viscosity decreases. The convenience of thixotropic ('jelly') paint depends on the same effect). The equation which defines η is usually written in the differential form

$$F = \eta A \frac{\mathrm{d}v_x}{\mathrm{d}y} \qquad \qquad ...(1)$$

and $\mathrm{d}v_x/\mathrm{d}y$ is called the *velocity gradient*; it indicates how rapidly the velocity is changing as we go from one layer to another.

Exercise 16.1

Find the SI units for η.

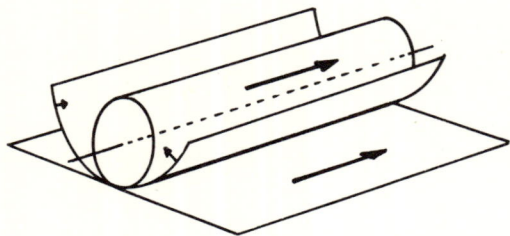

Edge effects can be avoided by working with cylindrical tubes rather than with plane sheets.

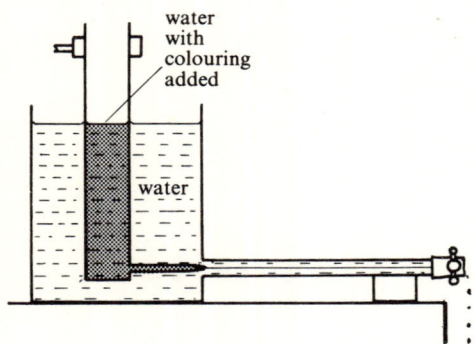

Figure 16.3

Streamline and turbulent flow

The viscous force between two parallel planes is simple to talk about in theory – that is why it is chosen as the basis of the definition of η – but it would not be easy to investigate experimentally because of edge effects. We should want to arrange the velocity to be the same for every point in a given plane, but the edge of the plane would have to be in contact with the container wall and therefore stationary. There is a way to get a layer with no stationary edges, and that is to curve it round into a cylinder moving at constant velocity in the direction defined by its axis. So the simplest experimental situation to consider is that of liquid flow down a tube with circular cross section; this gives us a straightforward method for determining η. We are going to imagine the liquid divided into thin cylinders all concentric with the tube axis, and all moving at different speeds. Perhaps it is expecting a lot to assume that the liquid will behave in such a well-organised fashion, but a simple experiment shows that this must actually be the case, providing the flow rate is not too great. The arrangement shown in Figure 16.3 can be used to inject a thin stream of coloured water into the tube, and it obediently flows in a straight line down the length of the tube, keeping to its own part of the cross section (Figure 16.4). This is an example of *streamline* flow. Strictly a streamline is defined in the following manner. If at every point in a vessel containing moving

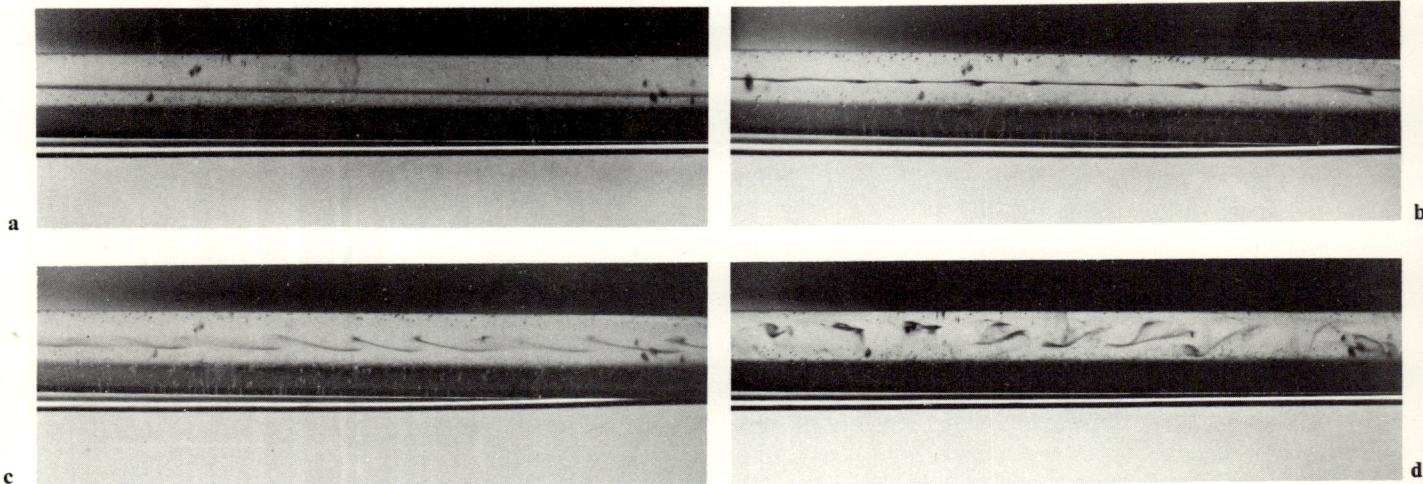

Figure 16.4

(a) shows streamline flow; (b), (c) and (d) show the result of successive increases of flow rate.

liquid we draw an arrow showing the direction of the velocity at that point, we can join up the arrows in continuous lines, just as we join up the directions given by a plotting compass at various places in a magnetic field to form lines of force. The lines we get are called the streamlines, and in general streamlines do not necessarily stay in the same place, because the velocity at any point may be continually changing. If the streamlines do *not* shift as time passes, however, they are then identical with the paths taken by individual parts of the liquid (the line taken by the coloured water in Figure 16.4a is a streamline); this is what is meant by *streamline flow*. If the flow rate down the tube is increased somewhat by opening the clip a little, the coloured line stays straight for a short distance and then breaks up into whirls and eddies which are never in the same place twice; the flow pattern has become *turbulent* (Figure 16.4b, c, d). Further increase in the flow rate removes all trace of streamline flow completely.

Though we are going to calculate the volume of liquid flowing down the tube per second in terms of the viscosity, we can do it only for a streamline flow situation; in fact turbulent flow is still an incompletely solved problem.

Rate of streamline flow in a tube of circular cross section

This is one problem for which dimensional analysis is very useful, and we shall establish the form of the equation before going on to a full derivation. We start by deciding what factors are likely to affect the volume Q flowing per second; certainly the radius r of the tube is involved, also its length l and the pressure difference P across its ends. We could perhaps make a case for including the density ρ of the liquid, but as we are specifically looking at a situation in which each part of the liquid travels down the tube at constant velocity we shall avoid introducing a term which has to do with the *inertia* of the liquid. Even so we have more than we can deal with; writing

$$Q = kP^{\alpha}l^{\beta}r^{\gamma}\eta^{\varepsilon}, \qquad \ldots(2)$$

we soon run into one of the common difficulties of dimensional analysis, four unknown numbers α, β, γ and ε and only three equations, for mass, length and time, with which to find them. What is more, r and l both have the dimensions of length so we should never be able to sort out the relative values of β and γ. We have to resort to a certain amount of inspired guesswork. We shall take it that the pressure difference P and length l of tube are related in such a way that it is the *pressure drop per unit length*, or pressure gradient, P/l which really matters; we can use the following argument to justify this assumption. We imagine the flow tube split up into a number of identical sections, each having length δl and pressure difference δP between its ends. If equation (2) is to represent the flow rate down the tube correctly it must be applicable to each small section, as well as to the entire length. The flow rate Q must be the same for each section because we cannot allow liquid to pile up in one place, so we must be able to write

$$Q = k(\delta P)^{\alpha}(\delta l)^{\beta}r^{\gamma}\eta^{\varepsilon} \qquad \ldots(3)$$

We can also apply equation (2) to, let us say, the first three sections, giving

$$Q = k(3\delta P)^{\alpha}(3\delta l)^{\beta}r^{\gamma}\eta^{\varepsilon} \qquad \ldots(4)$$

and equations (3) and (4) are consistent with each other only if we make the 3^{α} cancel with the 3^{β}, in other words $\beta = -\alpha$, and the pressure and length must be present in the equation as the ratio $\delta P/\delta l$, or P/l if we are talking about the entire length of tube.

The above reasoning enables us to write

$$Q = k(P/l)^{\alpha}r^{\gamma}\eta^{\varepsilon}$$

so that

$$[L^3T^{-1}] = [ML^{-2}T^{-2}]^{\alpha}[L]^{\gamma}[ML^{-1}T^{-1}]$$

The [M] equation then gives $\quad 0 = \alpha + \varepsilon$
the [L] equation gives $\quad\quad 3 = -2\alpha + \gamma - \varepsilon$
and the [T] equation gives $\quad -1 = -2\alpha - \varepsilon$.

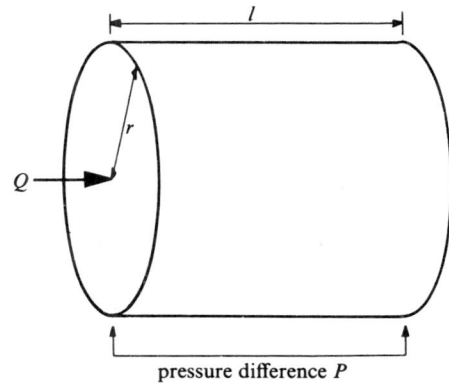

Flow rate Q down a cylindrical tube of radius r.

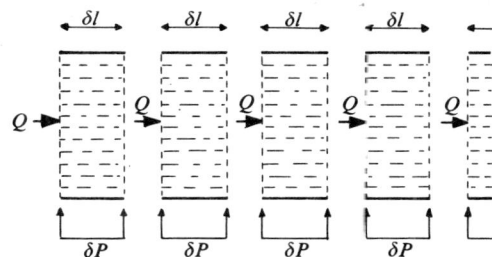

Flow tube split into small sections each of length δl.

Exercise 16.2

Solve these equations to find α, γ and ε and hence show that
$$Q = kPr^4/\eta l.$$

The full analytical approach goes as follows. We mentally divide the liquid in the pipe into three regions, separated by two coaxial cylindrical surfaces of radius y and $y + \delta y$ (Figure 16.5). We are going to look at all the external forces on the liquid within the

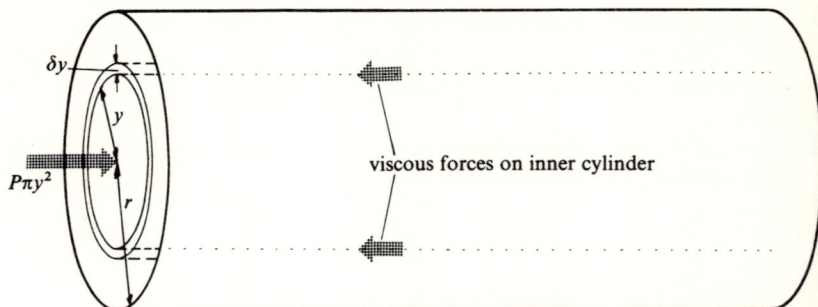

Figure 16.5

cylinder of radius y. We take it that the flow is steady and along streamlines, so no part of the liquid is accelerating; this means that the net force on our inner cylinder must be zero. The fluid in the cylinder is being pushed on owing to the pressure difference P across its ends, which produces a force of magnitude $P\pi y^2$, but this is exactly balanced by the viscous forces experienced from the surrounding liquid. If the velocity is v_x at radius y and $(v_x + \delta v_x)$ at radius $(y + \delta y)$, the viscous force F on a length l of the inner cylinder is given by

$$F = 2\pi y \eta l \frac{\delta v_x}{\delta y}$$

or in differential form

$$F = 2\pi y \eta l \frac{dv_x}{dy}.$$

So in conditions of steady flow,

$$P\pi y^2 + 2\pi y \eta l \frac{dv_x}{dy} = 0$$

$$\frac{dv_x}{dy} = \frac{-Py}{2\eta l}. \qquad \qquad \ldots(5)$$

The minus sign expresses the fact that the velocity decreases as we move away from the axis. Equation (5) tells us the velocity gradient at any radius y; we can use it to find the *velocity* at radius y by integrating:

$$\int dv_x = \frac{-P}{2\eta l} \int y \, dy$$

$$v_x = \frac{-Py^2}{4\eta l} + c, \qquad \qquad \ldots(6)$$

where c is a constant which must be fixed by boundary conditions, that is, by inserting a known value for v_x at some special value of y. The special value of y we know all about is the tube radius r, where v_x is zero. Using this set of values in equation (6) we get

$$0 = \frac{-Pr^2}{4\eta l} + c,$$

$$c = \frac{Pr^2}{4\eta l}.$$

Equation (6) now becomes

$$v_x = \frac{P}{4\eta l}(r^2 - y^2), \qquad \ldots(7)$$

which means that a graph of v_x against y is a parabola (Figure 16.6).

Now we have to find the volume of liquid flowing down the pipe per second, but each cylindrical shell of liquid must be treated separately because each has a different flow rate. Looking at the shell of radius y, thickness δy, where the velocity is v_x, we can say that the length of shell passing a fixed point per second is v_x, so that the volume flow per second, δQ, is given by

$$\delta Q = \text{cross-sectional area of shell} \times v_x$$
$$\approx 2\pi y \delta y \times v_x \qquad \text{if } \delta y \text{ is small.}$$
$$\approx \frac{\pi P y}{2\eta l}(r^2 - y^2)\,\delta y \qquad \text{using equation (7)}$$

So Q, the total volume flowing down the pipe per second, is the sum of all such contributions. In integral form

$$Q = \int_0^r \frac{\pi P y (r^2 - y^2)\,\mathrm{d}y}{2\eta l}$$
$$= \frac{P\pi}{2\eta l}\left[\frac{r^2 y^2}{2} - \frac{y^4}{4}\right]_0^r$$
$$Q = \frac{P\pi r^4}{8\eta l} \qquad \ldots(8)$$

Determination of η from flow rate down a narrow tube

Figure 16.7 shows the arrangement we need if we want to measure η using the above equation, known as Poiseuille's formula. The liquid is supplied from the constant-head apparatus shown and made to flow through the narrow horizontal tube; the rapid flow down the wide waste pipe ensures that the head of liquid is kept at height h for the whole experiment. The measurements needed are the length l of flow tube, its radius r, the pressure drop P across its ends, and Q the volume of liquid emerging per second. The length l gives no problem, and Q is most easily obtained by collecting the

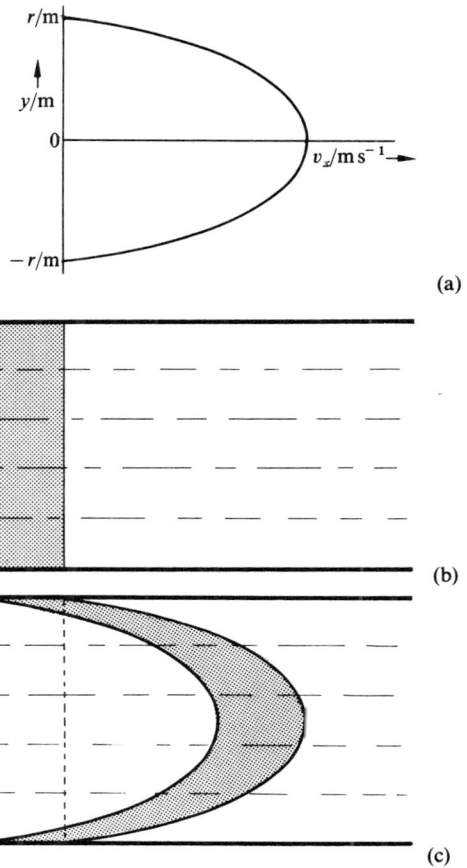

(a)

(b)

(c)

Figure 16.6

(a) Graph showing the velocity v_x at different distances y from the tube axis. The volume of liquid shaded in (b) moves so as to occupy the space shaded in (c).

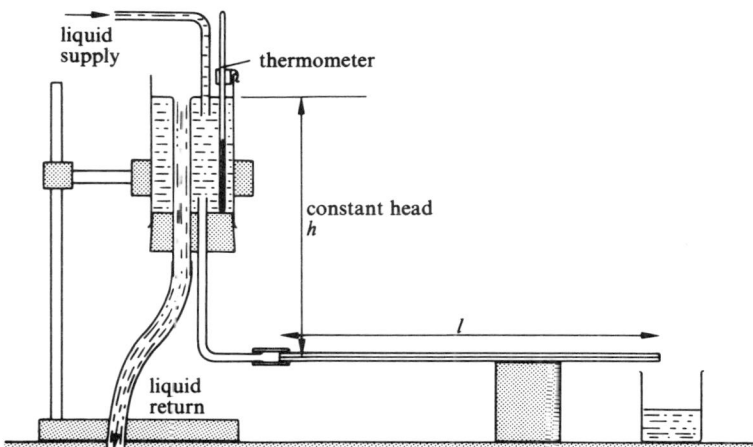

Figure 16.7

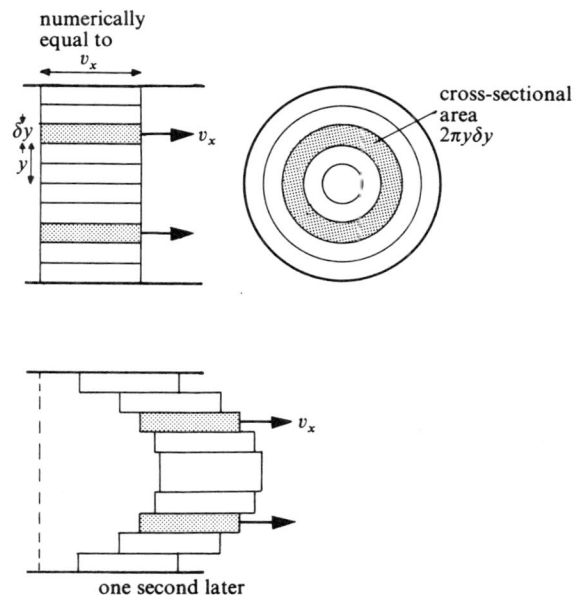

We consider the liquid split into concentric cylinders to evaluate the flow rate.

liquid in a measuring cylinder for several minutes. The radius r must be evaluated carefully because it appears raised to the fourth power in the formula: a travelling microscope offers one method but the tube must be cut and measured at several places to check for uniform bore and circular cross section. Alternatively if a pellet of mercury of known mass is introduced into the flow tube its length can be measured and hence the radius calculated, and by moving it to various parts of the tube a check can be made on the uniformity of cross section.

We have to think about the measurement of P a little more carefully. A small error is introduced if we simply express it as $\rho g h$ (where ρ is the liquid density) because of the kinetic energy acquired by the liquid entering the flow tube. If the capillary is narrow an additional error may arise from surface-tension effects at the outlet; there may be pressure differences across curved surfaces. We can correct for the former effect using Bernoulli's principle (p. 106) or instead we can measure P by the arrangement shown in Figure 16.8; this ensures that we are registering only the pressure gradient in the flow tube. The length l is then the distance between the T-junctions A and B. Note the analogy between this method of measuring pressure difference and a potentiometer method for measuring electric potential difference. No liquid flows at A and B; no current flows in the potentiometer leads.

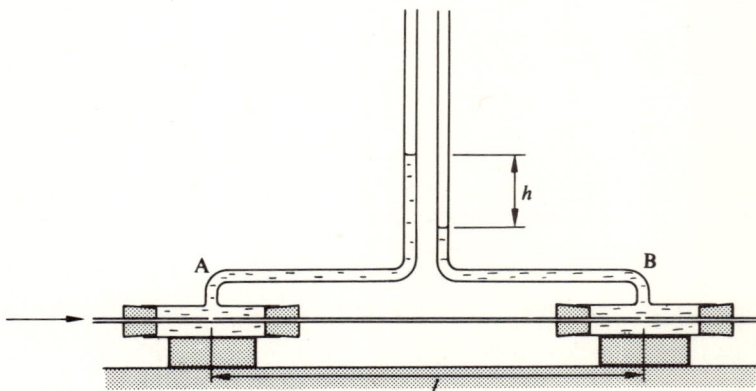

Figure 16.8

An arrangement which avoids the necessity for kinetic-energy and surface-tension corrections to the pressure gradient.

A graph of Q against P as the head of liquid is increased is shown in Figure 16.9. For small rates of flow the graph is linear, but as the velocity increases streamline flow breaks down and turbulent flow commences. If the results are to be used to determine η it is essential to use information from the linear part only; equation (8) is applicable only to streamline conditions.

Reynolds' number
Under what conditions can we rely on getting streamline flow? It turns out that the quantity to watch is

$$\frac{\text{liquid velocity} \times \text{tube diameter} \times \text{density}}{\text{viscosity}};$$

it is called *Reynolds' number R*. In symbols,

$$R = \rho d v / \eta.$$

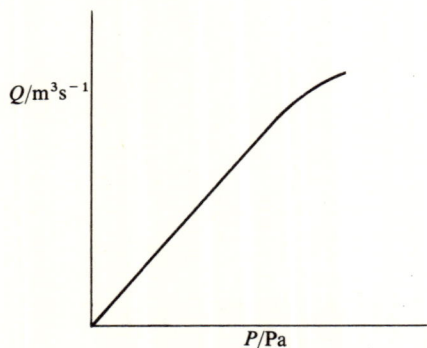

Figure 16.9

Graph of flow rate Q against pressure gradient for flow of liquid in a tube.

Exercise 16.3

Show that R is a number (that is, dimensionless).

If R is less than about 2200, the flow is streamline, but if it is much above this value there is turbulence.

Viscosity in a gas

The macroscopic equations we have derived to describe the viscous forces in a liquid are equally applicable to a gas; the only difference lies in the microscopic interpretation. In view of our microscopic view of gases we can hardly hold that viscous forces exist because of intermolecular attractive forces between adjacent layers; instead we must think in terms of molecules transferring from one region to another. If the gas is flowing it will simply mean that a steady drift velocity is superimposed on the much larger random velocities of the molecules. Suppose the molecules in slab A of Figure 16.10 have a larger drift velocity to the right than those in slab B. The random motion of molecules soon brings a transfer of some slab A molecules to B and vice versa: the molecules take their drift velocity with them, and in colliding with molecules in their new surroundings they transfer momentum from one layer to another. Newton's second law tells us that the viscous force exerted by one layer on the next is equal to the rate of transfer of momentum between the layers.

This argument can be extended to give a theoretical prediction of the *magnitude* of the coefficient of viscosity for a gas; what follows is merely a sketch of the ideas involved. A molecule leaving one layer does not contribute to the force on the adjacent layer until it suffers a collision, and it does not do that until it has travelled on average a distance λ called the *mean free path*; this is the name given to the average distance travelled by a molecule between collisions. We consider two thin layers of gas separated by λ; the difference δv_x between the drift velocities of the layers is given by

$$\frac{\delta v_x}{\lambda} = \frac{dv_x}{dy}. \qquad ...(9)$$

The average momentum transferred by one molecule of mass m in passing from one layer to the other will be $m\delta v_x$, and if we multiply this by the number of molecules moving across in unit time N, we obtain the total rate of transfer of momentum. It can be argued that N is proportional to the area A of one face of the layer, to the number n of molecules per unit volume, and to their average speed $\bar{c}$, so

rate of transfer of momentum $\propto A\bar{c}nm\delta v_x$.

We can replace nm, the mass of all the molecules in unit volume, by the density ρ, giving

viscous force F = rate of transfer of momentum
$$F \propto A\bar{c}\rho\delta v_x \qquad ...(10)$$

Exercise 16·4

By substituting for δv_x from equation (9) and comparing the result with equation (1), show that $\eta \propto \rho\bar{c}\lambda$.

We can expect the mean free path to decrease as the pressure increases, as there are more molecules with which any one molecule can collide in a given volume; quantitatively the kinetic theory of

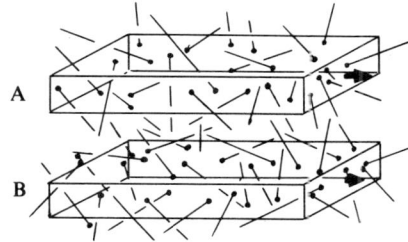

Figure 16.10.

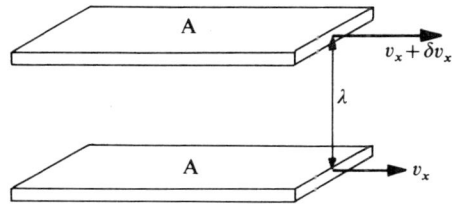

Two thin layers of gas separated by one mean free path.

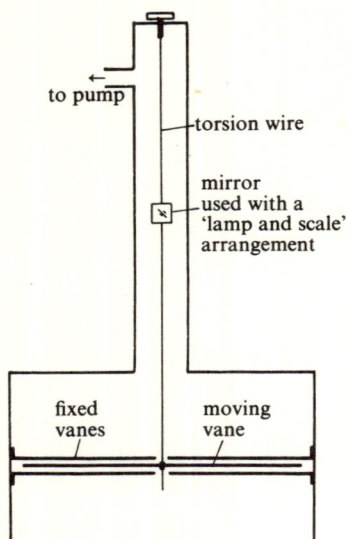

Figure 16.11

Essentials of Maxwell's apparatus for measurement of viscosity of a gas. The oscillations of the moving vanes are damped by viscous forces; the apparatus can be used to investigate the dependence of η on the pressure of the gas.

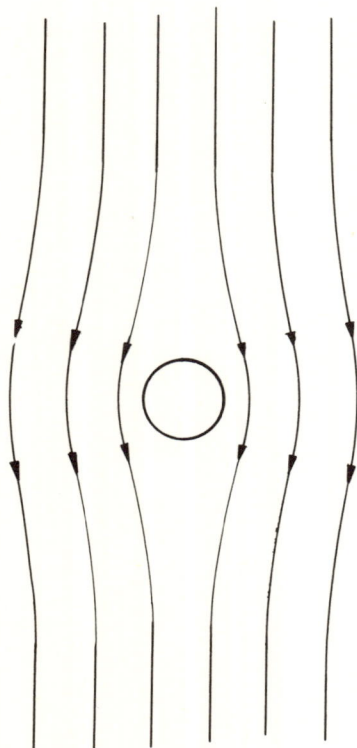

Figure 16.12

gases can be used to predict that $\lambda \propto 1/\rho$. This means that $\lambda\rho$ remains the same when the pressure is changed, and therefore η is *independent of pressure*. This prediction was confirmed experimentally by Maxwell (see Figure 16.11); it breaks down at low pressures when the mean free path is comparable with the gap between the moving surfaces, and also at high pressures when intermolecular attractions become significant.

Variation of viscosity with temperature

Viscous forces are highly temperature dependent. As Table 16.1 illustrates, liquid viscosities usually decrease as the temperature rises, whereas the viscosities of gases *increase*.

Table 16.1
Viscosities of water and air at various temperatures

Temperature/°C	Viscosity/Nsm^{-2}	
	Water	Air
0	17.9×10^{-4}	1.71×10^{-5}
20	10.1×10^{-4}	1.81×10^{-5}
40	6.6×10^{-4}	1.91×10^{-5}

We can find explanations for both phenomena in terms of our microscopic picture. We said that viscous forces occur in liquids because momentarily the liquid behaves like a solid under shear stress; then as slip commences the continuous random interchange of molecules removes all record of the previous arrangement. It is no surprise therefore to find that viscous forces decrease as the temperature rises, for as molecular activity increases we expect to see the shear stress 'melt away' even more quickly. By contrast, increase of molecular activity in a gas can be expected to increase the viscosity, because it speeds up the transfer of molecules between different layers and reduces relative drift velocities more rapidly. The increase is predicted by the formula quoted in Exercise 16.4 if we make use of the well-known result from kinetic theory that the average kinetic energy of a gas molecule is proportional to the absolute temperature T of the gas. This means that $\bar{c} \propto \sqrt{T}$, and therefore $\eta \propto \sqrt{T}$. Experimental results show poor agreement, indicating that the theory is oversimplified.

The drag force on an object moving through a fluid

This is a problem largely beyond our scope, in fact if the flow is turbulent it is a very difficult problem indeed. We shall simply look at the dimensions of the factors involved. We can expect the drag force to depend on the density ρ and viscosity η of the fluid; also the velocity v of the object and its shape and size; we shall have to be content with introducing a length r to represent the linear dimensions. If the object is a sphere, r is the radius, and if it has a less symmetrical shape, r must stand for some other significant length. Our method of dimensions is not adequate even for these four quantities because only three equations, for [M], [L] and [T], are available, so to make any progress we need to make some rather arbitrary choices. If we restrict ourselves to streamline flow we can fairly safely say that viscous forces are more important than forces needed to accelerate various parts of the fluid. Figure 16.12 indicates the flow pattern which results when a sphere moves slowly through a liquid; it can be established experimentally by keeping a sphere stationary in a tube and letting the liquid flow past it. There has got to be *some* acceleration of course as the liquid makes room for the sphere, but the impression the pattern gives is one of as little disturbance as possible. If the accelerations are negligible the liquid inertia should be unimportant, so we try leaving out the density ρ, and write

$$F = kr^\alpha v^\beta \eta^\gamma,$$

where k is a pure number. To get a dimensional balance we must have

$$[MLT^{-2}] = [L]^\alpha [LT^{-1}]^\beta [ML^{-1}T^{-1}]^\gamma.$$

Exercise 16.5

Form the [M], [L] and [T] equations and hence show that

$$F = k\eta rv. \qquad \qquad ...(11)$$

A full analysis of this situation shows that for a sphere moving under streamline conditions the drag force is given by

$$F = 6\pi\eta rv \qquad \qquad ...(12)$$

This formula is called *Stokes' law*, and we can use it to find how quickly small spherical particles will fall in a fluid. Figure 16.13 shows the forces on such a particle. If it starts from rest, the net force down is the weight mg minus the upthrust U, and this produces an acceleration a given by $(mg - U)/m$. As the speed builds up the drag force increases according to equation (12), so the net force decreases and so does the acceleration. The speed tends to a limiting value at which the drag force exactly compensates for $(mg - U)$ and then no further acceleration is possible; the particle has reached its *terminal velocity* v_T. In symbols, v_T is given by the expression

$$mg - U = 6\pi\eta rv_T. \qquad \qquad ...(13)$$

Exercises 16.6, 7

6 Show that if the particle has density σ,

$$v_T = \frac{2r^2(\sigma - \rho)g}{9\eta}.$$

7 Find the terminal velocity at room temperature for

a) Particles of chalk dust, density $2800\,kg\,m^{-3}$ and radius $0.01\,mm$ falling through air.

b) Air bubbles of radius $0.2\,mm$, $0.1\,mm$ and $0.01\,mm$ rising through water.

c) Water droplets of radius $0.001\,mm$ falling through air (as in a fog).

d) Particles of fat of density $900\,kg\,m^{-3}$ and radius $0.001\,mm$ rising through water (as in milk).

(Viscosities of water and air at room temperature are $1.0 \times 10^{-3}\,N\,s\,m^{-2}$ and $1.8 \times 10^{-5}\,N\,s\,m^{-2}$ respectively. The density of air can be taken as negligible compared with the other densities involved.)

If we now go to the other extreme and assume that at high velocities when the flow is completely turbulent the viscous forces are

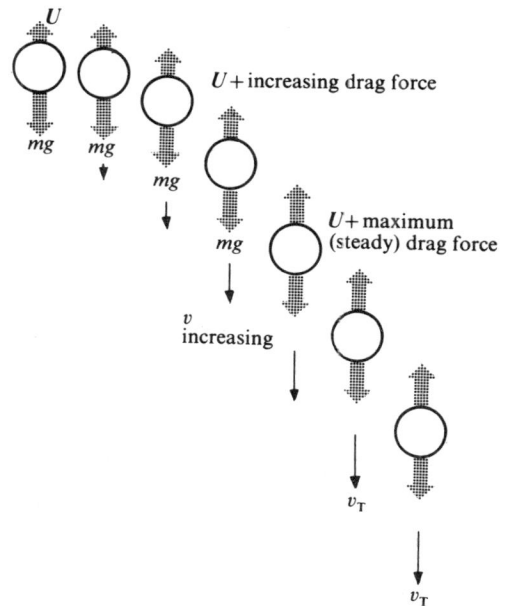

Figure 16.13
As the velocity of a falling object increases, the drag force increases, and the net force tends to zero. The velocity approaches a constant value, the terminal velocity v_T.

less important than the force needed to accelerate various parts of the fluid, we might try

$$F = kr^\alpha v^\beta \rho^\gamma,$$
$$[MLT^{-2}] = [L]^\alpha [LT^{-1}]^\beta [ML^{-3}]^\gamma.$$

Exercise 16.8

Show that these dimensions are consistent with

$$F = \rho k r^2 v^2.$$

The drag experienced by fast-moving objects cannot be fitted exactly with this equation; all we can say is that in many situations the drag is roughly proportional to the square of the speed and to the area of the object facing the flow. We can actually get an order of magnitude estimate for k if we are willing to make some very rough assumptions. Suppose, for example, we have a cube of side r moving face-on through a fluid of density ρ at speed v. It needs to clear a tunnel for itself; the volume of fluid which must be pushed away every second is $r^2 v$ (Figure 16.14). Let us take it that all of this fluid acquires the velocity of the cube; of course we know this will not happen in practice, so we can expect our estimate to be out by a factor of 2 or 3; this is an *order-of-magnitude* calculation, which means that we are content with finding the power of ten involved in the answer.

Momentum given to the fluid per second = new mass affected per second × velocity given to it.
$$= r^2 v \rho v$$

that is,

momentum lost by object per second $= \rho r^2 v^2$

In the earth's atmosphere where drag forces matter, vehicles often need to be streamlined...

...but near the moon, where there is no atmosphere and therefore no drag force, any shape will do.

and so, using Newton's second law,
$$\text{drag force} = \rho r^2 v^2.$$
Just as for small objects falling in conditions of streamline flow, so for larger objects falling in turbulent conditions, there is a terminal velocity, attained when the drag force is just equal to the weight minus the upthrust.

numerically equal to v

Figure 16.14

Exercise 16.9–11

9 Use the very approximate formula
$$mg = \rho r^2 v_\mathrm{T}^2$$
to estimate the terminal velocity of a man falling with and without a parachute. (A typical parachute radius is 5 m; take the man's 'radius' to be about 40 cm. The density of air is about 1·2 kg m^{-3}.)

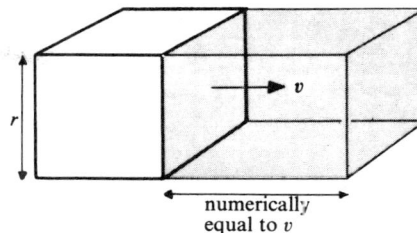

10 A heavy lorry has a front of area 10 m^2 and maximum power 100 kW. If the most important retarding force it experiences at high speed on a level road is the drag force due to the air, make an estimate of the top speed it can attain, given that the density of air is 1·2 kg m^{-3}.

11 Estimate the top speed attainable on a level road in still conditions by a cyclist who can develop a power of 500 W and who presents an area of 0·5 m^2 to the oncoming air.

Further exercises

Exercises 16.12–19

12 Explain what is meant by the *viscosity* of a liquid, and define the *coefficient of viscosity*. What phenomena suggest that viscosity is exhibited by gases as well as by liquids?

The following figures were obtained in an experiment to determine the coefficient of viscosity of water by measuring the rate of flow through a horizontal capillary tube:

diameter of capillary tube = 1·40 mm
length of capillary tube = 21·0 cm
pressure difference between ends of tube = 18·5 cm water
rate of flow of water = 0·74 cm^3 s^{-1}.

Give a diagram of the apparatus which would be used for this experiment, and explain how each measurement would be made.

Given that the diameter of the tube was measured to ±0·05 mm, the length to ±1 mm, the pressure difference to ±2 mm water, and the rate of flow to $\pm0.01\,\mathrm{cm^3\,s^{-1}}$, calculate the percentage error which the uncertainty in each of these factors introduces into the value obtained for the coefficient of viscosity. (C)

13 Confirm the dimensional consistency of the following statements in which η represents the coefficient of viscosity of a liquid, a and l are lengths, ρ a density, p a pressure difference, and v a speed:

a) The product $lv\rho/\eta$, known as Reynolds' number, is dimensionless.

b) According to Poiseuille, the volume of liquid flowing per second steadily through a capillary tube is $\pi p a^4/8\eta l$.

c) For a sphere of density ρ falling steadily under gravity through an expanse of liquid of density ρ',

$$6\pi\eta a v = \tfrac{4}{3}\pi a^3(\rho - \rho')g.$$ (O)

14 A solid sphere is suspended just below the surface of a viscous fluid. The sphere is released and falls through the fluid. What forces act on the sphere while it is falling through the fluid? Sketch graphs of (a) the velocity and (b) the acceleration of the sphere against time and explain the form of each graph. If the density of the solid is ρ_1 and of the liquid ρ_2, what is the initial acceleration of the sphere? (O & C)

15 Explain what is meant by the *velocity gradient* at a point in a fluid.

A small sphere when allowed to fall through a viscous medium eventually acquires a terminal velocity. Explain this qualitatively in terms of the forces concerned and discuss the energy transformations occurring when the terminal velocity has been reached.

A small steel ball-bearing falling through glycerine has a terminal velocity of $2.00\,\mathrm{cm\,s^{-1}}$. Find the terminal velocity of an air bubble of the same size rising through glycerine at the same temperature. Neglect the weight of air in the bubble and assume that the viscous force on each sphere is proportional to its terminal velocity.

(Density of steel $= 7700\,\mathrm{kg\,m^{-3}}$, density of glycerine $= 1260$ $\mathrm{kg\,m^{-3}}$.) (JMB)

16 Explain as fully as you can the phenomenon of viscosity, using the viscosity of a gas as the basis of discussion.

Show by the method of dimensions how the volume of liquid flowing in unit time along a uniform tube depends on the radius of the tube, the coefficient of viscosity of the liquid, and the pressure gradient along the tube.

The water supply to a certain house consists of a horizontal water main 20cm in diameter and 5km long to which is joined a horizontal pipe 15mm in diameter and 10m long leading into the house. When water is being drawn by this house only, what fraction of the total pressure drop along the pipes appears between the ends of the narrow pipe? Assume that the rate of flow of the water is very small. (O & C)

17 A shaft 4cm in diameter rotates inside a bearing 8cm long. The clearance between shaft and bearing is 0.02cm and is filled with oil of viscosity $1.5\,\mathrm{N\,s\,m^{-2}}$. Estimate the torque required to turn the shaft at a constant angular speed of 15 revolutions per second. (O & C)

18 Red cells are observed to sediment from blood at a rate of about 2cm in 1 hour. Assuming that they behave as spheres, density $1090\,\mathrm{kg\,m^{-3}}$, falling through a medium of density $1030\,\mathrm{kg\,m^{-3}}$

for which η has a numerical value $1\cdot5 \times 10^{-3}$ in SI units, calculate their average radius. (o & c)

19 A suspension of spherical particles in a fluid of viscosity η is rotated with an angular velocity ω in a centrifuge and a particle, of radius r, has a *radial* velocity v. Show that v becomes approximately constant when

$$v/R = m'\omega^2/6\pi\eta r,$$

where R is the distance of the particle from the axis of rotation and m' is the effective mass of the particle. You may assume that a particle of radius r moving with velocity v through a liquid of viscosity η experiences an opposing force equal to $6\pi\eta rv$.

Taking the effective mass, m', as $m(1 - \delta/\rho)$, where m is the true mass, δ is the density of the fluid and ρ is the density of the particle, calculate the value of this radial velocity for approximately spherical particles, radius 10^{-2} mm and density 1090 kg m^{-3}, suspended in a liquid for which $\eta = 0\cdot0015$ N s m^{-2} and whose density is 1030 kg m^{-3}, if it is attained 15 cm from the axis of a centrifuge rotating at 600 revolutions per minute.

(o & c)

17 Exponential decay

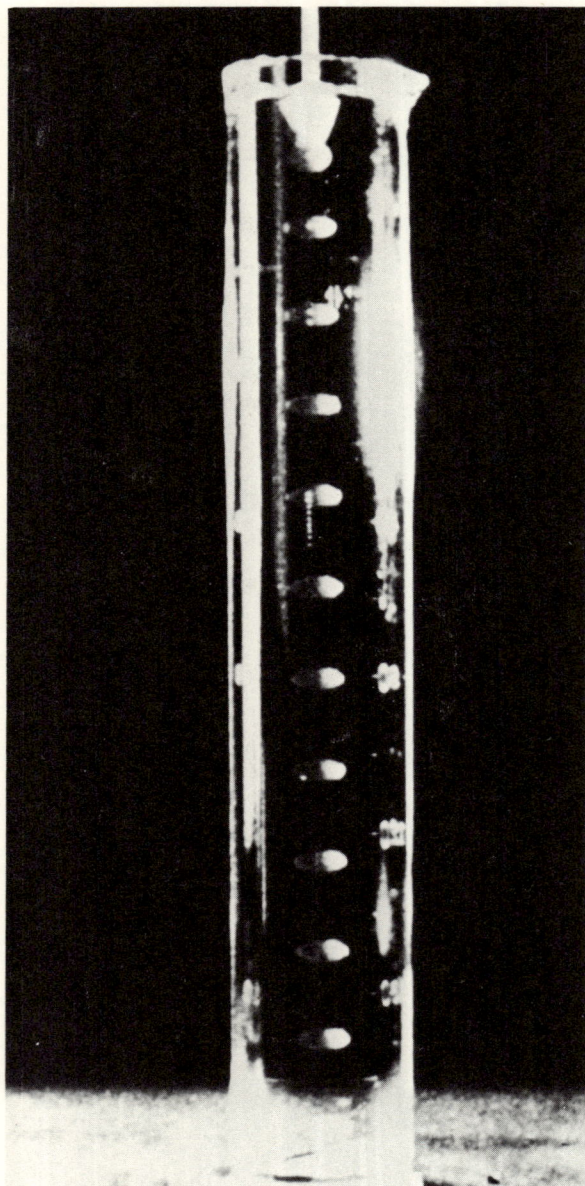

Drop of milk falling through liquid paraffin, illuminated by a stroboscopic lamp. Can you detect any evidence of the initial acceleration?

We have not yet finished with the falling sphere discussed on pp. 232–3. In the last chapter we concentrated on a qualitative description of the motion; quantitative treatment was confined to a calculation of the terminal velocity. We would now like to ask how the velocity and displacement of the sphere change as time goes on, in other words we shall aim for a complete mathematical description. We have to start with $F = ma$, so we must first write down the net downward force F on the sphere:

$$F = mg - U - 6\pi\eta rv, \qquad \ldots(1)$$

where v is the velocity and U the upthrust. The acceleration is most conveniently expressed as dv/dt, and $F = ma$ gives

$$mg - U - 6\pi\eta rv = m\frac{dv}{dt}$$

which is another differential equation. In more economical terms we can write it as

$$\frac{dv}{dt} = A - Bv \qquad \ldots(2)$$

where A and B are constants.

Exercise 17.1

Write down A and B in terms of the quantities appearing in equation (1).

A direct compact solution of this equation appears in Appendix 7; here we shall take a more leisurely look at what is involved. We shall start with an equation which is a little simpler to handle:

$$\frac{dv}{dt} = -Bv. \qquad \ldots(3)$$

Exercise 17.2

Show that this equation applies to the special case when the density of the sphere is the same as that of the medium. If left to itself therefore, the sphere will not move–we have to push it to get it going.

The velocity–time graph we get from equation (3) is sketched out in Figure 17.1. In words, the equation says that the gradient of the graph at every instant t is equal in magnitude to the corresponding value of v multiplied by a constant B, but if v is positive the gradient is negative. Of course this means that v must decrease as t increases, until v eventually dies out altogether; this is the reasoning which leads us to Figure 17.1. What we really want is a mathematical formula for the graph; we would like to be able to write v as a simple function of t. We might be forgiven for thinking the curve is part of a parabola, so we could try

$$v = a + bt + ct^2$$

as a solution; here a, b and c are constants. This will immediately commit us to putting

$$\frac{dv}{dt} = b + 2ct.$$

If these expressions are to constitute a genuine solution of equation (3) we shall need to insist that

$$-B(a + bt + ct^2) = b + 2ct$$

regardless of the value of t. We can see at a glance that this is not going to be possible; we could balance the terms independent of t by requiring that

$$-Ba = b$$

and the terms involving t by putting

$$-Bb = 2c,$$

but this still leaves the term involving t^2 unbalanced. The solution is *not* going to be the equation of a parabola.

Perhaps we need more terms. Let us try

$$v = a + bt + ct^2 + dt^3 + et^4$$

$$\frac{dv}{dt} = b + 2ct + 3dt^2 + 4et^3.$$

Fitting these into equation (3), we shall have to arrange things so that

$$-B(a + bt + ct^2 + dt^3 + et^4) = b - 2ct + 3dt^2 + 4et^3.$$

We can almost do it, but not quite. Remember we can choose a, b, c, d, e to be whatever values we like, as long as they are constant for all values of t.

Exercise 17.3

Show that the terms in t^3, t^2, t, and the term independent of t can be balanced if $-Bd = 4e$, $-Bc = 3d$, $-Bb = 2c$ and $-Ba = b$.

This still leaves the term involving t^4 unbalanced. Clearly if this type of approach is going to work we must use a trial solution which contains an *infinite* number of terms, because there is always one more term in v than there is in dv/dt.

Exercise 17.4

Show that if a solution of the form

$$v = a + bt + ct^2 + dt^3 + \ldots \qquad \ldots(4)$$

is to satisfy equation (3), the values of the various constants must be such that

$$b = -Ba, \quad c = \frac{+B^2a}{2}, \quad d = \frac{-B^3c}{2 \times 3}, \ldots,$$

and write down in terms of B and a the value of h, the coefficient of t^7 in the solution. Hence show that expression (4) can be written as

$$v = a\left[1 - Bt + \frac{(Bt)^2}{2} - \frac{(Bt)^3}{2 \times 3} + \frac{(Bt)^4}{2 \times 3 \times 4} \cdots\right] \quad \ldots(5)$$

So we can balance the terms in t, t^2, t^3, etc. as far as we like; but what about the extra term left over? Our only hope is that as the power of t increases the terms get progressively smaller, and that if we let the series go on long enough the last term will get so close to zero that it can be ignored. In mathematical terms we have to make sure that the infinite series (5) is *convergent* for all values of B and t

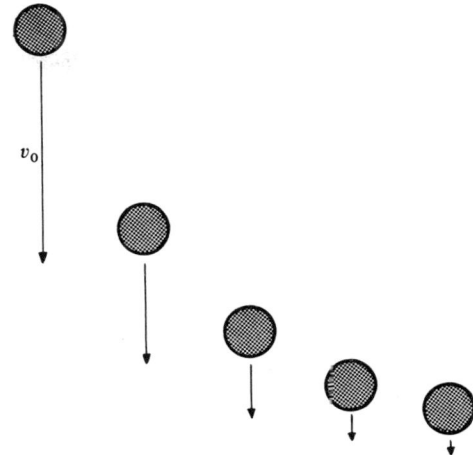

Motion of a sphere given an initial velocity v_0 in a resistive medium having a density equal to that of the sphere.

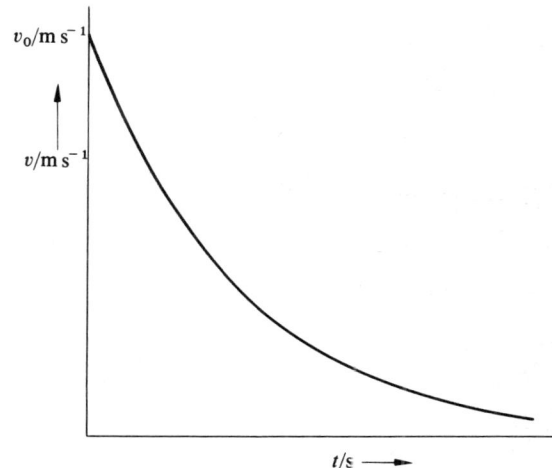

Figure 17.1

that we are likely to meet in a real situation. Mathematicians have rigorous tests for convergence, and this particular series passes the tests; we can safely assume that it is a satisfactory solution for our differential equation.

Strictly we could stop at this point; equation (5) is cumbersome but it is a genuine solution, and it would be possible to work out v for a given value of t to any required degree of accuracy by taking sufficient terms. We could reduce the labour involved by getting a computer to compile some appropriate tables, and it would help to devise some shorthand to save writing equation (5) every time we wanted to quote the solution. However we shall hardly ever have cause to use expression (5) because by a mathematical curiosity *it can be written in an entirely different form*. We can bring out the relationship with a little arithmetic.

Exercise 17.5

Taking both a and Bt to be numerically equal to 1, work out the approximate value of v obtained from the first six terms of equation (5). Repeat for Bt equal to 2 and 3. Confirm that these numbers are approximately equal to $1/2 \cdot 718$, $(1/2 \cdot 718)^2$ and $(1/2 \cdot 718)^3$ respectively.

This is the unexpected twist. It turns out that equation (5) can be written far more compactly as

$$v = a\left(\frac{1}{2 \cdot 718}\right)^{Bt}$$

The number $2 \cdot 718$ is not exact; to a few more significant figures it is $2 \cdot 71828$, but like π it cannot be written exactly in terms of a finite number of digits, and it is usually represented by the symbol e. We therefore write as the solution of equation (3),

$$v = a\left(\frac{1}{e}\right)^{Bt} \qquad \text{or} \qquad v = ae^{-Bt}, \qquad \ldots(6)$$

and it works for fractional values of Bt as well as for the integral values tried in Exercise (5). We say that equation (6) describes a velocity which *decays exponentially* as time goes on, and we find that the equation for exponential decay has many applications in all branches of physics.

Before going on we must take a look at a possible difficulty. An expression like $(2 \cdot 718)^{-2}$ we can understand – we could work it out by simple arithmetic if needed. Likewise we can no doubt picture a number like e^{-2}, where e is this strange number, close to $2 \cdot 71828$ but not exactly equal to it. Even $e^{-20 \cdot 6}$ we can perhaps visualise, though we should need logarithm tables at the very least to evaluate it to four significant figures. But e^{-Bt} would seem to be in a very different category, because B and t are not just numbers, they are *quantities* with units; how can we possibly raise e to a power which is a quantity! The answer is that we cannot; equation (6) makes no sense at all unless Bt happens to be dimensionless.

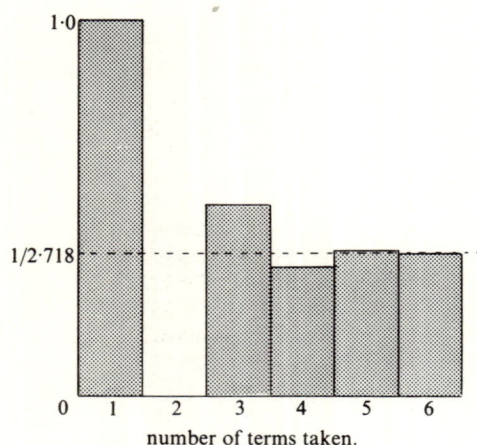

The series $1 - t + t^2/2 - t^3/(2 \times 3) + \ldots$, with t numerically equal to 1, rapidly approaches a value $1/2 \cdot 71828\ldots$

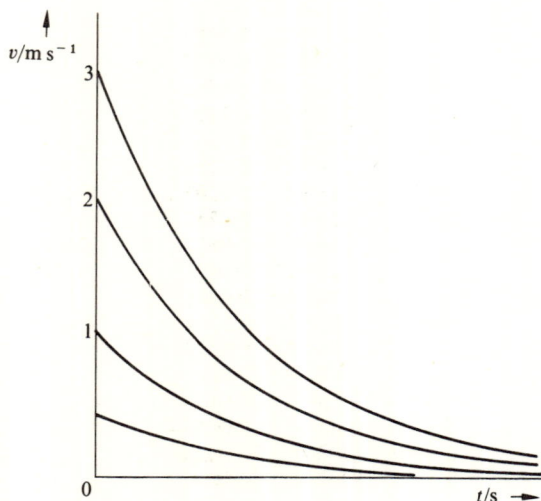

Several graphs corresponding to the formula $v = ae^{-Bt}$, with different values of a. We have to select the value of a we require by imposing boundary conditions, giving $a = v_0$.

Exercise 17.6

By referring to the result of Exercise 17.1, show that B does in fact have dimensions $[T]^{-1}$ and hence that Bt is a pure number.

Initial conditions

So far we can take it that the solution of

$$\frac{dv}{dt} = -Bv$$

is

$$v = ae^{-Bt},$$

but we have still not finished, because as yet a is completely arbitrary. We have met this situation before; we have to find a by looking for boundary conditions. Usually the conditions we know all about are those which apply at the start, when $t = 0s$. Our solution must give the right answer at that instant as at any other, so if we know that the initial velocity is v_0 we can write

$$v_0 = ae^{-B \times 0} = ae^0$$

Any non-zero number raised to power zero gives unity, and e is no exception, so

$$v_0 = a$$

and of course this value must apply for all possible values of t. The solution therefore becomes

$$v = v_0 e^{-Bt}.$$

Example

Given $v_0 = 0.05\,\text{ms}^{-1}$ and $B = 0.4\,\text{s}^{-1}$, find the value of v 6.0s after the start.

$$\begin{aligned}
v &= v_0 e^{-Bt} \\
&= 0.05 e^{-0.4 \times 6.0}\,\text{ms}^{-1} \\
&= 0.05 e^{-2.4}\,\text{ms}^{-1} \\
&= 0.05 \times 0.0907\,\text{ms}^{-1}
\end{aligned}$$

(Here we have used exponential tables: these are tables of e^x and e^{-x} for various values of x.)

$$v = 4.54 \times 10^{-3}\,\text{ms}^{-1}.$$

Exercise 17.7

Use Table 17.1 to find v at times 0.05s, 0.2s, 0.5s, 1.0s, 1.5s, 2.0s, 3.0s, 4.0s, 8.0s, 10.0s after the start, taking the values for v_0 and B from the example. Use your answers to plot a graph of v against t. Measure the slope of the graph at $t = 2.0s$ and check that it is equal in magnitude to $-Bv$ in accordance with the differential equation.

Example involving calculations of the time corresponding to a given velocity

Using the data for the sphere involved in Exercise 17.7, find the time taken for the velocity to drop to $0.01\,\text{ms}^{-1}$.

$$v = v_0 e^{-Bt},$$

so if we say that the velocity is $0.01\,\text{ms}^{-1}$ after Xs, X is given by

$$0.01 = 0.05 e^{-0.4X}$$
$$0.2 = e^{-0.4X}$$
$$e^{0.4X} = 5. \qquad \qquad \text{...(7)}$$

We now have to take logarithms of both sides to base e; these are called *natural* or *Naperian*, or sometimes *hyperbolic* logarithms. The

Table 17.1

Values of e^x and e^{-x}

x	e^x	e^{-x}
0.01	1.010	0.9900
0.02	1.020	0.9802
0.03	1.030	0.9704
0.04	1.041	0.9608
0.05	1.051	0.9512
0.06	1.062	0.9418
0.07	1.073	0.9324
0.08	1.083	0.9231
0.09	1.094	0.9139
0.10	1.105	0.9048
0.12	1.127	0.8869
0.14	1.150	0.8694
0.16	1.174	0.8521
0.18	1.197	0.8353
0.20	1.221	0.8187
0.3	1.350	0.7408
0.4	1.492	0.6703
0.5	1.649	0.6065
0.6	1.822	0.5488
0.7	2.014	0.4966
0.8	2.226	0.4493
0.9	2.460	0.4066
1.0	2.718	0.3679
1.1	3.004	0.3329
1.2	3.320	0.3012
1.3	3.669	0.2725
1.4	4.055	0.2466
1.5	4.482	0.2231
1.6	4.953	0.2019
1.7	5.474	0.1827
1.8	6.050	0.1653
1.9	6.686	0.1496
2.0	7.389	0.1353
2.2	9.025	0.1108
2.4	11.02	0.0907
2.6	13.46	0.0743
2.8	16.45	0.0608
3.0	20.09	0.0498
3.2	24.53	0.0408
3.4	29.96	0.0334
3.6	36.60	0.0273
3.8	44.70	0.0224
4.0	54.60	0.0183
5	148.4	0.00674
6	403.4	0.00248
7	1097	0.000912
8	2981	0.000335
9	8107	0.000123
10	22020	0.0000454
12	162800	0.00000615

symbol for 'logarithm of x to base e' is $\ln x$ (compare $\lg x$, the symbol for 'logarithm of x to base 10'). If we write

$$\ln x = y,$$

we mean that

$$x = e^y$$

so we see that the logarithm to base e of the left-hand side of equation (7) is simply $0.4X$.

So

$$0.4X = \ln 5$$
$$= 1.609 \text{ (using natural logarithm tables)}$$
$$X = 4.02$$

so that the velocity becomes $0.01\,\mathrm{m\,s^{-1}}$ $4.02\,\mathrm{s}$ after the start.

Exercise 17.8

Find how long it takes for the velocity to fall to (a) $0.04\,\mathrm{m\,s^{-1}}$, (b) $0.004\,\mathrm{m\,s^{-1}}$. [Hint: $\ln 12.5 = \ln(1.25 \times 10) = \ln 1.25 + \ln 10$.]

Acceleration and displacement

The whole point about the solution $v = v_0 e^{-Bt}$ is that it satisfies the relation $dv/dt = -Bv$. This means that the acceleration dv/dt is given by

$$\frac{dv}{dt} = -Bv_0 e^{-Bt}. \qquad \ldots(8)$$

The magnitude of the acceleration, like the velocity, decreases steadily to zero. Equation (8) lays down the rule for differentiating our exponential function; we merely write it down again and multiply it by $-B$. We can take it that to *integrate* therefore we have to *divide* by $-B$, so the displacement x is given by

$$x = \int v\,dt = \int v_0 e^{-Bt}\,dt$$
$$= \frac{-v_0 e^{-Bt}}{B} + c, \qquad \ldots(9)$$

where c is an arbitrary constant—we shall need some boundary conditions to find it. If we decide that $x = 0\,\mathrm{m}$ when $t = 0\,\mathrm{s}$, we can fit these special conditions into equation (9) and obtain

$$0 = \frac{-v_0 e^0}{B} + c$$

that is,

$$c = v_0/B.$$

Equation (9) then becomes

$$x = \frac{v_0}{B}(1 - e^{-Bt}). \qquad \ldots(10)$$

Figure 17.2 shows how we can interpret this formula. We know what a graph of $(v_0/B)e^{-Bt}$ looks like: it is just the exponential decay first sketched in Figure 17.1. So $-(v_0/B)e^{-Bt}$ is the same line turned upside down, or 'mirrored' in the t-axis. Finally $(1 - e^{-Bt})v_0/B$ is the mirrored line shifted up the axis by an amount v_0/B. The displacement x starts at zero, builds up at a rate which decreases steadily, and tends eventually to a constant value corresponding to a stationary sphere.

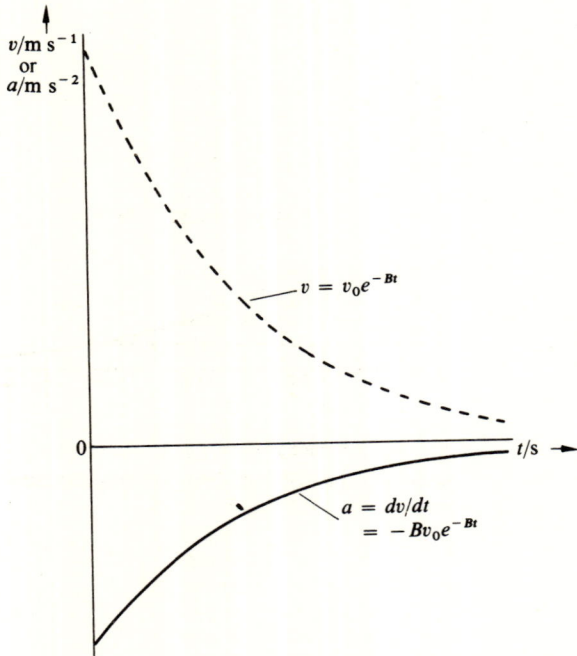

Graphs of velocity v and acceleration a against time t for the sphere moving in a resistive medium.

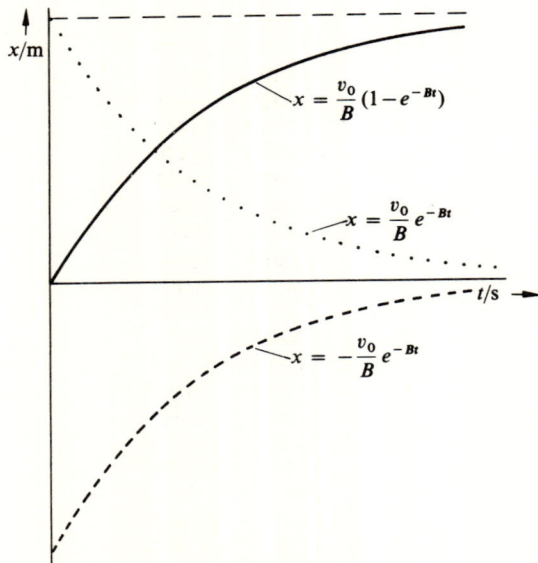

Figure 17.2

Exercise 17.9

Use the equations for acceleration and displacement to work out these quantities when $t = 1$ s, 5 s and 10 s for the sphere involved in Exercise 17.7. (Remember $B = 0.4$ s^{-1} and $v = 0.05$ m s^{-1} for this problem.)

Duration of exponential decay: half-life and relaxation time

The line sketched in Figure 17.1 never actually gets to the t-axis; it is asymptotic. Strictly, exponential decay takes an infinitely long time. In practice, however, the major part of the decay takes place in a finite time, sometimes a very short time indeed, and it is convenient to be able to quote a typical time interval at the end of which a certain fraction of the decay has taken place. Two criteria are commonly used; the first is the concept of *half-life*. In Figure 17.3 we have repeated the graph of v against t in order to bring out one of its special properties, namely the fact that the time taken for v to drop to half its value is *independent* of the value of v chosen. According to the graph, the velocity takes 1.5 s to fall from v_0 to $v_0/2$, a further 1.5 s to fall to $v_0/4$, and so on. This is a characteristic feature of exponential decay, and its half-life is a useful typical time to quote. At the end of five half-lives, for example, the velocity has dropped to $v_0/32$, or about 3% of the original value. The duration of the half-life is controlled by the value of B. Going back to the solution

$$v = v_0 e^{-Bt},$$

we get a value of t equal to the half-life $t_{1/2}$ when $v = v_0/2$:

$$v_0/2 = v_0 e^{-Bt_{1/2}},$$
$$2 = e^{Bt_{1/2}}.$$

Taking logarithms to base e

$$\ln 2 = Bt_{1/2}$$
$$t_{1/2} = 0.6931/B.$$

Often it is convenient to drop the factor 0.6931 and simply quote $1/B$ as a time interval representative of the rapidity of the decay; $1/B$ is called the *time constant* or in some contexts the *relaxation time* for the system.

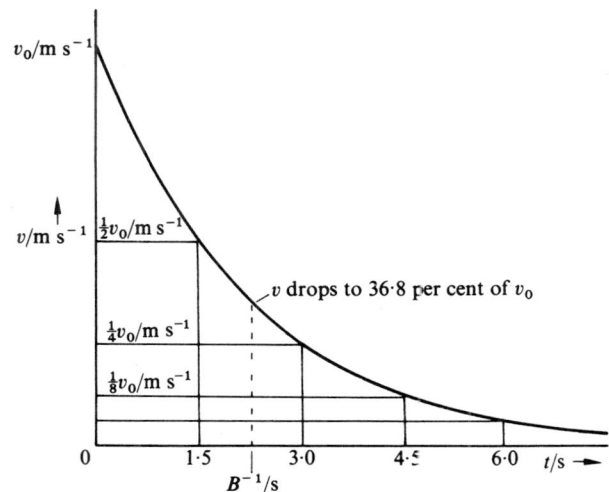

Figure 17.3

Exponential decay with a half-life of 1.5 s, showing the time constant $1/B$.

Exercises 17.10, 11

10 Show that the velocity drops to 36.79% of its original value in a time $1/B$, that is, about two thirds of the initial velocity has been lost. A further two thirds of what remains is lost in the time interval $1/B$ immediately following the first.
11 Show that at the end of an interval of duration $5/B$ less than 0.7% of the original velocity remains.

The solution of $dv/dt = A - Bv$

Now we can return to our original problem of the falling sphere in more general circumstances. A trial solution of the form

$$v = a e^{-Bt}$$

$$\frac{dv}{dt} = -aB e^{-Bt}$$

just fails, for if we substitute these expressions into the differential equation we have to try and arrange that

$$-aBe^{-Bt} = A - Bae^{-Bt}$$

and though the terms in t cancel the constant A remains unbalanced, so the equation is not satisfied. A slight modification puts things right; we try a solution

$$v = ae^{-Bt} + b, \qquad \qquad \ldots(11)$$

$$\frac{\mathrm{d}v}{\mathrm{d}t} = -aBe^{-Bt}, \qquad \qquad \ldots(12)$$

where b is another, as yet arbitrary, constant.

Exercise 17.12

Show that equations (11) and (12) satisfy the differential equation so long as $b = A/B$.

So the solution becomes

$$v = ae^{-Bt} + \frac{A}{B}$$

and we have simply to dispose of the arbitrary constant a. If we decide that $v = 0\,\mathrm{m\,s^{-1}}$ when $t = 0\,\mathrm{s}$, we get

$$0 = a + A/B \qquad \text{or} \qquad a = -A/B,$$

so that

$$v = (1 - e^{-Bt})A/B. \qquad \qquad \ldots(13)$$

This is very similar to the expression for the *displacement* obtained from the simpler differential equation; it is merely an exponential decay curve turned upside down and shifted up so as to start at the origin. After some time the velocity approaches the limiting value A/B, which is an old friend, the terminal velocity, and the time constant for this build-up is $1/B$ again (Figure 17.4).

Exercises 17.13–16

13 Recalling that $A = g - U/m$ and $B = 6\pi\eta r/m$, work out the terminal velocities and time constants for the following situations.
   a) A steel ball of density $7860\,\mathrm{kg\,m^{-3}}$ and radius $1\cdot0\,\mathrm{mm}$ falling through glycerine of density $1260\,\mathrm{kg\,m^{-3}}$ and viscosity $0\cdot85\,\mathrm{N\,s\,m^{-2}}$
   b) A steel ball of radius $4\cdot0\,\mathrm{mm}$ falling through glycerine
   c) A bubble of air of radius $0\cdot01\,\mathrm{mm}$ and density $1\cdot2\,\mathrm{kg\,m^{-3}}$ rising through water with viscosity $0\cdot0010\,\mathrm{N\,s\,m^{-2}}$
   d) The same bubble rising through glycerine.

14 Find the velocity of the ball in Exercise 17.13(a) $0\cdot0035\,\mathrm{s}$ after it is released from rest.

15 Show that in terms of the symbols used in equation (13) the acceleration of the sphere is $-Ae^{-Bt}$ and its displacement x from the point of release is given by

$$x = \frac{At}{B} + \frac{A}{B^2}(e^{-Bt} - 1).$$

Sketch the graph of x against t.
(Hint: show that the constant of integration is $-A/B^2$ if x is to be zero when $t = 0\,\mathrm{s}$.)

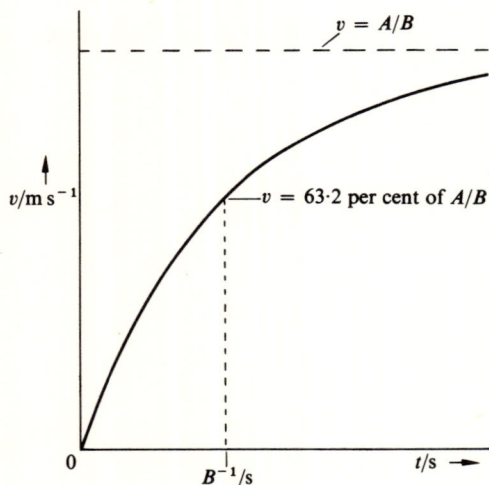

Figure 17.4
Graph of $v = (1 - e^{-Bt})A/B$ against t.

16 Figure 17.5 shows the essentials of an arrangement used by
   Millikan to determine the size of the fundamental electric charge.
   The method involves measuring the terminal velocity of a tiny
   charged oil drop, both with and without a potential difference
   between the horizontal plates.

   a) Find the weight W of an oil drop of radius $\lneq 0 \times 10^{-7}$ m,
      density $900\,\mathrm{kg\,m^{-3}}$, and its terminal velocity in air in the
      absence of an electric potential difference between the plates.
      (Viscosity of air at room temperature $= 1{\cdot}8 \times 10^{-5}\,\mathrm{N\,s\,m^{-2}}$.)
   b) A potential difference is suddenly set up between the plates so
      that the drop experiences an additional upward force of size
      $3W$. Using $F = ma$, write down the appropriate values of A
      and B in the differential equation $dv/dt = A - Bv$ which
      describes the subsequent motion. What is the new terminal
      velocity, and what is the size of the time constant involved in
      the velocity change?

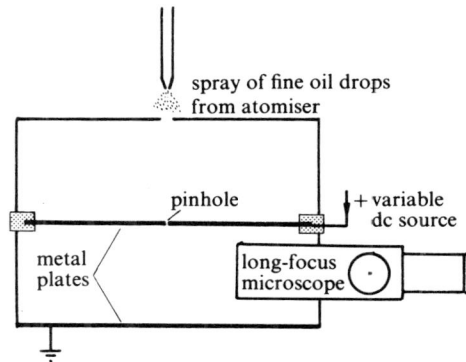

Figure 17.5 (Exercise 17.16)

Millikan's oil drop experiment. Light incident in a direction
perpendicular to the microscope axis is scattered by the
drops, which appear as tiny points.

The essential requirement for exponential decay

The equation

$$y = y_0 e^{-kx},$$

expressing the way y decreases exponentially as x increases, is
relevant to many situations in physics. Whenever the rate of decrease
of y with respect to x is proportional to the amount of y still left,
this is the equation which results, and that is such a simple requirement that it is satisfied in all sorts of unlikely situations. Suppose a
shop has N items of crockery on show. It is impossible to say when
a particular item will be broken, but we can make one valuable
generalisation: the more items on show, the more are likely to be
broken in a given time interval. In more mathematical language, the
rate at which N decreases, $-dN/dt$, is approximately proportional
to N, or

$$\frac{dN}{dt} \approx -kN,$$

leading to

$$N \approx N_0 e^{-kt},$$

where N_0 is the number of items present at the start. The approximation is best for very large values of N, becoming less and less reliable
as N decreases.

Radioactive decay

Very much the same reasoning controls the decay of a sample of
radioactive material. The nuclei responsible for the radioactivity
are unstable; they are liable to break up at any moment and emit
radiation. For example, if a sample of phosphorus is bombarded by
neutrons in a nuclear reactor, a proportion of its nuclei are transformed into a radioactive form, and the sample consists of a mixture
of ordinary phosphorus, ^{31}P, and radioactive phosphorus, ^{32}P.
During the first few hours after the sample is removed from the
reactor, it is found that for every microgram of ^{32}P present (that
is, about $1{\cdot}9 \times 10^{16}$ atoms), $1{\cdot}05 \times 10^{10}$ β particles are emitted per
second. Each β particle is emitted from a ^{32}P nucleus, which as a
result is transformed into a stable ^{32}S nucleus, the nucleus of a sulphur
atom. The only factor which alters the rate of decay into ^{32}S, and
the corresponding emission of β particles, is the amount of radioactive phosphorus present; if half of it is removed the rate of β

emission is halved. In symbols, if N ^{32}P nuclei are present at time t,

$$\mathrm{d}N/\mathrm{d}t = -\lambda N \qquad \ldots(14)$$

where λ is called the decay constant; it is equal to the fraction of N decaying in unit time.

Because this corresponds to the decay of the crockery on show in the shop, which is a *random* process, we believe that radioactive decay is also random, that is, the chance that any given nucleus decays in the next second remains constant for as long as it survives.

Equation (14) has the solution

$$N = N_0 e^{-\lambda},$$

where N_0 is the number of radioactive nuclei present when $t = 0\,\mathrm{s}$. This implies that the decay of the sample is exponential, with a definite half-life $t_{1/2}$ given by

$$t_{1/2} = (\ln 2)/\lambda.$$

It takes 14·5 days for half the ^{32}P nuclei in a sample to decay, and this means that the rate of β emission also halves in 14·5 days.

Exercise 17.17

Use the fact that a sample containing $1\cdot9 \times 10^{16}$ ^{32}P nuclei emits $1\cdot05 \times 10^{10}$ β particles per second to confirm that the half-life of phosphorus 32 is 14·5 days.

Exponential atmosphere

We can use the mathematics of exponential functions to give a picture of how the pressure P and density ρ of the atmosphere fall off with height h. Referring to Figure 17.6, the change in pressure δP over the small extra height δh is due to the weight of air between h and $(h + \delta h)$, and is given by

$$\delta P \approx \frac{-A\,\delta h\rho g}{A},$$

where A is the cross-sectional area of the column considered; it rapidly disappears from the calculation. We have assumed that δh is small enough to ignore any variation of density between h and $(h + \delta h)$.

$$\frac{\delta P}{\delta h} \approx -\rho g \qquad \text{or} \qquad \frac{\mathrm{d}P}{\mathrm{d}h} = -\rho g. \qquad \ldots(15)$$

Before we can solve this equation we need to be able to write ρ in terms of P. We can use Boyle's law to do this if we assume that the temperature does not vary with height. This is never realised in practice, so our treatment will give no more than an indication of the true situation.

The most convenient form of Boyle's law for us is

$$\frac{P}{\rho} = \frac{P_0}{\rho_0},$$

where P_0 and ρ_0 are the pressure and density at sea level. As a result we can write equation (15) in the form

$$\frac{\mathrm{d}P}{\mathrm{d}h} = \frac{-P\rho_0 g}{P_0}.$$

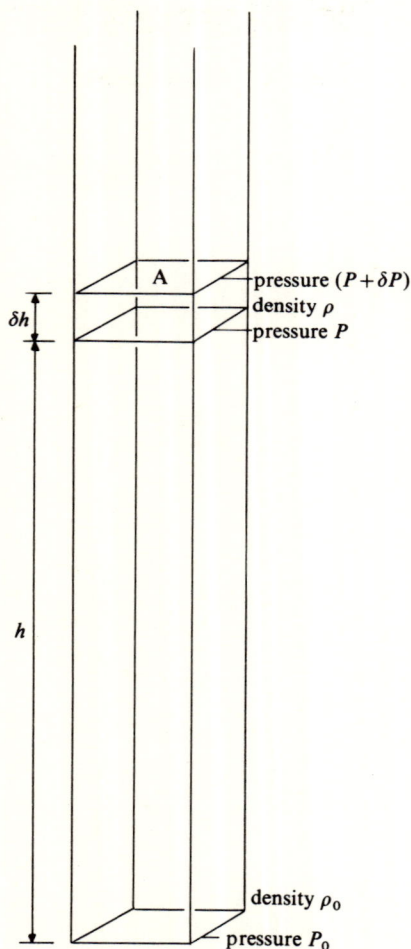

Figure 17.6

Exercises 17.18–20

18 Confirm by substitution that $P = Ke^{-\rho_0 gh/P_0}$ is a solution for this equation, and show that K must be put equal to P_0.

19 Given $P_0 = 10^5\,\mathrm{N\,m^{-2}}$ and $\rho_0 = 1{\cdot}2\,\mathrm{kg\,m^{-3}}$, find the pressure given by the equation at heights (a) 500 m, (b) 1 km, (c) 5 km, (d) 10 km, (e) 100 km above sea level.

20 Figure 17.7 shows a column of liquid of height h flowing out through a capillary tube of length l and radius r. Use Poiseuille's formula to show that if the flow is streamline

$$-\frac{dh}{dt} = \frac{\rho gh\pi r^4}{8A\eta l}, \qquad \ldots(16)$$

where ρ is the density and A the cross-sectional area of the wide tube. (This expression is not strictly accurate because some of the pressure head ρgh of liquid is needed to accelerate it.)

As a result show that

$$h = h_0 e^{-kt},$$

where h_0 is the height when $t = 0$ s, and identify k in terms of the symbols used in equation (16). Find how long it takes for the height to halve its value if $r = 0{\cdot}50$ mm, $l = 0{\cdot}50$ m, $A = 1{\cdot}0$ cm^2, and the liquid is water with viscosity $0{\cdot}001\,\mathrm{N\,s\,m^{-2}}$.

Further exercises

Exercises 17.21–28

21 An aluminium sphere is suspended by a thread below the surface of a liquid. Show on a sketch the forces acting on the sphere, and explain its equilibrium. (No formal proof is required.)

The thread is now cut. Show on a second sketch, or explain in words, the forces which act on the sphere when it is in motion.

Table 17.2 shows the values obtained for x, the total distance travelled by this sphere in the liquid at time t.

Table 17.2

t/s	1·0	2·0	3·0	4·0	5·0
x/cm	3·6	10·3	18·6	27·9	37·4

Draw a graph to display the relation between x and t, and explain its form. Find the terminal velocity of the sphere. Give a qualitative account of how you would expect the graph to be modified if the temperature of the liquid were increased.   (c)

22 A horizontal capillary tube is attached to the bottom of a vertical burette by means of rubber tubing. A liquid is then poured into the burette so that with the tap closed its meniscus stands at a height h above the axis of the capillary tube. The tap is then opened and the following readings of the height h at different times t are shown.

Table 17.3

h/cm	56·2	50·6	45·0	38·4	33·8	28·1	22·5	16·9
t/s	0	70	152	265	352	478	632	832

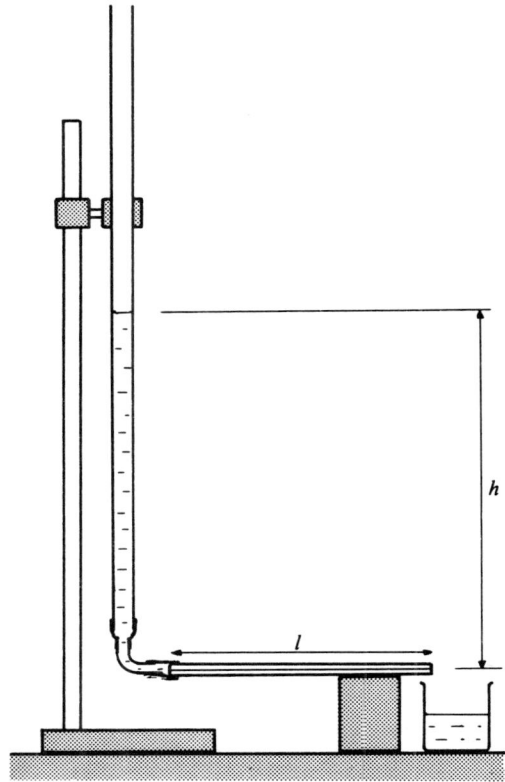

Figure 17.7 (Exercise 17.20)

Deduce how the rate of fall of the meniscus is related to its height above the capillary. (WJEC)

23 The force acting on a body of mass 2 kg varies with time according to the relation

$$F = 10e^{-0.01t},$$

where F is measured in newtons and t in seconds. Sketch the graph of F against t.

Write down an expression for the acceleration of the body at time t, and hence find an expression for its velocity, assuming that the body starts from rest when $t = 0$. Sketch a graph of the velocity of the body against time, and show that the kinetic energy ultimately gained by it is 2.5×10^5 J.

At what time is the force working at its maximum rate? (O)

24 A spherical soap bubble of surface tension T and radius R_0 is blown on the end of a capillary tube of length l and radius r and the air in the bubble is then allowed to escape through the open end of the tube. Find the time taken for the radius of the bubble to decrease to R, assuming that the volume rate of flow of the air through the tube is given by $\pi p r^4 / 8\eta l$, where p is the small pressure difference between the ends of the tube and η is the coefficient of viscosity of air.

Describe in detail how you would check your result experimentally in so far as the relation between radius of bubble and time is concerned. Indicate how you would make any other measurements necessary to enable you to determine a value for the coefficient of viscosity of air. (JMB)

25 A certain radioactive substance is thought to emit one type of radiation only. Its activity, in arbitrary units measured at weekly

Table 17.4

Week	0	1	2	3	4	5	6
Activity	2010	1776	1588	1382	1260	1076	974

intervals, is as shown in Table 17.4. Plot a suitable graph, and use it to find a value of the 'half-life' of the substance. (O)

26 A pure sample containing initially P_0 atoms of a radioactive element A with a decay constant λ_1 decays to give a second element B which is also radioactive and decays with a decay constant λ_2.

a) Write down an expression for the rate of disappearance of A.

b) Suppose that after time t the amount of B present is Q. What is the rate at which the amount of B is changing?

c) Assuming that Q varies with time according to the expression

$$Q = P_0(ae^{-\lambda_1} + be^{-\lambda_2 t})$$

where a and b are constants, show that after time t the number of nuclei of B present and the number of nuclei of the parent element A originally present are in the ratio

$$\frac{\lambda_1}{\lambda_2 - \lambda_1}(e^{-\lambda_1 t} - e^{-\lambda_2 t}).$$

d) At what time is the number of nuclei of B a maximum?

e) If element A is radium A with a half-life of 3.05 minutes, and element B is radium B with a half-life of 26.8 minutes, how long after obtaining a pure sample of radium A will the amount of radium B be a maximum? (WJEC)

27 A long, horizontal, gas-filled tube closed at both ends is rotated about a vertical axis through its centre normal to its length, the angular velocity being ω. Explain why a pressure gradient develops between the centre and either end and show that the pressure p at a point distance x from the centre is given by

$$\ln(p/p_0) = \omega^2 x^2/2rT,$$

where p_0 is the pressure at the centre, T the temperature in K and r the gas constant per unit mass. (The constant is defined by the relation $p_0 V = rT$, where V is the volume of unit mass of gas at pressure p_0 and temperature T.) (JMB)

18 Unbalanced force dependent on position and velocity: damped and forced oscillations

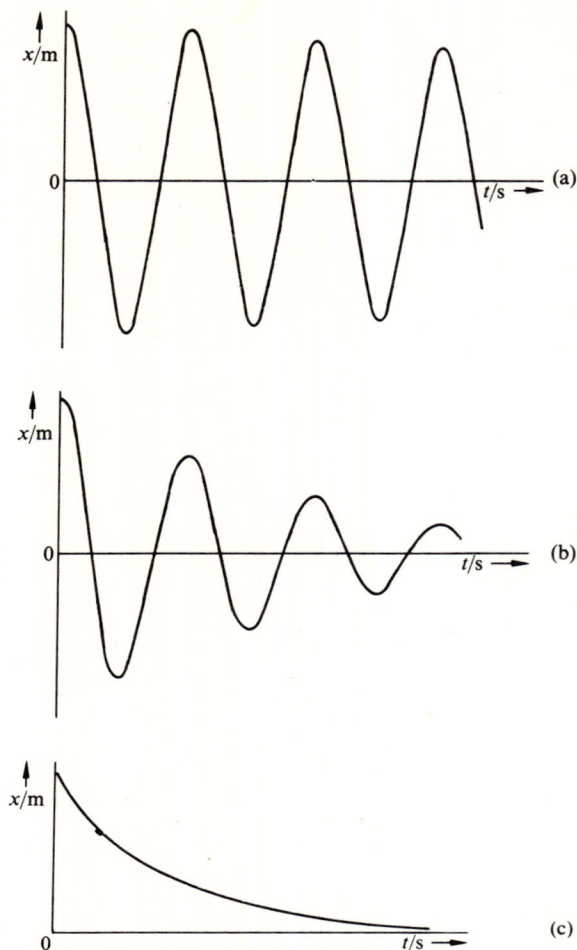

Figure 18.1

Displacement–time graphs for an oscillating mass; (a) lightly damped, (b) heavily damped, (c) very heavily damped.

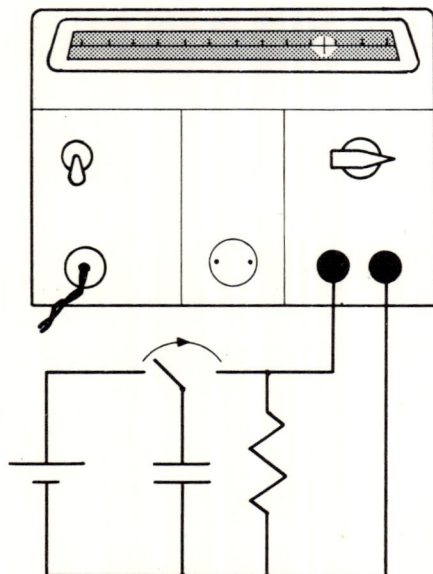

Figure 18.2

Simple harmonic motion, as a description of real oscillating systems, is often *too* simple; it does not give an adequate account of what happens in practice. The equations discussed in Chapter 15 represent oscillations which go on for ever with constant amplitude, whereas most vibrating systems are *damped*; they are subject to frictional and other dissipative forces which cause the vibrations to die away. Figure 18.1 indicates how the position of the vibrating mass changes as time goes on. The first graph corresponds to a *lightly damped* system; the oscillations persist for quite a long time compared with the period. A simple pendulum provides a good example of a lightly damped system. In Figure 18.1b we have a graph of a *heavily* damped system, perhaps a simple pendulum swinging in a bucket of water; the oscillations die away after only a few periods. Finally Figure 18.1c shows what can be expected from a simple pendulum swinging in a bucket of glycerine; not only has all trace of oscillatory behaviour vanished, but the bob takes a very long time indeed to reach the equilibrium position when displaced; this type of motion is said to be *dead beat*.

Critical damping

By mixing glycerine and water in different proportions we could get intermediate degrees of damping, and we should be able to find one particular situation for which oscillations are just, and only just, prevented. In this condition the system is said to be *critically damped*, and it is this degree of damping which we must supply if we want the system to get back to the equilibrium state as quickly as possible after a disturbance. Critical damping is of some importance in the design of galvanometers when rapid reading is needed; the idea can be illustrated very simply by means of a 'Scalamp' galvanometer on a sensitive range. The spot is disturbed most conveniently by discharging a capacitor through the Scalamp (Figure 18.2) and the damping can be varied by connecting different resistors across the terminals. Under such conditions damping due to mechanical causes – air resistance, for example – is relatively unimportant compared with the *electromagnetic damping*, which occurs as follows. When the moving coil rotates in the field of the permanent magnet, a voltage is induced across its ends. If the terminals are on open circuit that is the end of the matter, but if a resistor is connected across them a current flows, and this current also flows through the moving coil. Any conductor carrying a current in a magnetic field experiences a force, the *motor effect* force (unless the current and field are in the same direction) and the moving coil is no exception to this rule. The forces experienced by various parts of the coil oppose its motion and slow it down; this is an example of Lenz's law of electromagnetic induction, which tells us that the direction of the induced current is such that it opposes the change producing it. Readers who are familiar with the simple dynamo will recognise here an example of the 'back motor effect' of a generator. The smaller the resistance used, the greater the current and the greater the degree of damping, so that with a high resistance, oscillations persist for some time, while with a low resistance the motion is dead beat and the return to equilibrium can be very slow. There is some intermediate value of resistance which provides the critically damped condition, and this is the resistance to use if numerous or frequently repeated readings are required without waste of time waiting for oscillations to die away. When the Scalamp is used in series with a

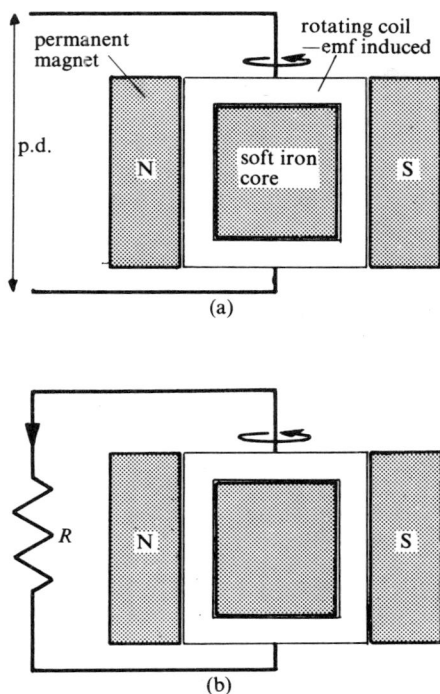

(a)

(b)

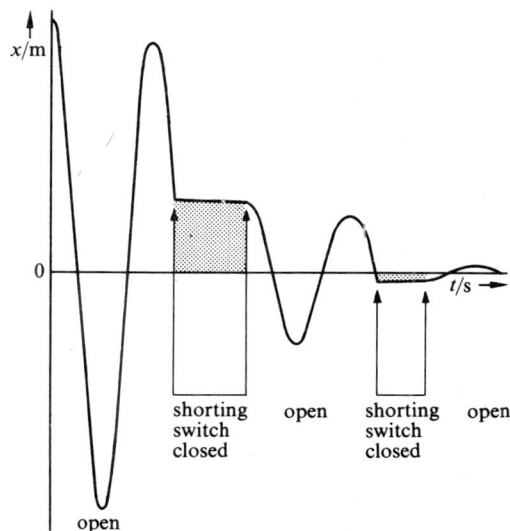

Figure 18.3

Electromagnetic damping. (a) The e.m.f. induced in the rotating galvanometer coil causes a p.d. to appear across the terminals. (b) If the circuit is completed externally, the p.d. drives a current through the rotating coil, producing 'motor effect' forces which slow it down. The rotational kinetic energy of the coil is converted into internal energy, and the temperature of the wire rises slightly.

high resistance, the time delay between readings can be reduced by adroit use of a shorting switch connected across the terminals; if the switch is closed just as the spot moves through zero the electro-magnetic damping causes the moving coil to stop dead. A 'near miss' can be corrected by repeated opening and closing of the switch (Figure 18.3).

Quantitative characteristics of damped harmonic motion

In Figure 18.4 a dotted line is drawn to show the way the amplitude of damped harmonic vibrations decays as time goes on. Is it possible that the dotted line is an *exponential* decay curve? It depends on the type of damping. If the damping force on the oscillating object happens to be proportional to the velocity, then the net force F on it is of the form

$$F = -kx - c\dot{x},$$

where k is the constant appropriate to the restoring force and c is a constant determined by the magnitude of the damping. Using $F = ma$ we then get

$$m\ddot{x} = -kx - c\dot{x}, \qquad \ldots(1)$$

which is a rather more formidable differential equation than those we have met so far. Appendix 8 shows that an expression of the form

$$x = Ae^{-\lambda t}\sin \omega t \qquad \ldots(2)$$

is one solution which can satisfy it, in other words we have a sinu-soidal motion with an amplitude which decays exponentially.

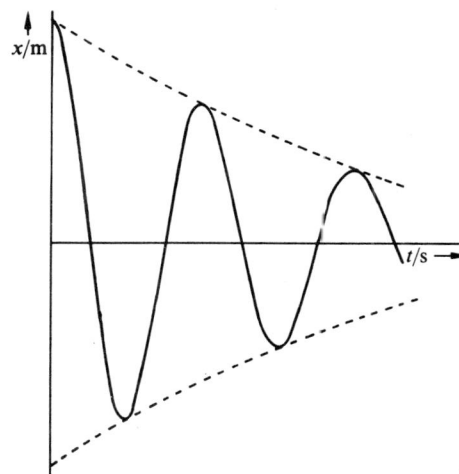

Figure 18.4

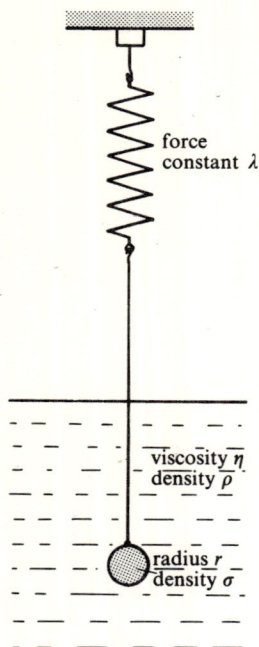

Damping for a loaded spring (Exercise 18.1)

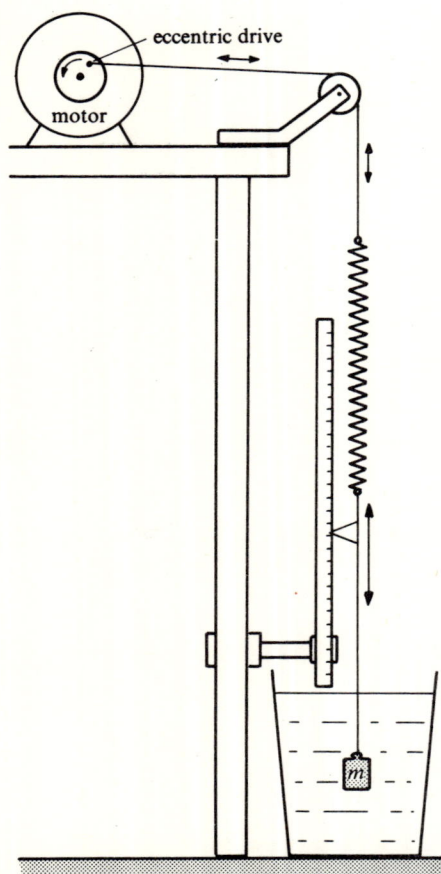

Figure 18.5

Appendix 8 also demonstrates that the frequency of the oscillations, $\omega/2\pi$, is a little less than that which would result if the damping were removed, and gives a derivation of the formula for the critically damped condition.

When can we expect the damping force to be proportional to the velocity, so that equations (1) and (2) follow? Electromagnetic damping in a uniform field can fulfil the requirement, because the force responsible for the damping is proportional to the current induced in the coil; this is proportional to the induced e.m.f., which in turn is proportional to the rate at which flux is cut by the rotating coil. The same applies to the drag produced by a resistive medium in conditions of streamline flow; for a sphere, for example, the force is $6\pi\eta rv$ (Stokes' law). If there is any turbulence, however, and this is quite likely, then equation (1) and its solution will not do, and the amplitude decay curve, though still of roughly the same shape, does not correspond to an exactly exponential decay.

Exercise 18.1

A sphere of radius r and density σ is suspended from a spring of force constant λ and immersed in a liquid of density ρ and viscosity η. Assuming that streamline conditions prevail when the sphere oscillates, find m, k and c [equation (1)] in terms of r, λ, ρ, σ, η, and g.

Forced oscillations

Figure 18.5 introduces a new aspect into the behaviour of oscillating systems. The mass m on the end of the spring is capable of performing damped harmonic motion if disturbed, and the frequency of the oscillations, the *natural* frequency f_0 of the system, is dictated by the inertia m, the stiffness of the spring, and, to a small extent, by the amount of damping. However, the upper end of the spring is fixed not to a rigid support, but to one which *vibrates* at some frequency f controlled by the speed of the motor; f is called the *driver frequency*. The motion which results can be split conveniently into two distinct sections. First, there is quite a long period of fairly irregular, almost chaotic, motion, during which it is rather meaningless to talk about any definite amplitude or frequency. What is happening is that the driver and the driven are fighting for supremacy; one wants the system to oscillate at frequency f, the other at frequency f_0. Gradually the driver gains the upper hand, and then the system settles down to steady oscillations at frequency f with constant amplitude; we call these *forced oscillations*.

Relaxation time

The time taken to establish steady oscillations at the driver frequency depends on the amount of damping. If the damping is heavy, the system settles down quickly, whereas a lightly damped system may take a very long time; in fact the interval required is roughly *that which would be needed for natural vibrations to decay away if the driver were removed* (see p. 243). It is convenient to quote a *relaxation time* for the system, which is the time needed for natural vibrations to decay to about one third of their original amplitude. An example of a very long relaxation time is provided by the arrangement shown in Figure 18.6; two simple pendulums A and B of roughly equal lengths and masses are suspended from the same non-rigid support, and one of them, A, is set swinging to act as the driver. It does not take many oscillations for the amplitude of A to disappear; all the energy is transferred to pendulum B which now swings with an amplitude almost as great as that given originally to A. At this point the pendulums exchange roles, and the energy of the oscillations is transferred back to A, B once more becoming stationary. The interchanges then continue repeatedly until eventually the whole system runs down.

Response curve

Going back to the situation pictured in Figure 18.5, where the driver is not materially affected by what happens to the driven system, we can make some important generalisations about the amplitude of the forced oscillations. It depends both on the damping and on how far f is from f_0, and the full picture is given by Figure 18.7, which is a graph of amplitude against driver frequency for various degrees of damping. The oscillations with largest amplitude occur when f is approximately equal to f_0, the natural frequency of the undamped system (the approximation is very good for lightly damped systems but less useful if the damping is heavy), and the change of amplitude with driver frequency is more pronounced with light damping. The lines in Figure 18.7 are called *response curves* for the system, and the phenomenon of maximum amplitude at a particular driver frequency is called *resonance*. The *resonant frequency*, at which the amplitude is maximum, can for most purposes be regarded as equal to the natural frequency of undamped oscillations of the driven system.

Notice that some way away from the resonant frequency the amplitude is rather insensitive to the degree of damping; this means that the *width* of the resonance peak is almost independent of the damping, whereas of course the height of the peak is *very* dependent on how much damping is present. Very lightly damped systems can have quite sharp resonance peaks, particularly at high frequencies. Readers who have met examples of electrical resonance in circuits containing capacitors and inductors will recognise the similarities; the addition of a resistor to the tuning circuit in a receiver can result in a 'flatter' response over a given frequency range.

Phase relationships in a resonant system

Although the oscillations of the driven system have the same frequency as the driver, they do not in general have the same *phase*: they are out of step with each other, they do not reach the end of their respective swings at the same instant. The only exception to this occurs when the driver frequency is very slow, as we might expect; if we move the end of the spring in Figure 18.6 *very* slowly up and down we can expect the mass on the other end to move exactly in step, just as though the spring were a solid rod.

Figure 18.8 shows how the phase difference between driver and driven varies with f in the case of a lightly damped system. For high values of f the two are exactly out of phase, that is, the phase difference is half a cycle, corresponding to a phase angle of π radians. This means that driver and driven stop at the same instant but are always moving in opposite directions. At resonance the phase difference is almost exactly a quarter of a cycle or $\pi/2$ radians (see Exercises 18.3 and 4 later).

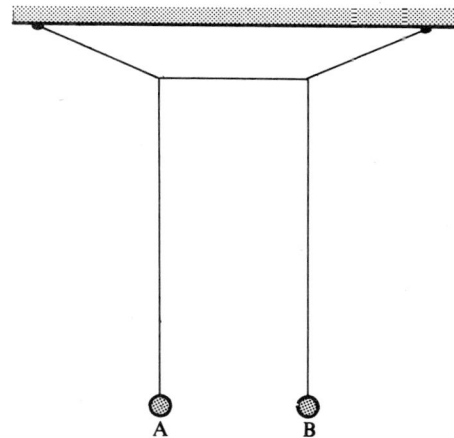

Figure 18.6

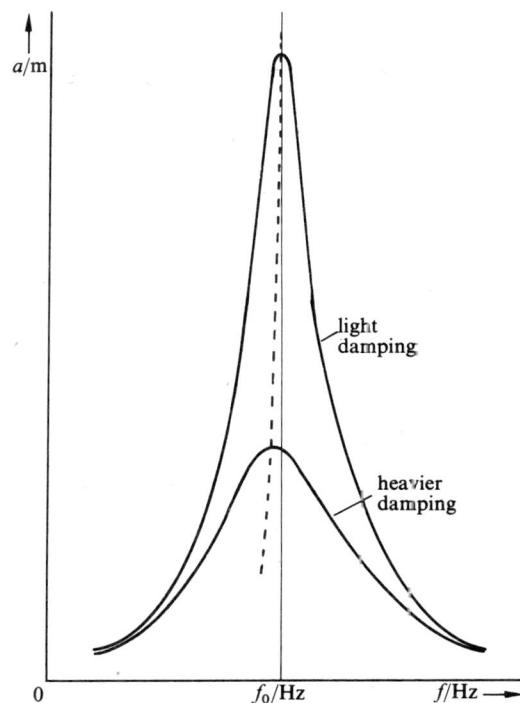

Figure 18.7

Response curves – graphs of amplitude a against driver frequency f for different degrees of damping.

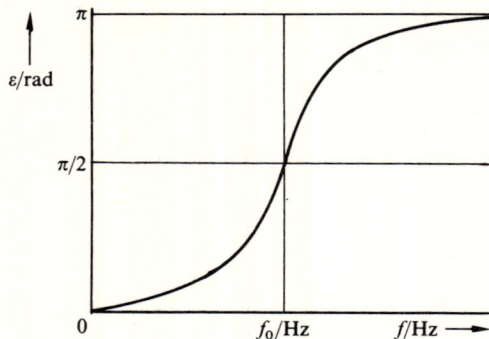

Figure 18.8

Variation of phase angle ε between driver and driven system with driver frequency f.

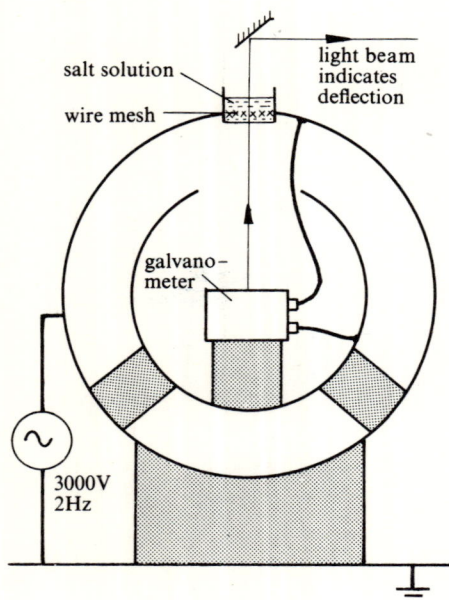

Figure 18.9

Everyday examples of resonance

Because most objects can vibrate with their own natural frequencies, resonance is a fairly common natural phenomenon, and is sometimes a nuisance, sometimes a boon. There are of course the stories of opera singers who can break wine glasses by singing notes of just the right frequency, but there are plenty of rather more down-to-earth examples. Loudspeakers and the cabinets enclosing them have to be carefully designed to be free from resonant frequencies–a poor record player may produce a rattle· whenever a particular note is sounded. Microphones for public address systems have to be carefully designed so as to have a 'flat' response over a wide frequency range (about 40–10 000 Hz), otherwise acoustic feedback between loudspeaker and microphone may well produce a continuous whistle at the resonant frequency. Car wheels which are not balanced properly, or poorly aligned steering, can produce annoying and possibly dangerous vibrations at a particular speed. The natural frequencies of pitch and roll of a boat must be thought out carefully in design so that they are well away from the wave frequency the boat is likely to meet. Resonance in electrical circuits is of prime importance in the field of communications of course, and in scientific work it can often be put to good use, especially in the detection of very small effects. When in 1936 Plimpton and Lawton repeated Cavendish's test of the inverse square law of force between electric charges, they used a resonance method to detect any variation from the law (Figure 18.9). The complete theory can be found in standard electricity textbooks, where it is shown that unless the inverse square law is exactly true, a voltage applied to the outer sphere shown in the figure causes a current to flow in the galvanometer. (The corresponding proof for gravitational effects inside a thin shell of matter appears on p. 254.) Plimpton and Lawton used a galvanometer with a natural frequency of 2 Hz and an alternating voltage of the same frequency, and although as a result the apparatus was sufficiently sensitive to detect a discrepancy of one part in 10^9, the galvanometer registered no response.

Differential equation for forced oscillations

Suppose we have a mass m on the end of a spring with force constant λ, so that the natural frequency f_0 of undamped oscillations is given by

$$f_0 = \frac{1}{2\pi}\sqrt{\left(\frac{\lambda}{m}\right)}. \qquad \ldots(3)$$

If the mass is subject both to a drag force equal to $c\dot{y}$, where $\dot{y}$ is the velocity, and to a driving force P of frequency f, so that

$$P = P_0 \sin 2\pi ft,$$

then the *net* force F it experiences a distance y from the equilibrium position can be written as

$$F = P_0 \sin 2\pi ft - \lambda y - c\dot{y}.$$

Using $F = ma$ we can then put

$$m\ddot{y} = P_0 \sin 2\pi ft - \lambda y - c\dot{y},$$

or

$$m\ddot{y} + c\dot{y} + \lambda y = P_0 \sin 2\pi ft. \qquad \ldots(4)$$

The ambitious reader may like to show that this equation can be

Collapse of Tacoma Narrows Suspension Bridge. Solid stiffening girders gave rise in transverse winds to eddies which were shed alternately from top and bottom edges of the girders. The resulting periodic forces produced resonant oscillations in the structure.

satisfied by a solution of the form

$$y = A \sin 2\pi ft + B \cos 2\pi ft$$

that is, by

$$y = C \sin(2\pi ft + \varepsilon),$$

where $C^2 = A^2 + B^2$ and $\tan\varepsilon = B/A$ (see p. 216).

The values of C and ε which emerge from this exercise are

$$C = \frac{-P_0}{4\pi^2 m\sqrt{[(f_0{}^2 - f^2)^2 + (c^2 f^2/4\pi^2 m^2)]}}, \qquad \ldots(5)$$

where f_0 is the natural undamped frequency [equation (3)], and

$$\tan \varepsilon = \frac{cf}{2\pi m(f^2 - f_0{}^2)}.$$

Exercises 18.2–4

2 Taking $P_0/4\pi^2 m$ to have unit magnitude and f_0 to be $100\,\text{Hz}$, work out C at $10\,\text{Hz}$ intervals and construct graphs of C against f for the following cases:
   (a) $c^2/4\pi^2 m^2 = 50\,\text{s}^{-2}$,   (b) $c^2/4\pi^2 m^2 = 200\,\text{s}^{-2}$.

3 Show that C is a maximum when $f_0{}^2 - f^2 = c^2/8\pi^2 m^2$, indicating that for light damping, that is when c is small, $f \approx f_0$ at resonance. (Hint: find when the denominator of equation (5) is a minimum.)

4 Show that when $f = f_0 = (1/2\pi)(\lambda/m)^{1/2}$, equation (4) is satisfied by $y = D\cos(2\pi f_0 t)$, giving a $\pi/2\,\text{rad}$ phase difference between driver and driven system, and find an expression for D.

Further exercises

Exercises 18.5–8

5 What is meant by (a) a forced vibration, (b) resonance? Give an example of each from (i) mechanics, (ii) sound.

Using the same axes, sketch graphs showing how the amplitude of a forced vibration depends upon the frequency of the applied force when the damping of the system is (a) light, (b) heavy. Point out any special features of the graphs.   (JMB)

6 Explain the terms *damped oscillation*, *forced oscillation* and *resonance*. Give one example of each.

Describe an experiment to illustrate the behaviour of a simple pendulum (or pendulums) undergoing forced oscillation. Indicate qualitatively the results you would expect to observe.

What factors determine (a) the period of free oscillations of a mechanical system, and (b) the amplitude of a system undergoing forced oscillation?   (O & C)

7 A spring is supported at its upper end. When a mass of 1 kg is hung on the lower end the new equilibrium position is 5 cm lower. The mass is then raised 5 cm to its original position and released. Discuss as fully as you can the subsequent motion of the system.   (O & C)

8 A simple pendulum of length l has a small bob made of lead. It is suspended in air from a support which can be moved horizontally in simple harmonic motion with frequency f and with a constant amplitude which is very small compared with l. At what frequency would you expect the pendulum bob to swing when its amplitude has reached a steady value? Sketch a rough graph to show how you would expect the steady amplitude of the bob to vary when it is measured at various values of the frequency f covering the range $f \ll f_0$ to $f \gg f_0$, where $1/f_0 = 2\pi \sqrt{(1/g)}$.

How would the graph differ if the bob were of the same dimensions as before, but made of cork?   (O & C)

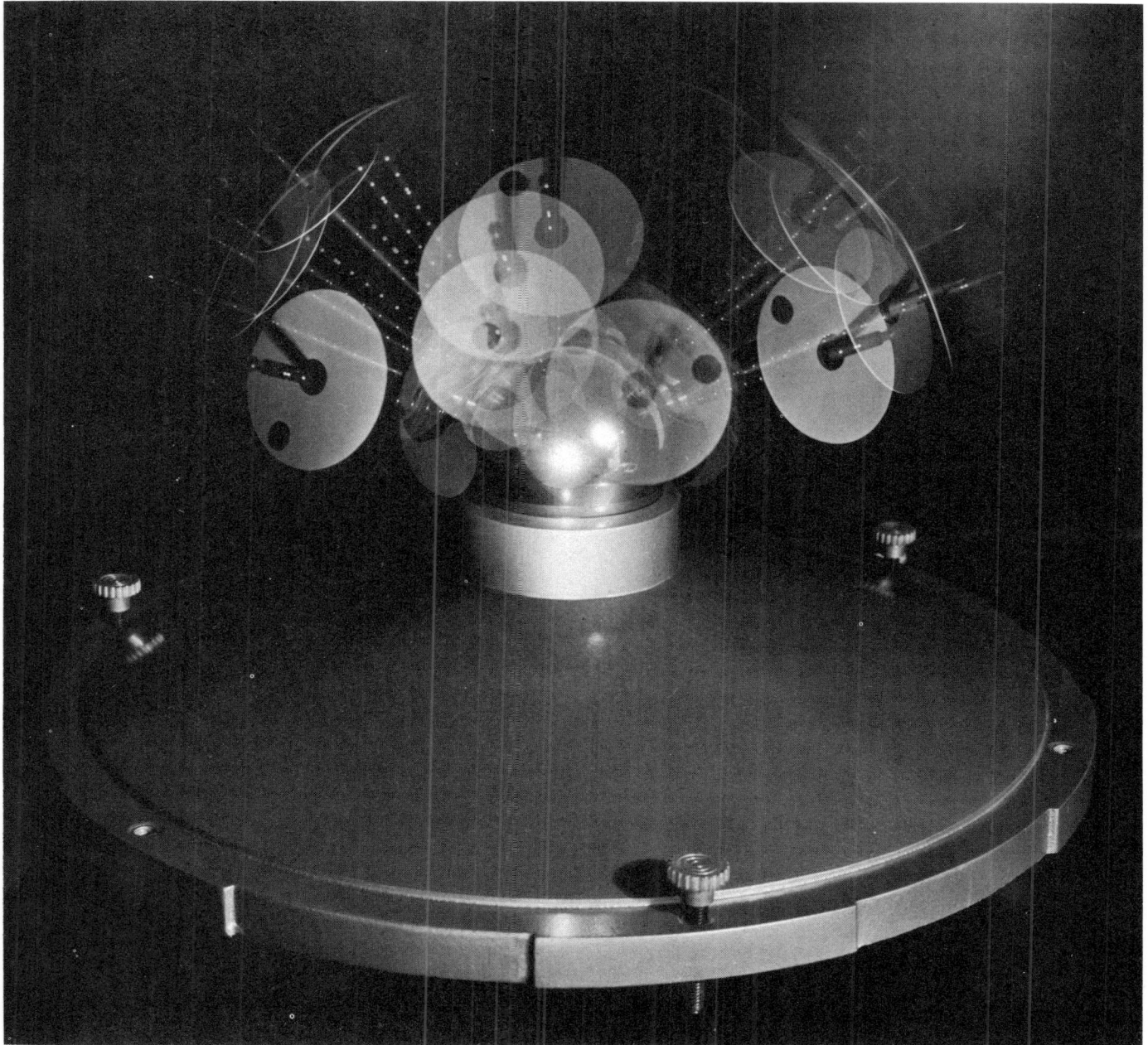

Gyroscope illuminated by a stroboscopic lamp.

We shall now look at how the equation $F = ma$ can be applied to a body which is *rotating*. We have only to think of a plate spinning on a table to realise that rotational motion can become quite complicated; in this book we shall deal only with cases for which the rotation is about an axis which does not wobble, that is, for one which remains perpendicular to a fixed plane.

Rotation about a fixed axis
Flywheels, rigid pendulums and rotating shafts of all kinds are important examples of bodies which rotate about an axis which is *fixed*. Figure 19.1 shows a rigid body able to rotate about a fixed axis through O, and we shall concentrate on just one of the many particles making up the body, a point mass δm distance r from O. Because it is part of the object and anchored firmly to its neighbours

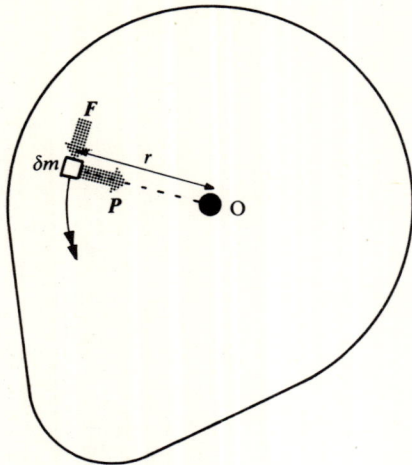

Figure 19.1

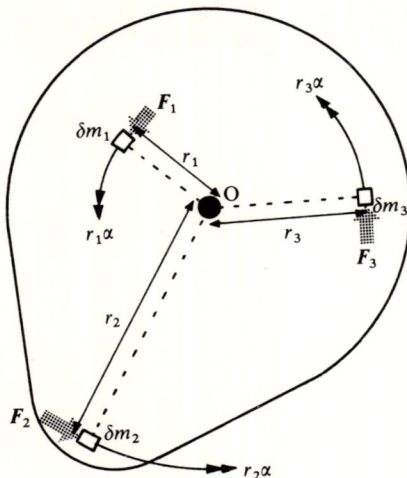

Figure 19.2

it experiences attractions and repulsions from all sides. If the body as a whole is stationary then these forces combine into a single force just sufficient to support the weight of the particle, but if there is any acceleration involved then there is a net force on δm sufficient to accelerate it with the rest of the body. It is this net force which we wish to consider; we shall assume that it acts in the plane of the paper, and resolve it into two components, one of magnitude P pointing towards O, and the other perpendicular to that direction and of magnitude F. We want to think about the sizes of P and F when the body is rotating about an axis through O perpendicular to the plane of the paper.

If it is rotating with uniform angular velocity we simply have the case considered in Chapter 14; F is zero and P must be great enough to provide δm with the necessary acceleration $\omega^2 r$ towards O. If, however, the angular velocity is *changing*, then F cannot be zero; it must supply δm with a tangential acceleration a as it moves round its circular path. Both F and a change direction continuously, but we shall nevertheless be able to write

$$F = \delta m a$$

as the link between them at every instant. Clearly the radial acceleration is still going on, and it can be shown that its magnitude is still $\omega^2 r$ where ω is the angular velocity at any given instant, but for the rest of the chapter we shall be concerned neither with this radial acceleration nor the force P producing it; we want to concentrate on the tangential acceleration.

Angular acceleration

Just as we progressed from the statement

distance along the arc of a circle $= r\theta$

to   speed along the arc $=$ rate of change of $(r\theta)$
$$= r\dot{\theta} \text{ or } r\omega,$$

where ω is the angular velocity, so we can go on to write

acceleration along the arc $=$ rate of change of $(r\dot{\theta})$
$$= r\ddot{\theta} \text{ or } r\dot{\omega} \text{ or } r\alpha,$$

where α is the rate of change of angular velocity, or the *angular acceleration*. It has units rad s^{-2}. The above formulae refer to an arc of constant radius, and since we are considering the rotation of a rigid body we can apply them to the motion of the mass δm. Now, because an arc is indistinguishable from its tangent if we look at a sufficiently small portion, we can safely say that the magnitude of the tangential acceleration is equal to $r\alpha$, enabling us to write

$$F = \delta m r \alpha$$

as the connection between the tangential force F and the acceleration it produces.

We could put down an equation like this for every small mass δm_1, $\delta m_2, \delta m_3 \ldots$ making up the body, and attempt some sort of summation (see Figure 19.2):

$$F_1 = \delta m_1 r_1 \alpha$$
$$F_2 = \delta m_2 r_2 \alpha$$
$$F_3 = \delta m_3 r_3 \alpha$$

Here we have used the fact that α is the same for all points in the body, because every point must take exactly the same time to complete a given angle of rotation. If we attempt a simple addition of all these expressions we get nothing useful at all, because $\bar{F}_1, F_2, F_3$ are the magnitudes of *vectors* which are all pointing in different directions. We can make progress however if we multiply each expression throughout by its respective value of r, to get

$$F_1 r_1 = \delta m_1 r_1{}^2 \alpha$$
$$F_2 r_2 = \delta m_2 r_2{}^2 \alpha$$
$$F_3 r_3 = \delta m_3 r_3{}^2 \alpha$$

Now there *is* a great deal of point in a simple algebraic addition of this latest set of equations. Each quantity on the left is the moment of the respective force about the axis of rotation, and we know that we *can* add moments to give the total torque about the axis concerned. In more advanced work the moment of a force is treated as another vector, of a type called an *axial* vector, and the direction associated with it is the direction of the axis about which the moment is evaluated. We are really adding vectors again, but this time they are all in the same direction. So we get

$$\sum Fr = \alpha \sum \delta m r^2,$$

where the sum is taken over all the particles making up the body. We need to think carefully about the meaning of $\sum Fr$. Remember F is the net tangential force on a particle; in most of the equations it is due to *internal* forces: interatomic attractions and repulsions which occur in equal and opposite pairs and which drop out of the sum. In fact the only contributions which do *not* drop out are those made by external forces acting on the body, that is, by whatever forces are producing the angular acceleration. So, $\sum Fr$ is simply the sum of the moments of the external forces about the axis of rotation; we shall write it as T, the total torque or moment about the axis concerned:

$$T = \alpha \sum \delta m r^2$$
or
$$T = I\alpha. \qquad \qquad ...(1)$$

The usual symbol for the quantity $\sum \delta m r^2$ is I, and it is a most important quantity called the *moment of inertia* of the body about the axis of rotation. Some of the significance of this term can be understood if we compare equation (1) with $F = ma$; it is in fact a form of $F = ma$ exactly tailored to suit rotational motion. In place of net force F we have net torque T, in place of acceleration a we have angular acceleration α, and in place of mass (or 'inertia') m we have the moment of inertia I, the 'rotational inertia' of the object. Notice that a given value of I is related to a *particular axis*: if we change the axis, the size of I changes also.

If we choose, we can solve any rotation problem we please simply by using $F = ma$ for every particle making up the body, but $T = I\alpha$ is a tremendously convenient and compact summary of the process we should have to work through. In one sense then, $T = I\alpha$ says nothing new, but in another sense it is a valuable step forward.

Exercise 19.1

An arm of length 2 m and negligible mass is pivoted at its centre so that it can rotate. It carries four masses of 0·5 kg each, one at each end and one on either side of the pivot and 0·5 m from it. Calculate

a) the moment of inertia of the arm.
b) the moment of the couple needed to give it an angular acceleration of 0·1 rad s^{-2}.

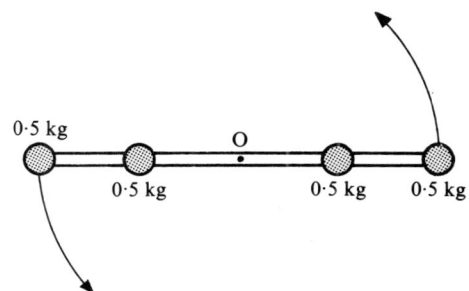

(Exercise 19.1)

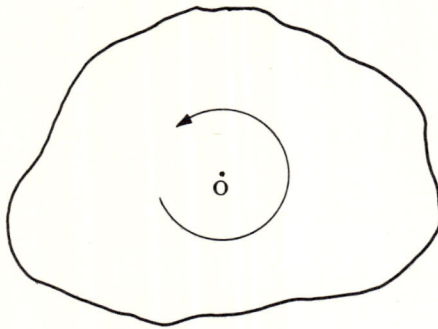

(a)

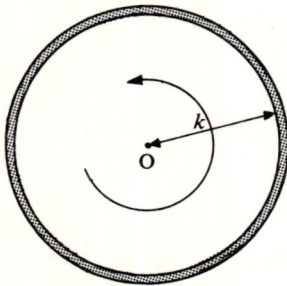

(b)

Figure 19.3

Radius of gyration. The thin ring or cylinder of matter of radius k, shown in (b), has the same mass and moment of inertia about an axis through O as the object shown in (a).

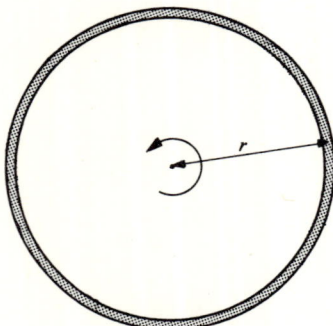

Figure 19.4

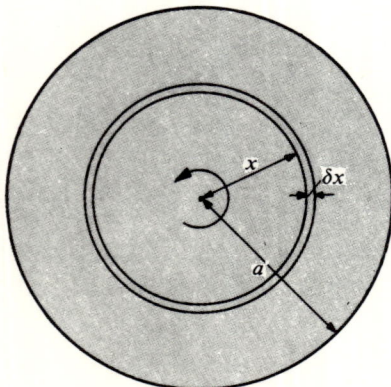

Figure 19.5

c) the time it would take to complete one revolution starting from rest, given the above acceleration.

Radius of Gyration

This is the name for the 'effective radius' of a rotating body with respect to the axis; it is defined by the formula

$$I = mk^2$$

where k is the radius of gyration and m is the total mass of the object. If we wanted to redistribute the mass of the object in such a way that every part of it should be equidistant from the axis of rotation, while still requiring that the body should respond to a given torque in the same way as before, then the ring or cylinder of matter so formed would have to be of radius k (Figure 19.3).

Calculation of moment of inertia

In principle we can work out $\sum \delta m r^2$ for any rigid body, but the process is straightforward only for objects of very simple shape, that is, for objects with a high degree of symmetry. The whole process depends on finding a section of the body with a size and shape such that every part of the section can be considered equidistant from the axis of rotation, and then writing the mass of the section in terms of its linear dimensions; this means introducing the density or mass per unit area or some other convenient quantity. In this way $\sum \delta m r^2$ can be turned into an integral and evaluated. It needs to be stressed that the moment of inertia about one axis is not equal to the moment of inertia about another, so that strictly *the* moment of inertia of a body is a term without meaning. We shall now work out some useful special cases.

Moment of inertia of a ring about an axis through its centre perpendicular to its plane

This is the simplest possible case; we consider a ring of mass m, radius r (Figure 19.4), and take it that the ring is so thin that we can regard every part of it as being equidistant from the centre. This being the case, we have no need to carve it into small pieces; $\sum \delta m r^2$ is simply mr^2, and of course the radius of gyration k is just r.

Moment of inertia of a circular disc about a perpendicular axis through its centre

We find it best to think of the disc, of mass m and radius a, as being made up of a large number of thin concentric rings (Figure 19.5). Looking at a typical thin ring of radius x and width δx, we can write its area as approximately $2\pi x \delta x$ and its mass as $2\pi x \sigma \delta x$, where σ is the mass per unit area, which we shall take to be constant. We know the moment of inertia of this ring about our axis, it is just (mass of ring $\times x^2$). (We are reckoning δx so small that all points on the ring are equidistant from the axis.) The moment of inertia is therefore the sum of all such contributions, that is

$$I = \sum 2\pi x \sigma x^2 \delta x$$

$$I = \int_0^a 2\pi \sigma x^3 \mathrm{d}x$$

So

$$I = \frac{2\pi \sigma a^4}{4},$$

and if we use the fact that
$$\sigma \times \text{disc area} = \text{total mass}$$
that is,
$$\sigma = m/\pi a^2$$
we get
$$I = \frac{2\pi a^4}{4} \times \frac{m}{\pi a^2}$$
$$= \tfrac{1}{2}ma^2. \qquad \qquad \ldots(2)$$

Exercise 19.2

Write down the radius of gyration of the disc.

Moment of inertia of a sphere about an axis through its centre

This time we must slice the sphere, of radius r and mass m, into a number of thin discs all perpendicular to the chosen axis. A typical disc is a distance x from the centre and of radius y (Figure 19.6). We cannot choose x and y independently; they are linked by the equation
$$x^2 + y^2 = r^2 \qquad \qquad \ldots(3)$$
If we make δx, the disc thickness, thin enough to ensure that its sloping edge makes no significant difference to its volume, then we can write

$$\text{mass of disc} = \text{area} \times \text{thickness} \times \text{density } \rho$$
$$= \rho \pi y^2 \delta x.$$

Its moment of inertia δI is therefore given by
$$\delta I = \tfrac{1}{2} y^2 \rho \pi y^2 \delta x,$$
where we have made use of equation (2), and the moment of inertia of the entire sphere is the sum of similar terms for all the discs which make it up. In integral form this becomes
$$I = \int_{-r}^{r} \tfrac{1}{2} \rho \pi y^4 \, dx,$$
where the limits are decided by the fact that x must be allowed to range from one edge of the sphere to the other in order to include all constituent discs. Now we make use of equation (3) to write the integral in terms of just one variable:
$$I = \int_{-r}^{r} \tfrac{1}{2} \rho \pi (r^2 - x^2)^2 \, dx.$$

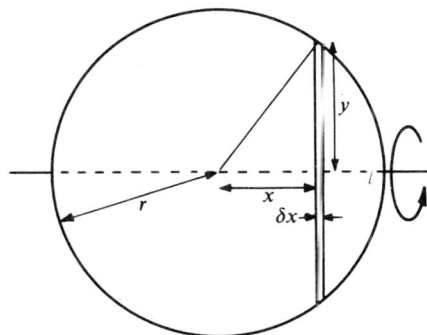

Figure 19.6

Exercise 19.3

Carry out the integration to obtain
$$I = 8\rho \pi r^5 / 15$$
and show that this can be written as $2mr^2/5$. Write down the radius of gyration in terms of r.

Moment of inertia of a thin plane rectangular sheet (i.e. a rectangular lamina) about an axis through its centre parallel to one edge

We take the rectangle to be of length a and width b (Figure 19.7), and the mass to be m. The axis chosen makes for the simplest arithmetic; the moment of inertia about an axis perpendicular to the plane of the lamina will be considered later.

We start by dividing the rectangle into narrow strips parallel to

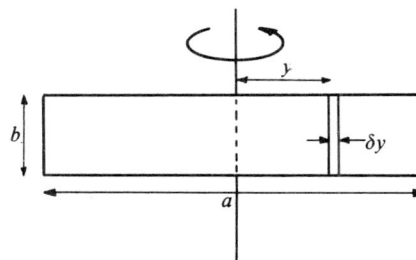

Figure 19.7

the chosen axis; a typical one is distance y from the axis and of thickness δy. If the rectangle has mass σ per unit area, the mass of this strip is $b\delta y\sigma$ and its moment of inertia about the axis selected is $by^2\delta y\sigma$. The total moment of inertia is therefore given by

$$I = \int_{-a/2}^{a/2} by^2\sigma\,\mathrm{d}y.$$

Exercises 19.4,5

4 Show that $I = ma^2/12$ and write down the moment of inertia with respect to an axis through the centre but parallel to an edge of length a.
5 Find the moment of inertia of the lamina with respect to an axis lying along one of the edges of length b.

Parallel-axis theorem

There are two rules which can often cut down the work involved in finding the moment of inertia about a given axis. The more useful of the two is the *parallel-axis theorem*; it applies to objects of any shape, regular or irregular, solid or hollow (unlike our next theorem, the perpendicular-axis theorem, which applies only to laminae–thin sheets of material). The parallel-axis theorem states that *if the moment of inertia about an axis through the centre of gravity is I_G, then the moment of inertia about a parallel axis displaced a distance a from the first is $I_G + ma^2$, where m is the mass of the body.* The proof is not difficult; in Figure 19.8, OO' is the axis through the centre of gravity and AA' is the displaced axis. A point mass δm at P, distance r from OO' and r_1 from AA', has moment of inertia δmr^2 with respect to OO' and δmr_1^2 with respect to AA', and if I_G and I' are the moments of inertia of the entire body about these axes respectively,

$$I_G = \sum\delta mr^2 \qquad \qquad \ldots(4)$$

and

$$I' = \sum\delta mr_1^2. \qquad \qquad \ldots(5)$$

But using the cosine rule in the triangle PQR, we obtain

$$r_1^2 = r^2 + a^2 - 2ar\cos\phi,$$

so that equation (5) can be written as

$$I' = \sum\delta mr^2 + \sum\delta ma^2 - 2\sum\delta mar\cos\phi.$$

Remembering that multiplying constants can be written either side of a summation sign, and making use of equation (4),

$$I' = I_G + a^2\sum\delta m - 2a\sum\delta mr\cos\phi. \qquad \ldots(6)$$

Now $\sum\delta m$ is simply the total mass m, and $\sum\delta mr\cos\phi$, though it looks a little formidable, is in fact an old friend. To interpret it we need to set up three perpendicular axes with origin at the centre of gravity as shown; we make the x-axis parallel to QR, and then we can see that $r\cos\phi$ is just the magnitude of the x-co-ordinate of P with respect to our reference frame. What is more, we can safely assert that

$$\sum\delta mx = 0$$

because the whole point about the centre of gravity is that for any axis passing through it the net moment of the weights of all the parts of the body is zero. In our case this means

$$\sum\delta mgx = 0$$

and so

$$\sum\delta mr\cos\phi = 0 \qquad \qquad \text{follows.}$$

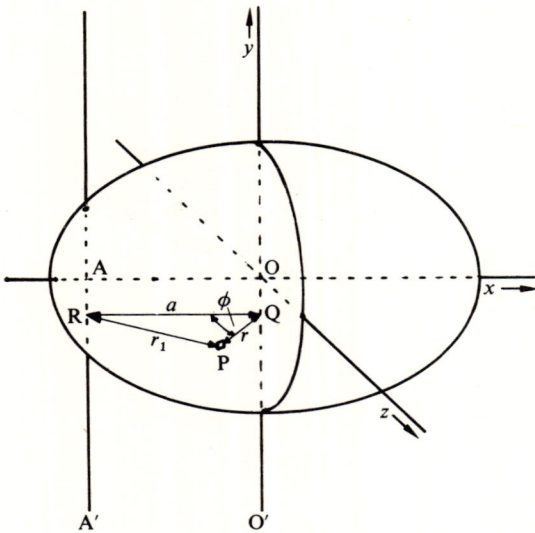

Figure 19.8

Equation (6) therefore reduces to

$$I' = I_G + ma^2. \qquad \text{...(7)}$$

Exercise 19.6

Use this formula to check your answer to Exercise 19.5.

It should be stressed that we can use this formula only if one of the axes involved passes through the centre of gravity. If a situation arises where this is not the case, then equation (7) must be used twice; the moment of inertia about an axis through the centre of gravity must be found as an intermediate step.

Perpendicular-axis theorem

This rule is rather more restricted: it applies only to a thin sheet in one plane, a lamina. In Figure 19.9, YY' is an axis perpendicular to the lamina and meeting it at any point P. The axes XX' and ZZ' are any two mutually perpendicular axes through P and in the plane of the lamina. The theorem says that *the moment of inertia of the lamina about YY' is equal to the sum of the moments of inertia about XX' and ZZ'*. The figure provides the key to the proof; the moments of inertia of the mass δm shown about YY', ZZ' and XX' are $\delta m r^2$, $\delta m x^2$ and $\delta m z^2$ respectively. Since

$$r^2 = x^2 + z^2,$$
$$\delta m r^2 = \delta m x^2 + \delta m z^2,$$

and what is true for one point mass is true for every other one in the lamina; we can write an equation like this for each one. If we add them all up we get

$$\sum \delta m r^2 = \sum \delta m x^2 + \sum \delta m z^2, \qquad \text{...(8)}$$

which is the result required.

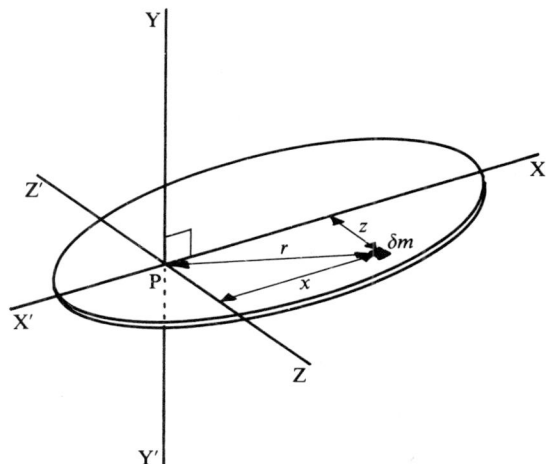

Figure 19.9

Exercise 19.7

Use equation (8) and the results of Exercise 19.4 to find the moment of inertia of a rectangular lamina with sides a and b and mass m about an axis through its centre perpendicular to its plane.

Centre of mass

The reference to the centre of gravity of the rotating body may seem a little strange when after all we are concerned primarily not with its weight but with its inertia. As the two properties of weight and inertia are linked so closely this need not worry us but for the fact mentioned on p. 67, that the term 'centre of gravity' is useful only when all the gravitational attractions are parallel. If they are *not* parallel the position of the centre of gravity can no longer be found from the equations

$$\sum \delta m x = \sum \delta m y = \sum \delta m z = 0. \qquad \text{...(9)}$$

In fact there may not even *be* a centre of gravity; it may not be possible to replace the body by an equivalent point mass. In such a situation however the relations (9) are still the ones we want for the proof of the parallel-axis theorem, and so the point they define is still significant. For this reason we find a different name for it, the *centre of mass*. For any object on the earth's surface small enough to walk round in an afternoon the centre of gravity and centre of mass

MOMENTS OF INERTIA FOR SOME UNIFORM OBJECTS

Table 19.1

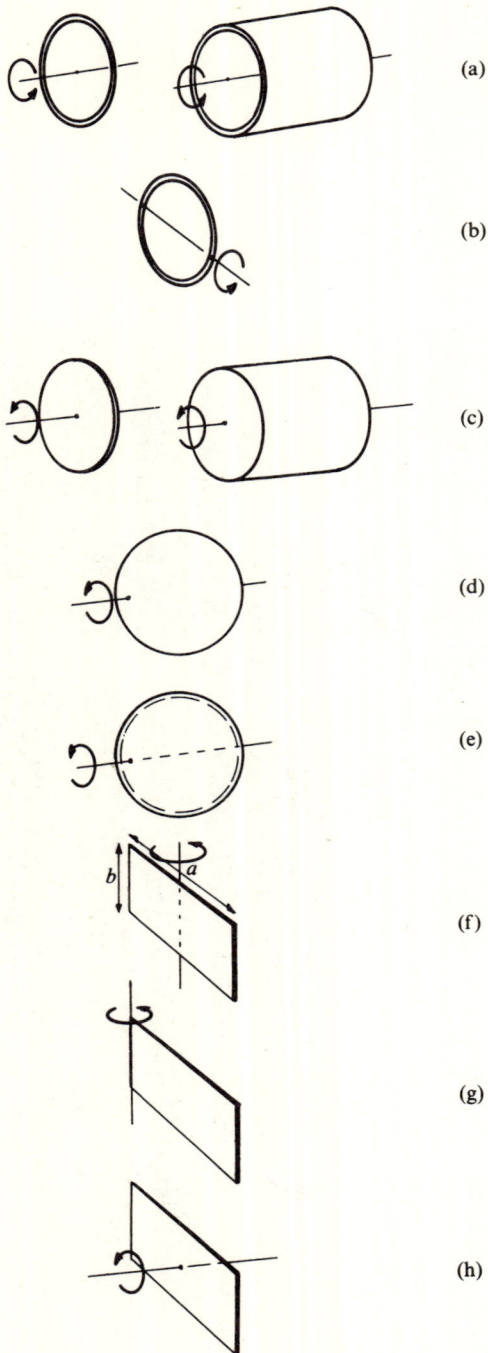

Description	Axis	Moment of Inertia
thin ring (or cylindrical shell) radius r mass m	through centre, perpendicular to plane of ring (or to plane containing edge of shell)	mr^2
thin ring radius r mass m	diameter of ring	$\frac{1}{2}mr^2$
uniform disc (or solid circular cylinder) radius r mass m	through centre, perpendicular to plane of disc (or to plane face of cylinder)	$\frac{1}{2}mr^2$
uniform sphere radius r mass m	through centre	$\frac{2}{5}mr^2$
thin spherical shell radius r mass m	through centre	$\frac{2}{3}mr^2$
rectangular lamina length a width b mass m	through centre, parallel to edge of length b	$\frac{1}{12}ma^2$
	edge of length b	$\frac{1}{3}ma^2$
	through centre, perpendicular to plane of lamina	$\frac{1}{12}m(a^2+b^2)$

(a)
(b)
(c)
(d)
(e)
(f)
(g)
(h)

are indistinguishable, and in any situation where they are *not* indistinguishable it is doubtful whether the centre of gravity has much significance anyway. We shall have cause to refer to the centre of mass again; it is a most useful concept in more advanced work.

Angular momentum

Equation (1) contains a surprising amount of information; take for instance a situation for which $T = 0$, that is, one for which there is no net torque on the body. This means that

$$I\mathrm{d}\omega/\mathrm{d}t = 0,$$

in other words the quantity $I\omega$ is constant. The corresponding statement for translational motion is that in the absence of any net force, mv is constant; linear momentum is conserved. By analogy $I\omega$ is called the *angular momentum*. Angular momentum remains constant providing there is no net external torque on the body. [An alternative name for angular momentum is *moment of momentum*, because $I\omega = \omega\sum\delta m r^2 = \sum\delta m(\omega r)r = \sum r\delta m v$, where δm is moving at speed v with respect to the chosen axis. Each term in this sum can therefore be thought of as the product of the momentum of an element of mass δm and its distance from the axis of rotation.] If there *is* an external torque, then somewhere else there is another object causing it and experiencing an equal and opposite torque, and the angular momentum change of the second body is equal and opposite to that of the first, just as in the case of linear momentum. We can as a result formulate the *law of conservation of angular momentum* as follows. *The total angular momentum of a system of bodies about any given axis remains constant in spite of any interaction between the bodies, providing they are not acted on by an external torque.*

One familiar consequence of the law is that the rate of rotation of the earth remains very nearly constant, providing us with a remarkably reliable clock. It is not *quite* constant (see p. 9), partly because the earth is not a perfectly rigid body and partly because it experiences a small external torque. The gradual lengthening of the day is mainly due to tidal friction, as a result of which kinetic energy of rotation is gradually converted into internal energy ('heat'). The forces responsible for the tides are discussed on p. 294.

Varying the moment of inertia

The law of conservation of angular momentum, like other conservation laws we have met, turns out to be of fundamental significance in its own right. What happens for example if we change I while the body is rotating, without touching anything else? Our derivation of the principle has not covered this possibility because we have always talked in terms of a rigid body, but it is occasionally very strange how some statements turn out to have a validity beyond their original terms of reference. In fact the angular momentum conservation principle is valid even when I is changing; what happens is that ω also changes to keep the quantity $I\omega$ constant. Watch a first-class ice skater working up to a rapid spin about a vertical axis; his first move is to achieve a much slower rotation with arms and one leg splayed out to get a large moment of inertia. Then when he straightens up and brings the whole of his body as near to the axis of rotation as possible, the resulting decrease in moment of inertia brings about a spectacular abrupt increase of angular velocity. It is difficult to avoid the impression that he has stabbed sharply at the

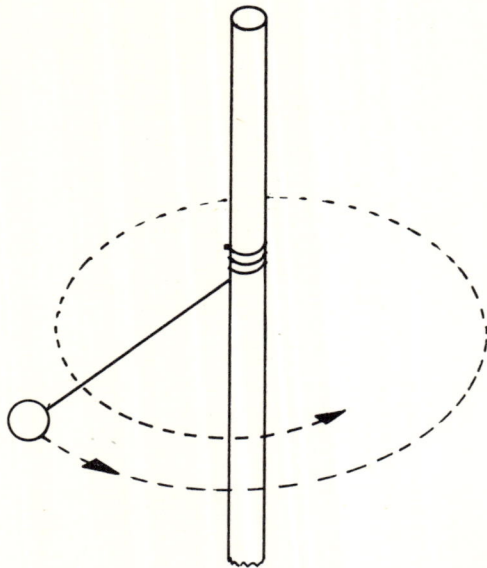

Figure 19.10

ice to spin faster, but that is a totally false notion; his increase in angular velocity all stems from the fact that $I\omega$ must remain constant. A high diver performing somersaults uses the same effect. He takes off from the board with a relatively small angular velocity, then tucks in limbs and head to reduce his moment of inertia, producing a significantly faster spin. Given the necessary skill, he can control the orientation with which he enters the water by straightening out at the correct instant, consequently reverting to a small angular velocity. We can find it easier to understand what is happening if we think of the situation in Figure 19.10; the mass is whirling round on the end of a string which winds on to the pole as the motion proceeds. Energy considerations tell us that the speed of the stone remains constant, since the tension in the string acts at right angles to the direction of motion and therefore does no work (we are ignoring air resistance and other dissipative forces). As the radius gets smaller therefore, the angular velocity must increase.

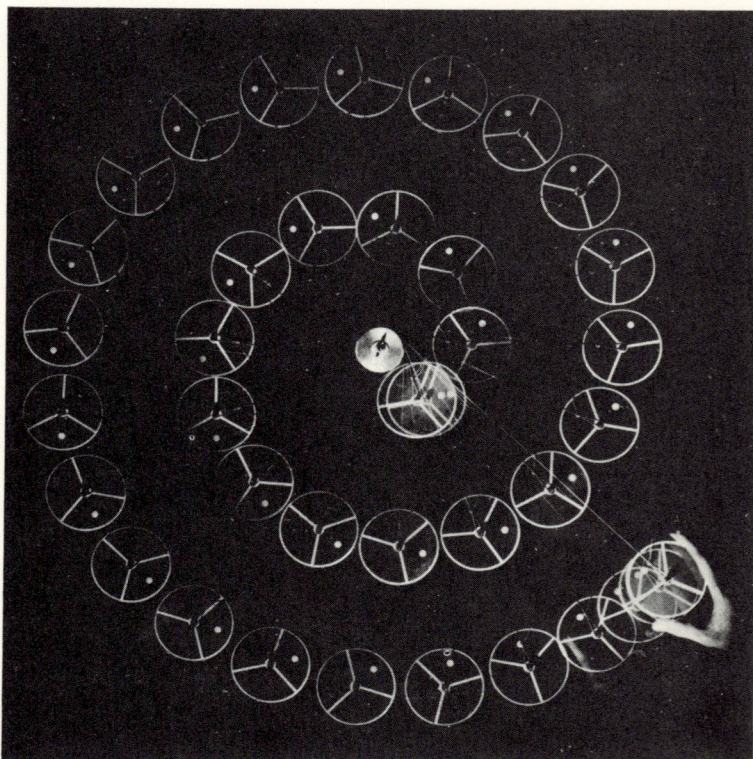

Angular momentum conservation has an important effect on winds. A mass of 'still' air at the equator naturally moves round with the land mass, at a speed of about 1700 km per hour with respect to the centre of the earth. By contrast, the air at the poles is stationary. If a pressure gradient causes air to commence moving north in the northern hemisphere therefore, it retains much of its angular momentum and is soon moving east faster than the land underneath it. Similarly a mass of air coming from the north veers west. For this reason, when a region of low pressure appears, the wind direction is not towards the centre of the 'low', as we might expect, but in a spiral round it (Figure 19.11).

Rotational kinetic energy

There is one more close correspondence between linear and rotational motion which we must consider, and this confirms again the role of moment of inertia as the quantity analogous to mass. Going back to our constituent point mass δm distance r from the fixed axis of rotation, we now wish to write down its kinetic energy:

$$\text{k.e.} = \tfrac{1}{2}\delta m \times (\text{speed})^2 = \tfrac{1}{2}\delta m r^2 \omega^2.$$

Energy is a scalar of course, and that means we can find the kinetic energy of the whole body merely by adding all the terms like this one.

$$\text{Total rotational k.e.} = \Sigma \tfrac{1}{2}\delta m r^2 \omega^2 = \tfrac{1}{2}\omega^2 \Sigma \delta m r^2,$$

since we are dealing with a rigid body and all the parts must go round with the same angular velocity. So,

$$\text{rotational k.e.} = \tfrac{1}{2}I\omega^2,$$

and the analogy with $\tfrac{1}{2}mv^2$ is clear. Rotational k.e. provides a useful approach to many problems, particularly those involving a moving axis of rotation, as we shall see later.

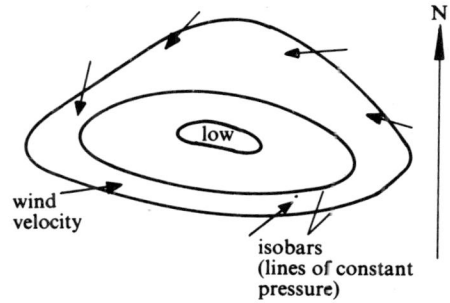

Figure 19.11

General pattern of winds around a cyclone (region of low pressure) in the northern hemisphere.

Table 19.2

Comparison of formulae for linear and rotational motion

Linear	Rotational
Displacement x	Angle θ
Velocity $v = \dot{x}$	Angular velocity $\omega = \dot{\theta}$
Acceleration $a = \ddot{x}$	Angular acceleration $\alpha = \ddot{\theta}$
$v = u + at$	$\omega = \omega_0 + \alpha t$
$x = ut + \tfrac{1}{2}at^2$	$\theta = \omega_0 t + \tfrac{1}{2}\alpha t^2$
Mass m	Moment of inertia I
Linear momentum mv	Angular momentum $I\omega$
force F	torque T
$F = ma$	$T = I\alpha$
Kinetic energy $\tfrac{1}{2}mv^2$	Rotational kinetic energy $\tfrac{1}{2}I\omega^2$

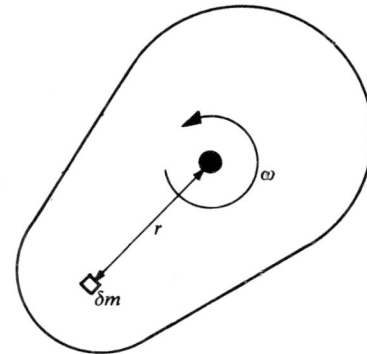

Kinetic energy of the small mass δm is $\tfrac{1}{2}\delta m r^2 \omega^2$.

Exercise 19.8

Find the energy stored in a flywheel consisting of a uniform disc of mass 4 kg and radius 10 cm rotating at 1200 revolutions per minute.

Measurement of moment of inertia of a body about a fixed axis

To find the moment of inertia of a body with a shape too complicated to allow easy calculation, we simply have to subject it to a torque acting about the required axis and measure the angular acceleration which results. The precise way in which this is done depends on the type of body concerned. Figure 19.12 illustrates a direct method suitable for a flywheel or similar piece of machinery arranged to rotate about an axis through the centre of mass; if this method is to be used for an axis *not* through the centre of mass then some arrangement with a vertical axis of rotation must be devised. The mass m is attached to one end of a cord wound round the axis of the flywheel; the other end is hooked over a peg projecting from the axle. The idea is that when the mass is allowed to fall the string will not slip off the axle until all of it has paid out, so that the tension in the string is responsible for a torque on the flywheel right up to the last moment. Then the mass drops to the floor and the flywheel slows down and stops. We could carry out the analysis using $T = I\alpha$, but we shall make use of energy to provide a quicker route. If the mass falls through a height h before the string is detached, it loses an amount of potential energy mgh up to the time the string becomes

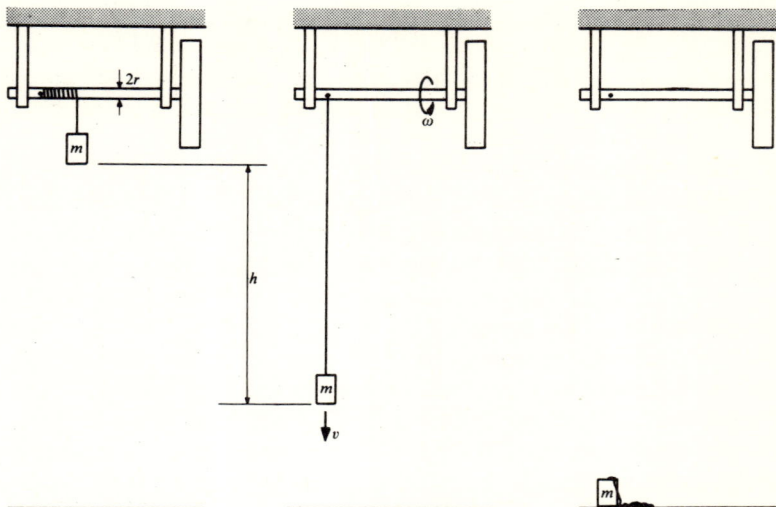

Figure 19.12

detached. This energy is used in three ways. First, the mass acquires kinetic energy $\frac{1}{2}mv^2$ where v is the velocity it has just before it drops off. Second, the flywheel picks up rotational kinetic energy $\frac{1}{2}I\omega^2$, ω being the maximum angular velocity attained. The quantities ω and v are directly connected because they both depend on how the string is paid out; clearly

$$v = r\omega, \qquad \ldots(10)$$

where r is the shaft radius (strictly, the sum of the radii of shaft and string if the string thickness cannot be neglected). Finally, because this is a real situation, there are energy losses on account of friction in the bearings, air resistance and so on. Though we cannot do much about air resistance, we can allow for friction in the bearings; we suppose that an amount of work W is lost per revolution in this way (as long as the frictional forces are reasonably independent of speed we are justified in regarding W as a constant). As a result we can write

$$mgh = \tfrac{1}{2}mv^2 + \tfrac{1}{2}I\omega^2 + n_1 W, \qquad \ldots(11)$$

where n_1 revolutions of the axle occur while the mass is still attached.

Once the string has dropped off, the energy $\frac{1}{2}I\omega^2$ is dissipated at the rate of W per revolution, so that

$$\tfrac{1}{2}I\omega^2 = n_2 W, \qquad \ldots(12)$$

where n_2 additional revolutions occur before the flywheel comes to rest. We also need to measure the time t taken for these n_2 revolutions. Assuming that the angular velocity decreases linearly from ω to zero we can write

$$n_2 = 2\pi \times (\text{average angular velocity}) \times t$$
$$= 2\pi \times \tfrac{1}{2}\omega \times t$$
$$\omega = n_2/\pi t. \qquad \ldots(13)$$

The four equations (10)–(13) contain four unknown quantities v, ω, W and I, and therefore enable us to find I in terms of the measurements available.

Exercises 19.9–11

9 Substitute equations (10) and (12) into (11) to remove v and W, and then use (13) to eliminate ω and obtain the expression

$$I = \frac{n_2 m}{(n_1 + n_2)}\left[\frac{2\pi^2 ght^2}{n_2{}^2} - r^2\right].$$

10 Show that n_1, r and h cannot be chosen independently of each other and find the relation between them.

11 It is also possible to measure the time t_1 taken for the first n_1 revolutions. Show that t_1 and n_1 are related by the expression $\omega = n_1/\pi t_1$. This equation can be used as an alternative to equation (13); find the corresponding expression for I.

Oscillations of a rigid pendulum

This provides an alternative way of measuring the moment of inertia about a non-symmetrical axis, such as the one through O in Figure 19.13. The object is freely pivoted from O and given a slight angular displacement θ. The only external force having a moment about the axis through O is the weight mg, so using the general equation

$$T = I\ddot{\theta}$$

we can write

$$mgl\sin\theta = -I\ddot{\theta},$$

where l is the distance between O and the centre of mass, and I is the appropriate moment of inertia. The minus sign indicates that the torque is acting so as to reduce θ. If we keep θ small we can put

$$mgl\theta = -I\ddot{\theta},$$

so that we have here a new case of simple harmonic motion, with period $2\pi\sqrt{(I/mgl)}$. (See p. 214: a distinction must be drawn here between the angular velocity $\dot{\theta}$ and the quantity ω referred to in Chapter 15, where $\omega/2\pi$ is the *frequency of oscillations*.)

We can take the argument a stage further however. If the radius of gyration of the body about an axis through its centre of mass is k, its moment of inertia about the axis is mk^2, and the parallel-axis theorem tells us that

$$I = mk^2 + ml^2.$$

The expression for the period therefore becomes $2\pi\sqrt{[(k^2 + l^2)/gl]}$. This varies with l in quite an interesting fashion; it has a minimum value when $l = k$, and though it increases only slightly for larger values of l it gets very big indeed when l is very small (Figure 19.14). Of course if l is zero the object is pivoted at the centre of gravity and there is no restoring torque at all.

Exercise 19.12

A certain metre rule has a mass of 120 g. Assuming that its width and thickness are negligible compared with its length, calculate

a) its moment of inertia about an axis passing through one end
b) the period of small oscillations if it is pivoted at one end
c) the mark at which it should be pivoted to give the most rapid oscillations, and the magnitude of the corresponding period
d) the period for small oscillations if it is pivoted at the 40 cm mark.

Torsional pendulum

Figure 19.15 shows an object suspended from a torsion wire so as to be able to rotate in a horizontal plane; we shall call the appropriate moment of inertia I. If it is given an angular displacement θ from equilibrium it experiences a restoring torque T given by

$$T = c\theta, \qquad \qquad \text{...(14)}$$

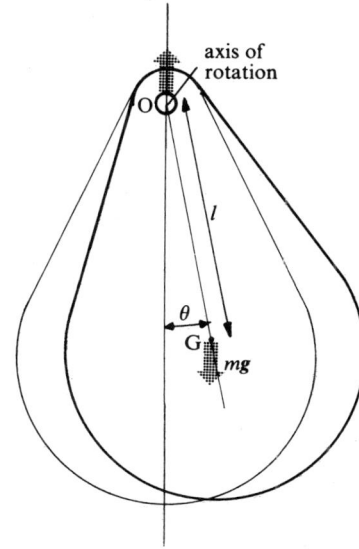

Figure 19.13

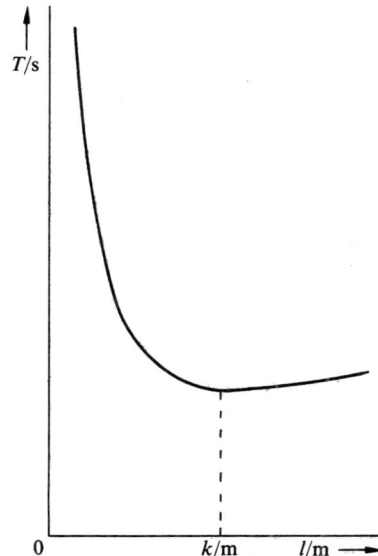

Figure 19.14

Graph of T against l for a metre rule pivoted a distance l from its centre and performing small-angle oscillations of period T.

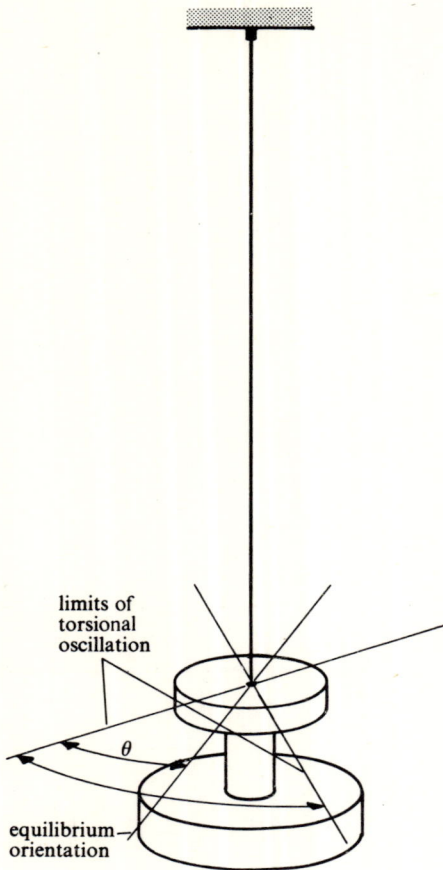

Figure 19.15

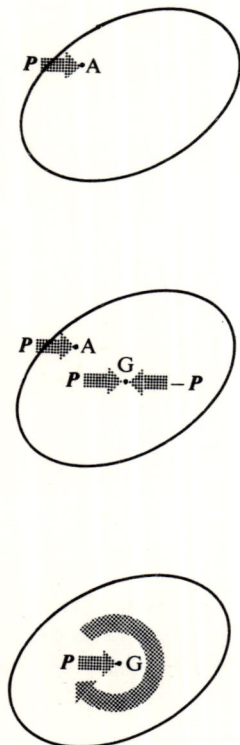

Figure 19.16

where c is the torsional constant for the wire (p. 146). Equation (14) is simply the appropriate form of Hooke's law. The equation of motion is

$$c\theta = -I\ddot{\theta}$$

which is another case of simple harmonic motion, with period $2\pi\sqrt{(I/c)}$. There is no restriction to small oscillations this time; so long as equation (14) still holds θ can be quite large.

Rotation about a moving axis

Hit a ball off-centre and it spins as it moves off. A single force applied to an extended object always produces a combination of translational and rotational acceleration if its line of action does not pass through the centre of mass, and we should like to be able to calculate how much translation and how much rotation occurs in any given case. The key to the problem lies in the fact that we can replace the single force by a rather more complex system of forces of our own choosing. Figure 19.16 illustrates the essential idea. Force P is the single force, acting at point A, and we imagine added to the system two forces P and $-P$ at G, the centre of mass. Because these additions are equal and opposite and have the same line of action, they cannot alter the translational or rotational motion.

However, now that we have three forces instead of one we can combine them in a very useful fashion. The force at A and the force $-P$ at G form a couple, producing pure rotation, and the single force P left over acts through G and therefore produces pure translation. We have succeeded in separating the translational and rotational effects of the original force, and what is more we have done it in such a way that both can be calculated without too much difficulty. In fact we can see that the translational acceleration of the centre of mass is that which would be produced by a force P acting at the centre of mass, and the rotational acceleration about the centre of mass is that which would be produced by a couple with moment equal to the moment of the original force P about an axis through the centre of mass. The way it works out in practice is illustrated in the following example.

Example

A uniform cylinder of mass m, radius r, starts from rest and rolls down an incline making angle β with the horizontal. Find the acceleration of the axis of the cylinder down the plane and the velocity after it has lost height h.

The forces on the cylinder are shown in Figure 19.17; they can in principle be combined into a single force P but we shall not need to go to that trouble, because in order to meet our target of replacing P by an equal force at G plus an appropriate couple we need just three pieces of information, all of which can be obtained without locating P precisely. First, we need to know the direction of P. Well, we can be sure from our previous argument that any translational motion occurs in the direction of P, and we can be equally sure that the translational motion is down the incline. Secondly, we need its magnitude. This too presents no problem; P after all is the resultant of the forces F, R and mg shown, and if its direction is down the plane then its magnitude must simply be the sum of the resolved components of F, R and mg down the plane. This gives

$$P = mg\sin\beta - F.$$

Incidentally we can be sure that there is no acceleration perpendicular to the plane, so

$$mg\cos\beta - R = 0.$$

Finally, we need the moment of P about an axis through G. But P is the force which entirely replaces F, R and mg; its moment about an axis through G must be equal to the sum of the moments of these three forces about the same axis. So

$$\text{moment of } P \text{ about } G = Fr.$$

Now we have enough information to replace P by our equivalent system of a force acting through G and a couple; the force producing translational acceleration of the centre of gravity down the plane is $mg\sin\beta - F$, and the moment of the rotating couple is Fr. For the translational part we use Newton's second law:

$$mg\sin\beta - F = m\frac{dv}{dt}, \qquad \text{...(15)}$$

where v is the velocity down the plane. For the rotational component of the motion we need to use $T = I\alpha$, giving

$$Fr = I\frac{d\omega}{dt}, \qquad \text{...(16)}$$

where I is the moment of inertia of the cylinder about the axis through G and ω is the angular velocity.

Both equations (15) and (16) can be integrated with respect to t to give v and ω respectively:

$$v = (g\sin\beta - F/m)t + c, \qquad \text{...(17)}$$

$$\omega = \frac{Frt}{I} + k, \qquad \text{...(18)}$$

and constants c and k are both zero if we start the cylinder from rest, since v and ω must be zero when $t = 0\,$s. Now we have to use the fact that v and ω are not independent. We get the connection between them by writing down, in two different ways, the velocity of that point B on the circumference which is momentarily in contact with the incline. For a start we can say that the velocity of B with respect to the incline = the relative velocity of B with respect to G + the velocity of G with respect to the incline. This is of course a vector equation and we must be careful about directions; the relative velocity of B with respect to G is just ωr up the incline, and the velocity of G is v down the incline. So

$$\text{velocity of } B = \omega r - v.$$

But the point B in contact with the plane is momentarily at *rest*; Figure 19.18 shows the path traced out by B as the cylinder rolls down the plane–it is called *a cycloid*. So

$$\omega r - v = 0, \qquad \text{...(19)}$$

and equation (18) can therefore be written as

$$\frac{v}{r} = \frac{Frt}{I}. \qquad \text{...(20)}$$

Elimination of v from equations (17) and (20) gives

$$\frac{Fr^2t}{I} = (g\sin\beta - F/m)t$$

that is

$$F = mg\sin\beta\left(\frac{I}{mr^2 + I}\right), \qquad \text{...(21)}$$

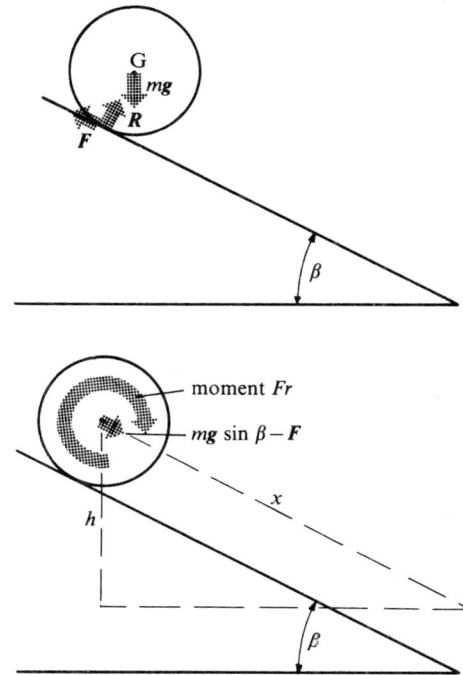

Figure 19.17

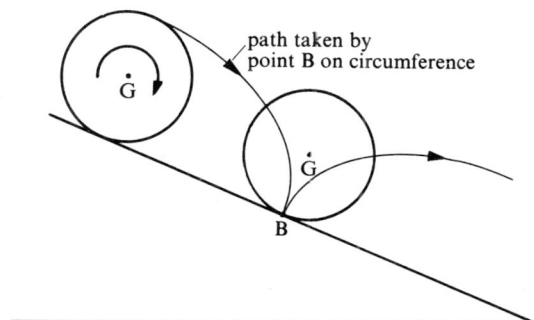

Figure 19.18

and using this value of F in equation (15) we finally get that the acceleration dv/dt is given by

$$\frac{dv}{dt} = g\sin\beta\left(\frac{mr^2}{mr^2 + I}\right). \qquad \ldots(22)$$

The moment of inertia for a cylinder is $\frac{1}{2}mr^2$ so that

$$dv/dt = (2g\sin\beta)/3.$$

Since this represents a constant acceleration we can use

$$2ax = v^2 - u^2$$

to get the velocity v a distance x down the plane, and x is related to the vertical drop h by

$$x = h/\sin\beta$$

giving

$$v^2 = 4gh/3.$$

Exercises 19.13, 14

13 Fill in the missing algebraic steps leading to equations (21) and (22).

14 Show that for a rolling sphere of mass m and radius r

$$dv/dt = (5g\sin\beta)/7, \qquad v^2 = 10gh/7$$

and that for a rolling cylindrical pipe of mass m, radius r and negligible thickness, $dv/dt = \frac{1}{2}g\sin\beta$, $v^2 = g\sin\beta$.

Combination of rotational and translational kinetic energy

Though the foregoing example is instructive, the method is rather long-winded, and energy considerations provide a far quicker route. Very conveniently it turns out that the kinetic energy of a rolling body such as we have been considering is simply the translational kinetic energy of an identical body moving with the same velocity v but not rotating, added to the rotational kinetic energy it would have if it were rotating at the same angular velocity ω but about a fixed axis (a proof of this proposition appears in Appendix 9). This means that as the cylinder drops through a vertical height h and loses potential energy mgh it picks up kinetic energy $\frac{1}{2}mv^2 + \frac{1}{2}I\omega^2$.

So ignoring frictional losses,

$$mgh = \tfrac{1}{2}mv^2 + \tfrac{1}{2}I\omega^2,$$

and making use of the arguments leading to equation (19) we obtain

$$v = \omega r,$$
$$\text{giving } v^2 = 2ghmr^2/(mr^2 + I) \text{ immediately.}$$

Exercise 19.15

Two uniform solids, with equal radii, one a sphere and the other a cylinder, are rolling on a plane horizontal surface, each with velocity $2 \cdot 0\,\mathrm{ms}^{-1}$. Find the vertical height each can climb up an incline.

Centre of percussion

In Figure 19.19 we have a rod, free to move, acted on by a force F for a short time δt. If the force is applied at the mid-point (Figure 19.19a) the rod simply moves off to the right without rotating, and in particular the end A also moves forward. If on the other hand F

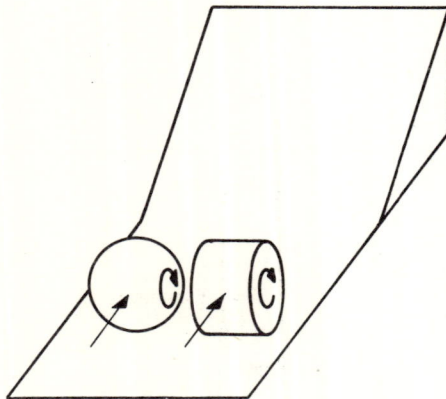

(Exercise 19.15)

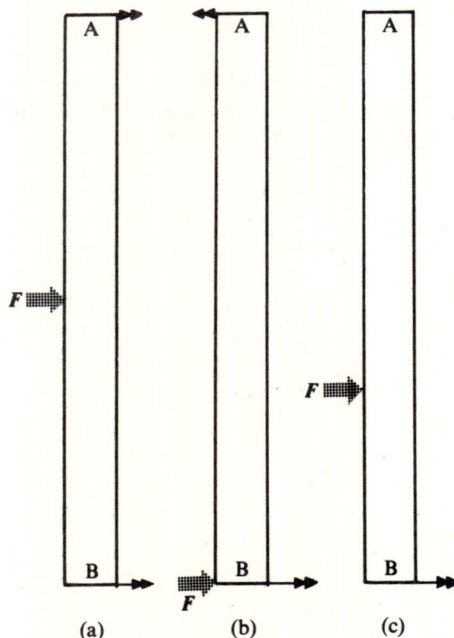

Figure 19.19

is applied at one end B, the combination of translation and rotation which results causes A to start moving to the left. There must be some point on the rod at which we can apply F so that, at least for a short time, A remains at rest. Figure 19.20 shows the problem generalised; the force F is applied at point Q a distance x from G the centre of mass, and as a result the point P a distance y from G on the other side remains stationary for a short time; it is the *instantaneous centre* of rotation. The first step in finding the connection between the positions of P and Q is to replace F by an equal force F at the centre of mass and an anticlockwise couple of moment Fx. The translational impulse is then $F\delta t$, and this gives the point G a velocity v dictated by

$$F\delta t = mv, \qquad \ldots(23)$$

where m is the mass of the body. In an analogous way the couple Fx acting for a time δt produces an angular velocity ω given by

$$Fx\delta t = I\omega, \qquad \ldots(24)$$

where I is the moment of inertia about an axis through G.
Now velocity of point P = relative velocity of P with respect to G

$$+ \text{ velocity of } G$$
$$= -\omega y + v$$

and we require this to be zero. Equation (24) can therefore be written in the form

$$Fx\delta t = Iv/y,$$

and if we use equation (23) to eliminate $F\delta t$ we obtain

$$mvx = Iv/y \quad \text{or} \quad xy = I/m.$$

The point Q is called the *centre of percussion* for the point P. If the body in question is a cricket bat gripped at P, then Q is the best point of impact for the oncoming cricket ball, in as much as it produces least jar to the hands. Figure 19.21 illustrates another everyday example.

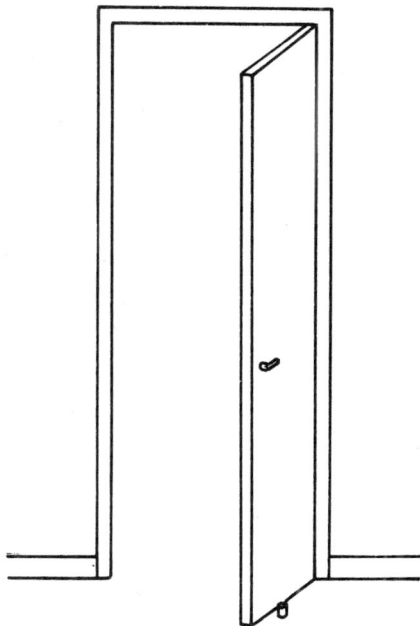

(a)

(b)

Figure 19.20

(a) Q is the centre of percussion for point P; (b) shows the equivalent force system used in the calculation.

Figure 19.21

Where is the best place for a doorstop?

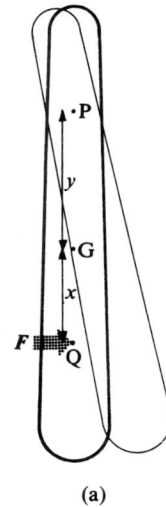

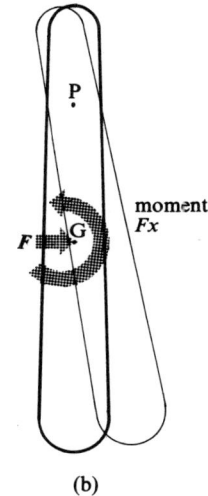

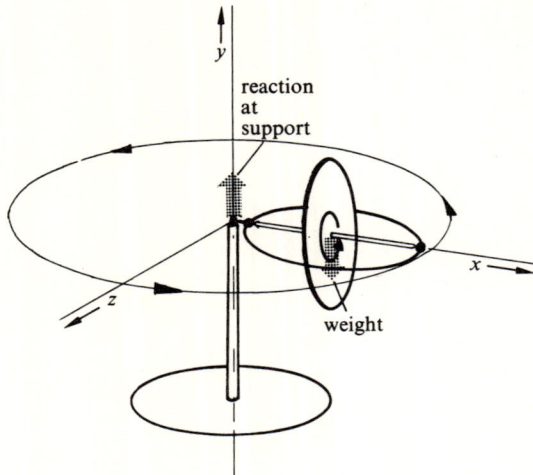

Figure 19.22

Qualitative description of the motion of a gyroscope

We conclude this chapter by looking at one of the less credible consequences of $F = ma$ applied to rotational motion. When the disc of a toy gyroscope is spun rapidly and the gyroscope is supported at one end, the other end *precesses*, that is, it rotates slowly round the path shown in Figure 19.22, apparently defying gravity. We cannot deal fully with this problem, which involves a study of rotation in three dimensions, but we can at least get a qualitative picture of how such behaviour comes about.

There is no denying that the external forces on the gyroscope in the diagram constitute a couple exerting a torque about the z-axis, and if the disc were not spinning the top would certainly fall off. Now, in order to get the disc spinning rapidly about the x-axis in the first place, a torque had to be applied about that axis – either a large torque for a small time, or a small torque for a long time. If the top were to behave as we instinctively expect it to behave, that is, fall off the support, it would have to end up spinning about the y-axis, and it cannot do so unless we supply a torque about that axis to provide the necessary angular momentum. We have no such torque; the torque we have is one which provides angular momentum about the z-axis, and so the system obediently *acquires* angular momentum about the z-axis; the plane of the disc commences to rotate into the xy-plane. What we are saying is that the couple needed to alter the axis of rotation of the spinning disc is far greater than the one we should need for a stationary disc, and, what is more, the plane in

Gyroscopic compass.

which it must act is different. The faster the disc spins the more important the effect becomes; naturally there is *some* tendency for the top to fall off, but there is far more tendency for it to precess. If the spinning top is positioned as shown in Figure 19.22 and released, it experiences a rather wobbly precession known as *nutation*.

Further exercises

Exercises 19.16–31

16 A bicycle wheel has a diameter of 0·5 m, a mass of 0·8 kg and a moment of inertia about its axle of $4·0 \times 10^{-2}$ kg m². Assuming $\pi = 22/7$ find the values of the following quantities when the wheel rolls, at 7 rotations per second without slipping, over a horizontal surface:

a) the angular velocity in radians per second
b) the linear velocity of the centre of gravity
c) the instantaneous linear velocity of the topmost point on the wheel
d) the total kinetic energy of the wheel.   (JMB)

17 Figure 19.23 represents a light rod of length 1 m with masses, each of small dimensions, distributed as shown. It is freely pivoted at the mid-point O. Find:

a) the moment of inertia of the system about O
b) the magnitude of the vertical force applied at the end B which just keeps the rod horizontal
c) the moment of the couple acting on the rod, when this force is removed
d) the angular acceleration with which the rod starts to rotate when the force at B is removed
e) the angular velocity with which it passes through the vertical position.

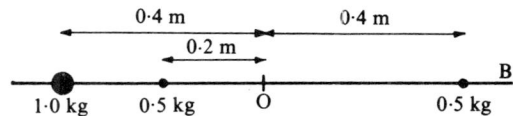

Figure 19.23 (Exercise 19.17)

Find the period of small oscillations of the rod about its (vertical) position of stable equilibrium when it is freely pivoted about a horizontal axis through its mid-point O.
(You may assume the formula $T = 2\pi\sqrt{(I/Mgh)}$ for the period, where I is the moment of inertia about the axis of suspension, M the mass of the whole system, and h the distance of the axis of suspension from the centre of gravity.)   (O)

18 How much kinetic energy is stored in a flywheel of moment of inertia 0·2 kg m² rotating at 1200 revolutions per minute? Find the value of the constant retarding couple which would bring the flywheel to rest in 5 minutes. Discuss briefly whether or not the flywheel is losing energy at a constant rate during this period.
           (O)

19 Define moment of inertia and explain the significance of this quantity in the description of the rotational movement of a rigid body. Explain briefly how you would determine the moment of inertia of an irregular solid about a given axis.

The moment of inertia of a solid flywheel about its axis is 0·1 kg m². It is set in rotation by applying a tangential force of 20 N with a rope wound round the circumference, the radius of the wheel being 0·10 m.

Calculate the angular acceleration of the flywheel.

What would the acceleration be if a mass of 2 kg were hung from the end of the rope?   (O & C)

20 A flywheel which can turn about a horizontal axis where there is a constant frictional couple is set in motion by means of a 0·5 kg mass which is hung from a light string wrapped around an axle of radius 12·5 mm. The mass falls from rest and the string unwinds without slipping so that the flywheel makes 10 revolutions before the string becomes detached from the axle. The flywheel then makes a further 50 revolutions before it finally comes to rest. If the velocity of the mass at the instant of detachment is 0·20 m s^{-1}, find the moment of inertia of the flywheel about its axis of rotation. (JMB)

21 Masses of 110 g and 90 g are hung over a frictionless pulley by a light inextensible thread. The pulley may be treated as a circular disc, whose moment of inertia about its axis of rotation is $\frac{1}{2}ma^2$, where m is the mass of the pulley and a its radius. If the mass of the pulley is 30 g, what will be the acceleration of the masses, and the tension in each vertical portion of the thread? (WJEC)

22 What do you understand by (a) angular momentum, (b) the law of conservation of angular momentum, and (c) moment of inertia?

A horizontal disc of radius r is pivoted freely about a vertical axis through its centre; its moment of inertia about this axis is I. A small insect A of mass M and a smaller insect B of mass m are initially at diametrically opposite points on the periphery of the disc; both insects and the disc are initially at rest. Insect A then walks round the periphery of the disc with a constant speed v relative to the disc until it reaches B, which remains at rest on the disc. A then eats B and returns to its starting point along the original path with the same speed relative to the disc. Find the angular velocity of the disc during the outward and return journeys and the net angular displacement of the disc from its initial position. (O & C)

23 Four point masses, each equal to m, are held by a rigid framework of negligible mass at the corners of a square of side l. Find the moment of inertia of the system (a) about a diagonal of the square, and (b) about an axis parallel to a diagonal which passes through only one of the masses.

The system is pivoted about a horizontal axis as described in (b) and is initially held at rest with all the masses in a horizontal plane. It is then released and swings freely under the action of gravity. What are the speeds of the masses when they pass through the position in which they are vertically below the axis? Suggest briefly a method for measuring their speeds and so checking your calculation. (O & C)

24 A uniform spherical ball starts from rest and rolls freely without slipping down an inclined plane at 10° to the horizontal along a line of greatest slope. Calculate its velocity after it has travelled 5 m. (The moment of inertia of a sphere about a diameter = $2Mr^2/5$.) (O & C)

25 Define *moment of inertia* and name a suitable unit in which it may be measured. What is the relationship in rotational motion which corresponds to '*force = mass × acceleration*' in linear motion?

Explain, in general terms and without numerical calculation, why it is reasonable that the moment of inertia of a solid cylinder about its own axis should be $\frac{1}{2}mr^2$, and that of a sphere about a diameter only $\frac{2}{5}mr^2$, when that of a thin cylindrical tube about its own axis is mr^2, where m and r represent the mass and radius in each case.

Find an expression for the velocity attained after a loss of height h by a solid sphere of radius r and density ρ which rolls without slipping down a uniform slope.

If the following pairs of objects were released simultaneously at the top of such a slope, which object, if either, would reach the foot of it first? Give reasons for your conclusions.

a) two steel balls of radii 1 cm and 2 cm;
b) a steel ball and a steel cylinder, of equal masses and radii;
c) a steel ball and a lighter aluminium ball of equal radius.

(SUJB)

26 A toy car contains a relatively massive flywheel which is geared to the rear wheels so that it revolves ten times faster than they do. In such a car of total mass 120 g, the flywheel has a moment of inertia of 50 gcm^2 and each of the four 'road' wheels are 1·0 cm in diameter and have a moment of inertia of 5·0 gcm^2. The moments of inertia are about the relevant axles in each case. Ignoring friction, find the speed of the car after travelling 1·0 m from rest down a slope of $\sin^{-1}(1/5)$ without slipping.   (JMB)

27 A horizontal disc is spinning freely about a vertical axis through its centre at 90 revolutions per minute. A piece of plasticine, of mass 20 g, is dropped vertically on to the rotating disc, to stick to it at a distance of 11 cm from the axis of rotation. The rate of revolution changes to 70 per minute. Calculate the moment of inertia of the disc about its axis of rotation.   (WJEC)

28 A small meteorite of mass m travelling towards the centre of the earth strikes the earth at the equator. Assuming that the earth is a uniform sphere of mass M and radius R, show that the length of the day is consequently increased by approximately $5mT/2M$ s, where T is the duration of the day, and that the rotational energy of the system is decreased by approximately $2\pi^2 mR^2/T^2$. How do you account for this loss of rotational energy?

What mass of meteorites and meteoric dust falling uniformly on the earth with no net angular momentum would increase the length of the day by one thousandth of a second?

(The moment of inertia of a uniform sphere of mass m and radius r about a diameter is $2mr^2/5$. Mass of earth $= 6\cdot0 \times 10^{24}$ kg.)   (O & C)

29 A gravity conveyor consists of two parallel bars at an angle of 5° to the horizontal, between which are mounted rollers of diameter 5 cm and moment of inertia $2\cdot5 \times 10^{-3}$ kgm^2 about their axes, evenly spaced at ten rollers per metre length of conveyor. The rollers, initially at rest, have rough surfaces, but may turn without friction on fixed bearings mounted on the bars, and their axes are horizontal and perpendicular to the bars. A long flat-bottomed box is placed at the top of the conveyor, exerting pressure on eight rollers, and is released so that it starts from rest and runs down under gravity. Given that the mass of the box is 40 kg, calculate its initial acceleration, by energy or other considerations.

Explain why the speed which the box can attain on a long conveyor is limited, and why this limiting velocity cannot be calculated by the energy method.

Discuss the effect of launching a second, similar box close behind the first.   (O & C)

30 The turntable of a record player has a moment of inertia of 0·0025 kg m². It is set spinning, with the driving motor disconnected, and its speed is measured at 10 s intervals. At $t = 50$ s a record is dropped on to the turntable and within a fraction of a second they appear to be moving together. From the data in Table 19.3 calculate the moment of inertia of the record and the frictional torques opposing the motion before and after $t = 50$ s.

Table 19.3

t/s	0	10	20	30	40	60	70	80	90	100
Angular velocity/rad s⁻¹	7·75	7·60	7·45	7·30	7·15	4·85	4·70	4·55	4·40	4·25

After the turntable has come to rest the record is removed and a light vertical barrier is fixed radially to the surface of the turntable in contact with the bob of a simple pendulum of mass 20 g, which is at rest in its equilibrium position. The pendulum is displaced and then released so that the bob strikes the barrier normally at 15 cm from the turntable axis with a velocity of 2·39 m s⁻¹. The pendulum rebounds with a velocity of 1·62 m s⁻¹. Calculate the initial angular velocity imparted to the turntable.

(o & c)

31 Figure 19.24 represents a cycle wheel with spokes of negligible mass and with a heavy rim of mass 10 kg. The wheel, of radius 0·30 m, is free to rotate about an axle of small diameter, and is supported with this axle horizontal by a boy sitting in a chair which is at rest but which can rotate freely about a fixed *vertical* axis. The wheel is then set in rotation by a constant torque of moment 4 N m applied for 3 s.

a) Find the angular velocity and the kinetic energy of the wheel when this torque is removed.

b) Explain why the chair begins to rotate about its own axis when the boy turns the axis of rotation of the wheel towards the vertical.

c) Does the magnitude of the angular velocity of the wheel about its axis change while the boy is turning this axis towards the vertical? If not, what is the source of the kinetic energy acquired by the boy and the chair? Explain.

d) When the boy has turned the axle of the wheel into a vertical position and the chair is still rotating, he stops the rotation of the wheel about its axis by pressing the rim against his chest. Will the chair be brought to rest or not? Explain. (o)

Figure 19.24 (Exercise 19.31)

**APPROXIMATE TIME-SCALE OF EVENTS IN THE
DEVELOPMENT OF THE SUBJECT OF GRAVITATION
SINCE A.D. 1500**

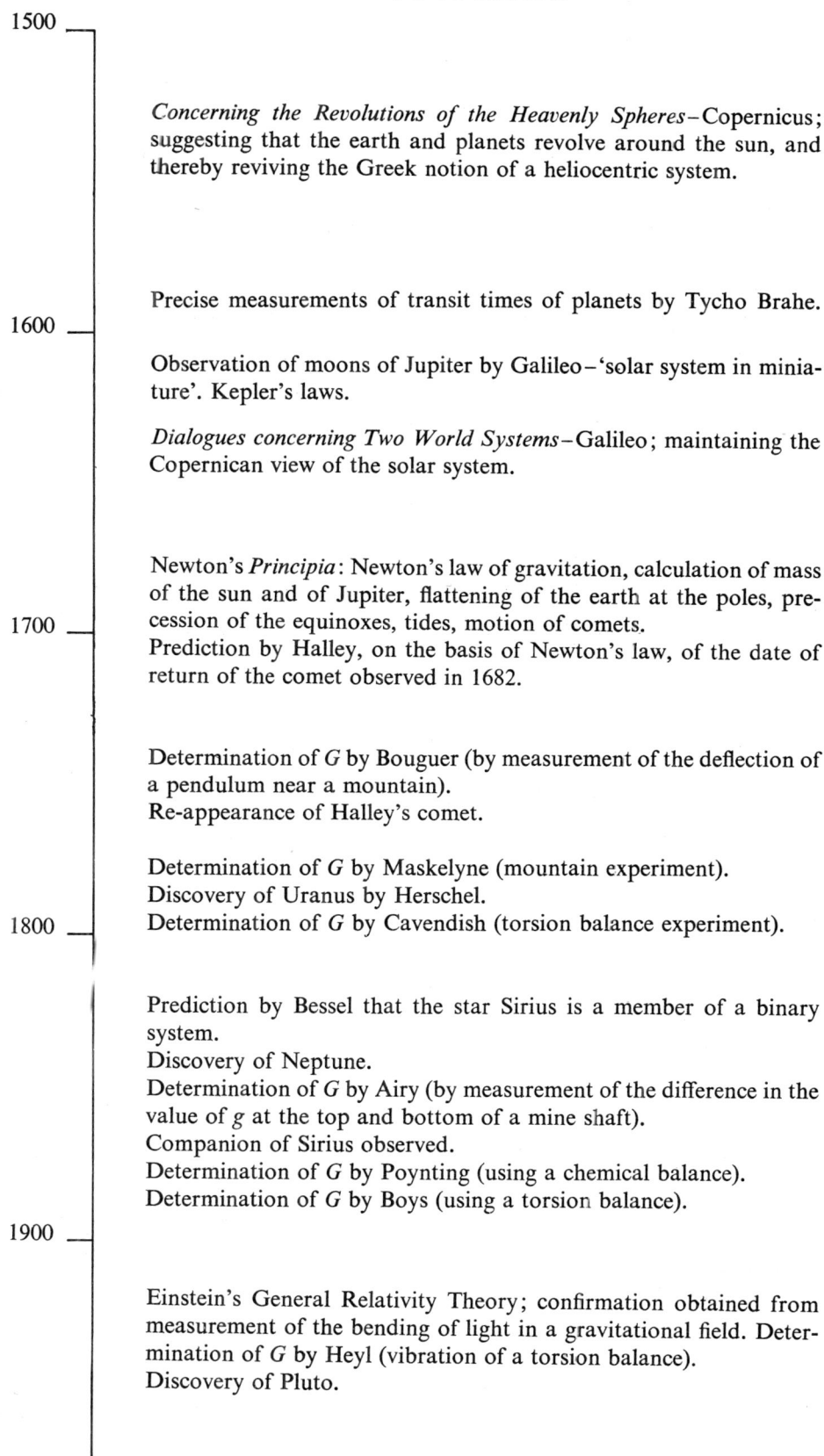

1500

Concerning the Revolutions of the Heavenly Spheres – Copernicus; suggesting that the earth and planets revolve around the sun, and thereby reviving the Greek notion of a heliocentric system.

1600

Precise measurements of transit times of planets by Tycho Brahe.

Observation of moons of Jupiter by Galileo – 'solar system in miniature'. Kepler's laws.

Dialogues concerning Two World Systems – Galileo; maintaining the Copernican view of the solar system.

Newton's *Principia*: Newton's law of gravitation, calculation of mass of the sun and of Jupiter, flattening of the earth at the poles, precession of the equinoxes, tides, motion of comets.
Prediction by Halley, on the basis of Newton's law, of the date of return of the comet observed in 1682.

1700

Determination of G by Bouguer (by measurement of the deflection of a pendulum near a mountain).
Re-appearance of Halley's comet.

Determination of G by Maskelyne (mountain experiment).
Discovery of Uranus by Herschel.
Determination of G by Cavendish (torsion balance experiment).

1800

Prediction by Bessel that the star Sirius is a member of a binary system.
Discovery of Neptune.
Determination of G by Airy (by measurement of the difference in the value of g at the top and bottom of a mine shaft).
Companion of Sirius observed.
Determination of G by Poynting (using a chemical balance).
Determination of G by Boys (using a torsion balance).

1900

Einstein's General Relativity Theory; confirmation obtained from measurement of the bending of light in a gravitational field. Determination of G by Heyl (vibration of a torsion balance).
Discovery of Pluto.

20 Gravitational forces

Nebula in Ursa Major. Gravitational forces are very important in controlling the behaviour of such huge systems.

As far as we can tell, every object in the Universe attracts every other object. This is not the conclusion we would come to from everyday experience; I can feel both my hands being pulled towards the earth but I do not get the impression that they are being attracted to each other at the same time. This is because the size of the attraction depends on the size of the objects concerned, and the tiny force which my left hand exerts on my right is entirely masked by much stronger forces.

Newton's law of gravitation

If we consider two *particles* of matter, that is, objects having sizes which are small compared with how far they are apart, then *the force of attraction between them is proportional to the mass of each and inversely proportional to the square of their separation.* This is *Newton's law of gravitation.* In symbols, the force F between particles of mass m_1 and m_2, separated by a distance r (Figure 20.1), is

$$F = \frac{Gm_1m_2}{r^2},$$

where G is a universal constant, the *gravitational constant.* The measurement of G by various methods is described on pp. 289–92.

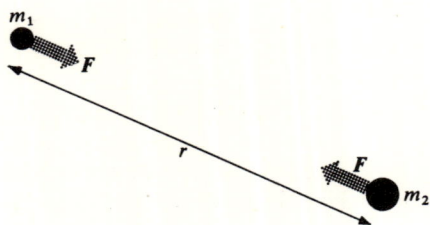

Figure 20.1

Exercise 20.1

Show that G has units $\mathrm{m^3\,kg^{-1}\,s^{-2}}$.

Application to objects of appreciable size

Newton's law can be applied to large objects, but only by treating them as aggregates of particles and adding up all the attractions and cross attractions involved. It would be very convenient if we could replace any object by an equivalent point mass at its centre whenever we wanted to work out the gravitational attraction on it, but generally speaking things are not that simple.

Exercise 20.2

Figure 20.2 shows two dumb-bells consisting of point masses m on the ends of light rods of length a, separated in such a way that the four masses occupy the corners of a square. Show that the cross attraction between A and C is of magnitude $Gm^2/2a^2$ and that this combines with the other cross attraction to give a force along EF of magnitude $Gm^2/\sqrt{2}a^2$. Hence show that the net force on either dumb-bell is $Gm^2(2 + 1/\sqrt{2})/a^2$ and compare this with the value that would be obtained if we could replace each dumb-bell by a mass $2m$ at its centre.

There are however two important cases where an object *can* be replaced by an equivalent point mass at its centre. Objects possessing *spherical symmetry* can be treated in this way – this is demonstrated theoretically on p. 301 – and so can any object which is very close to a *much larger* body, such as an object on the surface of the earth (Figure 20.3). The whole idea of centre of gravity depends on this second case; replacing the object by a point mass at the centre of gravity works only because the net attractive forces on the particles making up the object are all parallel and can therefore be combined according to the rules laid down on p. 67.

Exercise 20.3

Newton estimated the mean density of the earth as $5500\,\mathrm{kg\,m^{-3}}$. Using the assumption that all the mass of the earth acts as though it is concentrated at the centre, estimate the value of G. (Radius of earth $= 6400\,\mathrm{km}$, weight of $1\,\mathrm{kg} \approx 10\,\mathrm{N}$.)

Mt Schiehallion, Scotland. This cone-shaped mountain was used by Maskelyne for a determination of G by measurement of the angle of deflection of a pendulum near the base of the mountain.

Figure 20.2 (Exercise 20.2)

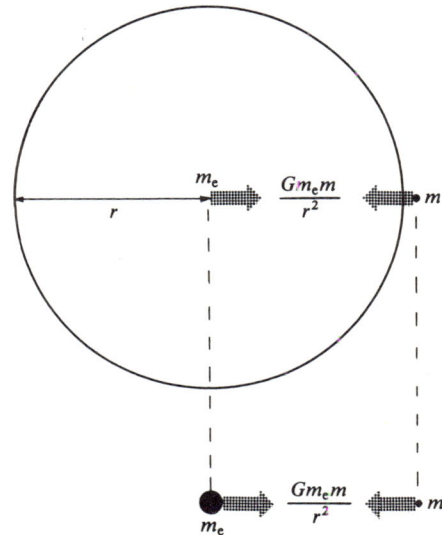

Figure 20.3
The gravitational attraction on the mass m due to the earth is the same as that which it would experience if the entire mass m_e of the Earth were concentrated at its centre.

Evidence for Newton's law: Kepler's laws

It was the close observation of a few dots of light that led to the discovery of Newton's law. The appearance of those dots in the sky at night has been a continual source of wonder and curiosity to man right down the ages, and his attempts to find some order in the way they change positions make a fascinating story. There is first the fact

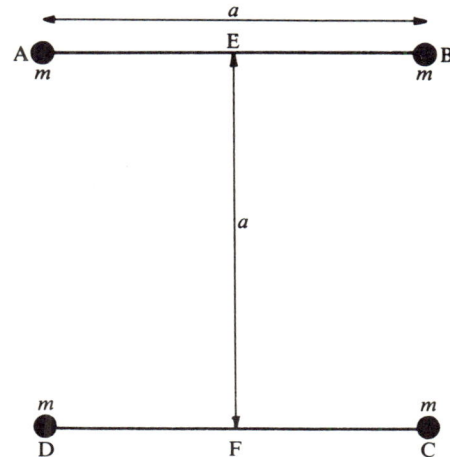

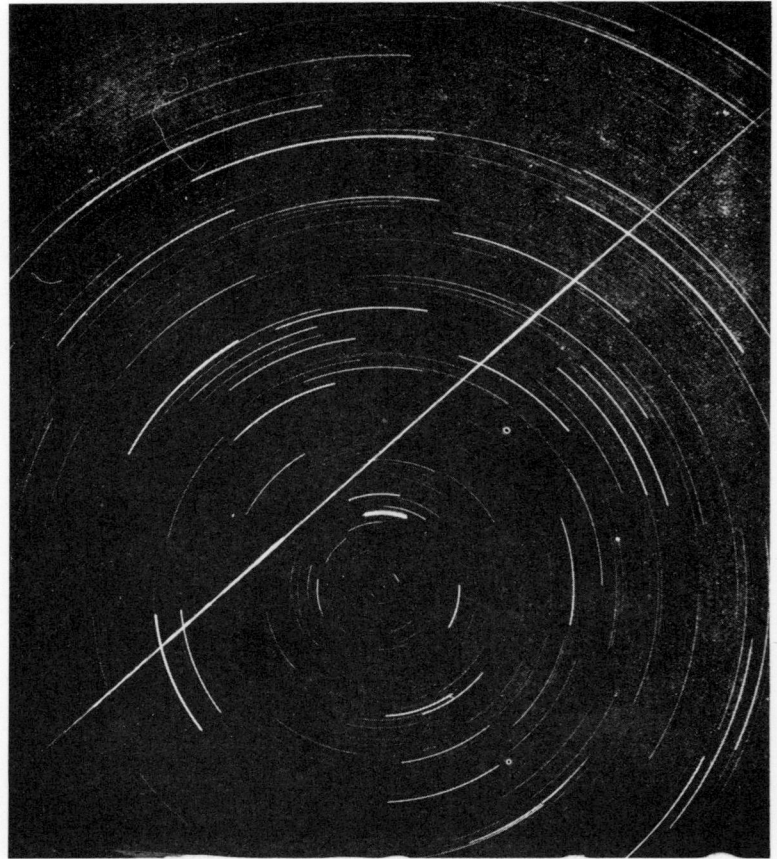

Figure 20.4
Apparent rotation of stars.

that as the night wears on, every one of the points of light appears to move in the arc of a circle, all the circles being concentric and centred near to one particular point of light which we call Polaris, the pole star (Figure 20.4). We might be inclined to think that the pole star exerted some peculiar influence on all the others, but it is simpler to take the view that *rotation of the earth* causes the stars to appear to rotate around it. With so many stars dotted in all directions around us it would be surprising if there was not one star more or less in line with the axis of rotation of the earth, and therefore not appearing to move, and this is how Polaris has come to have its rather special situation in the heavens.

If we want to look for any *relative* motion of the stars with respect to each other, the rotation of the earth is something of a nuisance – we have to 'freeze' it out. Most astronomical telescopes of any size have a so-called *equatorial mounting* for just this purpose; they are mounted on a rotating platform set at such an angle that its axis of rotation is parallel to the polar axis (Figure 20.5). The telescope can then be trained for long periods on any particular star merely by rotating the platform at a rate which compensates for the rotation of the earth.

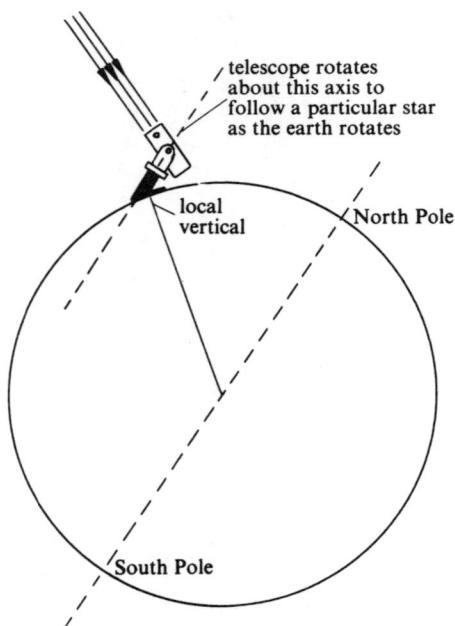

Figure 20.5

Exercise 20.4

At how many degrees per hour should a telescope with equatorial mounting be rotated about its axis in order to produce a stationary

Equatorially mounted telescope. The 98-inch Isaac Newton Telescope at the Royal Greenwich Observatory.

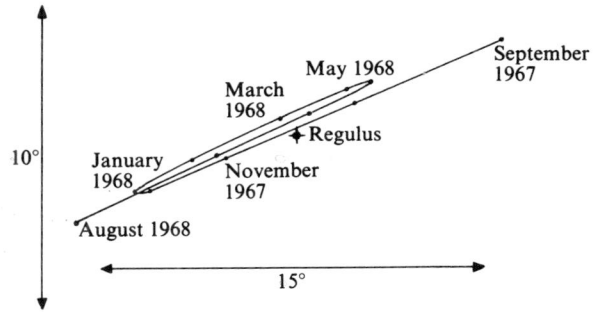

Figure 20.6
Path taken by Jupiter with respect to the fixed star Regulus as seen from the earth.

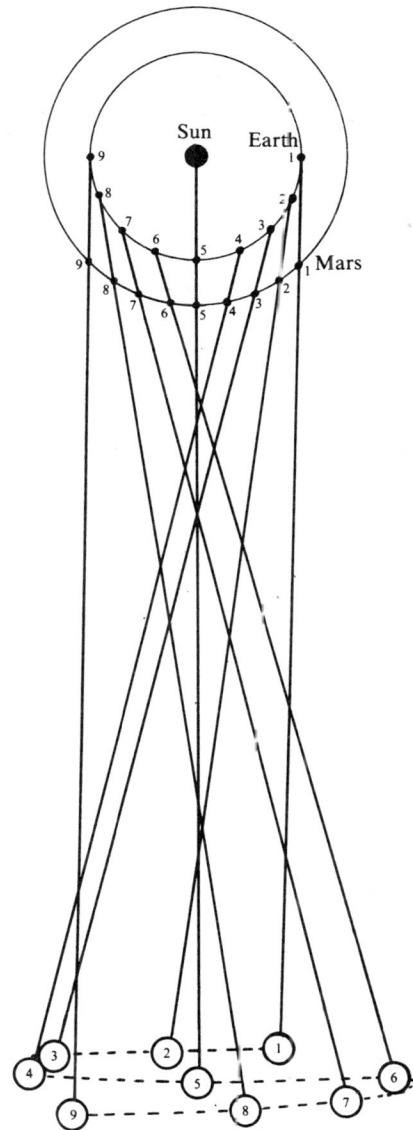

Figure 20.7
Successive positions of Mars as seen from the earth against the background of fixed stars.

pattern of stars in the field of view? What angle does the axis of rotation make with the vertical at latitude X degrees?

Once the rotation has been accounted for we find that with notable exceptions the points of light occupy substantially fixed positions in the sky. A very few stars do show movements which are perceptible over a matter of years, but for most purposes, including that of navigation, they can be regarded as fixed.

In marked contrast there is a mere handful of dots, some of them rather bright, which move around the sky in a far more active fashion: these are the planets. A very modest telescope shows that they have structure, that is, they appear as definite discs, or parts of discs, unlike the fixed stars which remain as points even when viewed by the largest telescopes. Figure 20.6 shows the path taken across the sky by Jupiter between September 1967 and August 1968; this apparently capricious behaviour is quite a common feature of planetary motions. The explanation is simple, though a good deal of careful observation and clear thinking was needed to discover it. The planets are moving around the sun in approximately circular orbits at more or less constant speed, and the occasional apparent back-tracking is caused by the fact that we are observing the situation not from the centre of the circle but from one of the planets, moving in its own orbit like the rest (Figure 20.7). Although the Greeks had suggested that planets might be moving round the sun, it was not until the sixteenth century that the notion began to gain favour again. A Polish monk named Copernicus gets the credit for re-introducing the idea, and Johannes Kepler, working with data collected painstakingly by Brahe, showed that the observed motions of the planets agreed with such an explanation to a high degree of precision. In fact Kepler was able to extract from his data three very simple and significant rules, known as *Kepler's laws.*

1 *The planets all move in elliptical orbits having the sun at one focus.* The foci of an ellipse have several properties; the most homely description we can give of them is in terms of the well-known exercise with two pins and a loop of cotton. The pins are stuck into the page and the cotton looped around them; a pencil point is used to pull it into a triangle. If we now move the pencil around the page in such a way that the cotton stays taut, it will trace out an ellipse, and the foci of the ellipse are the holes made by the pins (Figure 20.8). In fact the orbits of most of the planets are not far from being circles; the foci are quite close together and for many purposes it is a good approximation to think in terms of circular orbits with the sun at the centre.

2 *The radius vector sweeps out equal areas in equal time intervals.* The radius vector is the straight line between sun and planet, and according to this law it does not rotate with uniform angular velocity but sweeps round more rapidly when the planet is nearer the sun (Figure 20.9). We can find a useful mathematical form of Kepler's second law; if the radius vector, of length r, sweeps out a very small angle $\delta\theta$, measured in radians, then the area swept out is very near to that of a triangle of base $r\delta\theta$ and height r (Figure 20.10). The area is therefore approximately $\frac{1}{2}r^2\delta\theta$ and if it is covered in a short time δt then the rate at which the area is swept out is approximately $\frac{1}{2}r^2\delta\theta/\delta t$, which becomes exactly $\frac{1}{2}r^2\,\mathrm{d}\theta/\mathrm{d}t$ in the limiting case as we let δt tend to zero. Now, $r^2\,\mathrm{d}\theta/\mathrm{d}t$ is quite a significant quantity, in fact $mr^2\,\mathrm{d}\theta/\mathrm{d}t$ is the *angular momentum* of the planet, m being the mass. Kepler's second law says in effect then that the *angular momentum* of each planet is constant as it revolves round the sun.

3 *The squares of the periods of revolution of the planets are proportional to the cubes of their semi-major axes* (Figure 20.11). For a strictly circular orbit the semi-major axis becomes simply the radius of the circle. Symbols are a little more manageable than words in this case:

$$T^2 \propto r^3.$$

Kepler's laws provided Newton with a compact summary of the observed motion of the planets, and from them he was able to deduce what sort of force would be needed to provide such motion. In the first place, Kepler's second law shows that the force on the planet must be directed towards the sun at all times, otherwise there would be a net moment about an axis through the sun and we should not expect the angular momentum to remain constant. This in itself was a big step forward; the popular view at the time was that the planets had to be pushed round their orbits, as a car must be propelled

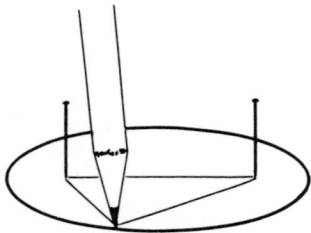

Figure 20.8

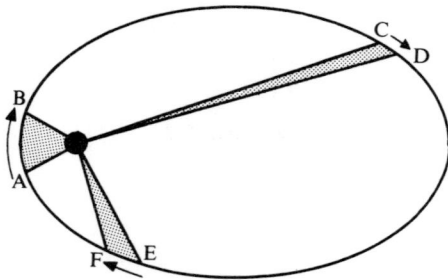

Figure 20.9

Kepler's second law. The three shaded portions all have equal areas; the planet takes equal time intervals to move through sections *AB*, *CD*, and *EF* of the orbit.

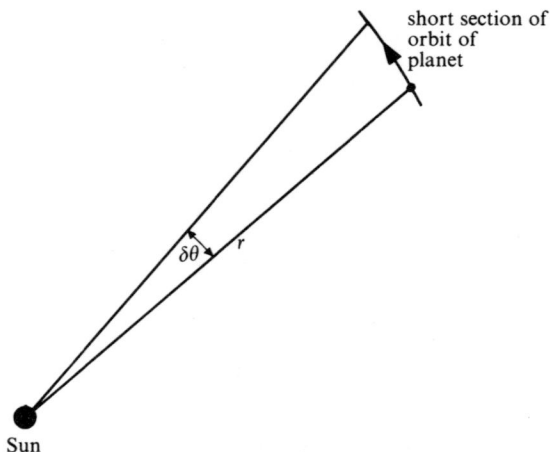

short section of orbit of planet

Sun

Figure 20.10

round a track. Kepler's third law then dictates that the force on the planet must be inversely proportional to the square of the orbit radius. We can work through the argument for the special case of a circular orbit as follows. If the planet is of mass m and travelling in an orbit r at angular velocity ω then the force F required to provide the necessary radial acceleration towards the sun is given by

$$F = m\omega^2 r$$

(see Chapter 14). The period of revolution T is given by

$$T = \frac{2\pi}{\omega},$$

so, writing F in terms of T we get

$$F = \frac{m4\pi^2 r}{T^2}.$$

If then $T^2 \propto r^3$ (law 3),

$$F \propto \frac{1}{r^2}.$$

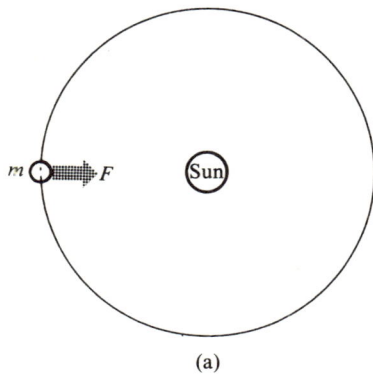

Figure 20.11
Axes of an ellipse.

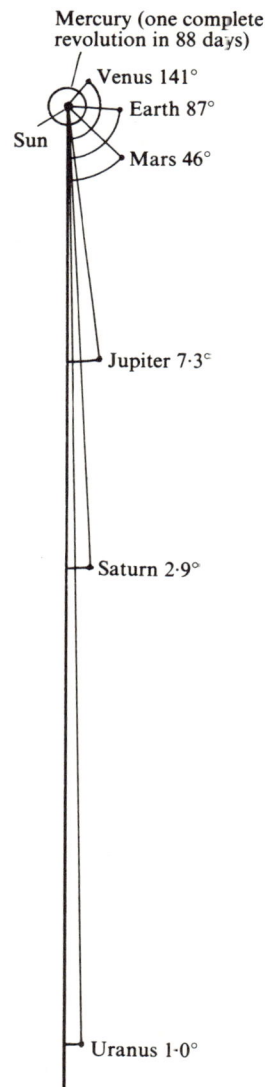

(a)

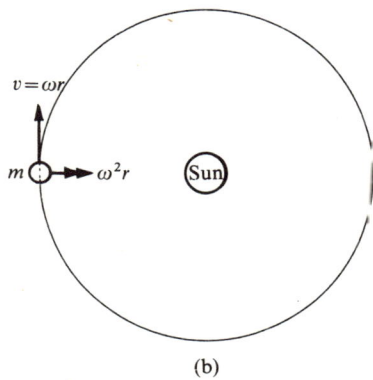

(b)

(a) Force on a planet in a circular orbit around the sun.
(b) Corresponding acceleration.

Exercise 20.5

Use $F = Gm_1m_2/r^2$ to show that if Kepler's third law is written in the form

$$T^2 = kr^3,$$

then

$$k = \frac{4\pi^2}{Gm_s},$$

where m_s is the mass of the sun.

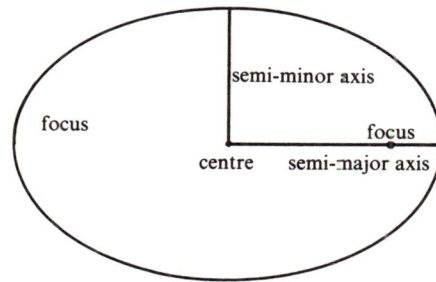

Relative distances of the inner planets from the sun, and the angular progress they make in orbit during one period of revolution of Mercury (approximately 88 earth days).

Newton was then able to show that an inverse square law force directed always towards the sun must necessarily produce an elliptical orbit. If we decided to launch a small artificial planet and wanted an exactly circular orbit for it we should have to be very careful about its velocity at the end of the rocket burn. Not only would it have to be moving perpendicular to its radius vector at that instant, it would also have to have exactly the right speed for the gravitational attraction of the sun to maintain it in a circular orbit. If either the size or direction of the velocity were wrong, the resulting orbit would be elliptical, assuming the planet did not escape from the solar system altogether or crash into the sun, but the mathematics needed to demonstrate that fact is beyond the scope of this book.

Tests of Newton's laws; the motion of the moon

It is one thing to discover a rule which accounts for the behaviour of the planets; it is quite another to claim that the rule applies to every particle in the universe, but this claim has turned out to be largely justified. One famous test of the law in new circumstances concerns the details of the orbit of the moon around the earth. Perhaps this may not seem a very different situation, but the moon is not a planet and if the rule governing the planet–sun system also works for the moon–earth system, this constitutes an extension of the terms of reference of the law. Assuming it is applicable in this case, the attraction F experienced by the moon is $Gm_e m_m/r^2$ where m_e and m_m are the earth and moon masses respectively, and r is the radius of the orbit, which we will take to be circular. If the angular velocity of the moon around the earth is ω, the attraction must be just sufficient to provide an acceleration $\omega^2 r$ towards the earth, in other words

$$\frac{Gm_e m_m}{r^2} = m_m \omega^2 r$$

or

$$Gm_e = \omega^2 r^3.$$

Putting ω equal to $2\pi/T$, where T is the period of revolution, we get

$$Gm_e = \frac{4\pi^2 r^3}{T^2}. \qquad \text{...(1)}$$

If we knew G and m_e we could check whether the orbit radius and period agreed with this prediction. We can in fact find a value for Gm_e if we are also allowed to apply Newton's law to an object on the surface of the earth. Replacing the entire mass of the earth by an equivalent point mass at its centre, we can write the gravitational force on a mass m on the surface as $Gm_e m/r_e^2$, where r_e is the radius of the earth. Of course we know another way of writing the gravitational force on the mass m; it is simply mg. So

$$\frac{Gm_e m}{r_e^2} = mg$$

giving $$Gm_e = gr_e^2 \qquad \text{...(2)}$$

If we use this relationship in equation (1) we get

$$gr_e^2 = 4\pi^2 r^3/T^2. \qquad \text{...(3)}$$

Exercise 20.6

Taking the radius of the earth as 6400 km and assuming the moon to be moving in a circular orbit of radius $3 \cdot 84 \times 10^5$ km, show that equation (3) gives a value for T of just over 27 days, consistent with the observed value.

Perturbations of the planetary orbits

According to Newton's law the planets should be attracted not only by the sun but also by each other, and although these cross attractions will be less important because of the smaller masses involved, they should nevertheless affect the orbits. The deviations from the simple elliptical paths can be calculated, and they agree very well with more precise measurements of planetary motions. In fact this exercise provided the law with perhaps its most spectacular successes. The planet Uranus, at one time the outermost known, did *not* behave properly even when the attractions of the other planets were taken into account. In 1845 Adams, a mathematician, suggested that the disagreement was due to the influence of another planet, as yet unknown, and worked out where it would have to be in order to produce the right effect. Leverrier carried out the calculation independently and sent his prediction to the Berlin Observatory. A detailed star map of the area involved had recently been completed, and when a telescope was pointed in the required direction it was at once apparent that a new body had moved in – the planet Neptune. Later it was found that Neptune itself was not obeying the rules quite as expected, and this led in 1930 to the discovery of yet another planet still further from the sun – Pluto.

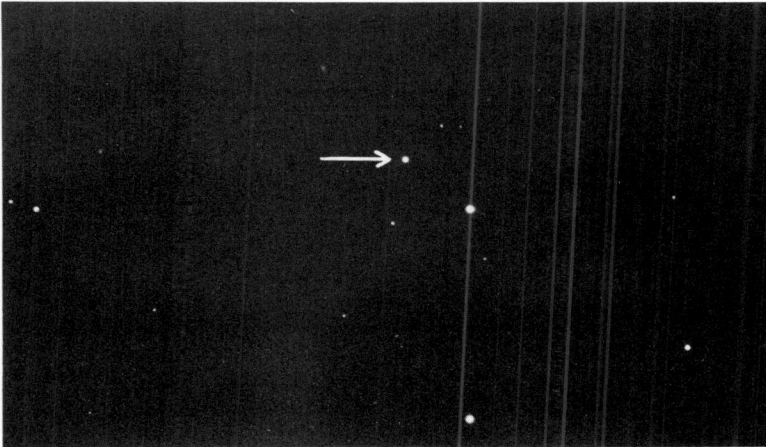

Motion of Pluto (arrowed) against the background of fixed stars.

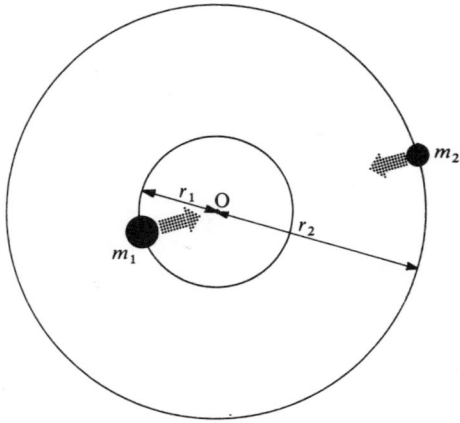

Figure 20.12

Binary systems

There is plenty of evidence to suggest that Newton's law applies well beyond the solar system; we get the clearest indications from the behaviour of a number of double star systems or *binaries*. These consist of two stars, quite close together and revolving around each other. We can work out how such a system should behave on the basis of Newton's laws, and the prediction stands up well to experiment. Usually the stars travel in elliptical orbits, but we shall work out the simplest case involving circular orbits. If the stars have masses m_1 and m_2 and are separated by a distance r they will each experience a force Gm_1m_2/r^2 towards the other. These forces cannot produce identical accelerations since they are acting on different masses, so if we are to have circular orbits they must certainly be of different radii – and they must be concentric otherwise the separation of the stars and the forces on them will change. We can therefore imagine the stars as though they were rotating on the ends of a huge tie-rod; the axis of rotation is not through the centre of the tie-rod but through some point on it which remains fixed, let us say, a distance r_1 from m_1 and r_2 from m_2 (Figure 20.12). We can apply Newton's second law of motion to each star in turn, giving

$$\frac{Gm_1m_2}{r^2} = m_1\omega^2 r_1 \qquad \ldots(4)$$

and

$$\frac{Gm_1m_2}{r^2} = m_2\omega^2 r_2,$$

where ω is the angular velocity of either star; it must be the same for both from our previous reasoning. One immediate conclusion from these equations is that we can put

$$m_1 r_1 = m_2 r_2, \qquad \ldots(5)$$

which fixes the axis of rotation very simply: it is in fact the *centre of mass* of the combination (p. 263).

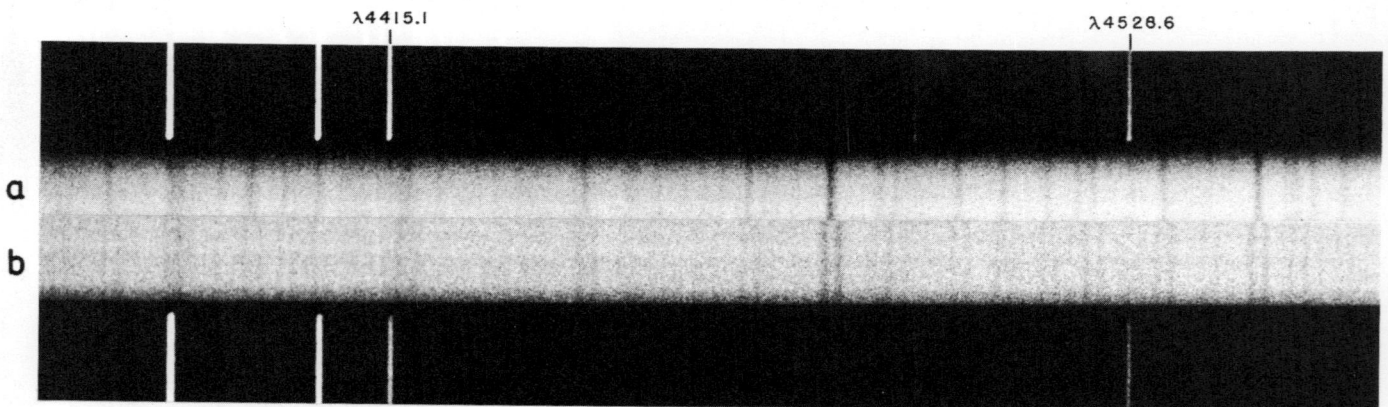

λ4415.1   λ4526.6

a

b

(a) June 11, 1927. (b) June 23, 1927. Each line in the upper spectrum appears twice in the lower one. The light received at the earth from a source moving relative to the earth does not have the same wavelength as the light from a similar terrestrial source (Döppler shift). Since the two stars are rotating around each other they have different relative speeds with respect to the earth, and the light they emit is Döppler-shifted to different extents. The line separation corresponds to an orbital speed of 140kms^{-1}. Why is there no separation in spectrum (a)?

The period of the binary system can be found with a little algebra; we need to be able to write r_1 or r_2 in terms of m_1, m_2 and r. Since

$$r = r_1 + r_2,$$

we can use equation (5) to put

$$r = r_1 + \frac{m_1 r_1}{m_2},$$

giving

$$r = \frac{m_2 r}{m_1 + m_2}.$$

Using this result in equation (4) we get

$$\frac{Gm_1 m_2}{r^2} = \frac{m_1 m_2 \omega^2 r}{m_1 + m_2}$$

or

$$\omega^2 = \frac{G(m_1 + m_2)}{r^3}. \qquad \text{...(6)}$$

Exercises 20.7–9

7 Strictly the earth–moon system should be treated in a similar manner. Find how far the centre of mass of the earth–moon system is from the centre of the earth using the following data:

$$\text{mass of earth} = 6\cdot 0 \times 10^{24}\,\text{kg}$$
$$\text{mass of moon} = 7\cdot 3 \times 10^{22}\,\text{kg}$$
$$\text{radius of moon orbit} = 3\cdot 8 \times 10^5\,\text{km}.$$

8 By comparing equations (6) and (1) estimate the fractional error involved in calculating the period of rotation of the moon when the earth is regarded as fixed.

9 It may be necessary to point out that the centre of mass is *not* the point at which a small mass must be placed in order that the separate gravitational attractions exerted on it by m_1 and m_2 cancel each other. The name for this point is the *neutral* point. Show that the neutral point of the binary system considered is $rm_1^{1/2}/(m_1^{1/2} + m_2^{1/2})$ from m_1 and calculate how far the neutral point of the earth–moon system is from the moon, using the data given in Exercise 20.7.

Determination of *G*: Boys' method

In principle G is very easy to measure. We have only to find two lumps of matter of known mass (preferably of spherical shape to make for easier arithmetic), hold them a known distance apart, measure the force between them and substitute the resulting values into the equation

$$F = \frac{Gm_1 m_2}{r^2}.$$

Unfortunately the gravitational attraction between any pair of objects with conveniently measurable masses is very difficult to detect, let alone measure precisely.

Exercise 20.10

Taking the density of lead to be $11\,000\,\text{kg m}^{-3}$, find the radius of a solid lead sphere of mass 200 kg. Use the value of G obtained by Newton's estimate (Exercise 20.3) to calculate the force expected between two such spheres not quite in contact.

The method used by Boys (1895) is indicated in principle in Figure 20.13; he simply suspended a small dumb-bell from a thin torsion wire and brought two masses sufficiently close to cause a measurable deflection. To understand the design of the apparatus we need to

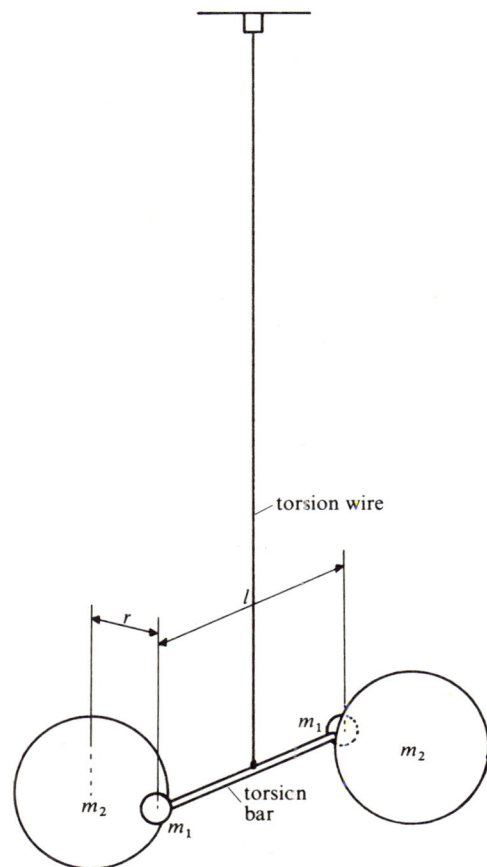

Figure 20.13

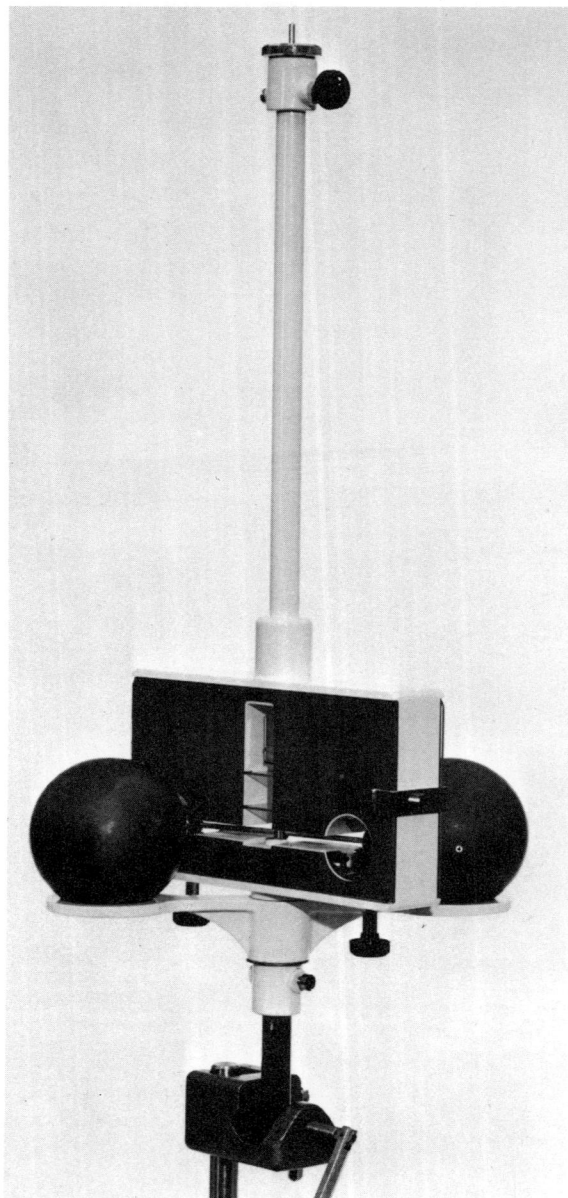

Torsion balance apparatus for determination of G.

look at the way the deflection depends on the various factors involved. If we consider spherical masses m_1 mounted at each end of the suspended horizontal bar (called the torsion bar) of length l, and bring near two spherical masses m_2 with centres distance r from the suspended masses, then the attractive force on each end of the torsion bar is Gm_1m_2/r^2. We arrange the attracting masses in such a way that the forces act in directions perpendicular to the torsion bar; this means we can write the deflecting couple as Gm_1m_2l/r^2, and the steady deflection θ is given by

$$c\theta = \frac{Gm_1m_2l}{r^2} \qquad \qquad ...(7)$$

where c is the torsional constant for the wire. Here we have ignored any attraction experienced by the torsion bar itself.

The main aim is to get θ as large as possible, and a preliminary glance at equation (7) may give the impression that we should go for large masses, a long torsion bar and a wire with as little rigidity as possible. Some of these aims are mutually exclusive however. For a start, large suspended masses need a relatively thick, and therefore stiff, torsion wire to support them. Both Boys and Cavendish, who performed a similar experiment in 1798, opted for small suspended spheres; those used by Boys were of mass 2·6 g.

There is some point in having the other masses large, though here too there is a conflicting requirement; r cannot possibly be less than the sum of the radii of large and small spheres, and is approximately equal to the radius of the larger sphere if the smaller radius can be ignored. This means that the mass m_2 is approximately proportional to r^3, so that all told the attractive force goes as r^3/r^2, or $(m_2)^{1/3}$. As the larger masses have to be moved with precision, however, too large a value is impracticable. Boys settled for 8 kg masses, and made them of cylindrical shape, sacrificing ease of computation in favour of a little more attraction and, more significantly, a simpler shape to form accurately. Of course it pays to use as dense a material as possible; Boys used lead for the large masses and an even denser material, gold, for the smaller ones.

Exercise 20.11

What value of m_2 would Boys have needed to double θ, other things being equal?

It is when we consider the best length for the torsion rod that the real surprise comes. Equation (7) suggests that l should be as long as possible, and in fact when Cavendish performed his experiment, using a silvered copper wire with as low a torsional constant as he could get, he used a wooden lathe about 2 m long for a torsion rod. Anything much longer than that would have run into two major difficulties; the wire would have to be thicker to support the weight, and the period of torsional oscillations of the system, from which c was calculated (p. 269), would become so long that it would be difficult to tell a true vibration from a stray deflection caused by draughts. In practice a period of about five minutes' duration is the largest which can reasonably be measured.

On the other hand, Boys' apparatus looked very different (Figure 20.14). The major improvement that he made depended on the use of a quartz fibre for the torsion wire, combining high tensile strength with low torsional restoring couple. As a result he was able to choose a value of c as low as he wished; it was no longer a limiting factor. In these circumstances it is more useful to replace c in equation (7) by the combination of quantities used to measure it. The period T of torsional oscillations is given by

$$T = 2\pi \sqrt{\left(\frac{I}{c}\right)},$$

where I is the moment of inertia of the torsion bar plus small masses about the axis of suspension, and is therefore approximately equal to $2m_1(l/2)^2$.

This gives
$$c = 2\pi^2 m_1 l^2 / T^2$$
and so equation (7) becomes

$$\theta = \frac{Gm_2 T^2}{2\pi^2 r^2 l}. \qquad \qquad ...(8)$$

Not only has m_1 dropped out of the expression, emphasizing the wisdom of using small suspended masses, but also θ is larger for a *small* value of l. This is not as strange as it looks; what has happened is that though c does not appear explicitly in equation (8) it is most certainly involved implicitly, and varying l without altering T implies varying c, that is, changing to a different torsion wire. The ultimate limit on θ is again set by the maximum measurable value of T, which is about five minutes, and Boys selected a fibre which produced this period with a torsion bar barely 2 cm long. The larger masses were arranged at different heights as shown to minimise the effects of cross attractions, and the whole torsion bar assembly was mounted in a cylindrical vessel which could be evacuated. The deflection θ was measured by making use of the torsion bar as the mirror for a lamp-and-scale arrangement, and the deflection obtained was doubled by placing the larger masses first on one side of the suspended masses and then on the other. The value Boys obtained for G was $6 \cdot 658 \times 10^{-11} \, \mathrm{m^3 \, kg^{-1} \, s^{-2}}$.

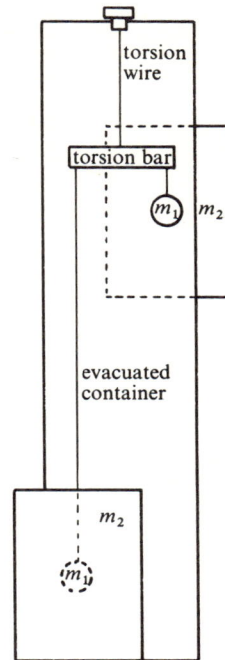

Figure 20.14
Arrangement of masses used by Boys.

Exercise 20.12

Using the values of m_1, m_2, T, l and G given in the text, estimate the value of θ Boys would have obtained. Take the large masses to be lead spheres (density $11\,000 \, \mathrm{kg \, m^{-3}}$).

Dynamic methods of determining G
The equation

$$T = 2\pi \sqrt{\left(\frac{I}{c}\right)}$$

for the period of oscillation of a torsional pendulum is applicable only when the restoring force is provided solely by the torsion wire. If there are any large masses in the vicinity they may well exert gravitational forces which are different at different stages of the oscillation, and produce a consequent modification of the period. The torsion bar indicated in Figure 20.15a for example has a shorter period of oscillation on account of the extra restoring forces provided by the masses m_2, whereas in Figure 20.15b the masses produce a *reduction* in restoring torque. Suitable measurements of the modified periods and the dimensions of the apparatus enable G to be calculated.

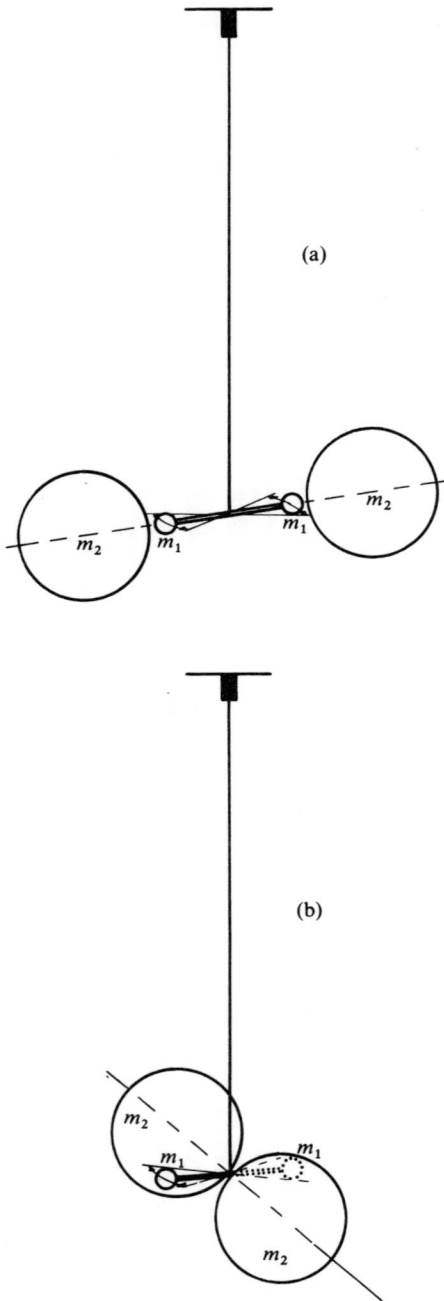

(a)

(b)

Figure 20.15

Exercise 20.13

In Figure 20.16, r is the separation of centres of large and small masses m_2 and m_1 in the equilibrium position, and l is the length of the torsion bar. A small deflection ϕ displaces m_1 to A, and, if l is very much greater than r, AO' can be considered perpendicular to OO'.

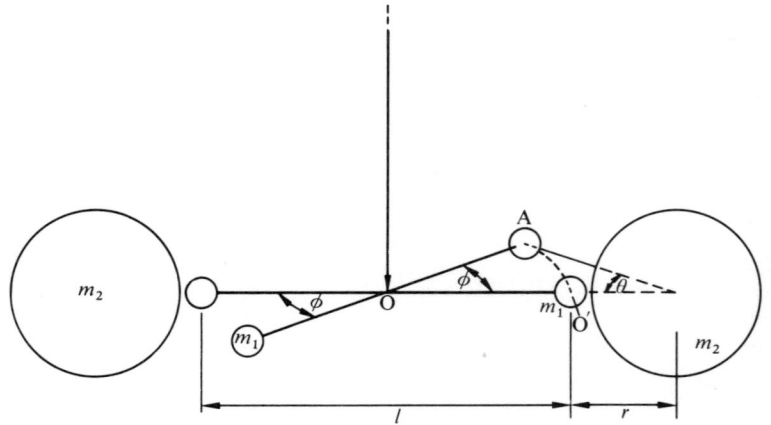

Figure 20.16 (Exercise 20.13)

a) Show that the gravitational attraction on m_1 at A is

$$Gm_1m_2 \cos^2\theta/r^2$$

and that the component of this in the direction AO' is

$$Gm_1m_2 \cos^2\theta \sin\theta/r^2.$$

For small displacements this reduces to $Gm_1m_2\theta/r^2$ approximately.

b) Show that the consequent period of oscillations T is related to the period T_0 in the absence of masses m_2 by

$$\frac{T}{T_0} = \left[\frac{(c + Gm_1m_2l/r^2)}{c}\right]^{1/2},$$

where c is the torsional constant for the suspending wire.

c) Confirm that the resulting increase in period is approximately $T_0Gm_1m_2l/2cr^2$.

Comparison of gravitational and electrical forces
There is a fascinating similarity between Newton's law of gravitation

$$F = \frac{Gm_1m_2}{r^2}$$

and the corresponding law for the force between electric charges q_1 and q_2,

$$F = \frac{1}{4\pi\varepsilon_0} \times \frac{q_1q_2}{r^2},$$

where ε_0 is the permittivity of a vacuum. The electric force can be repulsive or attractive according to whether like or unlike charges are involved, whereas as far as we know there is no such thing as a gravitational repulsion. Although the equations appear very similar, the magnitudes of the forces involved are very different; gravitational forces are very tiny indeed compared with corresponding electrical forces.

Exercise 20.14

Compare the electrical repulsion of two protons separated by a distance r with their mutual gravitational attraction, given the following information:

$$\text{proton mass} = 1\cdot67 \times 10^{-27}\,\text{kg}$$
$$\text{proton charge} = 1\cdot6 \times 10^{-19}\,\text{C}$$
$$\varepsilon_0 = 8\cdot85 \times 10^{-12}\,\text{F}\,\text{m}^{-1}$$
$$G = 6\cdot7 \times 10^{-11}\,\text{m}^3\,\text{kg}^{-1}\,\text{s}^{-2}.$$

In many ways the answer to Exercise 20.14 is surprising, because gravitational attraction dominates our lives, whereas the electrostatic forces we can demonstrate directly in the laboratory are relatively feeble. But comparison between the weight of a charged body and the electrostatic attraction it experiences in an electric field is scarcely fair, not only because the earth has such a large mass but also because even for an isolated body at several million volts potential the excess of positive charge it possesses represents a very tiny fraction indeed of the total positive charge present, most of which is delicately balanced by an equivalent amount of negative charge.

Exercise 20.15

A carbon atom has six orbital electrons and a nucleus comprising six protons and six neutrons, and 1 g of carbon contains about 5×10^{22} atoms. Estimate the force between two lumps of carbon, each of mass 1 g, separated by a distance of 100 m, (a) due to the gravitational attraction, (b) due to electrostatic repulsion, if by some means one electron could be removed from every atom in each lump. (This latter situation cannot be achieved in practice!)

Variations in g on the surface of the earth

In addition to the variation with latitude discussed on p. 199, the magnitude of the acceleration due to gravity varies slightly from place to place owing to local fluctuations in the composition of the crust of the earth and the shape of its surface. One way of prospecting for oil is to look for areas where g is a little smaller than usual; this betrays the fact that an appreciable amount of material below average density is nearby. Lunar astronauts have had to take note of local fluctuations too; regions of high mass concentration (mascons) on the moon have affected orbits to a measurable extent, and must be allowed for in orbit calculations.

Variation of g with height

Even if the earth were a completely homogeneous sphere we should expect g to vary with height above the surface. The force on a mass m at the surface is Gm_em/r_e^2 where m_e is the mass and r_e the radius of the earth, and if we call the corresponding acceleration g_0, we can write

$$mg_0 = Gm_em/r_e^2.$$

At a height h above the surface the acceleration has a different value g given by the equation

$$mg = Gm_em/(r_e + h)^2.$$

So

$$\frac{g}{g_0} = \frac{r_e^2}{(r_e + h)^2}$$
$$= (1 + h/r_e)^{-2} \approx 1 - 2h/r_e.$$

Exercise 20.16

What percentage reduction in g can be expected due to a change in height of 10000 m?

The tides

The old idea that tides are caused by the man in the moon pulling the water round with a fish hook is not a bad one; at least it recognises the link between the periodicity of tides and the motion of the moon. It has one basic snag however; it would mean there should be just one tide per day as the earth revolves under the moon, and there are two. Even when the fish hook is replaced by an inverse square law attraction the difficulty is not immediately resolved. The tide-generating forces are basically due to small differences in the attractive forces which the moon, and to a smaller extent the sun, exert on various parts of the earth. In the argument which follows we shall

The effect of the moon on the earth. These mud flats (Scolt Head Island) have been formed by tidal action.

forget initially the attraction which the earth exerts on a particle at its surface, and also the fact that the earth is rotating on its own axis. A particle m_O at O (Figure 20.17) experiences gravitational attraction towards the moon, but it also *accelerates* towards the moon as it moves in an orbit of radius r around the centre of mass P of the earth–moon system, with a period T of just over 27 days. The attractive force is the centripetal force for the circular motion; it has just the right magnitude to maintain the particle in orbit. A particle m_A at A is also moving in an orbit of radius r with period T. (Remember we are ignoring the rotation of the earth, and as the centre of the earth moves once around the centre of mass P every other point in the earth moves in a similar circle, with a different centre but with the same radius r (Figure 20.18).) The particle m_A also experiences gravitational attraction from the moon; it is slightly different from that experienced by m_O but is very nearly that required to keep m_A in its orbit. (Strictly speaking m_A experiences a very small tide-generating force towards O.) However, if we look at a particle m_B at B, it is definitely *not* experiencing the force it needs to keep in orbit: it is nearer the moon, and is attracted more than m_A. On the other hand the particle m_C at C is further away than m_A; it experiences *less* attraction than it needs to stay in orbit. If m_B or m_C form part of the solid crust they are restrained by interatomic forces, but if they form parts of the sea they tend to move, most of the movement being tangential to the surface, with the result that exact balance of forces and corresponding accelerations are not attained until bulges, of different sizes, are formed in the open sea at E and F. If we now think of the earth as rotating under the bulges with a period of about 24 hours, we can see that at any point on the surface of the sea there are two tides per day. More precisely, the period of rotation of the earth with respect to the moon is 24 hours 50 minutes, so that the tides generated by the moon are about $12\frac{1}{2}$ hours apart. The smaller tides generated by the sun are 12 hours apart; at new moon and full moon, when the sun, moon and earth are approximately collinear, the two effects combine to form tides higher than average (spring tides). The smaller neap tides occur at intermediate times, when the tide-generating forces due to sun and moon are opposed. This happens at half-moon, when the directions of the sun and moon from the earth are 90° apart.

There is a time lag involved in the formation of the bulges, so that they are formed somewhat east of the line joining the centres of earth and moon. This gives rise to the tidal friction torque referred to on p. 265.

The behaviour of tides in a given locality is influenced by the fact that large areas of sea can form *resonating systems*. Stationary waves can be set up with wavelengths perhaps hundreds or thousands of kilometres long, and with resonant frequencies matched by the frequencies of the tide-generating forces.

Gravitational field

We now want to describe a rather different way of thinking about the formula $F = Gm_1m_2/r^2$, one that not only turns out to be very useful but is actually more correct in circumstances where masses are moving rapidly. Instead of regarding mass m_1 as acting directly on mass m_2, we imagine the process split into two stages; we say that mass m_1 produces some sort of influence, called a *gravitational field*, everywhere around it, and that mass m_2, finding itself in this field,

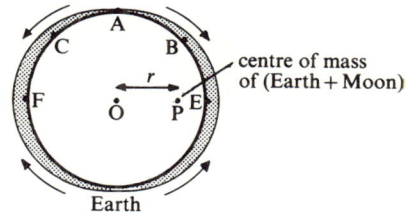

Figure 20.17

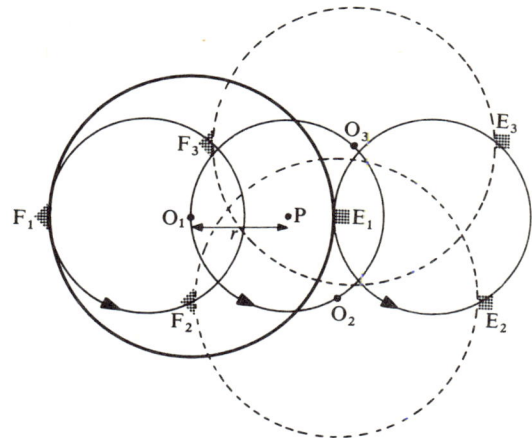

Figure 20.18

The motion of different points in the earth as it revolves around the centre of mass P of the earth–moon system; the rotation of the earth on its own axis is ignored. As the centre O moves through O_1, O_2, O_3, the triangle at F_1 moves through a similar circle F_1, F_2, F_3, and the square at E_1 moves through E_1, E_2, E_3.

experiences a force as a result. Similarly m_2 produces its own gravitational field which affects m_1. Admittedly this is all rather vague, in fact it appears to be just the sort of idea that physicists should reject; where are the operational definitions? We can certainly quote an operational definition for gravitational field, as follows: *the gravitational field strength E at a point is the force per unit mass on a small mass placed at that point.*

In symbols,

$$E = \frac{F}{\delta m} \quad \text{or} \quad F = E\delta m \qquad \qquad ...(9)$$

E is a vector, like the force from which it is defined, and has units $N kg^{-1}$.

Exercises 20.17–19

17 Write down the gravitational attraction on a point mass δm due to a point mass m distance r away and use it to find the magnitude and direction of the gravitational field strength due to a point mass m at a distance r from it.

18 Two stars, each of mass m, are separated by a distance r. Write down expressions for the magnitudes of the gravitational field strengths at the following points:

a) on the straight line joining the centres of the stars, distance $r/3$ from one of them
b) distance r from either star
c) midway between the stars.

19 By considering the symmetry of the contributions made by various parts, write down the magnitude of the gravitational field strength at the centre of a homogeneous sphere a very long way away from all other matter.

Gravitational potential

We are now going to think about the work needed to move a small mass δm around a gravitational field; we shall think in terms of a simple field produced by a point mass m at O (Figure 20.19). Wherever it goes, δm experiences a force F, and the work done is the product of force and distance moved, but because F varies from place to place we shall need to use formulae like

$$W = \int_{x_1}^{x_2} F_x dx,$$

where F_x is the component of the force in the direction of motion. Referring to the figure, suppose we move δm from A to C via B. First we evaluate the work done in getting to B; we need to write down the force F at some typical point D distance r from O:

$$F = \frac{Gm\delta m}{r^2}.$$

Because F is in the direction BA we do not need to worry about components; the work W_{AB} done against the force (we call it 'the work done on the field') is given by

$$W_{AB} = \int_{r_1}^{r_2} \frac{Gm\,\delta m\,dr}{r^2}$$

$$= Gm\delta m\left[\frac{1}{r_1} - \frac{1}{r_2}\right].$$

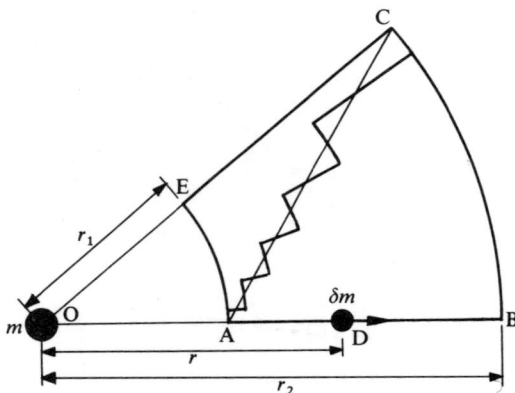

Figure 20.19

Now the work done in going from B to C along the arc of the circle of radius r_2 is zero, because the displacement is perpendicular to the force at every stage. So the work done in getting from A to C via B is $Gm\delta m[(1/r_1) - (1/r_2)]$. It is immediately apparent that the work done in getting from A to C via E is identical, since AE is another arc perpendicular to the force and EC is a line just like AB. What is a little more surprising is that *whatever* path is taken from A to C, the work done is the same. Consider for example the straight line AC as a route. We can think of it in the present context as the limiting case of a zigzag route like a staircase as shown; all the sections perpendicular to a radius require zero work and all the radial sections add up to make up a line equivalent to EC or AB. A field of force such as this, for which the work done between any two points is independent of the path taken, is called a *conservative field*. We have illustrated that this particular field is conservative, and we can show that *every* gravitational field must be conservative otherwise the law of energy conservation is contravened. For suppose X and Y are two points in a non-conservative gravitational field (Figure 20.20) and the work done in taking mass δm from X to Y along path 1, W_1, is different from that needed to go along path 2, W_2. If we go from Y to X along path 2 the field does work W_2 for us; we could waste that work of course but we could also make it do something useful–we could make it move another mass against a gravitational force in some other part of the field. Now if $W_2 > W_1$, then in going from X to Y via path 1 and back again via path 2 we could pick up a net amount of energy $W_2 - W_1$ and therefore use this as the basis of a perpetual motion machine. The inference is that W_2 must be equal to W_1. (Of course if δm were moving through a resistive medium, the work done along a given path would depend on how fast and in which direction the particle was moving, and a net amount of work would have to be done in getting it to travel round any closed path.)

The fact that the work done against the gravitational force is independent of the path has a very useful consequence. Referring to Figure 20.21, suppose it takes $5\cdot0\,\mathrm{J}$ to move a $1\,\mathrm{kg}$ mass from O to P, and $3\cdot0\,\mathrm{J}$ to move it from O to Q. We can immediately conclude that it takes $2\cdot0\,\mathrm{J}$ to move $1\,\mathrm{kg}$ from Q to P, since OQP is a perfectly good route from O to P and must therefore be a $5\cdot0\,\mathrm{J}$ route like any other. This is all quite familiar really; we are simply saying that if the *potential energy* of $1\,\mathrm{kg}$ is W_Q at point Q and W_P at point P, both measured with respect to a zero at point O, then the work needed to get the mass from Q to P is $W_P - W_Q$. The work done per unit mass in getting from Q to P is called the *gravitational potential difference* between the two points, and the potential energy possessed by unit mass at point P with respect to the zero at O is called the *gravitational potential* at P. If we are all to quote the same number for the gravitational potential we must all agree on the same zero, and the zero chosen is a point as far away as possible from all matter, in other words at infinity. The gravitational potential is really another name for the potential energy of unit mass with respect to a zero at infinity; its formal definition is as follows. *The gravitational potential V at a point is the work done per unit mass in bringing a small mass from infinity to that point.* In symbols

$$V = \frac{W}{\delta m} \quad \text{or} \quad W = V\delta m,$$

where W is the work done on mass δm. V is a scalar and has units $\mathrm{J\,kg^{-1}}$.

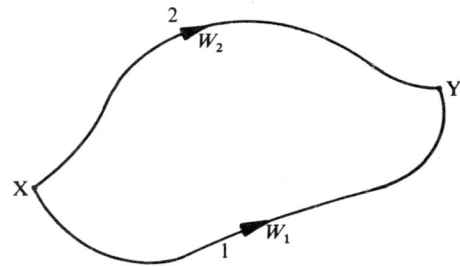

Figure 20.20

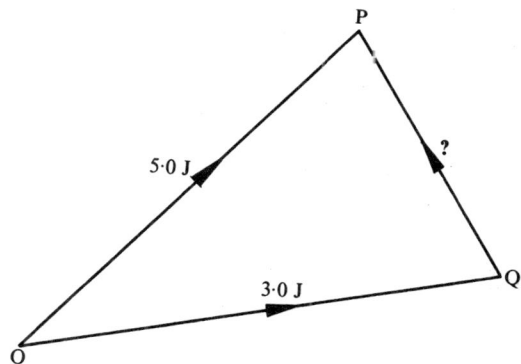

Figure 20.21

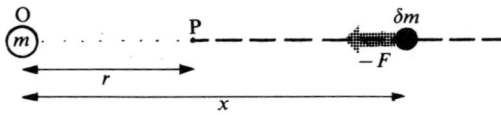

Figure 20.22

Gravitational potential due to a point mass

To evaluate the potential at a point P a distance r from O due to a point mass m at O we need to find the work done in bringing a small mass δm from infinity up to the point P (Figure 20.22). Because the work needed is independent of the path we can pick whatever path we like, so we choose the simplest, the straight line from O through P to infinity. The force on δm distance x from O has magnitude F given by

$$F = \frac{-Gm\,\delta m}{x^2},$$

where we have used the negative sign to indicate that $\boldsymbol{F}$ is in the $-x$-direction.

The work δW done in moving a little closer to P, that is increasing x by an amount $-\delta x$, is therefore

$$\delta W = \frac{Gm\,\delta m\,\delta x}{x^2},$$

and the total work W needed to get from infinity to P is

$$W = \int_{\infty}^{r} \frac{Gm\,\delta m\,\delta x}{x^2} = \frac{-Gm\,\delta m}{r},$$

so that the potential V at P is given by

$$V = -Gm/r.$$

Signs can be a little troublesome in this derivation; the main point is that the force is *attractive* so that there is a *loss* of potential energy in going from ∞ to P. Since all the gravitational forces we know of are attractive, it follows that *all gravitational potentials are negative.* Objection! What about the potential energy of mass m height h above the earth's surface? Surely that is a positive potential energy? It *is* positive with respect to a zero on the surface of the earth, but it is *not* positive with respect to a zero at infinity.

Exercise 20.20

Find the gravitational potential at a point on the surface of the earth, assuming that the earth produces the same gravitational field as an equivalent point mass at its centre. (Mass of earth = $6\cdot0 \times 10^{24}$ kg, radius = 6400 km.)

Equipotential surfaces

The point P in Figure 20.22 is not the only point for which $V = -Gm/r$, in fact this expression applies to every point on the surface of a sphere with centre O and radius r. This sphere constitutes an *equipotential surface*. Clearly no work is done in moving a mass between any two points in the same equipotential surface, and this means that the gravitational field at any point is perpendicular to the equipotential surface through that point. We can draw *gravitational field lines* which show the direction of the field at every point along them; in Figure 20.23 we have drawn two-dimensional representations of field lines and equipotential surfaces, both for a uniform sphere and for a binary star system.

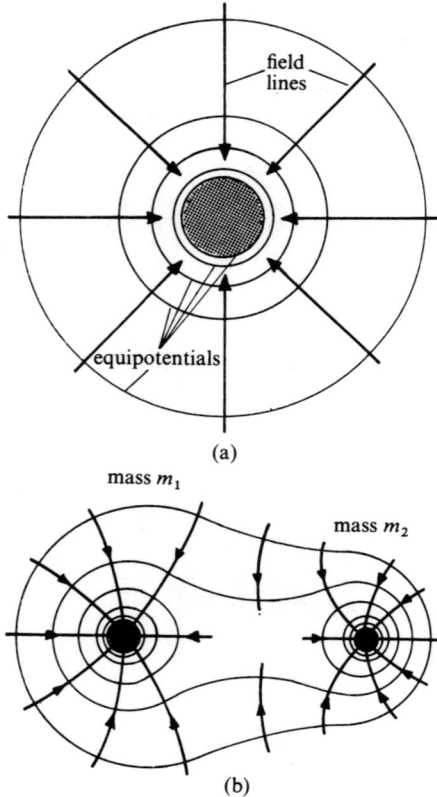

(a)

mass m_1

mass m_2

(b)

Figure 20.23

Gravitational field lines and equipotentials, (a) outside a uniform sphere, (b) for a binary star system with $m_1 > m_2$.

Connection between gravitational field and potential

We shall now derive a general formula linking E with V. Referring to Figure 20.24, the mass δm is in a region where the gravitational field strength is E in the direction shown, and we wish to move it from point O where the potential is V to an adjacent point P a distance δx along the x-axis, where the potential is $V + \delta V$. We shall take P and O to be so close that E does not change appreciably in going from one to the other. The work we need to do is $-E_x \delta m \delta x$ where E_x is the x-component of E; the minus sign is required because E is in fact doing the work for us–we are *gaining* energy. But in terms of the definition of potential we can also write down the work done as $\delta V \delta m$, so that

$$\delta V = -E_x \delta x,$$

and in the limit as δx tends to zero,

$$E_x = -\frac{dV}{dx}. \qquad \text{...(10)}$$

If we had moved in the y-direction instead we would have got

$$E_y = -\frac{dV}{dy} \qquad \text{...(11)}$$

and for the z-direction,

$$E_z = -\frac{dV}{dz}. \qquad \text{...(12)}$$

The potential at a point is a scalar; just one number is required to define it. The gravitational field is a *vector*, and to define its magnitude and direction completely at any point in space *three* numbers are needed: for example, three components E_x, E_y, E_z or the magnitude of E and two angles θ and ϕ (Figure 20.25). But equations (10) to (12) are saying that once we have specified the *potential* at every point in a region, one number per point, we shall find locked up in these numbers information to specify the field–three numbers at each point. We can draw a useful analogy with contour lines on a map. Contour lines, which join points of equal height, are in fact lines of equal gravitational potential, providing local fluctuations in

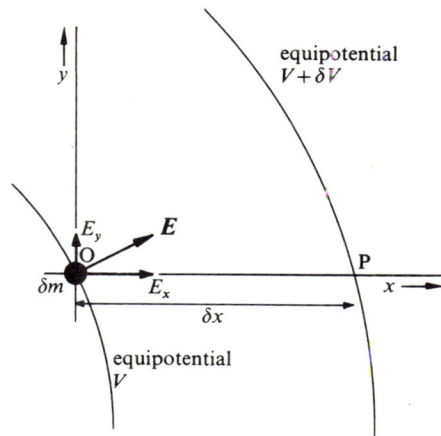

Figure 20.24

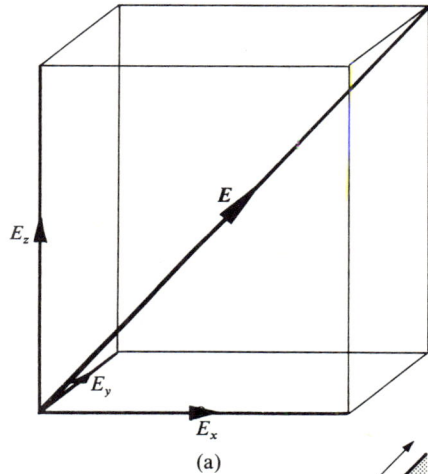

(a)

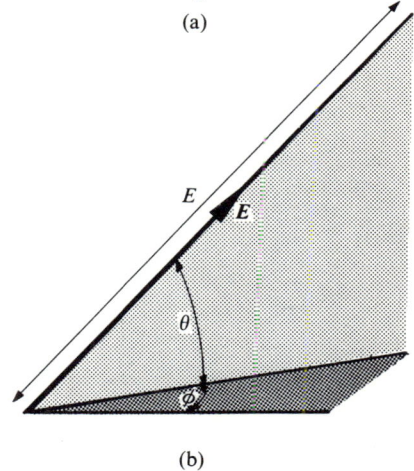

(b)

Figure 20.25
Two ways of specifying E in size and direction: (a) three mutually perpendicular components, (b) magnitude E, altitude θ, and azimuth ϕ.

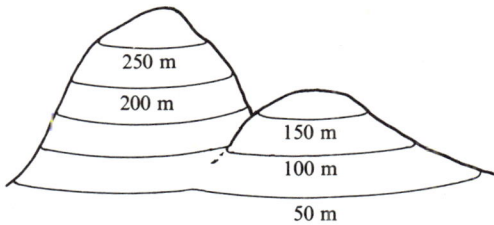

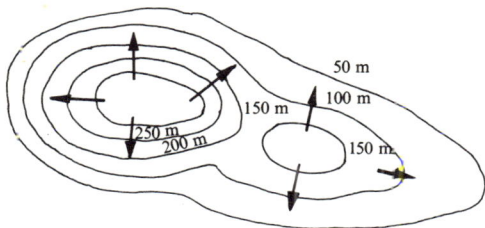

Contour lines. The arrows indicate lines of greatest slope, which are perpendicular to the contour lines.

g are negligible. They are defined on the map by just one number, 50m for example. One contour line by itself is of little value, but several at different heights give much more information than merely the height at a given spot. We can tell the *slope* from how close they are together, and they also show the direction of the line of *greatest* slope, which is clearly the line perpendicular to the contour lines at any point, since it is the shortest path between two given heights. The correspondence between slope and force is clear; the greater the slope the greater the tendency for an object to roll down it, and the direction in which it commences to roll from rest is the line of greatest slope.

Armed with the concepts of field and potential, and the link between them, we shall now tackle the problem of showing that the earth attracts an object on its surface just as though all its mass were concentrated at the centre.

Potential caused by a spherical shell at a point outside it

In principle the evaluation of the gravitational field at some point P distance r from the centre of the thin spherical shell of matter shown in Figure 20.26 is straightforward; we simply look at a small sample of mass δm, write down the contribution it makes to the field at P, and add up all such contributions from the entire shell. In practice, however, the summation is quite involved, especially as the individual contributions are *vectors*, all with slightly different directions. We can make the calculation a little simpler by evaluating the potential instead; the process of combining all the small contributions then involves the straight addition of scalar quantities.

We start by lumping together all those parts of the shell which are the same distance p from P; they constitute a thin ring of matter of radius y as indicated. If its mass is δm it makes a contribution δV to the potential at P given by

$$\delta V = -G\,\delta m/p.$$

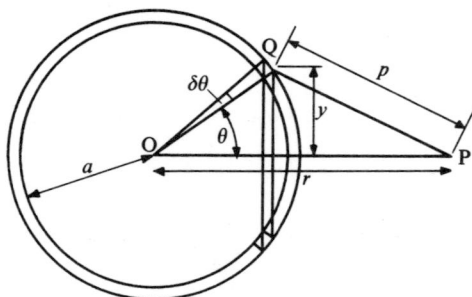

Figure 20.26

We need to write δm in terms of the ring dimensions; we shall say that the inner radius of the ring subtends an angle θ at the centre of the shell, and the outer radius an angle $\theta + \delta\theta$ where $\delta\theta$ is very small. Then the width of the ring is $a\,\delta\theta$, a being the shell radius, and the ring circumference is $2\pi y$ or $2\pi a\sin\theta$, so that if the shell has mass σ per unit area we can write

$$\delta m = \sigma \times 2\pi a\sin\theta \times a\,\delta\theta$$

and

$$\delta V = \frac{(-2\pi\sigma Ga^2\sin\theta)\,\delta\theta}{p},$$

$$V = -2\pi\sigma Ga^2 \int_0^\pi \frac{\sin\theta\,d\theta}{p}.$$

This integral is best written in terms of p; the limits are then from $p_1 = (r - a)$ at one edge of the sphere to $p_2 = (r + a)$ at the other.

$$V = -2\pi\sigma Ga^2 \int_{(r-a)}^{(r+a)} \frac{\sin\theta}{p}\frac{d\theta}{dp}\,dp, \qquad \ldots(13)$$

To evaluate it we have first to write down the cosine rule for triangle OPQ,

$$p^2 = a^2 + r^2 - 2ar\cos\theta.$$

and differentiate right through with respect to p

$$2p = 2ar\sin\theta\frac{\mathrm{d}\theta}{\mathrm{d}p}.$$

Substitution of this expression into equation (13) gives

$$V = -2\pi\sigma Ga^2 \int_{(r-a)}^{(r+a)} \frac{\mathrm{d}p}{ar}$$

$$= \frac{-2\pi\sigma Ga[(r+a) - (r-a)]}{r}$$

$$= \frac{-4\pi\sigma Ga^2}{r}.$$

Finally the total mass m of the shell can be written as

$$m = 4\pi a^2\sigma$$

so that

$$V = \frac{-Gm}{r}$$

which is the same as the potential due to a point mass m at O.

To get the field E, which is radial, we can use equation (10), but in the form $E_r = -\mathrm{d}V/\mathrm{d}r$, where E_r signifies the field in the direction of increasing r.

$$\text{Magnitude of } E = E_r = -Gm/r^2.$$

Potential and field caused by a spherically symmetric body at a point outside its surface

Once the potential for a thin spherical shell has been worked out the potential due to a solid sphere is quite simple. It can be regarded as a large number of thin shells with radii varying between zero and a the sphere radius. Each of these shells contributes to the potential at P distance r from O as though its mass were concentrated at O (Figure 20.27), in other words the entire sphere acts as though it were a point mass at its centre. It is not even necessary for the sphere to be homogeneous; it can be made up of spherical layers with different densities, as long as the density is constant for all points in a given layer. So, for a spherically symmetric body of mass m,

$$V = \frac{-Gm}{r} \quad \text{and} \quad E = \frac{-Gm}{r^2}$$

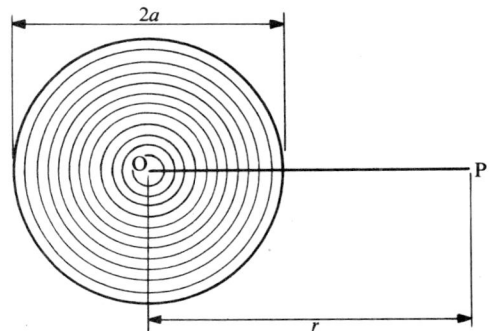

Figure 20.27

at a distance r from the centre, providing r is greater than or equal to the radius of the body.

In as much as a particle at P is attracted by the sphere as though the mass of the sphere were concentrated at O, and therefore the particle at P attracts such a mass as though it were concentrated at O, it follows that two spherically symmetric bodies attract each other as though their masses were concentrated at their centres.

Field and potential inside a spherical shell

Inside the shell the story is rather different. This time it is easier to work out the field, and it can most easily be done by dividing up the shell in the manner indicated in Figure 20.28. We shall calculate the field at point P; APB is a typical straight line through P meeting the shell at A and B. If we use this line as the axis of a cone of small angle, the cone defines areas δA_1 and δA_2 of the shell which are distances r_1 and r_2 from P. The attractions produced by δA_1 and

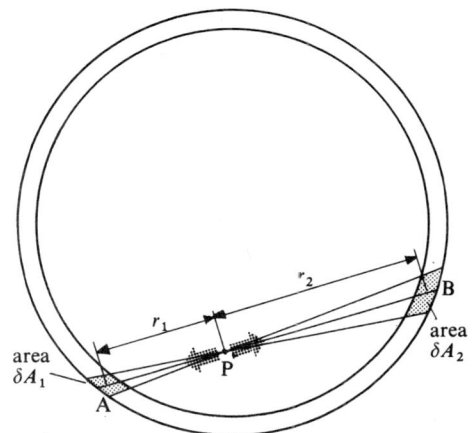

Figure 20.28

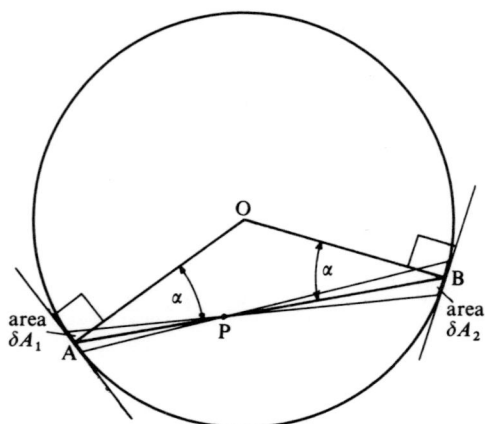

Figure 20.29

The normals to areas δA_1 and δA_2 are equally inclined to the line XPY, because $XO = OY$.

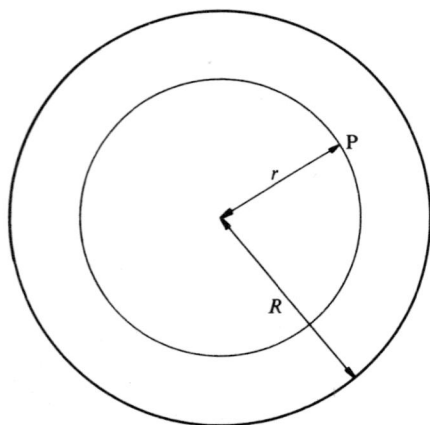

Figure 20.30

δA_2 both lie along APB but are in opposite directions, and the contribution δE they make to the field at P is given by

$$\delta E = G\sigma\left[\frac{\delta A_1}{r_1{}^2} - \frac{\delta A_2}{r_2{}^2}\right],$$

where σ is the mass per unit area of shell. A little solid geometry shows that

$$\frac{\delta A_1}{r_1{}^2} = \frac{\delta A_2}{r_2{}^2}$$

(see Figure 20.29) because the areas concerned are equally inclined to APB and the base of a circular cone of given angle has a radius proportional to its height.

So the field contributed by the areas δA_1 and δA_2 is exactly zero. But *every* small area of the shell can be paired up with a corresponding area on the other side of P in this way, and the result is that the field due to the entire shell is zero.

We have only to settle the question of V. If

$$E = 0,$$

$$\frac{\mathrm{d}V}{\mathrm{d}r} = 0$$

$$V = k, \text{ a constant.}$$

To decide the magnitude of k we go to the very edge of the inside: effectively we are applying boundary conditions. We already know the value of V at the surface; it is simply $-Gm/a$ where m is the total mass and a the radius of the shell. We cannot contemplate an abrupt change in V at the surface, for that would mean an infinite value of $\mathrm{d}V/\mathrm{d}r$ at some point, that is, an infinite field. So the constant k must be the value of V at the surface, $-Gm/a$.

Exercise 20.21

Figure 20.30 shows a sphere of radius R and uniform density ρ. Write down the contribution made to the field at P inside the sphere and distance r from the centre, (a) by the sphere of radius r, (b) by the shell lying outside r. As a result find the magnitudes of the field and potential at P.

Figure 20.31 shows how the field and potential due to a uniform sphere vary for points inside and outside it.

Long- and short-range forces

We are now in a position to discuss more fully what is meant by the term 'short range', applied to intermolecular forces (which are of course *not* gravitational–they arise from interactions between electric charges). We mean that the potential energy of a molecule or atom in bulk material does not depend on the volume of material involved, unless the sample is only a few atoms across. To ensure this, it is not good enough merely to have the force between the molecules decreasing with their separation; an attractive force obeying an inverse square law, for example, does not fall off sufficiently rapidly. The *gravitational* potential energy of a particle of mass δm at the centre of a homogeneous sphere of radius R is proportional to R^2 (see Exercise 20.21); in fact a thin shell of radius $2r$ contributes *more* potential at its centre than a shell of the same thickness but radius r. *Each particle* in the larger shell contributes less than a corresponding particle in the smaller shell, but in the larger shell there are *more particles*–four times as many. If we want the contribution from the larger shells to be negligible we must have a field of force which falls off *more rapidly* than $1/r^2$ to compensate. The critical case turns out to be a field falling off as $1/r^4$; any field which falls off more rapidly than this gives a potential at the centre of a homogeneous sphere which is *independent* of the radius of the sphere provided it is sufficiently large, and such a field of force is said to have 'short range'. Theory predicts that in a typical case the intermolecular attractive force falls off as $1/r^7$ and is therefore a short-range force. If the forces between molecules were *not* short range, the properties of matter would be very different from what we experience–for example, the elastic properties of a body would depend on its size. By contrast, gravitational forces are *long* range, so the larger the objects we consider, the more important become the gravitational effects.

Escape speed

How fast must a space vehicle be going in order to escape completely from the gravitational pull of a planet? If the astronauts are prepared to keep the rocket motors burning continuously, they can escape at any speed they please. This is not very practicable however. Fuel must be used to lift fuel, and it pays to use the bulk of it as quickly as possible and attain a high speed soon after launch; the vehicle then 'coasts' for most of the journey. So the question we must really ask is: what speed must a space vehicle acquire close to the surface of the planet in order that it can escape completely without further thrust from the motors? Energy provides the quickest answer; the potential at the surface of a planet with mass M and radius R is $-GM/R$ and the potential at infinity is zero. A rocket of unit mass must therefore gain potential energy numerically equal to GM/R to get clean away, and a rocket of mass m needs potential energy GMm/R. Within the terms we have set, all this must be provided by kinetic energy at the surface, so a speed v is needed such that

$$\tfrac{1}{2}mv^2 = GMm/R$$
$$v^2 = 2GM/R$$

and v, called the *escape speed*, is equal to $(2GM/R)^{1/2}$. Since $g = GM/R^2$ we can also write

$$v^2 = 2Rg.$$

Note that since kinetic energy is a scalar the *direction* of escape is immaterial.

Exercises 20.22, 23

22 Find the escape speed, (a) from the earth, (b) from the moon, using the following data. Ignore the help provided by the one in escaping from the other.

$$G = 6{\cdot}7 \times 10^{-11}\,\mathrm{m^3\,kg^{-1}\,s^{-2}}$$
$$\text{mass of moon} = 7{\cdot}3 \times 10^{22}\,\mathrm{kg}$$
$$\text{radius of moon} = 1740\,\mathrm{km}$$
$$\text{mass of earth} = 6{\cdot}0 \times 10^{24}\,\mathrm{kg}$$
$$\text{radius of earth} = 6400\,\mathrm{km}$$

23 A space vehicle is launched from near the surface of a planet, and acquires a speed of $10\,\mathrm{km\,s^{-1}}$ close to the surface before the

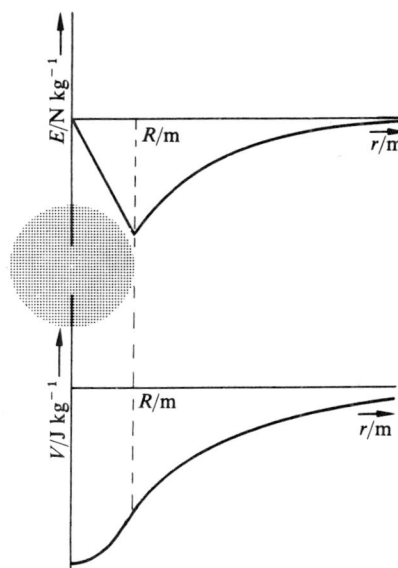

Figure 20.31
Variation of magnitude of gravitational field strength E and potential V with distance r from the centre of a uniform sphere of radius R.

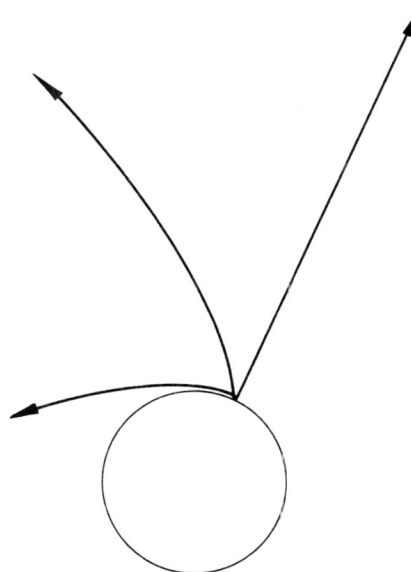

The speed of escape is independent of initial direction.

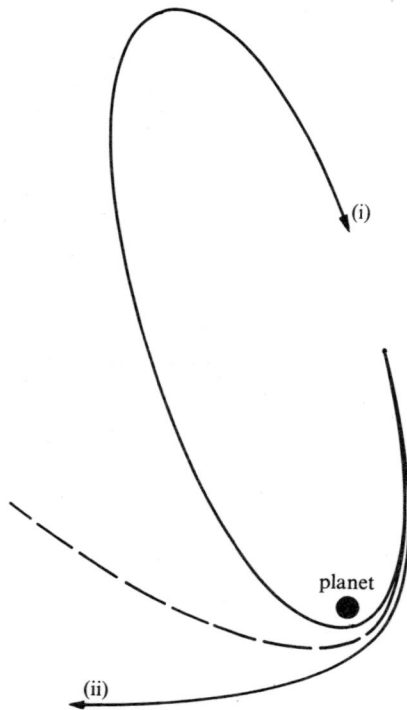

motors are switched off. The mass of the vehicle then remains constant at 40 000 kg. Taking the escape speed to be 9 km s^{-1} write down the potential energy at the surface of the planet and at a point very distant from it, and hence write down the kinetic energy and speed the vehicle will eventually attain when a very long way from the planet.

(Hint: write down the p.e. at the surface in terms of the escape speed.)

Space vehicles and angular momentum

A space vehicle with rocket motors switched off but under the influence of gravitational attraction from a nearby planet moves in an orbit which can be elliptical, parabolic or hyperbolic, according to its speed. If the sum of its k.e. and p.e. (the latter measured with respect to zero at infinity) is negative, the vehicle cannot escape from the gravitational pull of the planet and moves in an elliptical orbit with the centre of the planet at one focus of the ellipse. If the sum of its k.e. and p.e. is positive, it moves in a hyperbolic orbit and eventually escapes (Figure 20.32). An intermediate case occurs when the sum of k.e. and p.e. is zero; the orbit is then a parabola. [Strictly the 'parabolic' path of a small projectile launched from the surface of the earth, as discussed in Chapter 13, is an *ellipse* with the centre of the earth at one focus. This becomes an important consideration when the range is such that the curvature of the earth is relevant.] We cannot go into the full analysis of such orbits here, but we can get a considerable amount of information by applying the conservation laws to the situation. Energy must be conserved, and so must angular momentum about an axis through the centre of mass of the combination (planet plus vehicle), that is, for all practical purposes, about an axis through the centre of the planet. As we have seen

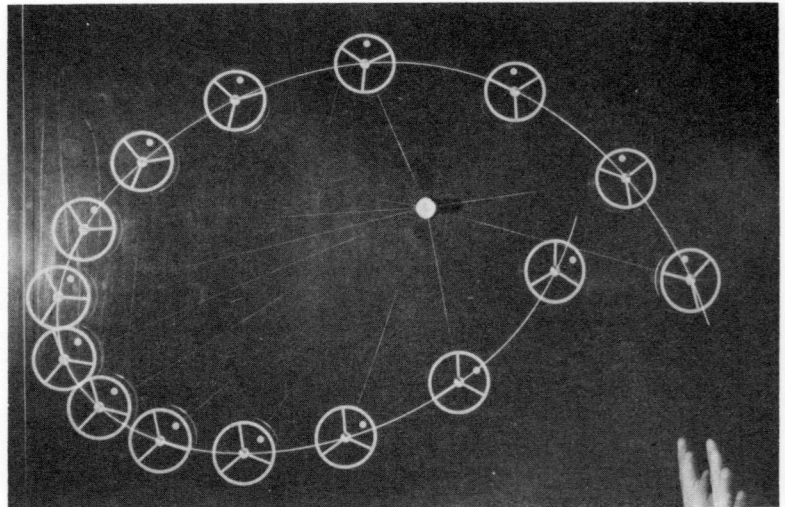

Figure 20.32

Possible trajectories for a space vehicle approaching a planet with motors switched off: (i) k.e. + p.e. negative; vehicle cannot escape, travels in elliptical orbit, (ii) k.e. + p.e. positive; vehicle eventually escapes, travelling in a hyperbolic orbit. The broken line shows the limiting case for which p.e. + k.e. = 0; the vehicle travels in a parabolic orbit, and its speed approaches zero a very long way from the planet.

Angular momentum is conserved in orbit. Low-friction puck illuminated by a stroboscopic lamp.

already, Kepler's third law is really a statement of the angular momentum conservation principle for a planet orbiting the sun. We shall give two examples of the way the conservation laws can be used to provide answers without going into the full details of the motion.

Example

A space vehicle is launched into a highly elongated elliptical orbit around the earth. At its perigee (closest point of approach) which is a negligible distance from the surface compared with the radius of the earth, it has a speed of $10 \, \mathrm{km \, s^{-1}}$. How far away from the centre of the earth is its apogee (furthest point) and how fast is it going there? (Radius of earth $= 6400 \, \mathrm{km}$, g at the surface $= 10 \, \mathrm{m \, s^{-2}}$.)

Writing the speeds at perigee and apogee as v_1 and v_2 respectively, and the corresponding distances from the centre of the earth as r_e and r, we can write for the energy conservation principle equation

$$\tfrac{1}{2}mv_1{}^2 - \frac{Gm_e m}{r_e} = \tfrac{1}{2}mv_2{}^2 - \frac{Gm_e m}{r}, \qquad \ldots(14)$$

where m is the mass of the vehicle and m_e the mass of the earth.

The problem is made easier by using the substitution

$$Gm_e = gr_e{}^2 \quad [\text{equation (2)}]$$

so that equation (14) can be written more compactly as

$$v_1{}^2 - 2gr_e = v_2{}^2 - \frac{2gr_e{}^2}{r}. \qquad \ldots(15)$$

Angular momentum conservation gives

$$mv_1 r_e = mv_2 r$$

(see Figure 20.33). The apogee and perigee are the only two places at which the angular momentum can be written quite so simply; at any other point in the orbit we should need to find the component of the velocity perpendicular to the radius vector. Substituting for v_2 in equation (15) we get

$$v_1{}^2 - 2gr_e = \frac{v_1{}^2 r_e{}^2}{r^2} - \frac{2gr_e{}^2}{r}$$

or

$$r^2(v_1{}^2 - 2gr_e) + 2gr_e{}^2 r - v_1{}^2 r_e{}^2 = 0.$$

The roots of this equation are $r = r_e$ or $v_1{}^2 r_e/(2gr_e - v_1{}^2)$; the first gives the perigee again and the second the apogee. So

$$r = \frac{v_1{}^2 r_e}{2gr_e - v_1{}^2}$$

$$= \frac{(10^4)^2 \times 6400 \times 10^3}{20 \times 6400 \times 10^3 - (10^4)^2} \, \mathrm{m}$$

$$= \frac{6400}{28} \times 10^2 \, \mathrm{km}$$

$$= 2 \cdot 3 \times 10^4 \, \mathrm{km}$$

and

$$v_2 = v_1 r_e / r$$

$$= \frac{10^4 \times 6400}{2 \cdot 3 \times 10^4} \, \mathrm{m \, s^{-1}}$$

$$= 2 \cdot 8 \, \mathrm{km \, s^{-1}}.$$

Notice that if $v_1{}^2 = 2gr_e$, r becomes infinite, giving us the escape-velocity condition again.

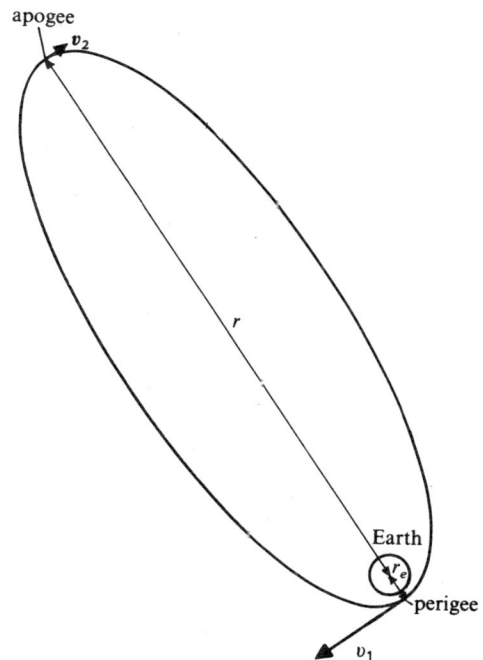

Figure 20.33

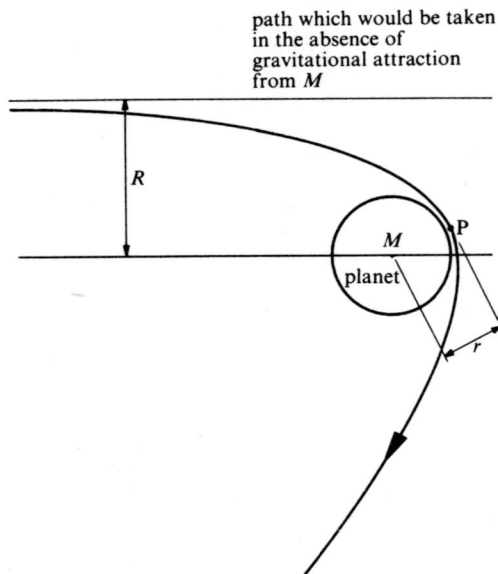

path which would be taken
in the absence of
gravitational attraction
from M

R

M

planet

P

r

Figure 20.34

Exercise 20.24

A space vehicle is moving with velocity v parallel to the surface of the earth and close to it. Find the value of v which produces (a) a circular orbit, (b) an orbit with an apogee 10^5 km from the centre of the earth.

Example

A space vehicle approaching a planet has a speed v a very long way out and is on a trajectory which would miss the centre of the planet by a distance R if it continued in a straight line (Figure 20.34). If the planet has mass M and radius r, what is the smallest value R can be in order that the resulting orbit will just miss the surface, and what is the velocity of the vehicle at the closest point of approach?

The energy a very long way from the planet is simply $\frac{1}{2}mv^2$ where m is the vehicle mass. If the velocity very near the surface is v_1, the total energy there can be written as $\frac{1}{2}mv_1{}^2 - GMm/r$, so that energy conservation gives

$$v^2 = v_1{}^2 - \frac{2GM}{r} \quad \text{or} \quad v_1 = \left(v^2 + \frac{2GM}{r}\right)^{1/2}.$$

The angular momentum equation is

$$mvR = mv_1 r$$

$$R = \frac{v_1 r}{v} = \frac{r}{v}\left[v^2 + \frac{2GM}{r}\right]^{1/2}.$$

Exercise 20.25

A space vehicle approaching the moon has a speed of $0 \cdot 5\,\mathrm{km\,s^{-1}}$ a long way out. Neglecting the effect of the earth, find how far it must be aimed from the centre of the moon in order to just avoid hitting the surface, and find the speed at closest approach.

mass of moon $= 7 \cdot 3 \times 10^{22}\,\mathrm{kg}$
radius of moon $= 1740\,\mathrm{km}$
$G = 6 \cdot 7 \times 10^{-11}\,\mathrm{m^3\,kg^{-1}\,s^{-2}}$.

If the vehicle described in Exercise 20.25 is to be put into a circular orbit round the moon, its speed must be reduced by turning the vehicle round and firing the rocket motors in the direction of motion. The motors must be switched on as the vehicle passes through the closest point of approach on the far side, which is the only part of the original orbit for which there is no radial component of velocity relative to the centre of the moon (Figure 20.34). The speed must be reduced as the vehicle passes point P until the correct speed for a circular orbit has been established.

Exercises 20.26, 27

26 Use the data provided in Exercise 20.25 to find the speed required for a circular orbit close to the surface of the moon.

27 The mass of the vehicle concerned is $40000\,\mathrm{kg}$ and the rocket thrust available to slow it down is $5 \times 10^5\,\mathrm{N}$. Estimate the duration of the rocket burn needed to establish a circular orbit when the vehicle is near P. (Hint: use the fact that impulse equals momentum change.)

Earth seen from lunar orbit.

Further exercises

Exercises 20.28–43

28 Calculate the radius of the orbit of an earth satellite whose period is equal to that of the earth on its axis. You may assume that the mass of the earth is $6\cdot0 \times 10^{24}$ kg and that the constant of gravitation is $6\cdot7 \times 10^{-11}$ SI units.   (JMB)

29 Explaining each step in your calculation and pointing out the assumptions you make, use the information below to estimate the mean distance of the moon from the earth.
Period of rotation of the moon around the earth = 27·3 days
Radius of earth = $6\cdot37 \times 10^3$ km
Acceleration due to gravity at earth's surface, $g = 9\cdot81$ m s^{-2}
                    (JMB)

30 Define moment of inertia and angular momentum.

A small planet, mass m, moves in an elliptical orbit round a large sun, mass M, which is at the focus F_1 of the ellipse (Figure 20.35).

Write down an expression for the force acting on the planet when it is at a position A, at a distance r from the sun. Indicate the direction of this force on a copy of Figure 20.35, and also mark on it the directions of the planet's velocity and acceleration when at A. What is the magnitude of this acceleration?

According to Kepler's second law, the planet is moving faster at B than at C. Account for this with reference to the principle of conservation of energy.

What is the moment of inertia of the planet, when at B, about F_1? The velocities at B and C are v_B and v_C. Use the principle of conservation of angular momentum to deduce the ratio v_B/v_C.
                    (SUJB)

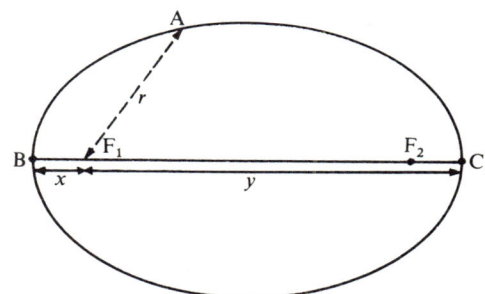

Figure 20.35 (Exercise 20.30)

31 Assuming that the earth is a uniform homogeneous sphere of radius R and density ρ, obtain expressions for the acceleration due to gravity:

a) at a pole of the earth
b) at a height h above the earth at the pole
c) at a point on the equator.

Any additional assumptions made should be clearly stated.

(JMB)

32 Assuming the earth to be perfectly spherical, give sketch graphs to show how (a) the acceleration due to gravity, (b) the gravitational potential due to the earth's mass, vary with distance from the surface of the earth for points external to it. If any other assumption has been made state what it is.

Explain why, even if the earth were a perfect sphere, the period of oscillation of a simple pendulum at the poles would not be the same as the period at the equator.

Still assuming the earth to be perfectly spherical, discuss whether the velocity required to project a body vertically upwards, so that it rises to a given height, depends on the position on the earth from which it is projected.

(C)

33 Show that the period of revolution of an earth satellite which is describing a circular orbit of radius x is given by

$$T = 2\pi \frac{x}{R} \sqrt{\left(\frac{x}{g}\right)},$$

where R is the radius of the earth (assumed to be a uniform sphere), and g is the intensity of gravity at the earth's surface.

Figure 20.36 represents, in plan and not to scale, a torsion balance of the Cavendish type. The arm AB is 2m long, the suspended lead spheres are of radius $r = 1\cdot3$cm, and the large lead spheres are of radius $R = 20$cm. Estimate the maximum couple which can be experienced with this arrangement by the suspended system. How would this be affected, (a) by doubling r, (b) by doubling R?

(Take the value of G, the universal gravitational constant, to be $6\cdot7 \times 10^{-11}$ in SI units, and the density of lead to be $11\,700$ kg m^{-3}.)

(O)

34 A binary star consists of two dense spherical masses of 10^{30} kg and 2×10^{30} kg whose centres are 10^7 km apart and which rotate together with a uniform angular velocity ω about an axis which intersects the line joining their centres. Assuming that the only forces acting on the stars arise from their mutual gravitational attraction and that each mass may be taken to act at its centre, show that the axis of rotation passes through the centre of mass of the system and find the value of ω. ($G = 6\cdot7 \times 10^{-11} \text{m}^3$ $\text{kg}^{-1}\text{s}^{-2}$)

(O & C)

35 Two spherical stars, each of mass M, with their centres distant D apart, revolve under their mutual gravitational attraction about the point midway between their centres. Find a formula for the period of revolution. Assume that each star behaves as if its mass were concentrated at its centre.

An artificial satellite is launched in an easterly direction into a circular orbit in the plane of the equator, concentric with the earth and at a height of 36000km above the earth's surface. Taking the radius of the earth as 6400km, and the value of g at

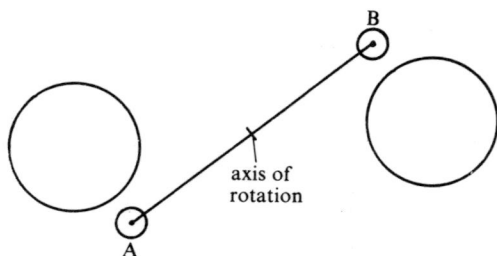

Figure 20.36 (Exercise 20.33)

the equator (corrected for the earth's own rotation) as $9.80\,\mathrm{m\,s^{-2}}$, find the period of revolution of the satellite in hours.   (o & c)

36 A sphere of radius $0.50\,\mathrm{m}$ made of metal of density $20000\,\mathrm{kg\,m^{-3}}$ is buried just below the surface of the earth. Calculate the percentage change in the acceleration due to gravity at the point immediately above the buried sphere. You may assume that the radius of the earth is $6000\,\mathrm{km}$, its density $5500\,\mathrm{kg\,m^{-3}}$ and that the material excavated to make room for the sphere has been removed to a great distance.   (JMB)

37 The earth may be regarded as a spherically uniform core of average density ρ_1, surrounded by a thin uniform shell of thickness h and density ρ_2. If the value of the acceleration due to gravity is the same at the surface as at depth h find the ratio of ρ_1 to ρ_2.   (JMB)

38 If the diameter of a planet is 11, its mean density $1/4$ and the mass of its atmosphere 10 in each case in relation to the corresponding quantities on earth, estimate (a) the ratio of the atmospheric pressure on the planet to that on the earth, (b) the mean height of a mercury barometer on the planet. Assume the effective height of the atmosphere to be very much smaller than the radius of the planet.   (JMB)

39 Show that the minimum velocity required to put a projectile into a circular orbit about the earth is $1/\sqrt{2}$ times the minimum velocity with which a projectile must be fired from the earth's surface so that it never returns.   (C)

40 Taking the earth to be a uniform sphere of radius $6.4 \times 10^6\,\mathrm{m}$ and mass $6 \times 10^{24}\,\mathrm{kg}$, the moon to be a uniform sphere of mass $7.3 \times 10^{22}\,\mathrm{kg}$ with its centre $3.8 \times 10^8\,\mathrm{m}$ from the centre of the earth, and the value of the universal constant of gravitation, G, to be $6.7 \times 10^{-11}\,\mathrm{m^3\,kg^{-1}\,s^{-2}}$, calculate

a) the value of g at the earth's surface, neglecting the effect of the moon
b) the greatest difference that the moon's effect can make to the value of g at the earth's surface
c) the period of rotation of the moon about the earth
d) the gravitational force between the earth and the moon.   (o)

41 The motors of a rocket launched from the earth are used only near the earth, in order to give the rocket just sufficient energy to reach the moon. Find where, subsequently, the velocity of the rocket is a minimum and calculate the speed with which it hits the moon's surface. (The motion of the moon may be neglected.)
    (Earth's mass $= 6.0 \times 10^{24}\,\mathrm{kg}$, moon's mass $= 7.3 \times 10^{22}\,\mathrm{kg}$, earth's radius $= 6.4 \times 10^3\,\mathrm{km}$, moon's radius $= 1.7 \times 10^3\,\mathrm{km}$, distance earth to moon $= 3.8 \times 10^5\,\mathrm{km}$, constant of gravitation $G = 6.7 \times 10^{-11}\,\mathrm{N\,m^2\,kg^{-2}}$.)   (C)

42 A satellite circling the earth over the equator is observed to complete one orbit in exactly 90 minutes. Show that the average height of the satellite above the earth's surface is very nearly $270\,\mathrm{km}$.

    It can be proved that if the next orbit takes exactly one second less the height decreases by $0.823\,\mathrm{km}$. Explain the energy changes accompanying this change in height and obtain a value for the average resistive force, expressing your answer in terms of the mass of the satellite. (Radius of earth $= 6400\,\mathrm{km}$, $g = 9.8\,\mathrm{m\,s^{-2}}$.)
                    (JMB)

43 Explain what is meant by *gravitational potential energy*.

A spherical star of uniform density ρ and radius R is formed by the condensation of interstellar dust from large distances due to gravitational forces. Find the energy change which would occur if the radius increased from r to $r + \mathrm{d}r$. Hence derive an expression in terms of G, R and the mass M of the star for the total loss of gravitational energy during the condensation.   (C)

Appendices

Appendix 1

Powers of ten

$$10^9 = 10 \times \underset{1}{10} \times \underset{2}{10} \times \underset{3}{10} \times \underset{4}{10} \times \underset{5}{10} \times \underset{6}{10} \times \underset{7}{10} \times \underset{8}{10} \times \underset{9}{10}$$

$$= 1\,000\,000\,000$$

$$10^6 = 1\,000\,000$$
$$10^3 = 1\,000$$
$$10^0 = 1$$
$$10^{-1} = 0{\cdot}1$$
$$10^{-3} = 0{\cdot}001$$
$$10^{-6} = 0{\cdot}000001$$

$$10^3 \times 10^2 = 1000 \times 100$$
$$= 100\,000$$
$$= 10^5$$

$$10^X \times 10^Y = 10^{(X+Y)}$$
$$10^X/10^Y = 10^{(X-Y)}$$
$$(10^X)^Y = 10^{(X \times Y)}$$

Appendix 2

Differentiation

[The arguments in Appendices 2 and 3 are illustrative rather than rigorous, and are intended to help those readers with no previous experience of calculus.]

In what follows, y is a function of x, that is, the value of y depends on the value of x in some definite way. The symbol δx represents a small increase in x; it does not mean 'delta times x', The symbol δy stands for the corresponding change in the value of y (remember x and y are connected).

The ratio $\delta y/\delta x$ depends in general both on x and on δx (note that the δ's cannot be cancelled). We are interested in the value this ratio takes when we make δx very small; we speak of δx as *tending to the limit* zero. Usually δy also tends to the limit zero, but $\delta y/\delta x$ tends to a *definite value* which depends on the value of x. We take as an example the function $y = x^2$, and work out the value of $\delta y/\delta x$ when $x = 2$, for progressively decreasing values of δx.

Table A2.1

δx	$x + \delta x$	$y + \delta y$	δy	$\delta y/\delta x$
0·1	2·1	4·41	0·41	4·1
0·01	2·01	4·0401	0·0401	4·01
0·001	2·001	4·004001	0·004001	4·001

Clearly the smaller we make δx, the more nearly $\delta y/\delta x$ approaches the value 4·0 exactly.

The quantity '$\delta y/\delta x$ as δx tends to zero' is called the *differential coefficient* of y with respect to x, and is given the symbol dy/dx. This is a single symbol: the d's do not stand for quantities and they cannot be cancelled according to the normal rules of algebra.

A graphical interpretation is shown in Figure A2. The ratio $\delta y/\delta x$ is the gradient or *slope* of the straight line AB, and as δx tends to zero, that is, as the vertical line DC moves towards A, the line AB

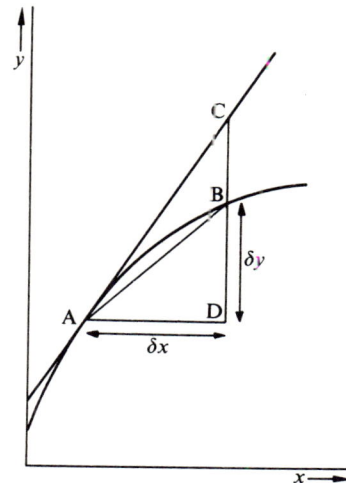

Figure A2

approaches the tangent AC. The slope of the tangent, that is, the slope of the curve at the point A, is dy/dx.

The differential coefficient dy/dx can be calculated in terms of x for many functions of x.

If for example
$$y = x^2$$
then
$$y + \delta y = (x + \delta x)^2$$
$$\delta y = (x + \delta x)^2 - y$$
$$= (x + \delta x)^2 - x^2$$
$$= 2x\,\delta x + (\delta x)^2$$
$$\delta y/\delta x = 2x + \delta x$$

and as δx tends to zero $\delta y/\delta x$ tends to $2x$, so

$$\frac{dy}{dx} = 2x.$$

Similar methods can be used to establish the differential coefficients in Table A2.2 as quoted in this book:

Table A2.2

y	dy/dx	Example
ax^n (a and n constant, $n \neq 0$)	nax^{n-1}	$y = 3x^{1/2}$ $dy/dx = \dfrac{3x^{-1/2}}{2}$
$\sin(bx + c)$ (b and c constant, x in radians)	$b\cos(bx + c)$	$y = \sin 2x$ $dy/dx = 2\cos 2x$
$\cos(bx + c)$	$-b\sin(bx + c)$	$y = \cos(x + \pi/4)$ $dy/dx = -\sin(x + \pi/4)$
ae^{bx}	abe^{bx}	$y = 3e^{-4x}$ $dy/dx = -12e^{-4x}$
$f(z)$ where z is a function of x	$\dfrac{dy}{dz} \times \dfrac{dz}{dx}$	$y = (x^2 + 9)^{-1/2}$ $dy/dx = \frac{1}{2}(x^2 + 9)^{-1/2} \times 2x$

Appendix 3

Integration

If y and z are both functions of x such that $y = dz/dx$, then z is called the *integral* of y with respect to x, and written as

$$z = \int y\,dx;$$

in other words, integration is the reverse of differentiation.

Generally speaking there is more than one expression z for a given function y. For example,

if $z = 3x^2$, then $y = 6x$
if $z = 3x^2 + c$, where c is any constant, then $y = 6x$ again.

So we write

$$\int 6x\,dx = 3x^2 + c$$

and the constant c is included to provide for all the possible expressions for z; it is called the constant of integration.

The following integrations are used in the text:

$$\int ax^n\,dx = \frac{a}{n+1}x^{n+1} + c \qquad (n \neq 1).$$

$$\text{Example:} \int 3x^4\,dx = \frac{3x^5}{5} + c$$

$$\int \sin(ax+b)\,dx = -\frac{1}{a}\cos(ax+b) + c.$$

$$\text{Example:} \int \sin 3x\,dx = -\frac{1}{3}\cos 3x + c$$

$$\int \cos(ax+b)\,dx = \frac{1}{a}\sin(ax+b) + c.$$

$$\text{Example:} \int \cos(x+\pi/4)\,dx = \sin(x+\pi/4) + c$$

$$\int e^{ax}\,dx = \frac{e^{ax}}{a} + c.$$

$$\text{Example:} \int e^{0 \cdot 01x}\,dx = 100e^{0 \cdot 01x} + c$$

and the following are quoted in Appendices 6 and 7 respectively:

$$\int \frac{dx}{\sqrt{a^2 - x^2}} = \cos^{-1}(x/a) + c$$

$$\int \frac{dx}{a + bx} = \frac{1}{b}\ln(a + bx) + c.$$

The integral as the limit of a sum; graphical interpretation

As the vertical line AB in Figure A3 moves towards larger values of x it sweeps out more and more area under the curve shown, which is the graph of y against x. We can show that the area S swept out in this manner is given by

$$S = \int y\,dx$$

(we are regarding S as a variable depending on x). We can think of S as the sum of the areas of a large number of narrow strips like $EFGH$, of area δS:

$$S = \sum \delta S.$$

If $EFGH$ has height y and width δx,

$$\delta S \approx y\,\delta x,$$

enabling us to write two useful expressions:

$$S \approx \sum y\,\delta x \quad \text{and} \quad y \approx \delta S/\delta x.$$

The above expressions are not *exactly* true, because $EFGH$ is not quite a rectangle: the line EF is inclined to the x-axis. But the smaller we make δx, the smaller is the error involved in the approximation, and as δx approaches zero, $\sum y\,\delta x$ approaches S more and more closely. Moreover, $\delta S/\delta x$ approaches dS/dx more and more closely, so that we can write $y = dS/dx$, or, by definition,

$$S = \int y\,dx.$$

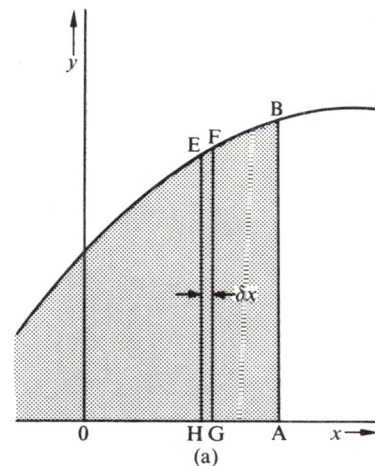

(a)

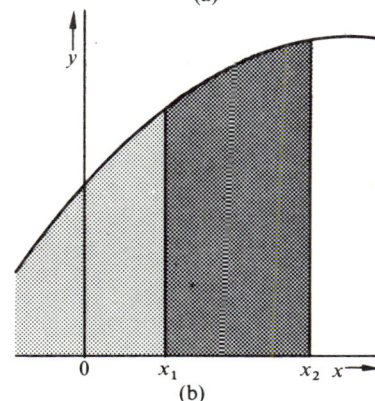

(b)

Figure A3

So,

$$\text{area under the graph of } y \text{ against } x \text{ as far as } x = \underset{\text{as } \delta x \to 0}{\text{limit}} (\Sigma\, y\delta x) = S = \int y\mathrm{d}x.$$

Though we have defined the right-hand limit of the area under the curve, no mention has yet been made of the left-hand limit, that is, the value of x from which measurement of S commences. We can get a whole range of values for S depending on where we start to measure from. This reminds us of the fact that the integration of y with respect to x involves the arbitrary constant c in the answer.

Definite integration

Often we require to evaluate the area under the curve between two definite values of x, x_1 and x_2. We shall consider a specific example, the area under the graph of $y = 3x^3$.

The formula for the area S from an arbitrary starting value of x is

$$S = \int y\mathrm{d}x = \int 3x^3\,\mathrm{d}x$$
$$= 3x^4/4 + c.$$

That is, the area as far as x_1 is $3x_1{}^4/4 + c$
and the area as far as x_2 is $3x_2{}^4/4 + c$.
So the area from x_1 to x_2 is

$$(3x_2{}^4/4 + c) - (3x_1{}^4/4 + c) = (3x_2{}^4/4) - (3x_1{}^4/4),$$

and the arbitrary constant c disappears.
The above process is usually written as follows:

$$\text{Area between } x_1 \text{ and } x_2 = \int_{x_1}^{x_2} y\mathrm{d}x = \int_{x_1}^{x_2} 3x^3\mathrm{d}x$$
$$= \left[3x^4/4\right]_{x_1}^{x_2} = (3x_2{}^4/4) - (3x_1{}^4/4).$$

Appendix 4   (see p. 69)

If the force R is the resultant of forces P and Q, as defined by the parallelogram law, then the moment of R about any axis perpendicular to the plane containing P, Q and R is equal to the sum of the moments of P and Q about that axis.

Proof

Moment of P about the axis through $A = P \times AX$ (Figure A4)
$$= P \times OA\cos\alpha$$
$$= OA \times \text{ component of } P \text{ in direction } OK$$

(where OK is perpendicular to OA).

Similarly, moment of Q about the axis through A
$$= OA \times \text{ component of } Q \text{ in direction } OK$$
and   moment of R about the axis through A
$$= OA \times \text{ component of } R \text{ in direction } OK$$

But because R is the resultant of P and Q,

component of R along $OK =$ component of P along OK
$$+ \text{ component of } Q \text{ along } OK.$$

(In terms of the geometry of the figure, this is equivalent to the relation $OD + OB = OC$.)

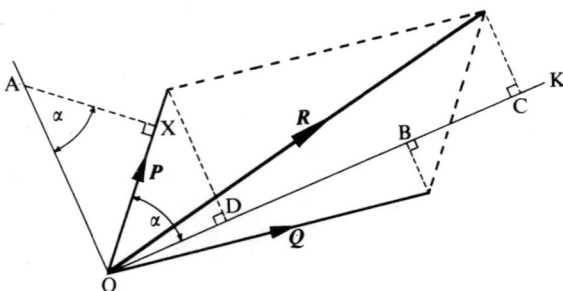

Figure A4

Appendix 5

The force R is chosen so as to have the same translating effect as the system of forces $P_1, P_2, P_3, \ldots$, and its line of action chosen so that its moment about an axis through O perpendicular to the plane of the forces equals the sum of the moments of $P_1, P_2, P_3, \ldots$, about that axis. We shall show that the moment of R about an axis through a different point O' is equal to the sum of the moments of P_1, P_2 $P_3, \ldots$, about the axis through O'.

Proof
Using the result proved in Appendix 4,
moment of a typical force P about the axis through O

= moment of x component of P about O + moment of y
    component of P about O
= $P_x y - P_y x$ (See Figure A5; we are counting clockwise
    moments as positive.)

So sum of moments of $P_1, P_2, P_3, \ldots$, about the axis through O

$$= \Sigma (P_x y - P_y x),$$

and moment of R about the axis through $O = R_x y' - R_y x'$.
So

$$\Sigma (P_x y - P_y x) = R_x y' - R_y x'.$$

Sum of moments of $P_1, P_2, P_3, \ldots$, about a parallel axis through O'

$$= \Sigma [P_x(y - b) - P_y x]$$
$$= \Sigma (P_x y - P_y x) - \Sigma P_x b$$
$$= \Sigma (P_x y - P_y x) - b \Sigma P_x$$
$$= R_x y' - R_y x' - bR_x$$
$$= R_x(y' - b) - R_y x'$$
$$= \text{moment of } R \text{ about the axis through } O'.$$

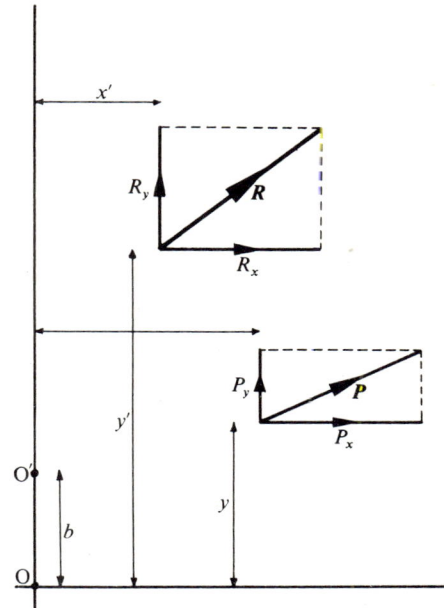

Figure A5

Appendix 6

Solution of $d^2x/dt^2 = -\omega^2 x$ by integration (p. 209)

The equation can be written as

$$dv/dt = -\omega^2 x,$$

where v is the velocity in the x-direction. So

$$\frac{dv}{dx} \times \frac{dx}{dt} = -\omega^2 x$$

$$v dv/dx = -\omega^2 x$$

$$\int v dv = -\omega^2 \int x \, dx$$

$$\tfrac{1}{2}v^2 = -\tfrac{1}{2}\omega^2 x^2 + c,$$

where c is a constant which must be decided by boundary conditions. If we specify that $x = a$ when $v = 0$, we obtain

$$0 = -\tfrac{1}{2}\omega^2 a^2 + c,$$

that is,

$$c = \tfrac{1}{2}\omega^2 a^2.$$

Hence

$$\tfrac{1}{2}v^2 = -\tfrac{1}{2}\omega^2 x^2 + \tfrac{1}{2}\omega^2 a^2$$

and

$$dx/dt = v = \omega\sqrt{(a^2 - x^2)},$$

$$\int \frac{dx}{\sqrt{(a^2 - x^2)}} = \int \omega \, dt.$$

Using standard results of integration,

$$\cos^{-1}(x/a) = \omega t + \varepsilon,$$

where ε is a constant (i.e., 'the angle having a cosine equal to x/a is $\omega t + \varepsilon$').

$$x/a = \cos(\omega t + \varepsilon),$$
$$x = a\cos(\omega t + \varepsilon).$$

Appendix 7

Solution for $dv/dt = A - Bv$ by intergration (p. 238)

$$\int \frac{dv}{A - Bv} = \int dt,$$

$$-\frac{1}{B}\ln(A - Bv) = t + c,$$

where c is a constant.

$$\ln(A - Bv) = -Bt - Bc.$$

Taking antilogarithms of both sides,

$$A - Bv = e^{-Bt}e^{-Bc}$$
$$= ke^{-Bt},$$

where k is another constant.

$$v = \frac{A}{B} - \frac{ke^{-Bt}}{B}$$

and k can be shown to equal A by inserting boundary conditions, as on p. 244.

Appendix 8

Trial solution for the equation $m\ddot{x} + c\dot{x} + kx = 0$ (p. 251)

We shall try as a solution

$$x = Ae^{-\lambda t}\sin\omega t$$

implying

$$\dot{x} = -\lambda Ae^{-\lambda t}\sin\omega t + \omega Ae^{-\lambda t}\cos\omega t$$

and

$$\ddot{x} = -\lambda\omega Ae^{-\lambda t}\cos\omega t + \lambda^2 Ae^{-\lambda t}\sin\omega t - \omega\lambda Ae^{-\lambda t}\cos\omega t$$
$$- \omega^2 Ae^{-\lambda t}\sin\omega t.$$

Substituting these expressions into the differential equation and collecting terms in $\sin\omega t$ and $\cos\omega t$, we have to insist that

$$Ae^{-\lambda t}[\sin\omega t(m\lambda^2 - m\omega^2 - c\lambda + k) + \cos\omega t(-2m\lambda\omega + c\omega)] = 0$$

for all values of t if the trial solution is to be successful. This can be achieved only if the coefficients of $\sin\omega t$ and $\cos\omega t$ are both zero, that is, if

$$2m\lambda\omega = c\omega$$

and

$$m(\lambda^2 - \omega^2) = c\lambda - k.$$

The first condition requires that $\lambda = c/2m$ and the second that

$$\omega^2 = \lambda^2 - (c\lambda - k)/m$$
$$= (k/m) - (c^2/4m^2)$$

using the expression for λ.

When c is small, corresponding to light damping, $\omega^2 \approx k/m$, which is the value of ω^2 for undamped (simple harmonic) oscillations. For larger values of c, ω decreases, that is, the frequency of damped oscillations is smaller than that of undamped oscillations. When $c^2/4m^2 > k/m$, ω^2 is negative, so that ω is no longer a real number. In this situation the solution $x = Ae^{-\lambda t}\sin\omega t$ is not suitable; there are no oscillations and the motion is dead beat. The condition for critical damping is given by $\omega^2 = 0$, that is, $c^2/4m = k$.

Appendix 9

Kinetic energy of rotation and translation (p. 272)

If a rigid body of mass m is moving in a straight line with velocity v and at the same time rotating about an axis through its centre of mass with angular velocity ω, its total kinetic energy is $\frac{1}{2}mv^2 + \frac{1}{2}I\omega^2$, where I is the moment of inertia about the axis of rotation.

Proof

The velocity of the mass δm_1 at P (Figure A9) is the sum of two velocities, with magnitudes v and ωr_1 in the directions shown. The magnitude of the resultant velocity is

$$\left[v^2 + \omega^2 r_1^2 - 2v\omega r_1 \cos\left(\frac{\pi}{2} - \theta_1\right)\right]^{1/2},$$

or $(v^2 + \omega^2 r_1^2 - 2v\omega r_1 \sin\theta_1)^{1/2}$.

So the kinetic energy δE_1 of mass δm_1 is given by

$$\delta E_1 = \frac{1}{2}\delta m_1(v^2 + \omega^2 r_1^2 - 2v\omega r_1 \sin\theta_1).$$

Similarly the kinetic energy of mass δm_2 is given by

$$\delta E_2 = \frac{1}{2}\delta m_2(v^2 + \omega^2 r_2^2 - 2v\omega r_2 \sin\theta_2)$$

and so on. The total kinetic energy is the sum of all such contributions:

$$E = \sum \frac{1}{2}\delta m(v^2 + \omega^2 r^2 - 2v\omega r \sin\theta)$$
$$= \frac{1}{2}v^2 \sum \delta m + \frac{1}{2}\omega^2 \sum \delta m r^2 - 2v\omega \sum \delta m r \sin\theta$$
$$= \frac{1}{2}mv^2 + \frac{1}{2}I\omega^2 - 2v\omega \sum \delta m r \sin\theta.$$

But $\sum \delta m r \sin\theta$ is the sum of moments of masses about an axis parallel to v and *passing through G, the centre of mass*; this sum is zero by definition.

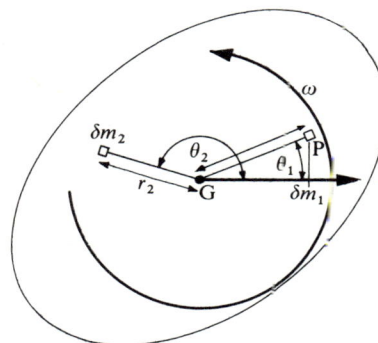

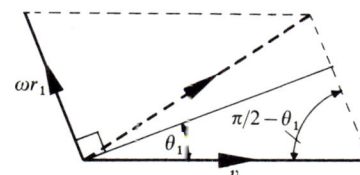

Figure A9

Answers to exercises

1.1 (a) 10^{16} (b) 10^{24}
1.3 $\frac{8}{9}$
1.5 10^{19} or 10^{20}
1.6 about 10^{-25} kg
1.7 about 10 atomic diameters

2.1 0.25 nm
2.2 (a) 1 year (b) stars in view would never set
2.3 10 cm
2.4 (a) 10^4 (b) 10^6
2.5 (a) 10^6 (b) 10^9
2.6 (a) 1000 kgm^{-3} (b) 13600 kgm^{-3} (c) 1.2 kgm^{-3}, 36 kg
(d) $1000X$ kgm^{-3}
2.7 10^{18} kgm^{-3}
2.8 $V^{1/3}$, $V^{2/3}$, dV
2.9 $r \times$ (angle in degrees) $\times \pi/180$
2.10 (a) (i) AC/r, (ii) AB/r, (iii) AB/DB, (iv) DB/r
(b) $\sin\theta \approx \tan\theta \approx \theta$ (in radians), $\cos\theta \approx 1$
(c) (i) 0.0013, (ii) 0.0051 (d) 1.04 (e) (i) 0.0002, (ii) 0.0038
2.11 1990 km
2.12 6400 km

3.1 $\pm 1\%$, $\pm 0.1\%$, $\pm 3\%$
3.2 (a) 0.129 (b) 50.00, 0.168
3.4 0.029, 0.036, about 280
3.6 $1/500$
3.7 $y = \frac{1}{2}x$, $y = x + 1$, $y = -\frac{1}{3}x + 4$, $y = x - 1$ $y = 5$,
$x = 3$
3.8 0.02 s, after
3.9 (a) $(p/\text{m})^2$ against $(q/\text{s})^3$, k numerically equal to m
(b) (p/m) against $(q/\text{s})^2$, k numerically equal to m, r
numerically equal to c
(c) $(p/\text{m})^2$ against $(q/\text{s})^2$, k numerically equal to $-m$,
r numerically equal to c.
(d) $(p/\text{m})^{-2}$ against $(q/\text{s})^{-2}$, k numerically equal to $-1/m$,
r numerically equal to c
3.10 (a) 0.49 (b) 0.43, (0.46 ± 0.03) m$^{-1/2}$ s, $(0.02 = 0.03)$ s
3.11 $y = 1.58x + 0.07$
3.12 $n = \frac{1}{2}$, $k = 2$m$^{-1/2}$ s
3.14 $x + z + \delta x + \delta z$, $x + z - \delta x - \delta z$
3.15 $x - z + \delta x + \delta z$, $x - z - \delta x - \delta z$
3.16 1%
3.18 (a) 2.2%, perhaps better quoted as 2%
(b) 8.5%, perhaps better quoted as 10%
3.19 1.7%, perhaps better quoted as 2%
3.22 $\pm 6\%$
3.23 $(T_1 - T_2)/200$

4.1 (a) 4.35 ms^{-1} (b) 9.1 ms^{-1}
4.2 9.1 ms^{-1}
4.3 6.3
4.5 1.2 ms^{-1}, 0.7 ms^{-1}
4.6 (a) 2.4 ms^{-2} (b) 1.6 ms^{-2}
4.7 1.1 ms^{-1}, 1.25 ms^{-1}, 1.9 ms^{-2}
4.8 3.8 ms^{-2}
4.10 2.3 ms^{-2}
4.11 2.5 ms^{-2}
4.14 $x = vt - \frac{1}{2}at^2$
4.15 2.5 ms^{-2}
4.16 40.3 ms^{-2}
4.17 (a) 0.45 s (b) 1.41 s
4.18 45 m
4.19 4.0 s
4.20 38.8 ms^{-1}, 35
4.21 3.58 m, S$53.6°$E
4.22 33 s, 45 s
4.23 6.9 ms^{-1}, 1.22 ms^{-1}

4.24 245 ms^{-1}
4.25 915 m, 10.6 ms^{-1}
4.26 30 ms^{-1}, 30 ms^{-1}, after 4 s, 40 m above ground

5.4 1.5 ms^{-2}
5.5 Increased by a factor 3.5
5.6 3 N
5.7 (b) $P/5m$, $P/5$, $2P/5$
5.8 17.2 ms^{-1}
5.9 (a) M

6.3 $5l/16$
6.4 $2h/3$

7.1 15 cm
7.3 10 m
7.5 2300 Pa
7.6 (a) (i) 180 mm, (ii) 230 mm (b) (i) 225 mm, (ii) 290 mm
7.7 7.5 N
7.8 10^6 N
7.9 0.01%
7.10 58.7 cm^3
7.11 55 mm
7.14 40.7 N
7.15 40.2 N to 41.2 N
7.16 0.009 N
7.17 (a) $\frac{1}{2}$ (b) $\frac{1}{10}$ (c) 0
7.18 0.08 m^3
7.19 $1000Y/(Z + X - Y)$ kgm^{-3}
7.20 5.0 m
7.21 0.96 N, 0.46 cm
7.23 1460 N
7.24 1.74×10^6 N, 170 m^3
7.25 940 kgm^{-3}, 875 kgm^{-3}
7.26 (a) $M/d_1 A$ (b) $Md_2 g/d_1 A$ (c) $Md_2 g/d_1$
7.27 $l = 1000(V + aL)/\rho a - V/a$, 0.076 m^3

8.2 (a) 30 Ns (b) $10\sqrt{13}$ Ns, $\tan^{-1}(3/2)$ to the horizontal
8.3 (a) 0.067 ms^{-1} (b) 0.2 ms^{-1} (c) 0.33 ms^{-1}
8.4 (a) before, 0.1 ms^{-1}, after, 0
(b) 0.6 m from C, 0.3 m from C
8.5 (a) before, 0.133 ms^{-1} and 0.067 ms^{-1}, after, 0 and 0
(b) 0.4 m from C, 0.2 m from C
8.6 $\sqrt{17}/3$ ms^{-1}, $\tan^{-1}(4.0)$ to the horizontal
8.7 (a) 196 N (b) 196 N (c) 294 J (d) 294 J
8.9 2.67, 4, 66.7%
8.10 y/x, $mg/F = y/x$
8.12 5 J
8.14 (a) $mg\sin\alpha$ (b) $g\sin\alpha$ (c) $(2gl\sin\alpha)^{1/2}$
8.15 6 ms^{-1}, $2\sqrt{5}$ ms^{-1}
8.16 (a) 15 J (b) 6 J (c) 9 J (d) $2\sqrt{3}$ ms^{-1} (e) 6 J
8.18 (a) u, 0 (b) $u/2$, $u/2$ (c) $u/2$, 50%
8.19 $v = \sqrt{3}u/2$, $V = u/2$
8.24 (a) 0.25 ms^{-2} (b) 0.42 ms^{-2}
8.25 $(b - kv^2)/m$; $(b - kv^2 - \frac{1}{20}mg)/m$, $v^2_{max} = (b - \frac{1}{20}mg)/k$
8.26 1875 N, 8 s, 15000 Ns, 112500 J
8.27 (a) 0.49 m (b) 9.8 J (c) 7.84 J
8.29 $\frac{1}{2}mg/N$
8.30 $2E/103$
8.31 $3U$, (a) $\frac{1}{2}MU^2$ (b) $7MU^2/6$, 2.67×10^9 J
8.32 $u(m - M)/(M + m)$, $2um/(M + m)$; fractional share of
energy: (a) neutron 0, nucleus 1 (b) neutron $\frac{1}{9}$, nucleus $\frac{8}{9}$
8.33 2.59×10^5 ms^{-1}
8.34 36%
8.35 mv/M
8.36 $\frac{1}{4}v$, at $60°$ to the original velocity of the first sphere.

8.37 (a) $m_1 = m_2$ (b) $m_1 < m_2$ (c) $m_1 > m_2$
8.38 $M/2ApV$
8.40 (a) $3.33\,\text{N}$ (b) $0.56\,\text{W}$ (c) $0.28\,\text{W}$
8.41 $1.0\,\text{ms}^{-1}$, $600\,\text{J}$
8.43 $x^2 = 4y(z - y)$, $\tfrac{1}{2}z$, z

9.2 $\alpha = \tfrac{1}{2}$, $\beta = -\tfrac{1}{2}$
9.5 $7.2 \times 10^{-2}\,\text{Nm}^{-1}$
9.6 0.07%
9.7 $x = 2$, $y = 1$, $z = 2$
9.8 $v \propto \sqrt{(E/\rho)}$
9.10 $T \propto \sqrt{(a^3\rho/\gamma)}$

10.1 (a) $10^{-8}\,\text{J}$ (b) 5×10^{16} (c) 5×10^{12} (d) $5 \times 10^{-13}\,\text{kg}$
   (e) $50\,\text{K}$
10.2 (a) $14\,\text{N}$ (b) $3.5\,\text{N}$ (c) 0.25 (d) $8.5\,\text{N}$
10.3 0.29
10.7 (a) slide if $\mu < a/2b$, topple if $\mu > a/2b$
   (b) slide if $\mu < a/(2b + a/\sqrt{3})$,
      topple if $\mu > a/(2b + a/\sqrt{3})$
10.8 $\tfrac{1}{2}W$ vertical, $\sqrt{3}W/2$ horizontal
10.9 $T = \sqrt{3}W/4$
10.10 $30\sqrt{2}\,\text{N}$
10.11 $57.5\,\text{N}$
10.12 $77°$
10.13 $\cos\theta = \dfrac{(m + M)r\sin\phi}{Ml}$, giving two values of θ between $-\tfrac{1}{2}\pi$ and $+\tfrac{1}{2}\pi$ as long as the expression is less than 1
10.15 Stable if $r > \tfrac{1}{2}t$

11.1 (a) $10\,\text{cm}$ (b) $0.1\,\text{cm}$ (c) $2.5\,\text{cm}$
11.2 $8.2 \times 10^{10}\,\text{Nm}^{-2} \pm 2\tfrac{1}{2}\%$; diameter contributes most error
11.6 $1000\pi\,\text{N}$
11.14 $1.4 \times 10^{12}\,\text{Nm}^{-2}$, 0.35
11.15 1.9×10^{-4} degrees
11.17 Ratio $a:b:c = 625:369:81$
11.20 $6, 2, 2$
11.22 $4, 2$
11.23 $0.083\,\text{J}$
11.24 (a) $2.13 \times 10^{11}\,\text{Nm}^{-2}$ (b) $2.49 \times 10^{-2}\,\text{J}$
11.25 5%
11.26 $66.7\,\text{N}$
11.27 $4F/\pi d^2$, e/L, $2.04 \times 10^{11}\,\text{Nm}^{-2}$, 4%, $10\,\text{cm}$
11.28 $628\,\text{N}$
11.29 (a) $1.8 \times 10^4\,\text{N}$ (b) $5.4\,\text{J}$
11.30 $2.16\,\text{mm}$, brass $53\,\text{N}$, steel $113\,\text{N}$
11.31 Copper $2.9\,\text{N}$, steel $4.9\,\text{N}$, $9.6 \times 10^{-5}\,\text{J}$
11.32 (a) $0.4\,\text{J}$ (b) $0.2\,\text{J}$ (c) $0.2\,\text{J}$
11.33 $10\,\text{m}$
11.34 $4800\,\text{Jm}^{-3}$
11.35 $0.09\,\text{m}$
11.36 $1.08 \times 10^4\,\text{Jkm}^{-1}$
11.37 $1.22 \times 10^7\,\text{Pa}$

12.2 $0.072\,\text{Nm}^{-1}$
12.11 $39\,\text{mm}$
12.13 About $10^{-12}\,\text{N}$
12.14 2000 atmospheres
12.15 $7.1 \times 10^{-2}\,\text{Nm}^{-2}$
12.16 $1.8\,\text{mm}$
12.18 $\pi r^2 h\rho g - 2\pi r\gamma\cos\alpha$
12.20 $7.8 \times 10^{-5}\,\text{J}$
12.21 $1300\,\text{Pa}$
12.22 $2.0\,\text{mm}$

12.23 $5.7\,\text{cm}$
12.24 $5.0\,\text{cm}$
12.26 $220\,\text{N}$
12.28 $1\,\text{mm}$
12.29 2%

13.1 $v = c$
13.2 $y = k$
13.3 $k = 3\,\text{m}$, $y_{max} = 7\,\text{m}$, $y = 0$ when $t = 4.65\,\text{s}$
13.5 (a) $57.2\,\text{m}$ (b) $65.5\,\text{m}$
13.6 $15°$ and $75°$
13.8 $60\,\text{m}$
13.9 (a) $0.245\,\text{s}$ (b) $12.2\,\text{ms}^{-1}$ (c) $12.4\,\text{ms}^{-1}$
13.10 (a) $31.3\,\text{ms}^{-1}$ $25.2°$ above horizontal and $29.1\,\text{ms}^{-1}$ $13.4°$ below horizontal
   (b) $4\,\text{Ns}$ (c) $2\,\text{N}$
13.11 $125\,\text{m}$, $870\,\text{m}$; $125\,\text{m}$, $1070\,\text{m}$
13.12 $9.3\,\text{cm}$

14.1 $6 \times 10^4\,\text{N}$
14.4 $29°$
14.7 $27\,\text{ms}^{-1}$
14.10 $\sqrt{(8gr)}$, $\sqrt{(4gr)}$, $3mg$, $9mg$
14.11 (a) $0.085°$, 0.085% (b) $0°$, 0.34%
14.12 84 minutes
14.13 $(5\lambda l/16m)^{1/2}$, $5:1$
14.14 (a) $1.3\,\text{N}$ horizontal   (b) $1.1\,\text{s}$
14.15 $18.7\,\text{ms}^{-1}$
14.16 $11\,\text{cm}$, $1.0\,\text{N}$
14.17 (a) $3.4\,\text{rads}^{-1}$ (b) $6.8\,\text{N}$ (c) $1.85\,\text{s}$ (d) $0.85\,\text{J}$
14.18 0.145 revolutions per second
14.19 $29.4\,\text{km}$ per hour
14.20 (a) mgl (b) $\sqrt{2gl}$ (c) $2g$ (d) $3mg$
14.21 $28\,\text{cm}$
14.22 $9.79\,\text{ms}^{-2}$
14.23 $2.4\,\text{N}$, $2.0\,\text{N}$, $1.2\,\text{N}$
14.24 $9.79281\,\text{ms}^{-2}$

15.1 $g(m/\lambda)^{1/2}$
15.3 s^{-1}. Later we shall use rad s^{-1}.
15.5 $9.2\,\text{m}$
15.6 All three are solutions
15.7 $0.184\,\text{m}$, $-0.131\,\text{m}$
15.8 $2\pi\sqrt{(m/\lambda)}$, $2\pi\sqrt{(m/\rho Ag)}$, $2\pi\sqrt{(l/g)}$
15.9 $(1/4\pi)\,\text{m}$, $x = (1/4\pi)\sin 8\pi Z\,\text{m}$, $v = 2\cos 8\pi Z\,\text{ms}^{-1}$, where $t = Z\,\text{s}$
15.10 (a) $0.042\,\text{s}$, $0.208\,\text{s}$, $0.292\,\text{s}$, $0.458\,\text{s}$, $\ldots$
   (b) $0.083\,\text{s}$, $0.167\,\text{s}$, $0.333\,\text{s}$, $0.417\,\text{s}$, $\ldots$
15.11 $\pm 7.55\,\text{ms}^{-1}$
15.14 $\pi/2\,\text{rad}$, $\pi/2\omega\,\text{s}$, $\tfrac{1}{4}$ cycle
15.16 $\tfrac{1}{2}m\omega^2 a^2$
15.22 $1.58 \times 10^{-4}\,\text{J}$, 0
15.23 (a) $0.63\,\text{s}$ (b) $1.25 \times 10^{-4}\,\text{J}$
15.24 (a) $10\pi\,\text{cms}^{-1}$ (b) $5\pi^2\,\text{cms}^{-2}$ towards A (c) $\tfrac{1}{3}\,\text{s}$
   (d) $\tfrac{1}{2}\pi^2 \times 10^{-3}\,\text{J}$ at A, zero top and bottom
   (e) $\tfrac{1}{2}\pi^2 \times 10^{-3}\,\text{J}$
15.25 $10^{-4}\,\text{J}$
15.26 (c) $2.09\,\text{s}$, $1.11\,\text{m}$
15.27 Shortened by $2\,\text{mm}$
15.28 (a) $16\,\text{Hz}$ (b) $0.1\,\text{N}$ (c) $0.1\,\text{ms}^{-1}$
15.29 $1.6\,\text{Hz}$; at highest point
15.30 $0.05\pi^2\,\text{ms}^{-2}$, $1.42\,\text{Hz}$, 0.5
15.31 (a) $a\cos\theta$ (b) $v\sin\theta$ (c) $(v^2/a)\cos\theta$; $2\pi\sqrt{(h/g)}$
15.32 $8.33\,\text{cm}$, $4 \times 10^{-2}X\,\text{N}$, $5 \times 10^{-5}\,\text{J}$
15.33 (a) $4\pi^2\,\text{rads}^{-1}$ (b) $-2\sqrt{3}\pi^2\,\text{rads}^{-1}$ (c) $-4\pi^3\,\text{rads}^{-1}$; F_{max} numerically equal to $16\pi^4 mr$

16.1 Nsm^{-2}

16.7 (a) $3.5\,cms^{-1}$ (b) $8.9\,cms^{-1}$, $2.2\,cms^{-1}$, $0.22\,mms^{-1}$
(c) $0.124\,mms^{-1}$ (d) $2.2 \times 10^{-4}\,mms^{-1}$

16.9 $5\,ms^{-1}$, $60\,ms^{-1}$

16.10 $20\,ms^{-1}$

16.11 $10\,ms^{-1}$

16.12 Diameter 14%, length 0.5%, pressure 1.1%, rate of flow 1.4%

16.14 $(\rho_1 - \rho_2)g/\rho_1$

16.15 $3.9\,mms^{-1}$

16.16 98.5%

16.17 $2.84\,Nm$

16.18 $8 \times 10^{-3}\,mm$

16.19 $0.53\,mms^{-1}$

17.1 $A = g - (U/m)$, $B = 6\pi\eta r/m$

17.4 $-B^7 a/2 \times 3 \times 4 \times 5 \times 6 \times 7$, written as $-B^7 a/7!$

17.8 (a) $0.556\,s$ (b) $6.31\,s$

17.9 $-1.34 \times 10^{-2}\,ms^{-2}$ and $0.0412\,m$ when $t = 1\,s$;
$-2.71 \times 10^{-3}\,ms^{-2}$ and $0.108\,m$ when $t = 5\,s$;
$-3.66 \times 10^{-4}\,ms^{-2}$ and $0.123\,m$ when $t = 10\,s$

17.13 (a) $1.73 \times 10^{-2}\,ms^{-1}$, $2.06 \times 10^{-3}\,s$
(b) $0.276\,ms^{-1}$, $3.3 \times 10^{-2}\,s$
(c) $2.2 \times 10^{-4}\,ms^{-1}$, negligible time constant
(d) $3.3 \times 10^{-7}\,ms^{-1}$, negligible time constant

17.14 $1.41 \times 10^{-2}\,ms^{-1}$

17.16 (a) $2.4 \times 10^{-15}\,N$, $1.8 \times 10^{-5}\,ms^{-1}$
(b) $A = 2g = 20\,ms^{-2}$, $B = 6\pi\eta rg/W = 5.6 \times 10^5\,s^{-1}$,
new (upward) terminal velocity $= 3.6 \times 10^{-5}\,ms^{-1}$,
time constant $= 1/B = 1.8 \times 10^{-6}\,s$

17.19 (a) $9.4 \times 10^4\,Pa$ (b) $8.9 \times 10^4\,Pa$ (c) $5.5 \times 10^4\,Pa$
(d) $3.0 \times 10^4\,Pa$ (e) $0.615\,Pa$

17.20 $k = \rho g\pi r^4/8A\eta l$, $140\,s$

17.23 $5e^{-0.01t}$ in ms^{-2}, $500(1 - e^{-0.01t})$ in ms^{-1}, $69.3\,s$

17.24 $t = 2\eta l(R_0^4 - R^4)/r^4 T$

17.26 (a) $\lambda_1 P_0 e^{-\lambda_1 t}$ (b) $\lambda_1 P_0 e^{-\lambda_2 t} - \lambda_2 Q$
(d) $(\ln\lambda_1 - \ln\lambda_2)/(\lambda_1 - \lambda_2)$ (e) 10.8 minutes

18.1 $\frac{4}{3}\pi r^3\rho$, λ, $6\pi\eta r$

18.4 $D = -P_0/2\pi f_0 C$

18.8 f

19.1 (a) $1.25\,kgm^2$ (b) $0.125\,Nm$ (c) $11.2\,s$

19.2 $a/\sqrt{2}$

19.3 $r\sqrt{2/5}$

19.4 $mb^2/12$

19.5 $ma^2/3$

19.7 $m(a^2 + b^2)/12$

19.8 $158\,J$

19.10 $h = 2\pi r n_1$

19.11 $I = \dfrac{n_2 m}{n_1 + n_2}\left[\dfrac{2\pi^2 g l t_1^2}{n_1^2} - r^2\right]$

19.12 (a) $0.04\,kgm^2$ (b) $1.62\,s$ (c) 15.4 cm or 84.6 cm mark
(d) $1.8\,s$

19.15 28 cm, 30 cm

19.16 (a) $44\,rads^{-1}$ (b) $22\,ms^{-1}$ (c) $44\,ms^{-1}$ (d) $232\,J$

19.17 (a) $0.26\,kgm^2$ (b) $6\,N$ (c) $3\,Nm$ (d) $11.5\,rads^{-1}$
(e) $4.8\,rads^{-1}$; $1.85\,s$

19.18 $1580\,J$, $8.4 \times 10^{-2}\,Nm$

19.19 $20\,rads^{-2}$, $16.7\,rads^{-2}$

19.20 $2.55 \times 10^{-2}\,kgm^2$

19.21 $0.93\,ms^{-2}$, $1.0\,N$, $0.98\,N$

19.22 $Mrv/(I + mr^2 + Mr^2)$, $-rv(M + m)/(I + mr^2 + Mr^2)$,
$\pi r^2 m/(I + mr^2 + Mr^2)$

19.23 (a) ml^2 (b) $3ml^2$; $0.97\sqrt{(gl)}$ and $1.94\sqrt{(gl)}$

19.24 $3.5\,ms^{-1}$

19.25 $\sqrt{(10gh/7)}$

19.26 $0.16\,ms^{-1}$

19.27 $8.47 \times 10^{-4}\,kgm^2$

19.28 $2.8 \times 10^{16}\,kg$

19.29 $0.48\,ms^{-2}$

19.30 $0.001\,kgm^2$, $3.75 \times 10^{-5}\,Nm$, $5.25 \times 10^{-5}\,Nm$,
$4.81\,rads^{-1}$

19.31 $13.3\,rads^{-1}$, $80\,J$

20.2 $4Gm^2/a^2$

20.3 $6.8 \times 10^{-11}\,m^3 kg^{-1} s^{-2}$

20.4 15° per hour, $(90 - X)°$

20.7 $4600\,km$

20.8 0.6%

20.9 $3.8 \times 10^4\,km$

20.10 $0.163\,m$, $2.56 \times 10^{-5}\,N$

20.11 $64\,kg$

20.12 $5 \times 10^{-2}\,rad$

20.14 $1.24 \times 10^{36}:1$

20.15 (a) $6.7 \times 10^{-21}\,N$ (b) $5.8 \times 10^{13}\,N$

20.16 0.31%

20.17 $E = Gm/r^2$

20.18 (a) $27Gm/4r^2$ (b) $\sqrt{3}Gm/r^2$ (c) 0

20.19 0

20.20 $-6.3 \times 10^7\,Jkg^{-1}$

20.21 (a) $4G\pi r\rho/3$
(b) 0; $E = 4G\pi r\rho/3$, $V = -2G\rho\pi(R^2 - r^2/3)$

20.22 (a) $11.2\,kms^{-1}$ (b) $2.4\,kms^{-1}$

20.23 $-1.62 \times 10^{12}\,J$, 0, $3.8 \times 10^{11}\,J$, $4.36\,kms^{-1}$

20.24 (a) $8.0\,kms^{-1}$ (b) $11.0\,kms^{-1}$

20.25 $8400\,km$, $2.42\,kms^{-1}$

20.26 $1.67\,kms^{-1}$

20.27 $60\,s$

20.28 $4.24 \times 10^4\,km$

20.29 $3.84 \times 10^5\,km$

20.30 GMm/r^2 towards F_1, acceleration $= GM/r_1^2$ towards F_1;
mx, y/x

20.31 (a) $4\pi RG\rho/3$ (b) $4\pi R^3\rho G/3(R + h)^2$ (c) $R(\frac{4}{3}\pi G\rho - \omega^2)$
where ω = angular velocity of rotation of the earth

20.33 $1.2 \times 10^{-7}\,Nm$; (a) $7.1 \times$ larger (b) $2.1 \times$ larger

20.34 $1.42 \times 10^{-5}\,rads^{-1}$

20.35 $2\pi\sqrt{(D^3/2GM)}$, 18.8 hours with respect to the fixed stars,
87 hours with respect to an observer on the earth

20.36 2.2×10^{-5} %

20.37 $\frac{3}{2}$

20.38 (a) $\frac{10}{44}$ (b) $\frac{10}{121}$

20.40 (a) $9.81\,ms^{-2}$ (b) $3.50 \times 10^{-5}\,ms^{-2}$
(c) $2.32 \times 10^6\,s$ (d) $2.03 \times 10^{20}\,N$

20.41 $3.8 \times 10^4\,km$ from moon, $2.27\,kms^{-1}$

20.42 $9 \times 10^{-5}\,Nkg^{-1}$

20.43 $3GM^2/5R$

Index